The Grenville Event in the Appalachians and Related Topics

Edited by

Mervin J. Bartholomew
Montana Bureau of Mines and Geology
Montana Tech
Butte, Montana 59701

Associate Editors
Eric R. Force
United States Geological Survey
National Center
Reston, Virginia 22092

A. Krishna Sinha
Department of Geological Sciences
Virginia Polytechnic Institute and State University
Blacksburg, Virginia 24061

Norman Herz
Department of Geology
The University of Georgia
Athens, Georgia 30602

SPECIAL PAPER
194

THE GEOLOGICAL SOCIETY OF AMERICA
3300 Penrose Place, P.O Box 9140, Boulder, Colorado 80301

Published by The Geological Society of America, Inc.
3300 Penrose Place, P.O. Box 9140, Boulder, Colorado 80301

Printed in U.S.A.

Library of Congress Cataloging in Publication Data

Main entry under title:

The Grenville event in the Appalachians and related
   topics.

   (Special paper ; 194)
   Includes papers presented Mar. 27, 1982 as part of a
symposium at the combined northeastern and southeastern
sectional meeting of the Geological society of
America, held in Washington, D.C.
   Includes bibliographies and index.
   1. Geology, Stratigraphic—Pre-Cambrian—Congresses.
2. Geology—Appalachian Mountains—Congresses.
I. Bartholomew, Mervin J. II. Geological Society of
America. III. Series: Special paper (Geological Society
of America) ; 194.
QE 653.G74   1984   551.7'1'0974   83-25503
ISBN 0-8137-2194-6

Contents

iv   *Contents*

Preface

Many classic studies in geology have been done in the Grenville province of Canada and the adjacent Appalachian province of the United States of America and Canada. Moreover, for many years it has been generally recognized that rocks which were affected by the "Grenville event," about 1 billion years ago, formed the crystalline "basement" on which uppermost Precambrian (younger than about 750 m.y.) and Paleozoic sedimentary and volcanic rock accumulated, prior to the middle to late Paleozoic orogenic pulses that resulted in tectonic development of the Appalachian orogen as we know it today. This Grenville basement, concealed beneath flat-lying Paleozoic rocks of the eastern mid-continental portion of the United States (Fig. 1), rises to the surface in the Adirondack Mountains of New York and the Grenville province of Canada, but is exposed in the Appalachian province only because post-Paleozoic erosion has cut deep into lower levels of some structurally higher crystalline thrust sheets.

Grenville-age rocks are distributed longitudinally in the Appalachians in two discreet belts. Forming a discontinuous belt near the center of the Appalachians, the western chain of Grenville massifs includes: the Chain Lakes, Green Mountains and Berkshire Massifs of New England; the Reading Prong of the central Appalachians; the core of the Blue Ridge anticlinorium of Virginia; and a major part of the Blue Ridge province in North Carolina and Tennessee, as well as smaller gneiss domes of Tennessee, North Carolina and Georgia. Smaller, but no less important massifs form a second chain to the east. Here, Grenville rocks form the Hudson Highlands near New York; the Honey Brook Upland and associated terrane near Philadelphia; the Baltimore Gneiss domes of Maryland; and the Sauras Massif of North Carolina. Still farther east are the isolated Grenville rocks of both the Goochland area in eastern Virginia and the Pine Mountain window in Georgia and Alabama.

To be able to treat all of these Grenville terranes comprehensively in one volume is desirable; however, a number of them have yet to receive anything but cursory work as geologists have pursued what seemed like more timely and momentous geologic problems of the Appalachians. Timely and momentous: the pursuit of the suture zone of the Appalachians—the real boundary between the craton of North America and that of the African-European mass with which it collided to form the Appalachian

orogen. Such a suture zone likely should separate different basement rocks of the opposing cratons, e.g., the Grenville rocks of the North American craton from probably more ancient Precambrian rocks of the African craton (Fig. 1). Timely and momentous: indeed the discovery of 2-billion-year or older rocks in eastern North America or, alternatively, Grenville rocks in northwestern African and adjacent Europe would stimulate considerably more interest in the basement rocks of the Appalachians.

Alas, this volume does not announce such discoveries. It does, however, provide the first synopsis ever done on basement rocks of the Appalachians. Thus, it lays the foundation for future investigations dealing with searches for similar rocks in Africa and Europe or for comparison of suspected fragments of the African craton in North America. Neither could be accomplished without a standard for comparison, and that is the purpose of this special paper—to provide a state-of-the-art treatise on the Grenville rocks of the Appalachian orogen.

The papers are presented in a geographical order, starting with the exhaustive summary of the Adirondacks of New York and ending with a discussion of the tectonics of the Pine Mountain window of Georgia and Alabama. In addition, a variety of special topics are treated, including a comprehensive study of the Reading Prong and detailed petrologic investigations of the Honey Brook Upland and the Roseland District. Six papers present new geochronological data on both Grenville-age rocks and spatially associated, younger Precambrian plutons. The newly discovered Grenville rocks of eastern Virginia are described. Contrasting viewpoints on some topics have been included—compare the papers on the age of the Robertson River Pluton and those dealing with the origin of Grenville rocks in the Blue Ridge of central Virginia. Most of the papers, particularly those not already mentioned, deal largely with either regional or local tectonics as well as petrology.

Like any scientific work, the papers presented here were developed on geologic foundations provided by many pioneers who made the first geologic maps or descriptions of rocks we now know to have been affected by the Grenville event. Thus, this volume is dedicated to all the geologists who did pioneer work on the Grenville terranes of the Appalachians. A few of these early geologists, whose vintage works speak for them, are

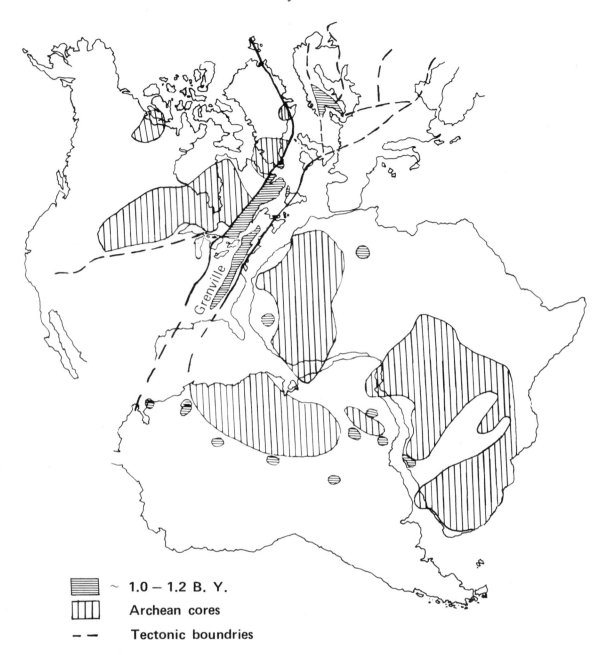

~ 1.0 – 1.2 B. Y.

Archean cores

– –   Tectonic boundries

Figure 1. A predrift reconstruction, modified from Bullard, Everett, and Smith (1965), showing the Grenville Province of eastern North America relative to other billion-year-old rocks, Archean cores, and selected tectonic boundaries for geologic provinces of the cratons. Ages and distribution of tectonic provinces modified after Baragar and McGlynn (1976); Chumakov and Semikhatov (1981); Engel (1963); Gibb and others (1983); Hurley and others (1967); Hurley and Rand (1969); Kröner (1977); Lidiak and others (1966); Muehlberger and others (1967); Skjernaa and Pedersen (1982); Turek and Robinson (1982); Wynne-Edwards (1972); Wynne-Edwards and Hasan (1970).

listed below in special recognition of their contributions. Just a glance at the reference lists within this volume provides a fair appreciation of the value and longevity of their works.

Florence Bascom
William S. Bayley
Eleanora Bliss (Knopf)
Arthur F. Buddington

Ernst Cloos
C. Willard Hayes
Anna I. Jonas (Stose)
Arthur Keith

Most of the papers in this book were presented orally on March 27, 1982, as part of the symposium on "Grenville terranes of the Appalachians" at the combined Northeastern and Southeastern sectional meeting of the Geological Society of America held at Washington, D.C. It is certain that without the solid geological base, built through the prodigious research efforts of these pioneers, neither the symposium nor this compendium volume on the Grenville terranes could have been possible at this point in time. This special paper is our thanks for their efforts.

Finally, completion of this special paper depended on the patience and cooperation of the people who served as critical reviewers for papers selected to be included here. The following list is a way of acknowledging their efforts to improve the scientific quality of this volume.

J. Robert Butler
James W. Clarke
William A. Crawford
R. David Dallmeyer
George W. Fisher
Paul D. Fullagar
M. Charles Gilbert
Lynn Glover, III
David R. Gray
Robert D. Hatcher, Jr.
David A. Hewitt
Allan Kolker
Sharon E. Lewis

Bruce R. Lipin
Peter T. Lyttle
John M. Moore, Jr.
Louis Pavlides
Nicholas M. Ratcliffe
Douglas W. Rankin
Edwin S. Robinson
John J. W. Rogers
Gail S. Russell
J. Alexander Speer
Thomas W. Stern
Mary Emma Wagner

REFERENCES CITED

Baragar, W.R.A. and McGlynn, J. C., 1976, Early Archean basement in the Canadian Shield: a review of the evidence: Geological Survey of Canada, Paper 76-14, 21 p.

Bullard, E. C., Everett, J. E., and Smith, A. G., 1965, The fit of the continents around the Atlantic: Philosophical Transactions of the Royal Society of London, A-258, p. 41–51.

Chumakov, W. M., and Semikhatov, M. A., 1981, Riplean and Vendian of the USSR: Precambrian Research, v. 15, no. 314, p. 229–253.

Engel, A.E.J., 1963, Geologic evolution of North America: Science, v. 140, p. 143–152.

Gibb, R. A., Thomas, M. D., Lapointe, P. L., and Mukhopadhyay, M., 1983, Geophysics of proposed proterozoic sutures in Canada: Precambrian Research, v. 19, no. 4, p. 349–384.

Hurley, P. M., and Rand, J. R., 1969, Pre-drift continental nuclei: Science, v. 164, p. 1229–1242.

Hurley, P. M., de Almeida, F.F.M., Melcher, G. C., Cordani, V. G., Rand, J. R., Kawashita, K., Vandoros, P., Pinson, W. H., and Fairbairn, H. W., 1967, Test of continental drift by comparison of radiometric ages: Science, v. 157, p. 495–500.

Kröner, A., 1977, The Precambrian geotectonic evolution of Africa: Plate accretion versus plate destruction: Precambrian Research, v. 4, no. 2, p. 163–213.

Lidiak, E. G., Marvin, R. F., Thomas, H. H., and Bass, M. N., 1966, Geochronology of the midcontinent region, United States—Part 4, eastern area: Journal Geophysical Research, v. 71, p. 5427–5438.

Muehlberger, W. R., Denison, R. E., and Lidiak, E. G., 1967, Basement rocks in continental interior of United States: American Association of Petroleum Geology Bulletin, v. 51, p. 2351–2380.

Skjernaa, L. and Pedersen, S., 1982, The effects of penetrative sveconorwegian deformations on Rb-Sr isotope systems in the Römskog—Aurskog-Höland area, SE Norway: Precambrian Research, v. 17, nos. 3/4, p. 215–243.

Turek, A. and Robinson, R. N., 1982, Geology and age of the Precambrian basement in the Windsor, Chatham, and Sarnia area, southwestern Ontario: Canadian Journal Earth Sciences, v. 19, no. 8, p. 1627–1634.

Wynne-Edwards, H. R., 1972, The Grenville Province *in* Variations in tectonic styles in Canada, eds. R. A. Price and R.J.W. Douglas: Geological Association of Canada, Special Paper 11, p. 263–334.

Wynne-Edwards, H. R. and Hasan, Z., 1970, Intersecting orogenic belts across the North Atlantic: American Journal of Science, v. 268, p. 289–308.

Geological Society of America
Special Paper 194
1984

Stratigraphy and structural geology of the Adirondack Mountains, New York: Review and synthesis

Richard W. Wiener*
Geological Survey
New York State Museum
The State Education Department
Albany, New York 12230

James M. McLelland
Department of Geology
Colgate University
Hamilton, New York 13346

Yngvar W. Isachsen
Geological Survey
New York State Museum
The State Education Department
Albany, New York 12230

Leo M. Hall
Department of Geology and Geography
University of Massachusetts at Amherst
Amherst, Massachusetts 01003

ABSTRACT

A new stratigraphic and structural synthesis is presented for Precambrian rocks of the Adirondack Mountains, New York, an amphibolite-granulite facies terrane in the 1.1-b.y.-old Grenville province exposed in a dome on the North American craton. The geology of the Adirondacks appears to be explicable in terms of a stratigraphic sequence that has been subjected to multiple folding, metamorphism, and intrusive activity. This stratigraphic sequence is correlated across the entire width of the Adirondacks and westward into Ontario. Recognition of the widespread nature of this stratigraphic sequence has resulted in a coherent structural framework for the Adirondacks, consisting of two stages of nappe formation followed by three stages of upright to overturned folding. Both widespread intrusive activity and subsequent mylonitization of intrusive rocks occurred mostly during the second phase of nappe formation.

The stratigraphy of the Adirondacks is interpreted to consist of an older granitic basement, referred to as the Piseco Group, overlain unconformably by a metamorphosed clastic/carbonate sequence, referred to as the Oswegatchie Group in the

*Present address: ST 4112, Exxon Production Research Co., P.O. Box 2189, Houston, Texas 77001.

northwest and the Lake George Group in the east. The oldest recognized formation in the Piseco Group is the Pharaoh Mountain Gneiss, consisting of charnockitic and granitic gneiss.^{t} This unit is overlain in many places by the Alexandria Bay Gneiss, consisting of pink leucogranitic gneiss. The Alexandria Bay Gneiss is equivalent to the Brant Lake Gneiss in the eastern Adirondacks.

The basal formation of the metasedimentary rocks of the Oswegatchie Group is the Baldface Hill Gneiss. This thin and discontinuous unit consists of garnet-sillimanite gneiss and quartzite. The overlying Poplar Hill Gneiss consists of biotite-quartz-plagioclase gneiss that contains granitic portions. Rocks of the Baldface Hill and Poplar Hill may represent metamorphosed basal quartz sand and conglomerate, shale, shaly arkose, and possibly reworked Fe- and Al-rich regolith that was formed by weathering of the basement prior to deposition of the cover rocks. The Baldface Hill and Poplar Hill are equivalent to the Eagle Lake Gneiss of the Lake George Group.

Overlying these thin basal clastic deposits of the Oswegatchie Group is the Gouverneur Marble that consists of five members, two of which contain three sub-divisions within them. Member A at the base consists of thick, calcitic, dolomitic, and siliceous marbles. Member B is a thin, pyritic biotite schist. Member C consists of interbedded siliceous marbles, quartzites, and calc-silicate rocks. Member D consists of well-layered calcareous gneiss, and Member E, only locally present, is a quartz-feldspar granulite. To the east, the Gouverneur Marble correlates with carbonate rocks of the Cedar River, Blue Mountain Lake, and Paradox Lake Formations, and correlates via facies changes to metamorphosed calcareous clastics of the Cranberry Lake, Sacandaga, Tomany Mountain, and Springhill Pond Formations. The previously defined upper and lower marble may be stratigraphically equivalent in the Northwest Lowlands and possibly in the Adirondack Highlands.

The Pleasant Lake Gneiss overlies the Gouverneur Marble and consists largely of migmatitic gneiss equivalent to the Treadway Mountain Formation of the Lake George Group. K-feldspar megacrystic granitic gneisses overlie the Pleasant Lake Gneiss. These rocks are equivalent to the Lake Durant Formation in the Lake George Group and probably represent intrusive sheets.

Anorthosite, charnockite, hornblende granite, and gabbro successively intruded the metamorphosed sedimentary rocks and themselves were later metamorphosed and deformed. Mangerite-charnockite suites that mantle anorthosite contain xenoliths of anorthosite and are thought to be produced by partial melting of Pharaoh Mountain Gneiss by heat from the anorthosite. Megacrystic hornblende granitic gneisses intrude various formations of the Lowlands and Highlands but show gross structural concordance.

Five phases of folding affected all stratigraphic units, but only the last four phases affected the intrusive rocks. The first phase of folding resulted in northwest-directed nappes and formation of regional foliation and lineation. A second phase of isoclinal folding folded the regional foliation and lineation. Intrusion of anorthosite, charnockite, hornblende granite, and gabbro accompanied second-phase folding, as well as local mylonitization of charnockite and local thrusting. The third-phase folds are upright to overturned and responsible for the "grain" of the Adirondacks. The axial traces of these folds form an arc convex to the north that swings continuously from N70°W to east-west to N45°E from south to northwest. Peak 1.1 to 1.02-b.y.-old granulite facies metamorphism outlasted third-phase folding in the Adirondack Highlands and second-phase folding in the Lowlands. Fourth-phase, northwest-trending folds are open and best developed in the northwestern Adirondacks, where they are associated with retro-grade metamorphism and possibly with intrusion of diabase dikes at mid-amphibolite facies. Fifth-phase, north-northeast-trending folds are open and best developed in the Adirondack Highlands, where they are associated with retrograde metamorphism and with 930-m.y.-old pegmatite dikes. The fourth- and fifth-phase folds interfere with third-phase folds to produce dome and basin map patterns. Hook and heart and anchor

map patterns result from interference of the later folds with first- or second-phase isoclinal folds.

INTRODUCTION

In this paper we present a review and regional synthesis of the stratigraphy and structural geology of the Adirondack Mountains, New York with brief treatment of plutonic rocks and metamorphism. The rocks of the Adirondack Mountains constitute an extension of the Grenville province of Ontario and Quebec connected via the Frontenac Arch or Axis (Fig. 1). These Precambrian terranes were metamorphosed and deformed 1.1 to 1.0 b.y. ago (Silver, 1969) and are overlain along their margins with profound angular unconformity by the Upper Cambrian to Lower Ordovician Potsdam Sandstone and other unmetamorphosed lower Paleozoic rocks (Fisher and others, 1971). The Adirondack dome is defined by the dip of overlying Paleozoic strata, is slightly elongate, and has a north-northeast dimension of about 200 km and an east-west dimension of about 140 km (see tectonic-metamorphic map in Fisher and others, 1971; Fig. 1 of this paper). The Adirondack dome is extensively broken by north-northeast-trending block faulting along its eastern and southern flanks, where the overlying Upper Cambrian through Middle Ordovician strata are downfaulted against Proterozoic rocks (Fisher and others, 1971) (Fig. 1). The downfaulted Appalachian foreland lies east of the Adirondacks, and the Middlebury

synclinorium and the Green Mountain anticlinorium are successively eastward from the foreland.

The Adirondack dome is an anomalous feature on the eastern edge of the North American craton differing from other uplifts in the Interior Lowlands not only in its greater structural relief ($\geqslant$2,000 m), but also in the present height of its mountains (1,600 m) (Isachsen, 1975). This high elevation and consequent stream drainage pattern indicate that the dome is relatively young (Tertiary?), and releveling profiles suggest that it is undergoing current doming of several millimeters per year (Isachsen, 1975; Isachsen and others, 1978; Barnett and Isachsen, 1980).

The metamorphic and igneous basement rocks exposed in the core of the Adirondack dome lie west of the Appalachian orogen and differ from the 1.1-b.y.-old Grenvillian basement rocks of the Appalachians in that they are not known to have been subjected to any Paleozoic metamorphism or compressive deformation. Furthermore, the roughly circular map pattern of the Adirondacks related to doming contrasts with the elongate map patterns of Grenville terranes in the Appalachians formed by Paleozoic orogenies. Thus, with the exception of postmetamorphic high-angle fault blocks, the basement rocks in the Adiron-

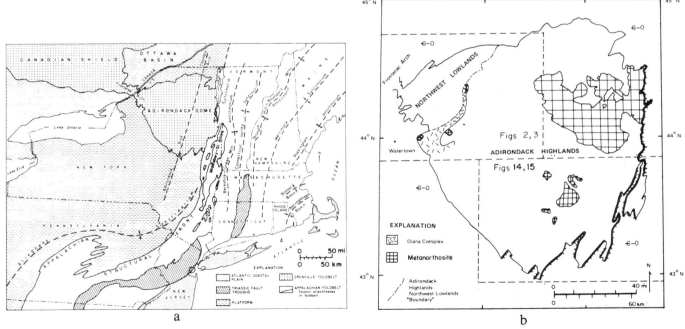

a   b

Figure 1. (a) Location of the Adirondacks in New York and with respect to the Paleozoic Appalachian foldbelt to the east (from Fisher and others, 1971). (b) Map of Adirondacks showing extent of Adirondack Highlands and Northwest Lowlands, anorthosite bodies, Diana Complex, and location of map area of the northwestern Adirondacks (Fig. 3), and southern and central Adirondacks (Fig. 14). Fault boundaries of Adirondacks shown by heavy line with tick marks on downthrown block €O = Cambrian and Ordovician sedimentary rocks.

TABLE 1. AREAL EXTENTS OF PRECAMBRIAN LITHIC UNITS IN THE ADIRONDACKS
AS SHOWN ON GEOLOGIC MAP OF NEW YORK (FISHER AND OTHERS, 1971)

Lithologic type	Total Adirondacks			Adirondack Highlands			Northwest Lowlands		
	Area (%)	Area (km^2)	Area (mi^2)	Area (%)	Area (km^2)	Area (mi^2)	Area (%)	Area (km^2)	Area (mi^2)
Metagabbro	0.1	42	18	0.1	35	15	0.3	7	3
Leucogranitic gneiss	3.1	825	320	1.9	465	180	16.2	360	140
Hornblende-(biotite) granitic gneiss	23.8	6,410	2,470	24.1	5,970	2,300	19.8	440	170
Syenitic gneiss (mangerite in Highlands)	2.9	772	297	3.0	740	285	1.4	32	12
Charnockitic gneiss	23.8	6,413	2,475	25.9	6,400	2,470	0.5	13	5
Meta-anorthosite	14.0	3,750	1,450	15.2	3,750	1,450	--	--	--
Metasedimentary rocks, amphibolite, mixed gneisses	32.4	8,740	3,370	29.8	7,360	2,840	61.8	1,380	530
Totals	100.0	26,952	10,400	100.0	24,720	9,540	100.0	2,232	860

Note: Accuracy of planimetric measurements is slightly better than 1% (by summing areas).

dacks retain the metamorphic mineral assemblages and structural configurations imparted to them 1.1 b.y. ago during the Grenville orogenic cycle. The Adirondacks, therefore, provide an ideal "protolith terrane" against which the "poly-orogenic" Grenvillian basement of the Appalachian foldbelt can be compared.

The Proterozoic core of the Adirondack dome has an area of about 27,000 km^2, and the areal extents of the constituent rock types are shown in Table 1. The Adirondacks have been divided physiographically into the Adirondack Highlands province, consisting mostly of metamorphosed intrusive rocks, and the Northwest Lowlands province, consisting mostly of metamorphosed sedimentary and volcanic rocks (Buddington, 1939; Engel and Engel, 1953a; Buddington and Leonard, 1962) (Fig. 1). Metasedimentary rocks are subordinate but present in the Highlands and are here interpreted to be stratigraphically equivalent to the metasedimentary formations in the Northwest Lowlands.

Although stratigraphic analysis in the Adirondacks is complicated by multiple folding and intrusive events, coherent parallel lithic sequences can be traced on the ground, around major fold hinges, for distances exceeding 100 km, without disruption by major thrusts. This was shown in the southern and central Adirondacks (McLelland and Isachsen, 1980), the eastern Adirondacks (Walton, 1961), the central Adirondacks (deWaard, 1962), and the western Adirondacks (Buddington and Leonard, 1962). No reliable top and bottom criteria are present in the metamorphosed sedimentary rocks of the Adirondacks. However, isotopic age data, a possible unconformity, and the structural configuration suggest the direction of younging adopted in this paper.

We present evidence for correlation of a coherent stratigraphic sequence throughout the Adirondacks and part of contiguous Ontario. This sequence consists of a basement of granitic gneisses of the Piseco Group overlain unconformably by metamorphosed sedimentary rocks referred to as the Oswegatchie Group in the western Adirondacks and the correlative Lake George Group in the central, eastern, and southern Adirondacks.

The metamorphosed sedimentary and intrusive rocks of the Adirondacks have undergone polyphase deformation, as noted by many workers. Folding accompanied regional metamorphism during the Grenville orogeny 1.1 b.y. ago (Silver, 1969). Metamorphism in the Northwest Lowlands attained amphibolite to hornblende granulite facies, and hornblende-granulite to pyroxene granulite facies metamorphism occurred in the Highlands toward the southeast (deWaard, 1969) (Fig. 2).

The boundary between the Adirondack Highlands and Northwest Lowlands (Fig. 1) (Buddington, 1939; Engel and Engel, 1953a; Buddington and Leonard 1962; Fisher and others, 1971; Wiener, 1977, 1981) coincides with the northwest border of a 1-m to 5-km-wide mylonite zone (Geraghty and Isachsen, 1981) and is parallel to a smooth metamorphic gradient to the southeast (Fig. 2). In its southern extent, the Highlands-Lowlands "boundary" is an intrusive contact between mylonitic rocks of the Diana Complex and metasedimentary rocks of the Northwest Lowlands (Figs. 1, 2). The contact has undergone multiple folding but no major fault displacement (Wiener, 1977, 1983). Although the amount of displacement both along the boundary to the northeast of the Diana Complex and also within the mylonite zone to the southeast of the "boundary" is unknown, correlation of stratigraphy from the Northwest Lowlands to the Adirondack Highlands suggests a lack of major displacement along the mylonite zone.

STRATIGRAPHY OF THE NORTHWEST LOWLANDS

The Northwest Lowlands form the northwestern corner of the exposed Precambrian high-grade metamorphic rocks in the Adirondacks (Fig. 1; western half of Fig. 3). This terrane is characterized by broad carbonate valleys and intervening ridges underlain by various metasedimentary, metavolcanic, and plutonic gneisses. The stratigraphy and structure of parts of the Northwest Lowlands have been mapped in detail at 1:24,000 and larger

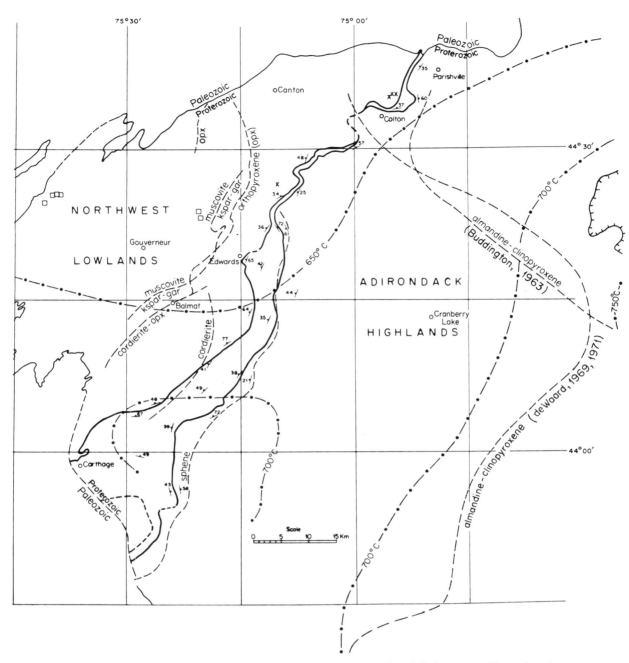

Figure 2. Map showing Carthage-Colton mylonite zone (heavy outline), foliation, mapped isograds and paleoisotherms. K-feldspar-garnet isograd is composite from Engel and Engel (1962), deWaard (1969, 1971), and Buddington (1963). Major orthopyroxene (opx) isograd is composite from same sources and Nielsen (1971); short opx isograd is from Bloomer (1969). Isolated opx occurrences (squares) are from Guzowski (1979, western cluster) and Foose (1974, eastern cluster). Cordierite isograds are from deWaard (1969, 1971). Isolated cordierite occurrences (X) are from Buddington (unpub. guidebook, south of Canton) and Stoddard (1976), p. 67 and Pl. 2, north of Colton). Sphene isograd from Buddington (1963). Almandine-clinopyroxene isograds are from deWaard (1969, 1971) and Buddington (1965). Paleoisotherms are from Bohlen and others (1980). Hachured line is boundary of meta-anorthosite to the east.

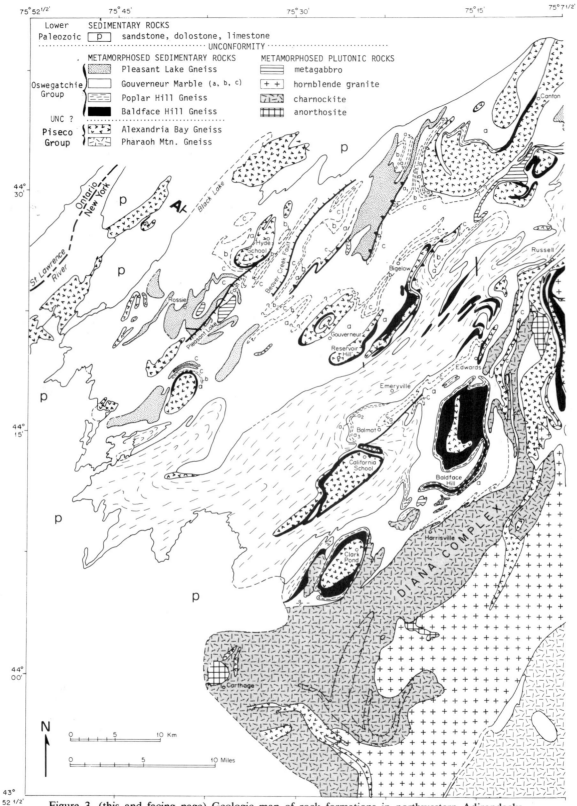

Figure 3. (this and facing page) Geologic map of rock formations in northwestern Adirondacks. Principal sources are Fisher and others (1971, Geologic Map of New York), Foose and Brown (1976, the Northwest Lowlands); and Buddington and Leonard (1962, the Adirondack Highlands to the

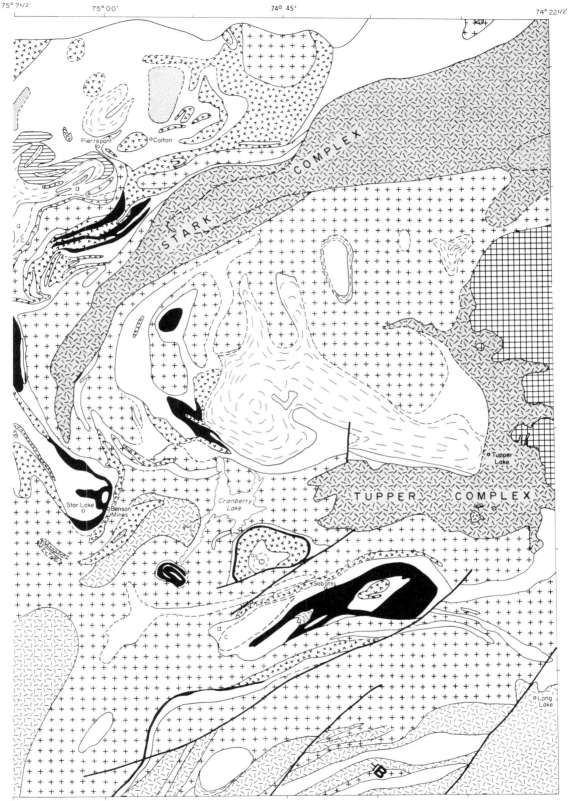

southeast). Other sources include Lewis (1969), Guzowski (1979), Isachsen (1953), Bannerman (1972), Foose (1974), Brown and Engel (1956), Engel and Engel (1953b), Buddington (1929), Wiener (1981), and Leavell (1977).

scales by several workers (Buddington, 1929, 1934; Brown, 1936; Gilluly, 1945; Isachsen 1953; Engel and Engel, 1953a; Brown and Engel, 1956; Buddington and Leonard, 1962; Lea and Dill, 1968; Bloomer, 1969, and unpub. maps; Lewis, 1969; Brown, 1969; Brocoum, 1971; Bannerman, 1972; Foose, 1974; Foose and Brown, 1976; Stoddard, 1976; Foose and Carl, 1977; Guzowski, 1979; deLorraine, 1979; Tyler, 1979; Wiener, 1979, 1981; and Romey and others, 1980).

The following stratigraphic sequence is proposed for this terrane (Fig. 4). Leucogranitic gneiss of the Alexandria Bay Gneiss is at the top of the Piseco Group and is unconformably overlain by the Oswegatchie Group. The Baldface Hill Gneiss is the basal formation of the Oswegatchie Group and consists mostly of garnet-sillimanite gneiss. This formation grades upward into the overlying Poplar Hill Gneiss, which is mostly biotite-quartz-plagioclase gneiss. These rocks are in turn overlain by the Gouverneur Marble with five members, and the Pleasant Lake Gneiss (Fig. 4). The major intrusive rock suites include, from oldest to youngest, anorthosite and related rocks, quartz syenite gneiss and related rocks, hornblende granite gneiss, and metagabbro. Present-day formation thicknesses are indicated below.

Alexandria Bay Gneiss (≧1,200 m) (of Piseco Group)

The Piseco Group, formally defined in the section on the central and eastern Adirondacks, consists of the Pharaoh Mountain Gneiss overlain by the Alexandria Bay Gneiss (Fig. 3, 4). The Alexandria Bay Gneiss is the basal formation exposed in the Northwest Lowlands.

The Alexandria Bay Gneiss consists mostly of fine-grained, pink, leucogranitic gneiss (Fig. 5). We arbitrarily select the type area as that within about 1 km of the Alexandria Bay. Characteristic exposures also include the series of low outcrops in the pasture on the east side of the Rock Island Road, 2 to 3 mi (3–5 km) north-northwest of Gouverneur, in the classical Gouverneur phacolith. Rocks in this unit originally were named Laurentian granite (Cushing and others, 1910, p. 36) for their similarity to batholithic rocks in Canada. Buddington (1929, p. 53) applied the name Alexandria granite to this unit because of exposures of pink or light gray, fine- to medium-grained leucogranitic to quartz dioritic gneisses near Alexandria Bay, St. Lawrence River, New York (Fig. 3), and this name has been used by many subsequent workers.

The Alexandria Bay Gneiss consists mostly of pink or gray, fine- to medium-grained, equigranular, leucogranitic to quartz dioritic gneisses characterized by low mafic mineral content, typically less than 10%. The pink alaskitic gneisses make up most of the Alexandria Bay Gneiss and contain an average of 34% quartz, 11% plagioclase, 50% microcline perthite, 2% pyroxene and hornblende, 1% biotite, and 1.5% magnetite (Carl and Van Diver, 1975). Pyroxenes include both clinopyroxene and, locally, orthopyroxene (Cushing and Newland, 1925; Carl and Van Diver, 1975). Dark gray, plagioclase-rich gneisses, which are interlayered with the thicker, pink to light gray alaskitic gneisses,

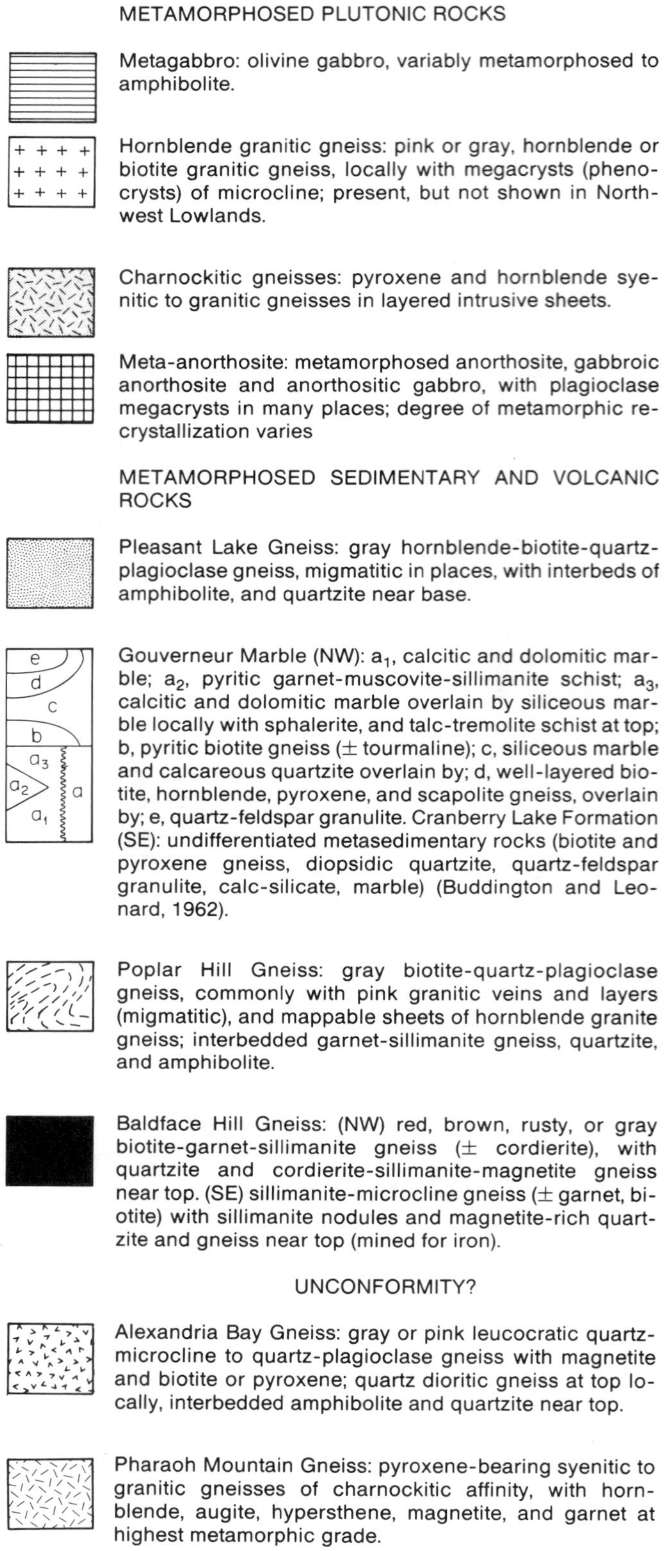

METAMORPHOSED PLUTONIC ROCKS

Metagabbro: olivine gabbro, variably metamorphosed to amphibolite.

Hornblende granitic gneiss: pink or gray, hornblende or biotite granitic gneiss, locally with megacrysts (phenocrysts) of microcline; present, but not shown in Northwest Lowlands.

Charnockitic gneisses: pyroxene and hornblende syenitic to granitic gneisses in layered intrusive sheets.

Meta-anorthosite: metamorphosed anorthosite, gabbroic anorthosite and anorthositic gabbro, with plagioclase megacrysts in many places; degree of metamorphic recrystallization varies

METAMORPHOSED SEDIMENTARY AND VOLCANIC ROCKS

Pleasant Lake Gneiss: gray hornblende-biotite-quartz-plagioclase gneiss, migmatitic in places, with interbeds of amphibolite, and quartzite near base.

Gouverneur Marble (NW): a_1, calcitic and dolomitic marble; a_2, pyritic garnet-muscovite-sillimanite schist; a_3, calcitic and dolomitic marble overlain by siliceous marble locally with sphalerite, and talc-tremolite schist at top; b, pyritic biotite gneiss (± tourmaline); c, siliceous marble and calcareous quartzite overlain by; d, well-layered biotite, hornblende, pyroxene, and scapolite gneiss, overlain by; e, quartz-feldspar granulite. **Cranberry Lake Formation (SE):** undifferentiated metasedimentary rocks (biotite and pyroxene gneiss, diopsidic quartzite, quartz-feldspar granulite, calc-silicate, marble) (Buddington and Leonard, 1962).

Poplar Hill Gneiss: gray biotite-quartz-plagioclase gneiss, commonly with pink granitic veins and layers (migmatitic), and mappable sheets of hornblende granite gneiss; interbedded garnet-sillimanite gneiss, quartzite, and amphibolite.

Baldface Hill Gneiss: (NW) red, brown, rusty, or gray biotite-garnet-sillimanite gneiss (± cordierite), with quartzite and cordierite-sillimanite-magnetite gneiss near top. (SE) sillimanite-microcline gneiss (± garnet, biotite) with sillimanite nodules and magnetite-rich quartzite and gneiss near top (mined for iron).

UNCONFORMITY?

Alexandria Bay Gneiss: gray or pink leucocratic quartz-microcline to quartz-plagioclase gneiss with magnetite and biotite or pyroxene; quartz dioritic gneiss at top locally, interbedded amphibolite and quartzite near top.

Pharaoh Mountain Gneiss: pyroxene-bearing syenitic to granitic gneisses of charnockitic affinity, with hornblende, augite, hypersthene, magnetite, and garnet at highest metamorphic grade.

Figure 4. Expanded stratigraphic column of Precambrian rocks in northwestern Adirondacks to accompany Figure 3.

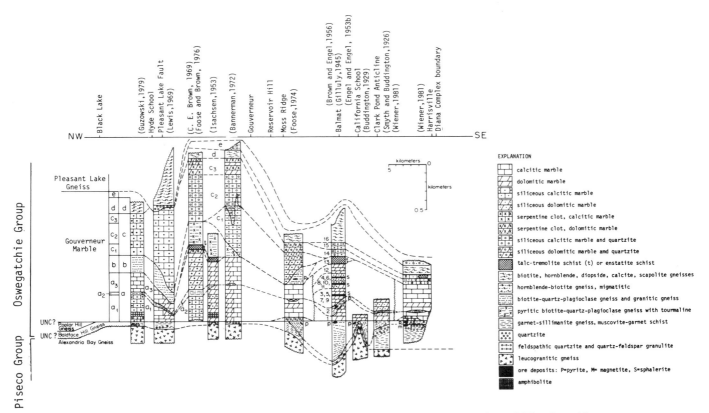

Figure 5. Stratigraphic correlation of rocks of the Northwest Lowlands shown prior to folding, but with observed thicknesses. Reinterpreted and modified from Guzowski (1979), Lewis (1969), Foose and Brown (1976), Brown (1969), Isachsen (1953), Bannerman (1972), Foose (1974), Brown and Engel (1956), Gilluly (1945), Engel and Engel (1953b), Buddington (1929), Smyth and Buddington (1926), and Wiener (1981). Units 1–16 at Balmat are units 1–16 of Brown and Engel (1956).

occur at the top and near the lowest exposed parts of the unit (Buddington, 1929; Carl and Van Diver, 1975). The plagioclase-rich gneisses contain an average of 20% quartz, 60% plagioclase, 10% potassium feldspar, 5% biotite, 2% magnetite, 2% pyroxene and hornblende, and 1% zircon, apatite, and other accessories (Carl and Van Diver 1975). Ubiquitous magnetite is responsible for aeromagnetic highs associated with this unit (Foose and Brown, 1976).

Thin amphibolites and quartzites are interbedded with the pink leucogranitic gneisses and are especially abundant near the top of the Alexandria Bay Gneiss (Cushing and others, 1910, p. 37; Buddington, 1929; Lewis, 1969; Carl and Van Diver, 1975). Carl and Van Diver (1975) also recognized garnet-sillimanite gneiss, calc-silicate granulite, carbonate, and biotite schist as interbeds in the leucogranitic gneiss unit, but these may be infolds of the overlying stratigraphic units.

The Alexandria Bay Gneiss occupies the cores of 14 domes in the Northwest Lowlands (Fig. 3), and Buddington (1929) concluded that each is a phacolith consisting of granite that was emplaced during folding and then deformed after initial crystallization. Engel and Engel (1963) emphasized metasomatic altera-

tion of a metasedimentary unit, possibly arkose, for the origin of this granitic gneiss, as did Bloomer (1969). Subsequent workers (Silver, 1965; Lewis, 1969; Carl and Van Diver, 1975; Foose and Carl, 1977; Foose and Brown, 1976; Romey and others, 1980; Wiener, 1981) have interpreted the unit as stratiform metamorphosed felsic volcanic rocks. Carl and Van Diver (1975) showed that internal mineralogical and chemical layering exist within the Alexandria Bay Gneiss in the classical Hyde School phacolith, which is now thought to be a dome. Mappable units of dark gray trondhjemitic and other plagioclase-rich gneiss occur at the top and near the lowest exposed sections of the Hyde School dome (Carl and Van Diver, 1975) and may be repeated by early folding (Foose and Carl, 1977). Carl and Van Diver (1975) interpreted the alaskitic and trondhjemitic units as representing metamorphosed rhyolitic ash-flow tuffs with a cap rock of quartz latite, and they suggested that these deposits may be partly water lain. Sedimentary reworking of this unit is suggested by the high percentage of quartz (25%–50%, Carl and Van Diver, 1975, p. 1695) in alaskitic gneisses of the Northwest Lowlands and also in the Adirondack Highlands where 32% to 41% quartz is present in alaskitic gneisses (Buddington and Leonard, 1962, p. 67). Re-

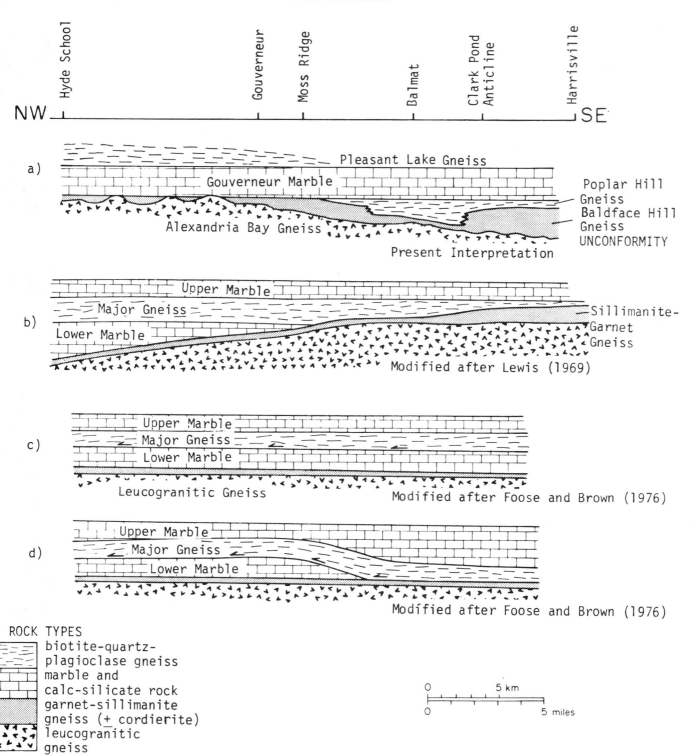

Figure 6. Four restored stratigraphic interpretations of the Northwest Lowlands: (a) This paper. (b) Lewis (1969), modified herein. (c) Foose and Brown (1976), Foose and Carl (1977). (d) Foose and Brown (1976), modified herein. Note unconformity at top of Alexandria Bay Gneiss suggested in this paper (a). See text for additional discussion.

working, or sedimentary (arkosic) parentage, is further suggested by rounded zircon grains in the leucogranitic gneiss noted by Carl and Van Diver (1975) and Eckelmann (1976).

Oswegatchie Group

Metamorphosed sedimentary rocks of the Oswegatchie Group unconformably overlie granitic gneisses of the Piseco Group. The Oswegatchie Group consists of four formations: Baldface Hill Gneiss, Poplar Hill Gneiss, Gouverneur Marble, and Pleasant Lake Gneiss, in order of decreasing age. The Oswegatchie Group is named for the Oswegatchie River, which courses across all the formations of the group in the Northwest Lowlands.

Baldface Hill Gneiss (0–1,000 m).
The Baldface Hill Gneiss (new name) overlies the Alexandria Bay Gneiss and consists mostly of garnet-sillimanite gneiss. The type locality of the Baldface Hill Gneiss is the thick section exposed along the rocky face of Baldface Hill, 3 to 5 km north of Harrisville, New York, in the Harrisville 7½-minute quadrangle (Wiener, 1981) (Fig. 3). The Baldface Hill Gneiss is up to 30 m (100 ft) thick and apparently discontinuous in the central and northwest part of the Northwest Lowlands, whereas it thickens to the southeast to >300 m (Figs. 3, 5, 6a). Although apparently discontinuous, this unit may actually be continuous but not exposed everywhere. Further mapping in the region should involve specific efforts to find the highly aluminous rocks of this unit.

The Baldface Hill Gneiss consists mostly of well-layered garnet-sillimanite gneiss, biotite-garnet-sillimanite gneiss, and biotite-cordierite-garnet-sillimanite gneiss (Buddington, 1929; Lewis, 1969; Foose and Brown, 1976; Wiener, 1981). These rocks are gray- or rusty brown-weathering and commonly display lenses, plates, or nodules that are rich in sillimanite and quartz or sillimanite, biotite, and cordierite, and stand out in relief on weathered surfaces. Near Harrisville, these rocks consist of biotite, garnet, sillimanite, cordierite, quartz, plagioclase, and perthite in highly varied proportions, along with local kyanite, spinel, hypersthene, pyrite, and magnetite, which locally form up to 10% of the rock (Wiener, 1981). Near the top of the unit, quartzite interbeds are abundant, whereas Al-, Fe-, Mg-rich magnetite-sillimanite-cordierite rock is locally present (Wiener, 1981).

We interpret rocks in the Baldface Hill Gneiss as representing metamorphosed regolith and basal clastics. A very thin, basal, Al-, Fe-, Mg-rich regolith (sillimanite-cordierite-magnetite rock) formed more or less in place in a subaerial environment and was mixed with deposits of shale (garnet-sillimanite gneiss) and mature quartz sands (quartzite), with a thicker clastic section to the southeast. Previous interpretations of the origin of the garnet-sillimanite gneiss emphasized its position adjacent to granitic rocks as indicative of origin as a contact metamorphic aureole developed by metasomatic alteration of amphibolite (Buddington, 1929). However, the expected contact metamorphic minerals formed between granite and calcitic or dolomitic marble would

be wollastonite, diopside, tremolite, and other calc-silicate minerals, and not garnet, sillimanite, and cordierite.

Poplar Hill Gneiss (0–1,000 m):
The Poplar Hill Gneiss overlies the Baldface Hill Gneiss and consists mostly of biotite-quartz-plagioclase gneiss. This unit is absent in the northwest and thickens to the southeast (Figs. 3, 5, 6a). The unit is named for a distinctive long roadcut on Route 58 and exposures on adjacent Poplar Hill, 1.5 mi west of Emeryville, between Balmat and Gouverneur, New York (Fig. 3). This unit was previously named the Major Paragneiss (Engel and Engel, 1953a, 1953b), the Major Gneiss (Lewis, 1969; Foose and Brown, 1976), and is here given a place name.

The typical rock in this unit is a rusty-brown- or dark-gray-weathering, gray, fine- to medium-grained biotite-quartz-oligoclase gneiss with thin layers and veins of pink to white granitic gneiss or granite (Engel and Engel, 1953b; Foose, 1974; Wiener, 1981). The granitic layers and veins may be extensively or sporadically distributed and are garnet-bearing in many places. Where granitic components of the rocks are extensively developed, the rock typically is migmatitic. Interbeds of amphibolite, quartzite, augite-hypersthene-hornblende-biotite-quartz-plagioclase-gneiss, garnet-sillimanite-biotite-quartz-plagioclase gneiss, and marble and calc-silicate rock are present (Engel and Engel, 1953b; Foose, 1974, Wiener, 1981). The garnet-sillimanite-biotite-quartz-plagioclase gneiss is similar to rocks in the interbedded and underlying Baldface Hill Gneiss. Stratabound, pyrite-rich biotite-quartz-plagioclase gneiss and nodular and layered pyrite-chlorite-quartz ores are present in many places at the top of the Poplar Hill Gneiss (Fig. 5) (Buddington and others, 1969; Wiener, 1981). These pyrite deposits are isotopically distinct from pyrite deposits described by Buddington and others (1969) in the pyritic schist members of the overlying Gouverneur Marble. Very thin but persistent marble and calc-silicate rock occur near the top of the Poplar Hill Gneiss near Emeryville (Foose and Brown, 1976; Foose and Carl, 1977) and Harrisville (Wiener, 1981) (Fig. 3). Also, quartzite and amphibolite beds are especially abundant near the top of the Poplar Hill Gneiss, as at Baldface Hill (Wiener, 1981).

The origin of the biotite-quartz-plagioclase gneiss in the Poplar Hill Gneiss has been the subject of much study. Engel and Engel (1953a, 1953b) studied the mineralogy and chemistry of this gneiss and concluded that the granitic portions were introduced and responsible for extensive metasomatism that increased toward the Adirondack Highlands. This interpretation refers to the gray biotite-quartz-oligoclase gneiss as the "least altered" gneiss, which has an enigmatic origin suggested to be graywacke, dacitic tuff, or sodic shale, based on the composition of the gray biotite gneiss alone, and not including the granitic portions of the gneiss. The biotite-quartz-plagioclase gneiss has since been studied by Carl (1978) who took whole-rock channel samples of both the gray biotite-quartz-plagioclase gneiss and the associated granitic gneiss. This sampling procedure considered the granitic portions of the gneiss to be locally derived by in situ partial melting, so that the bulk composition of the protolith of the gneiss includes both rock types and not only specific layers. The result of this study indicates that the protolith of this gneiss is argillaceous (illitic) arkose (Carl, 1978).

Gouverneur Marble (100–2,200 m). The Gouverneur Marble overlies the Baldface Hill Gneiss in the northwest and Poplar Hill Gneiss in the southeast (Figs. 3, 5). The primary sedimentary nature of this contact is indicated by the interbedded nature of the base of the Gouverneur Marble with the underlying rocks (Brown and Engel, 1956; Lewis, 1969; Foose and Brown, 1976; Wiener, 1981). We propose a type locality for the Gouverneur Marble, which is the long, high roadcut and surrounding hills on Route 58, 2.2 km southeast of the Oswegatchie River at Gouverneur, in the vicinity of Hailesboro. The exposures display layered calcitic marbles interbedded with amphibolitic and calc-silicate rocks, and are typical of much of Member A and Member C of the Gouverneur Marble. The name Gouverneur Marble is adapted from the name Gouverneur limestone applied by Buddington (1934, p. 137) to the apparently thick calcitic marble belt near Gouverneur. Other marble-rich stratigraphic units named in the Northwest Lowlands include the Black Lake metasedimentary belt and Balmat-Edwards marble belt (Engel and Engel, 1953a) and the lower marble and upper marble (Lewis, 1969; Foose and Brown, 1976). All of these named units are interpreted to be correlative in this paper and are included in the Gouverneur Marble.

The Gouverneur Marble is divided into five members in accordance with the work of Foose and Brown (1976), who used studies of Lewis (1969), Bannerman (1972), Brown (1969), Foose (1974), and Brown and Engel (1956). These members are here discussed in order of decreasing age with additional data included from other areas (Smyth and Buddington, 1926; Buddington, 1929; Gilluly, 1945; Isachen, 1953; Guzowski, 1979; Wiener, 1981) and with slight modifications (Figs. 3, 5). Member A is mostly marble and divided into three submembers in much of the area as follows: A_1, calcitic and dolomitic marble, locally siliceous, with amphibolite; basal quartzite and calc-silicate in places; A_2, thin, discontinuous, quartz-feldspar schist or gneiss, which is gray-, buff- or rusty-weathering and contains varied amounts of graphite, biotite, garnet, muscovite or sillimanite, and local pyritic ore deposits; A_3, siliceous marble with clots or layers rich in diopside or tremolite, dolomitic and calcitic marbles, and talc-tremolite schist locally at top of member.

Submembers A_1, A_2, and A_3 are tentatively correlated with units 1, 2, and 3 to 13, respectively, described by Brown and Engel (1956) for the Balmat area (Figs. 3, 5). Units 3, 5, 7, and 9 (Brown and Engel, 1956) may be a single unit of dolomite marble repeated by isoclinal folding and overlain by a single siliceous dolomite marble unit consisting of units 4, 6, 8, 10, and 11 repeated by isoclinal folding, as suggested by map patterns near Balmat (Brown and Engel, 1956). Alternatively, units 1 to 16 may represent a continuous stratigraphic sequence (de Lorraine, 1982, written commun.). The sequence siliceous dolomite, dolomite, talc-tremolite schist at Balmat (units 11, 12, and 13, Brown and Engel, 1956) may be correlative with a similar se-

quence present near Richville (Brown, 1969) at the top of submember A_3 of the Gouverneur Marble (Fig. 5). Also, stratabound sphalerite occurs in submember A_3 in the Balmat area (Brown and Engel, 1956) and near Beaver Creek and Bigelow to the northwest in the Gouverner Marble (Brown, 1969; Foose, 1981), further supporting connection of the Gouverneur and Balmat Marbles near Pierrepont (Fig. 3).

Member B is thin, discontinuous, and consists of gray- or rusty-weathering, tourmaline-bearing, biotite-quartz-plagioclase gneiss, which is migmatitic in places and locally contains pyritic ore. Quartzite, micaceous quartzite, and quartz-feldspar granulite are interbedded with biotite-quartz-plagioclase gneiss in Member B.

Member C contains siliceous calcitic and dolomitic marble, diopside or tremolite quartzite, pure diopside or tremolite granulite, microline-rich gneiss, and tremolite or enstatite-rich schists. Locally, Member C is divided into three submembers as follows: C_1, siliceous calcitic and dolomitic marble; C_2, thinly interbedded calcareous quartzite and either siliceous marble or calc-silicate granulite ("quartz mesh rock"); C_3, commonly pyritic (and rusty-weathering) marble, siliceous marble, calc-silicate granulite, and microcline granulite. Submembers C_2 and C_3 appear to correlate with units 14 and 15, respectively, in the Balmat area (Brown and Engel, 1956) and the Siliceous Marble and Rusty Calc-Silicate in Harrisville (Wiener, 1983).

Member D consists of well-layered quartz-feldspar gneisses with varied amounts of biotite, hornblende, diopside, scapolite, and calcite. Pronounced layering results in slabby exposures with ribs and troughs. Marble and calc-silicate rock are locally present at the base. Member D is here correlated with unit 16, the Median Gneiss, in the Balmat area (Brown and Engel, 1956) and the Diopside Gneiss near Harrisville (Wiener, 1983).

Member E is discontinuous and consists of pink to buff quartz-microcline granulite or gneiss and feldspathic quartzite.

Pleasant Lake Gneiss (0–1,050 m). The Pleasant Lake Gneiss overlies the Gouverneur Marble and is exposed only in the northwest part of the Northwest Lowlands, in the area around Pleasant Lake and Rossie, and also east of Canton, New York (Fig. 3). Good exposures can be seen within 1 km of Rossie, along the county road that leads north-northeast from the hamlet. This unit was previously named the Major Gneiss (Lewis, 1969; Foose and Brown, 1976), but is renamed to emphasize its distinction from the Poplar Hill Gneiss, to which it was previously correlated.

The Pleasant Lake Gneiss consists of a thin, basal, feldspathic quartzite, which is not everywhere present, overlain by medium- to coarse-grained migmatitic hornblende-biotite-quartz-microcline-plagioclase gneiss (Lewis, 1969) (Fig. 5). Thin interbeds of quartzite, rusty gneiss, and siliceous marble are present in the migmatitic gneiss (Lewis, 1969), which itself grades upward into orthoclase augen gneiss.

Rocks in this unit are similar in composition to the Poplar Hill Gneiss, but hornblende is present in the Pleasant Lake Gneiss, whereas garnet or sillimanite are commonly present in the

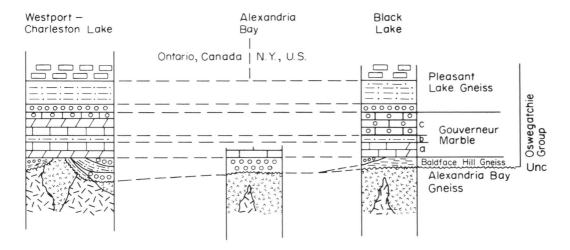

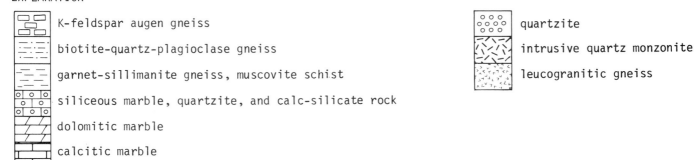

EXPLANATION

K-feldspar augen gneiss

biotite-quartz-plagioclase gneiss

garnet-sillimanite gneiss, muscovite schist

siliceous marble, quartzite, and calc-silicate rock

dolomitic marble

calcitic marble

quartzite

intrusive quartz monzonite

leucogranitic gneiss

Figure 7. Stratigraphic correlation of Westport map area in southeastern Ontario with Alexandria Bay and Northwest Lowlands. Not to horizontal or vertical scale. Modified from Wynne-Edwards (1967, Westport), Lewis (1969, Black Lake), and Fisher and others (1971, Alexandria Bay).

Poplar Hill Gneiss. Furthermore, the base of the Poplar Hill Gneiss is in contact with the Baldface Hill Gneiss (Fig. 5) and, locally, the Alexandria Bay Gneiss (Buddington, 1929), whereas the base of the Pleasant Lake Gneiss is in contact with Gouverneur Marble. The Pleasant Lake Gneiss is correlated with unnamed and identical sequence of quartzite, biotite gneiss, and augen gneiss in the Westport area (Wynne-Edwards, 1967) (Fig. 7).

Basal Oswegatchie Group Unconformity

We suggest that the metamorphosed volcanic (and plutonic?) rocks of the Piseco Group may be unconformably overlain by metamorphosed basal clastic and carbonate rocks of the Oswegatchie Group and correlative Lake George Group. This unconformity predates 1.1 to 1.0-b.y. metamorphism of the Grenville orogenic cycle (Silver, 1969).

In the Northwest Lowlands, the Alexandria Bay Gneiss of the Piseco Group is 1.27 b.y. old (Grant and others, 1981). A 1.22-b.y. U-Pb zircon age from the Poplar Hill Gneiss (Major Paragneiss) (Silver, 1965) may be a relict zircon from Alexandria Bay Gneiss deposited in the Oswegatchie Group. The age of

rocks of the Oswegatchie Group is poorly constrained. A dike in the Gouverneur Marble is 1.05 b.y. old (Foose and others, 1981, giving a minimum age for the marble. Correlative rocks of the Lake George Group are 1.17 to 1.02 b.y. old (Bickford and Turner, 1971; Silver, 1969). Evidence of an angular unconformity is preserved in the Hyde School dome of the Northwest Lowlands (Fig. 3), where contacts between dioritic and leucogranitic gneisses in the Alexandria Bay Gneiss are truncated by the overlying Baldface Hill Gneiss and Gouverneur Marble (Foose and Carl, 1977).

The unconformity is thought to have developed on an older, deformed terrane of felsic volcanic rocks (Alexandria Bay Gneiss). Subaerial weathering of the felsic volcanic terrane produced a very thin basal iron-, aluminum-rich regolith which was probably transported and deposited as sediment in many places. These rocks are represented by sillimanite-rich and cordierite-rich gneiss in the Baldface Hill Gneiss (Wiener, 1981, 1983). Succeeding deposits of shale (garnet-sillimanite gneiss), shaly arkose (biotite gneiss), and clean quartz sands (quartzite) were deposited above the transported regolith in a southeastward-thickening clastic wedge (Figs. 5, 6a). The pure quartz sands were the final weathering and erosional product of the buried terrane, which had a fairly thin carapace of clastic deposits. In places such as the St. Lawrence River area (Fisher and others, 1971), mostly sands remained on top of the Alexandria Bay Gneiss due to reworking of the other deposits. Carbonate deposition followed, forming limestone, dolomite, and local evaporitic deposits in closed basins developed on the underlying thin sands, arkoses, and shales. Eventually the carbonates were succeeded by deposition of illitic arkose as represented by biotite-quartz-plagioclase gneiss of the Pleasant Lake Gneiss.

Previous Interpretations of Stratigraphy and Structure of the Northwest Lowlands

There is a long history of investigation into the stratigraphy, structure, and petrology of rocks of the Northwest Lowlands. However, due to complex multiple folding and uncertainty regarding the intrusive or stratiform nature of certain units, significantly different regional stratigraphic syntheses have been presented.

Broadly classed, two hypotheses have been proposed concerning the stratigraphy and structure of the Northwest Lowlands: (1) The rocks of the Northwest Lowlands occupy the northwest-dipping limb of a large-scale fold and are thus in a homoclinal sequence if considered alone. (2) The continuous stratigraphic section of the Northwest Lowlands is repeated several times due to complex multiple folding.

The homoclinal sequence hypothesis was proposed by Buddington (1939) who interpreted these rocks to be on the southeast limb of a synclinorium and also to be thrust southeastward over the Adirondack Highlands (Fig. 8a). This interpretation suggests that the stratigraphic section is younger to the northwest (Fig. 3). Engel and Engel (1953a) interpreted the rocks to occupy the southeast limb of an overturned anticline and thus to be progressively older to the northwest (Fig. 8b). They (Engel and Engel, 1953a) named five stratigraphic belts in their interpretation, which are from oldest to youngest (Fig. 8b): Black Lake metasedimentary belt, Gouverneur Marble belt, Antwerp-Hermon Major Paragneiss belt, Balmat-Edwards upper marble belt, and Harrisville-Russell upper feldspathic granulite. Walton and deWaard (1963), Jacobi (1964), and Wynne-Edwards (1967) agreed with Buddington's (1939) synclinorium interpretation, but discounted the presence of a thrust fault at its southeast margin (Fig. 8c). They suggested that the stratigraphic section in the Northwest Lowlands unconformably overlies the Diana Complex (Walton and de Waard, 1963; Jacobi, 1964) and, furthermore, that the Diana Complex, along with other metamorphosed plutonic rocks of the Adirondack Highlands, constitutes a basement complex (Fig. 8c). Wynne-Edwards (1967) proposed that stratigraphy of the Northwest Lowlands is right-side-up based on primary facing criteria near Westport, Ontario.

The hypothesis that the stratigraphic section in the Northwest Lowlands is repeated by folding was presented by Brown (1936), who worked in the Balmat-Edwards mining district. He correlated the Balmat-Edwards marble belt with the Gouverneur

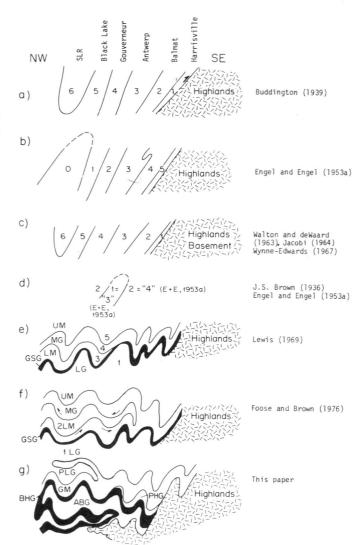

Figure 8. Schematic NW-SE cross-sectional diagrams, not to scale, showing generalized structural/stratigraphic theories of Northwest Lowlands. Numbers indicate relative age of stratigraphic units. SLR = St. Lawrence River. (e) and (f) UM = Upper Marble, MG = Major Gneiss, LM = Lower Marble, GSG = garnet-sillimanite-gneiss, LG = Lower Gneiss. (g) PLG = Pleasant Lake Gneiss, GM = Gouverneur Marble, PHG = Poplar Hill Gneiss, BHG = Baldface Hill Gneiss, ABG = Alexandria Bay Gneiss.

marble belt and interpreted this single marble unit to overlie the biotite-quartz-plagioclase gneiss (Poplar Hill Gneiss in this paper) (Fig. 8d). Brown's concept is an essential part of the stratigraphic model adopted in this paper as well as ones proposed by Lewis (1969) and by Foose and Carl (1977). Lewis (1969) proposed that the Black Lake belt and Gouverneur Marble belt of Engel and Engel (1953a) are equivalent and that the Major Paragneiss of Engel and Engel (1953a) is equivalent to the Major Gneiss in the Black Lake area (Figs. 6b, 8e). He also pointed out the stratigraphic sequence of leucogranitic gneiss overlain successively by garnet-sillimanite gneiss and carbonates, which is a key to present interpretations (Figs. 8e, 8f, 8g). This stratigraphic sequence was also mapped in the Bigelow area by Foose (1974) and shown to be persistent throughout much of the Northwest Lowlands (Foose and Brown, 1976; Foose and Carl, 1977). Foose and Brown (1976) proposed that the Gouverneur and Balmat marble belts are correlative (Fig. 8f) and further subdivided the Gouverneur Marble into five members. In addition, they proposed that the biotite-quartz-plagioclase gneiss (Poplar Hill Gneiss in this paper) is thrust over the Gouverneur and Balmat Marbles and is correlative with the Major Gneiss in the Rossie area studied by Lewis (1969) (Pleasant Lake Gneiss of this paper) (Figs. 6b, 8f). They also discussed the effects of three periods of folding on these rocks.

Stratigraphic interpretations consistent with recent advances in understanding of the multiply folded rocks of the Northwest Lowlands (Foose and Carl, 1977; deLorraine, 1979; Wiener, 1981) and the stratigraphic nature of the contact between the Alexandria Bay Gneiss and overlying Baldface Hills Gneiss (Carl and Van Diver, 1975; Foose and Carl, 1977) are presented in Figure 6. Lewis (1969) suggested the presence of five stratigraphic units (Fig. 6b) in the Black Lake area near Hyde School where he did his detailed mapping. In his interpretation (Lewis, 1969), the lower marble pinches out stratigraphically below the Major Gneiss southeast of Gouverneur, and the marbles and calc-silicates near Balmat are equivalent to the upper marble near Hyde School (Fig. 6b). However, the Balmat-Harrisville ("upper") marble belt appears to be the same stratigraphic unit as the Gouverneur ("lower") marble belt, since the two apparently connect near Pierrepont (Fig. 3). Furthermore, stratabound sphalerite occurs in the Gouverneur Marble in both belts (Foose, 1981).

Also, the "upper marble" and "lower marble" near Rossie (Lewis, 1969) are interpreted as one stratigraphic unit in this paper because they apparently connect around the Major Gneiss (Pleasant Lake Gneiss) ~7 km south of Rossie, and also ~6 km east-southeast of Rossie (Fig. 3). In further support of this hypothesis, the lower and upper marbles (Lewis, 1969) are lithically similar, consisting of diopsidic calcite marble with thin beds of rusty biotitic gneiss with quartzite beds at the Major Gneiss contact (Fig. 9a). Thus, the lower and upper marbles are interpreted to be the same units and correlative to the Gouverneur Marble, and to be successively overlain by thin discontinuous quartzite and the Pleasant Lake Gneiss (Fig. 9b).

A stratigraphic sequence consisting of five units, as shown in Figure 6c, has been suggested (Foose, 1974; Foose and Brown, 1976; Foose and Carl, 1977). This interpretation involves a thrust fault between the Major Gneiss and the lower marble (Fig. 6c). This thrust or slide is based on the truncation of stratigraphic units on opposite sides and on the presence of slabby augen gneiss near the lower contact of the Major Gneiss (Foose, 1974) in the vicinity of Bigelow, New York (Fig. 3). Local thrusting is undoubtedly present along this contact, but a major regional thrust fault along this contact throughout the Northwest Lowlands is less certain. One problem is that the sequence Alexandria Bay Gneiss, Bald-

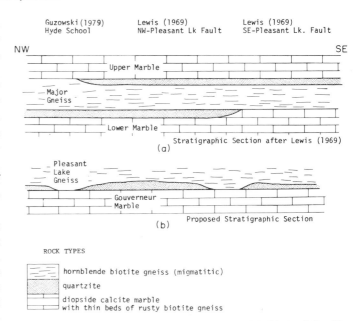

Figure 9. Correlation of Upper Marble and Lower Marble (as defined by Lewis, 1969). (a) Stratigraphic sequence proposed by Lewis (1969). Note repetition of quartzite and diopside calcite marble above and below the biotite gneiss (Major Gneiss), suggesting the presence of isoclinal fold with axial surface in Major Gneiss. (b) Stratigraphic sequence proposed in this paper as a result of the correlation of the Upper Marble and Lower Marble.

face Hill Gneiss, Poplar Hill Gneiss (Major Gneiss), and Gouverneur Marble (lower marble) is continuously exposed near Baldface Hill, Clark Pond, and at the California School antiform (Buddington, 1929; Gilluly, 1945; Wiener, 1981) (Fig. 3). This sequence would seem to suggest the configuration of Figure 6d rather than Figure 6c. However, the critical relations where the thrust cuts across the section to the garnet-sillimanite gneiss (Fig. 6d) are not exposed anywhere. Instead, the sequence in Figure 6a where the Gouverneur (and Balmat) marble belts are in contact with the top of the Baldface Hill Gneiss and Poplar Hill Gneiss is observed (Fig. 6a). Also, the thrust contact configuration fails to explain, unless one appeals to thrust slices, the interbedding of garnet-sillimanite gneiss and biotite-quartz-plagioclase gneiss observed in many localities. The intimate interbedding of the Baldface Hill Gneiss and Poplar Hill Gneiss in these areas, as well as their position in mutual contact throughout the Balmat-Harrisville region (Fig. 3), indicates that these two formations are in normal stratigraphic contact there (Fig. 6a) and not in thrust contact (Fig. 6d). Also, the similarity of protoliths of the Baldface Hill Gneiss (shale and sandstone) and Poplar Hill Gneiss (shaly arkose) suggests a primary sedimentary contact between these two formations.

Another problem with the interpretation of Figure 6d—that a regional thrust is at the base of the Poplar Hill Gneiss—is the presence of interbeds of marble and quartzite at the top of the Poplar Hill Gneiss and interbeds of biotite gneiss and garnet-sillimanite gneiss (commonly rusty-weathering) near the base of

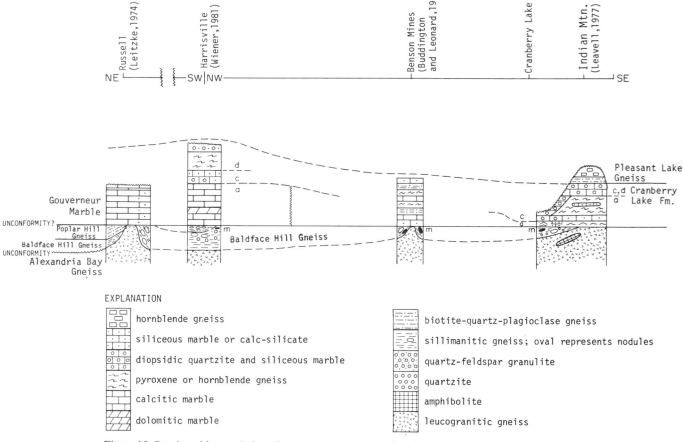

Figure 10. Stratigraphic correlation of southeastern Northwest Lowlands and northwestern Adirondack Highlands. Modified from Leitzke (1974), Wiener (1981), Buddington and Leonard (1962), and Leavell (1977); a, c, d = members of Gouverneur Marble; m = magnetite.

the Gouverneur Marble. This interbedding of the basal Gouverneur Marble with the Baldface Hill Gneiss and Poplar Hill Gneiss, reported by many workers in the Northwest Lowlands, suggests its lower contact is a primary sedimentary contact.

Workers near Balmat suggest the configuration of Figure 6c without the regional thrust, that is, the Harrisville and Gouverneur (lower) marbles are correlative and overlain by the Major Gneiss, which is in turn overlain by the Balmat (upper) marble (deLorraine, 1982, written commun.). One problem with this hypothesis is the apparent connection of the Harrisville and Balmat marbles around the northeast nose of the doubly plunging anticline near Baldface Hill (Fig. 3). Additional field work needs to be done to resolve the various stratigraphic hypotheses of the Northwest Lowlands.

STRATIGRAPHY OF WESTERN ADIRONDACK HIGHLANDS

The western Adirondack Highlands is the area west of the Marcy anorthosite and east of the Northwest Lowlands (Fig. 1, eastern half of Fig. 3). The stratigraphy and structure of the west-

ern Adirondack Highlands is less well known than that of the Northwest Lowlands due to the large intrusive complexes, poor exposure of metasedimentary rocks, small-scale mapping, and lack of detailed stratigraphic mapping. The main source of information is Buddington and Leonard's (1962) geologic map of St. Lawrence County (1:62,500). Recently, some parts of this region have been the subject of detailed mapping by graduate students at the University of Massachusetts. Leavell (1977) mapped a portion of the Darning Needle syncline in the Cranberry Lake 15-minute quadrangle; Bran Potter is currently mapping part of the Loon Pond syncline near Sabbatis, New York, and Page Fallon is mapping part of the Little Tupper Lake area, both in the Tupper Lake 15-minute quadrangle.

Piseco Group

Pharaoh Mountain Gneiss. This formation is formally defined in the section on the stratigraphy of the central and eastern Adirondacks. In the western Adirondack Highlands, it consists of interlayered charnockitic and granitic gneisses exposed in the cores of domes and anticlines.

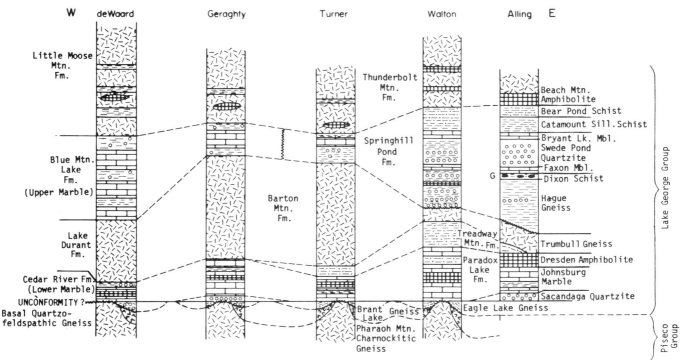

Figure 11. Stratigraphy and correlation of central and eastern Adirondacks prior to folding, as proposed by Geraghty and others (1975), with slight modifications. Sections from Walton and deWaard (1963), Turner, (1979), Geraghty (1978), and Alling (1917, 1927). Not to horizontal or vertical scale. Note possible early isoclinal fold cored by Lake Durant Formation, which may repeat the stratigraphy (see Fig. 13). Symbols same as in Fig. 5.

Alexandria Bay Gneiss. Pink leucogranitic gneisses in the western Adirondack Highlands overlie the Pharaoh Mountain Gneiss and are correlative with the Alexandria Bay Gneiss of the Northwest Lowlands. A. F. Buddington considered the pink, equigranular, leucogranitic (alaskitic) gneisses of the Northwest Lowlands to be the same rock series as those in the Adirondack Highlands (Buddington and Leonard, 1962; Buddington, 1929). Although Buddington believed these rocks to be intrusive, we believe them to be metamorphosed felsic volcanic rocks equivalent to the Alexandria Bay Gneiss of the Northwest Lowlands. In the Highlands these rocks are characteristically rich in quartz (32%–41%) and poor in mafic minerals (about 2%), which are mainly biotite and magnetite (Buddington and Leonard, 1962).

A 2,300-ft thickness of leucogranitic gneisses is present in the Cranberry Lake area (Fig. 3) and has been divided into three parts by Leavell (1977). These subdivisions consist of homogeneous leucogranitic gneiss in the middle that is physically overlain and underlain by interlayered leucogranitic gneiss and amphibolite. This physical repetition of rock types suggests that the axial surface of an isoclinal fold is present in the homogeneous gneiss. This fold is interpreted to be a stratigraphic anticline on the basis of regional correlation with the Alexandria Bay Gneiss (Fig. 10).

Oswegatchie Group

Baldface Hill Gneiss. A discontinuous unit mapped by Buddington and Leonard (1962) as sillimanite-microcline granite gneiss ("gms") overlies the Alexandria Bay Gneiss in the western Adirondack Highlands and is correlated with the Baldface Hill Gneiss (Fig. 11) on the basis of position and rock types. Although Buddington and Leonard (1962) considered these rocks to be a contaminated granite, the peraluminous and quartz-rich gneisses seem to clearly represent metamorphosed sedimentary rocks. There are also granitic gneisses in the "gms" unit (Buddington and Leonard, 1962), and their origin is open to interpretation, although we think they represent metamorphosed sedimentary rocks for the most part.

The sillimanite-microcline gneisses ("gms") are rich in quartz and microcline and contain sillimanite, biotite, garnet, muscovite (after sillimanite), and an average of 2.5% magnetite (Buddington and Leonard, 1962). Modes of sillimanite-microcline gneiss shown by Buddington and Leonard (1962) may selectively represent the least aluminous facies, since they wanted to sample what they believed to be granite that is least contaminated. The sillimanite-microcline gneisses are also characterized by aggregates consisting of varied proportions of sillimanite and quartz, and minor magnetite, that occur as sillimanite-rich eyes, discs, lenticles, platy lenses, and quartz-rich nodules up to several inches long and stand out in relief on weathered outcrop surfaces (Buddington and Leonard, 1962). The nodules are 1 to 6 in. long and 0.5 to 1 in. wide and may form up to 25% of the rock, giving it a "conglomerate appearance" (Buddington and Leonard, 1962,

p. 73). Similar quartz-sillimanite nodular gneisses of the Baldface Hill Gneiss overlie Alexandria Bay Gneiss in the Russell area (Fig. 10) (Leitzke, 1974).

In the Cranberry Lake area (Fig. 3), Leavell (1977) mapped magnetite-rich feldspathic quartzite in many places near or at the top of the Baldface Hill Gneiss (Fig. 10). These rocks give rise to magnetic anomalies of remarkable stratigraphic consistency and have been mined for iron at various places including Benson Mines (Buddington and Leonard, 1962; Leonard and Buddington, 1964). Magnetic anomalies are present in many places at the contact of the Alexandria Bay Gneiss with the Gouverneur Marble (Cranberry Lake Formation of Leavell, 1977). These anomalies occur where no intervening Baldface Hill Gneiss ("gms" of Buddington and Leonard) has been previously mapped, suggesting that the Baldface Hill may be present although thin and/or not exposed. These magnetite-rich quartzites and sillimanite-cordierite gneisses are also present at the top of the Baldface Hill Gneiss in the Northwest Lowlands (Wiener, 1981).

Poplar Hill Gneiss. The rocks at the base of the Poplar Hill Gneiss are interbedded with those at the top of the Baldface Hill Gneiss in the Northwest Lowlands. This situation likely prevails in the western Adirondack Highlands where the Poplar Hill Gneiss may be the biotite-rich facies of "gms" and much of "gma" mapped by Buddington and Leonard (1962). Therefore, although not shown throughout much of the western Adirondack Highlands on Figure 3, the Poplar Hill Gneiss may be present, but not separately mapped. Furthermore, Poplar Hill Gneiss may be present in the "ms" map unit of undifferentiated metasedimentary rocks (Buddington and Leonard, 1962). Biotite-quartz-plagioclase gneiss, garnet-biotite migmatitic gneiss, and biotite-sillimanite gneiss, at the base of the "ms" unit of Buddington and Leonard (1962), are similar to rocks in the Poplar Hill Gneiss and in a comparable stratigraphic position (Fig. 10). Biotite gneisses that have been mapped separately in the Sabbatis area (Fig. 3) (Buddington and Leonard, 1962) are correlated with the Poplar Hill Gneiss on the basis of the lithic similarity and their position above sillimanite gneiss. A large terrane of sillimanite or biotite microcline granite gneiss interlayered with amphibolite is also exposed in the area north of Cranberry Lake. These rocks are tentatively correlated with the Poplar Hill Gneiss owing to their stratigraphic position and the prevalence of biotite-rich rocks.

Cranberry Lake Formation. The Cranberry Lake Formation was originally described by Leavell (1977), and it is proposed for formal usage here. Typical exposures of the Cranberry Lake Formation are present at the base of Indian Mountain near the south end of Cranberry Lake (Fig. 3). The lower part of the Cranberry Lake Formation consists of well-layered pyroxene-hornblende gneiss, calc-silicate gneisses and granulites, and marble. The upper part consists mostly of rusty-weathering diopsidic quartzites and granulites, siliceous marbles, biotite-hornblende gneiss, and, at the top of the unit, calc-silicate rocks containing large crystals of diopside, hornblende, and scapolite (Leavell, 1977).

The lower part of the Cranberry Lake Formation as defined by Leavell (1977) is correlated here to the Whippoorwill Corners Formation of Leitzke (1974) and to undifferentiated metasedimentary rocks mapped as unit "ms" by Buddington and Leonard (1962). Map units "ms" consists of various biotite and pyroxenic gneisses, marbles, quartzites, and calc-silicate rocks. The upper part of the Cranberry Lake Formation (Leavell, 1977) is here correlated to a quartz-rich map unit ("msq") and a pyroxene quartzite ("mspg") map unit of Buddington and Leonard (1962). These units consist of calcareous quartzite, calcareous quartz-feldspar granulites and gneisses, and lesser calcitic marble, tremolite schist, and diopside granulite, all of which are variably pyritic.

Cranberry Lake Formation, including map units "ms," "msq," and "mspq," is correlated with the Gouverneur Marble, which is interpreted to be its facies equivalent. The rocks mapped as "ms" by Buddington and Leonard (1962), excluding those parts rich in biotite gneiss, are probable facies equivalents of Member A of the Gouverneur Marble. In support of this correlation, Buddington and Leonard (1962) suggested that marble is much more prevalent in unit "ms" than scant exposures indicate. Although only two exposures of marble were encountered in their mapping of several hundred square miles, marble was intersected in eight of ten drill holes in pyroxene skarn iron ore deposits and in two holes out of six in granite iron ore deposits. In none of these instances was marble observed in the related rocks exposed at the surface. Therefore, Buddington and Leonard (1962) concluded that marble is a widespread, major member of the metasedimentary rocks (map unit "ms") as a whole. Thus, it is reasonable to correlate the lower part of the Cranberry Lake Formation with Member A of the Gouverneur Marble (Fig. 10), despite the lack of surface exposures of marble. However, a facies change within the lower Cranberry Lake Formation is necessary to explain the abundance of pyroxene-hornblende gneisses in the western Adirondack Highlands. Rocks mapped as "msq" by Buddington and Leonard (1962) are probably equivalent to Members C and D, and "mspq" is probably equivalent to Member E of the Gouverneur Marble. Biotite-quartz-plagioclase gneisses, garnet-biotite migmatitic gneisses, and sillimanite-biotite gneisses in the "ms" map unit (Buddington and Leonard, 1962) probably represent undifferentiated parts of the underlying Poplar Hill Gneiss and Baldface Hill Gneiss, where these rocks occur at the base of the "ms" unit, but not where they are interlayered with the calcareous rocks in the Cranberry Lake Formation. Other rock types in the "ms" map unit, which are part of the Cranberry Lake Formation, include marble, calc-silicate granulite, amphibolite, hornblende gneiss, pyroxene gneiss, quartzite, feldspathic quartzite, and quartz-feldspar granulite (Buddington and Leonard, 1962).

Pleasant Lake Gneiss. Biotite and hornblende gneisses and amphibolites overlie the Cranberry Lake Formation. These rocks correlate with hornblende-biotite gneisses with interbedded quartzites and amphibolites of the Pleasant Lake Gneiss, exposed near Rossie, New York, in the Northwest Lowlands (Fig. 3). In the Cranberry Lake area (Leavell, 1977), the Pleasant Lake Gneiss consists of fine- to medium-grained, gray, well foliated,

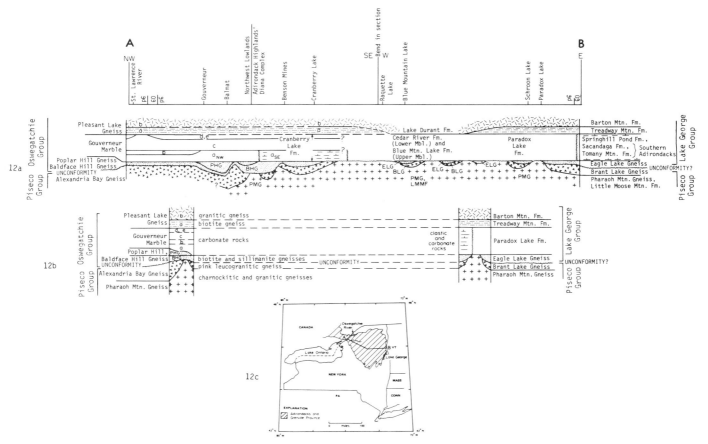

Figure 12. Stratigraphic correlation of northwestern, central, and eastern Adirondacks. (12a) Correlation assuming Upper Marble is equivalent to Lower Marble and Little Moose Mountain Formation equivalent to Piseco Group. Vertical saw tooth lines indicate facies change. (12b) Correlation of lithic types of Oswegatchie Group and Lake George Group. (12c) Location of line of restored section.

biotite gneisses interbedded with feldspathic quartzites that are overlain by black- to brown-weathering biotite-hornblende gneisses with porphyroblasts of hornblende and feldspar. Similar hornblende megacrystic rocks are present at the top of the Pleasant Lake Gneiss in the Northwest Lowlands near Rossie (Lewis, 1969).

STRATIGRAPHY OF THE CENTRAL AND EASTERN ADIRONDACKS AND THEIR CORRELATION

The central and eastern Adirondacks comprise the northern and eastern part of the area outlined in Figure 1. A geologic map of part of this area is shown in Figure 14. Pioneering stratigraphic studies in this area were done by Harold Alling (1917, 1927) during his research on Adirondack graphite deposits in the eastern Adirondacks. Subsequent detailed geologic mapping was done in the eastern Adirondacks in the Paradox Lake, Elizabethtown, Port Henry, and Ticonderoga 15-minute quadrangles by Matt Walton (Walton, 1961; Walton and deWaard, 1963) and by his students (Hills, 1964; Berry, 1961; Turner, 1979; Walton

and Turner, 1963), and in the central Adirondacks by Dirk de-Waard and his students (Lettney, 1969; Geraghty, 1978, 1979). Walton and deWaard (1963) established a stratigraphic sequence of five supracrustal formations consisting of metasedimentary rocks, which they interpret to overlie a basement complex of anorthosite, charnockitic and granitic gneisses, and paragneisses (Fig. 11). Geraghty and others (1975) extended the mapping of Walton and deWaard (1963) and proposed a correlation of the five stratigraphic units of the central and eastern Adirondacks, herein named the Lake George Group.

We redefine the basement complex (Walton and deWaard, 1963) to consist only of charnockitic and granitic gneisses of the Piseco Group (named herein). Furthermore, we suggest that the anorthosite represents a younger intrusive and that the paragneisses are the basal formation of the overlying Lake George Group. Therefore, we interpret the anorthosite and the paragneisses to be younger than the basement (Piseco Group) instead of part of it, as originally thought by Walton and deWaard (1963). In addition, we present the stratigraphy and correlation of the Lake George Group of the central and eastern Adirondacks as proposed by Geraghty and others (1975) and also suggest correla-

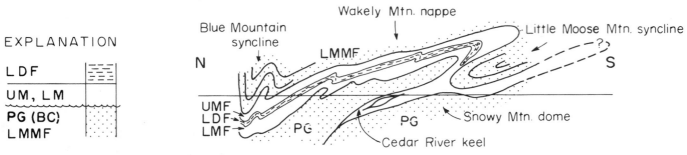

Figure 13. Structural cross section in central Adirondacks, after Geraghty and deWaard (written commun.). Interpretation here shows possible early F_1 isoclinal fold in Lake Durant Formation, which results in connection and correlation of upper marble (UMF) and lower marble (LMF) and also of Piseco Group (PG) and Little Moose Mountain Formation (LMMF). F_1 isoclinal fold in Lake Durant Formation refolded by F_2 Wakely Mountain nappe and Little Moose Mountain syncline, and F_3 Snowy Mountain dome and Blue Mountain syncline.

tions with the stratigraphy of the northwestern Adirondacks (Fig. 12). Two modifications of the stratigraphy of the central and eastern Adirondacks include tentative proposals that the lower marble and upper marble (Fig. 11) are the same formation, and that the Little Moose Mountain Formation is equivalent to granitic gneisses of the Piseco Group. In this interpretation, both the marble and Piseco Group are repeated about an early isoclinal fold in the Lake Durant Formation (Fig. 13).

Piseco Group

Basement in the central and eastern Adirondacks consists of the Pharaoh Mountain Gneiss and overlying Brant Lake Gneiss, which together constitute the Piseco Group. The Pharaoh Mountain Gneiss consists of interlayered charnockitic and granitic gneisses, and the Brant Lake Gneiss consists of pink leucogranitic gneisses. Resistant rocks of the Piseco Group are exposed in most of the ridges and mountain ranges of the Adirondacks, excepting the High Peaks. The Piseco Group is named for exposures of interlayered charnockitic, granitic, and leucogranitic gneisses in the Piseco dome near Piseco Lake, New York, in the southern Adirondacks (Figs. 15, 16). Several rock units were mapped near Piseco Lake by Cannon (1937) and Glennie (1973), and they are here considered subdivisions of the Piseco Group. The Piseco Group is equivalent to the Basal Quartzo-Feldspathic Gneiss (McLelland and Isachsen, 1980) (Fig. 17), but is given a place name in accordance with stratigraphic principles and to emphasize its separateness from the overlying Lake George Group.

Pharaoh Mountain Gneiss. Interlayered charnockitic and granitic gneiss constitutes the lowest stratigraphic unit in the central and eastern Adirondacks (Walton and deWaard, 1963; Geraghty, 1978). Charnockitic gneisses in large domical massifs in the eastern Adirondacks were referred to as the Pharaoh Mountain and Ticonderoga massifs (Walton, 1961). Therefore, the name Pharaoh Mountain Gneiss is adopted for this unit for exposures on Pharaoh Mountain in the Paradox Lake 15-minute quadrangle. Typical exposures can be observed on Pharaoh Mountain, on the east side of Schroon Lake, and in the woods on

the east side of the town road north from Adirondack toward Schroon Lake Village.

The charnockitic and mangeritic gneisses in the Pharaoh Mountain Gneiss are tan to maple-sugar-brown weathering and gray to gray-green on fresh surfaces. Orthopyroxene, clinopyroxene, hornblende, and locally garnet are present in these mesoperthite-rich gneisses, which contain subordinate interlayered amphibolite (Geraghty, 1978). Blue-gray xenocrysts of andesine and xenoliths of anorthosite are present in the charnockitic gneisses adjacent to anorthosite plutons (deWaard and Romey, 1963, McLelland and Isachsen, 1980), suggesting melting of charnockite in proximity to anorthosite plutons, resulting in intrusion of remelted charnockite into partially solidified anorthosite (McLelland and Isachsen, 1980). Granitic gneisses in the Pharaoh Mountain Gneiss are gray, pink, or tan, and contain hornblende and, locally, garnet and interlayered amphibolite (Geraghty, 1978). The charnockitic and granitic gneisses in this unit are thinly interlayered in places, as can be seen in maps of the Bolton Landing 7½-minute quadrangle (McConnell, 1965) and in the Skiff Mountain area of the Paradox Lake 15-minute quadrangle (Walton and Turner, 1963; Walton, 1961).

Exposures of the basal charnockitic and granitic gneisses in the central Adirondacks include the Thirteenth Lake dome (Lettney, 1969), Snowy Mountain dome, and Wakely Mountain nappe (McLelland and Isachsen, 1980). In the eastern Adirondacks, characteristic domical exposures include the Pharaoh Mountain and Ticonderoga massifs (Walton, 1961), Lake George Pluton (Hills, 1964), and "Whitehall-type" gneiss (Berry, 1961).

Brant Lake Gneiss. The Brant Lake Gneiss (named by Bickford and Turner, 1971) overlies the Pharaoh Mountain Gneiss in many places in the eastern Adirondacks. The Brant Lake Gneiss is equivalent to the Alexandria Bay Gneiss and consists of pink to gray, fine-grained leucogranitic gneiss interlayered with similar appearing pink quartz-albite granulite or gneiss, both containing sparse biotite, pyroxene, and magnetite. Within the leucogranitic gneisses of the Brant Lake Gneiss are magnetite iron ore deposits. Quartzites are interlayered with the

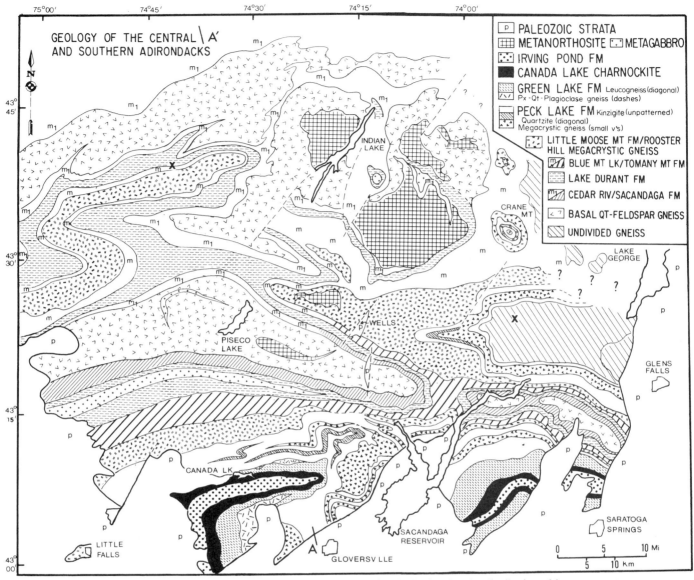

Figure 14. Geologic map of the central and southern Adirondacks showing the distribution of formational units throughout the area. The two X's locate bodies of metagabbro referred to in the text. Most of the faults have been omitted for the sake of clarity, but several are shown as dashed lines where their presence is essential to interpreting the geology. Generalized from field maps at 1:24,000 and 1:62,500 (on open file at New York State Geological Survey) and from New York State Geological Map, 1:250,000 (Fisher and others, 1971). From McLelland and Isachsen (1980).

leucogranitic gneiss near its top (Geraghty, 1978), as is the case with the Alexandria Bay Gneiss in the northwestern Adirondacks. The leucogranitic gneisses have locally undergone extensive melting, as indicated by their minimum melt composition and the "synmetamorphic" 1.1-b.y. age of the Brant Lake Gneiss in the Brant Lake dome (Bickford and Turner, 1971). However, not all the rocks in the Brant Lake Gneiss have melted, and whatever melting did occur in this unit was in situ, as indicated by its stratigraphic coherence (Turner, 1979). The protolith of the Brant Lake Gneiss may have been felsic volcanics, as is the case

for the Alexandria Bay Gneiss. These rocks underwent anatexis at the high metamorphic grade in the eastern Adirondacks.

The Brant Lake Gneiss of the eastern and central Adirondacks is correlated with the Lyon Mountain granite (gneiss) (Postel, 1952) of the northeastern Adirondacks and the Alexandria Bay Gneiss of the northwestern Adirondacks (Fig. 12) on the basis of similar lithology, high magnetic character (and magnetite deposits), and stratigraphic position. The 1.1-b.y. ages in the Brant Lake Gneiss are equivalent to the younger ages reported from the Alexandria Bay Gneiss (Maher and others, 1981).

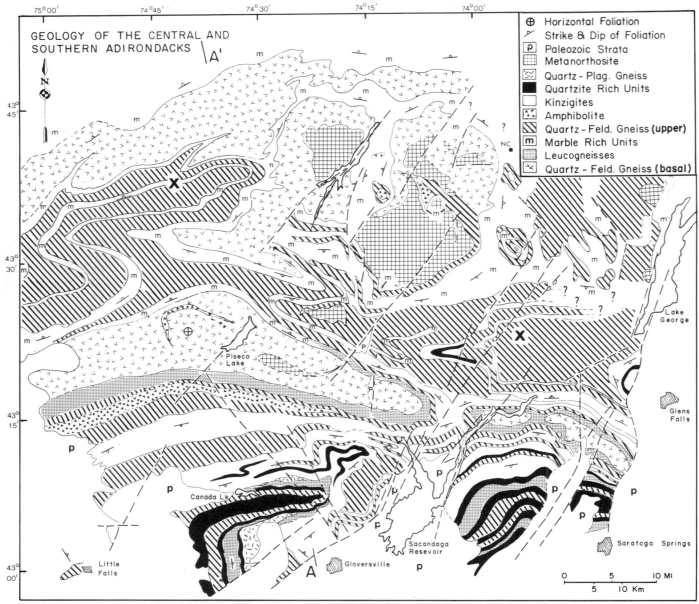

Figure 15. Geologic map of the central and southern Adirondacks showing the distribution of lithologies throughout the area. The two X's refer to bodies of metagabbro discussed in the text. Major faults are shown by dashed lines. Generalized from field maps at 1:24,000 and 1:62,500 (on open file at New York State Geological Survey) and from New York State Geological Map, 1:250,000 (Fisher and others, 1971). Most of the faults have been omitted for simplicity; for fault data, see Isachsen and McKendree (1977). From McLelland and Isachsen (1980).

Lake George Group

Metamorphosed sedimentary rocks of the Lake George Group unconformably overlie granitic gneisses of the Piseco Group. The Lake George Group is correlative with the Oswegatchie Group and consists of the Eagle Lake Gneiss, Paradox Lake Formation, Treadway Mountain Formation, and Lake Durant Formation, in order of decreasing age. The Springhill Pond Formation and Thunderbolt Mountain Formation physically overlie

the Lake Durant Formation, but may be stratigraphic equivalents of the Paradox Lake Formation and Piseco Group, respectively. The Lake George Group is named for exposures of metamorphosed sedimentary rocks in the area of Lake George and the region north and west of there.

Eagle Lake Gneiss. The Piseco Group is overlain in places by the thin, discontinuous Eagle Lake Gneiss, the basal formation of the Lake George Group. This unit was formerly named the Older Paragneiss (Walton and deWaard, 1963) and is renamed

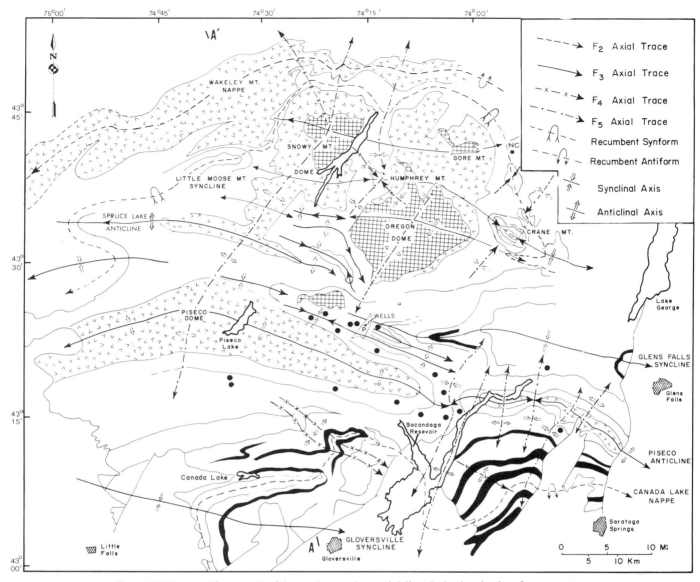

Figure 16. Structural framework of the southern and central Adirondacks showing interference patterns produced by four regional fold sets. Map units are generalized from Figures 14 and 15. Dots locate small meta-anorthositic intrusives referred to in the text. P refers to Paleozoic strata in Wells inlier. Lithologic symbols from Figure 15. From McLelland and Isachsen (1980).

since it is now thought to be part of the younger Lake George Group. The Eagle Lake Gneiss is named for exposures on the north shore of Eagle Lake in the Paradox Lake 15-minute quadrangle in Walton's (1961) map area. This unit consists of biotite-quartz-plagioclase gneiss with granitic layers and veins, migmatitic in places, as well as quartzite and quartz-sillimanite-rich gneisses. The quartz-sillimanite-rich biotite gneisses are nodular in places and locally contain kyanite as well (Geraghty, 1978), as in the Baldface Hill Gneiss of the northwestern Adirondacks (Wiener, 1981). Owing to these lithic similarities and its stratigraphic position, the Eagle Lake Gneiss is correlated with both

the sillimanite-rich rocks of the Baldface Hill Gneiss and migmatitic biotite-quartz-plagioclase gneisses of the Poplar Hill Gneiss of the northwestern Adirondacks (Fig. 12).

Paradox Lake Formation. The Paradox Lake Formation (Walton and deWaard, 1963) overlies the Eagle Lake Gneiss and rocks of the Piseco Group where the Eagle Lake Gneiss is not present (Fig. 11). The Paradox Lake Formation consists of calcitic and dolomitic marble interlayered with amphibolite, quartzite, plagioclase gneiss, graphitic schist, garnet granulite, and calc-silicate granulite, some of which are locally mappable units (Walton and deWaard, 1963; Hills, 1964; Geraghty, 1978). Al-

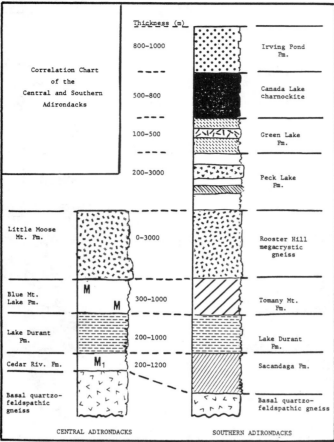

Figure 17. Correlation chart for the southern Adirondacks from McLelland and others (1978) and central Adirondacks from Walton and deWaard (1963) with thicknesses in meters. From McLelland and Isachsen (1980).

ling (1927) named the calcitic marbles the Johnsburg Limestone, and he also named the mappable amphibolite the Dresden amphibolite.

The Paradox Lake Formation in the eastern Adirondacks has been correlated with the Cedar River Formation (Lower Marble of Walton and deWaard, 1963) in the central Adirondacks (Geraghty and others, 1975) (Fig. 11). We suggest that the Paradox Lake Formation and Cedar River Formation correlate with the Cranberry Lake Formation and Gouverneur Marble (Fig. 12). A facies change on the southeast and northwest margins of the Cranberry Lake Formation may be necessary to explain the apparently smaller percentage of marble in the Cranberry Lake Formation (Fig. 12), although this may be the result of poor exposure. In either case, the lithic character and stratigraphic position of the Gouverneur Marble are identical to the Cedar River Formation and Paradox Lake Marble (Fig. 12).

Treadway Mountain Formation. The Treadway Mountain Formation (Walton and deWaard, 1963) overlies the Paradox Lake Marble in the eastern Adirondacks (Fig. 11). This

formation consists of migmatitic biotite-quartz-plagioclase gneiss, quartz-rich graphite-garnet-sillimanite-gneiss, quartzite, and garnet-biotite gneiss (Walton and deWaard, 1963; Geraghty, 1978). The Treadway Mountain Formation correlates to the Pleasant Lake Gneiss (Fig. 12).

Lake Durant Formation. In the central Adirondacks, the Lake Durant Formation (deWaard, 1964) occupies the position of much or all of the Treadway Mountain Formation (Geraghty and others, 1975) (Fig. 11). Biotitic and garnetiferous rocks similar to the Treadway Mountain Formation are present in the Lake Durant Formation of the central Adirondacks, suggesting that the Treadway Mountain Formation may be present but thinner in the central Adirondacks. The Lake Durant Formation consists of pink and green layered charnockitic and leucogranitic gneisses containing thin amphibolite layers as well as minor quartzite and carbonate rocks. In the eastern Adirondacks, the Barton Mountain Formation is equivalent to the Lake Durant Formation and consists of hornblende, pyroxene, and garnet-bearing charnockitic and granitic gneisses interbedded with amphibolite, calc-silicate granulite, and marble (Turner, 1979). Rocks of this unit are intimately associated with and overlie the Treadway Mountain Formation (Fig. 11).

The Treadway Mountain Formation and overlying Lake Durant Formation are correlated with the Pleasant Lake Gneiss of the northwestern Adirondacks (Fig. 12) owing to lithic similarity and stratigraphic position.

Springhill Pond Formation. The Springhill Pond Formation (Alling, 1917) structurally overlies the Lake Durant Formation in the eastern Adirondacks (Fig. 11). The Springhill Pond Formation consists of garnet-sillimanite gneiss (locally graphitic), quartzite, and marble, and was subdivided by Alling (1917, 1927) into various units in his mapping of Adirondack graphite deposits. These units are, from base to top (Fig. 11): (1) Hague gneiss, a garnet-sillimanite gneiss; (2) Dixon Schist, a graphitic biotite quartz-feldspar schist; (3) Faxon Limestone, a thin discontinuous marble; (4) Swede Pond Quartzite, a thick, massive, vitreous quartzite; (5) Bryant Lake Limestone, a thin discontinuous marble; (6) Catamount Sillimanite Schist, a rusty red sillimanite schist; and (7) Bear Pond Schist, a biotite-quartz-feldspar schist. This detailed stratigraphy is only known to be present in the eastern, southeastern, and southern Adirondacks.

The upper marble (Walton and deWaard, 1963) or Blue Mountain Lake Formation (Geraghty, 1979, written commun.) has been correlated with the Springhill Pond Formation of the eastern Adirondacks (Geraghty and others, 1975; Geraghty, 1978) (Fig. 11). The Blue Mountain Lake Formation consists of siliceous marble, calcareous quartzite, and calc-silicate granulite (Walton and deWaard, 1963). Thus, the calcareous and siliceous rocks of the Blue Mountain Lake Formation undergo a facies change to correlative aluminous and siliceous rocks of the Springhill Pond Formation (Geraghty and others, 1975) (Fig. 11).

Thunderbolt Mountain Formation. The Thunderbolt Mountain Formation structurally overlies the Springhill Pond

Formation in the eastern Adirondacks (Walton and deWaard, 1963). This formation consists of gray-green, charnockitic gneisses, and pink granitic gneisses interlayered with amphibolite, garnet granulite, garnet-biotite-quartz-plagioclase gneisses, and quartzite (Walton and deWaard, 1963; Geraghty, 1978). The Thunderbolt Mountain Formation correlates with the Little Moose Mountain Formation in the central Adirondacks (Geraghty and others, 1975) (Fig. 11).

Correlation of Paradox Lake–Cedar River Formation (Lower Marble) to Blue Mountain Lake–Springhill Pond Formation (Upper Marble)

It is possible that the upper marble and lower marble of the central Adirondacks are stratigraphic equivalents. These units may connect on the ground around an early isoclinal fold in the Lake Durant Formation (Geraghty, 1978, personal commun.). If the upper marble and lower marble are the same stratigraphic unit, then the aluminous Springhill Pond Formation (upper marble) would appear to undergo a facies change to the calcareous Paradox Lake Formation (lower marble) in the eastern Adirondacks. In support of this correlation, McLelland and Isachsen (1980) have demonstrated similar facies changes *within* the lower marble, between the aluminous Sacandaga Formation and calcareous Paradox Lake–Cedar River Formations, and also *within* the upper marble, between the aluminous Tomany Mountain Formation and Springhill Pond Formation and calcareous Blue Mountain Lake Formation. A facies change *between* the upper and lower marble would occur around a high amplitude, postulated but not proven to be present, early isoclinal fold cored by the Lake Durant Formation (Fig. 13). This earliest postulated fold is then interpreted to be refolded by the F$_2$ Wakely Mountain nappe and Little Moose Mountain syncline (Fig. 13). Although we believe that the upper and lower marble may be the same stratigraphic unit, we have described them separately and shown them separately on the correlation chart (Fig. 11) for ease of description and also because folds which may repeat the marble units are not obvious from presently available maps.

Correlation of Little Moose Mountain Formation and Piseco Group

According to the interpretation presented above, with an early isoclinal fold in the Lake Durant Formation (Fig. 13), charnockitic and other gneisses of the Little Moose Mountain Formation are equivalent to the Piseco Group. This correlation is supported by similarity of rock types: in the Little Moose Mountain Formation gray-green charnockitic gneisses may correlate with Pharaoh Mountain Gneiss, pink leucogranitic gneisses with Brant Lake Gneiss, and garnet-biotite-quartz-plagioclase gneiss, garnet granulite and quartzite with the Eagle Lake Gneiss, which overlies the Piseco Group.

Unconformity between the Piseco Group and the Lake George Group

We suggest an unconformity that predates the Grenville orogeny (1.1 b.y.) may be present above the granitic gneisses of the Piseco Group and beneath the metasedimentary rocks of the Lake George Group (Fig. 11). This unconformity would correlate with that interpreted to be present at the base of the Oswegatchie Group in the Northwest Lowlands (Fig. 12). The presence of the unconformity is based on regional correlation and local evidence of truncation of the contact between the Pharaoh Mountain Gneiss and Brant Lake Gneiss of the Piseco Group by the overlying Eagle Lake Gneiss and Paradox Lake Marble of Lake George Group (Fig. 11) (Walton, 1961). Charnockitic gneisses of the Pharaoh Mountain Gneiss range from 1.39 b.y. (Spooner and Fairbairn, 1970) to 1.13 b.y. old (Silver, 1969). The anatectic Brant Lake Gneiss is 1.10 to 1.12 b.y. old, coinciding with the time of metamorphism, whereas Lake George Group rocks have been dated as 1.18 b.y. (Bickford and Turner, 1971) to 1.02 b.y. old (Silver, 1969). These age data are permissive of an unconformity but not conclusive, inasmuch as some of the younger ages are metamorphic. It is possible that parts of an older basement terrane (Piseco Group) were remelted during peak 1.1-b.y.-old Grenville metamorphism, giving a spread of ages from older primary igneous crystallization to younger metamorphic-anatectic ages.

Walton and deWaard (1963) proposed a major unconformity between the "Basement Complex" and the base of the Paradox Lake Marble and overlying metasedimentary rocks. This interpretation was based on the presence of many lithic types in the Basement Complex in contact with the base of the Paradox Lake Marble. We suggest a modification of this, by considering the Eagle Lake Gneiss ("Older Paragneiss") to be at the base of the stratigraphic sequence *above* the unconformity, rather than including it in the Basement Complex as suggested by Walton and deWaard (1963). According to the present interpretation, the Eagle Lake Gneiss represents a thin, discontinuous unit of metamorphosed, reworked, aluminous, weathered horizons and basal clastics deposited above an unconformity on granitic basement. This is then succeeded by widespread carbonate deposition, as in the northwestern Adirondacks.

STRATIGRAPHY OF THE SOUTHERN ADIRONDACKS

The central and southern Adirondacks are outlined in Figure 1 and shown in Figures 14 and 15. Figure 14 is a formational map, whereas Figure 15 is a lithologic map. The approximate boundary between the southern and central Adirondacks is arbitrarily placed along the Piseco anticline (Fig. 16).

The earliest research in the southern Adirondacks consists of regional studies done primarily by Miller (1909, 1911, 1916, 1920, 1923) and, to a lesser extent, by Balk (1932, 1944) and Kreiger (1937). Most of this early mapping has either been redone or reevaluated in later studies by McLelland (1969), McLelland and others (1978), and McLelland and Isachsen, (1980). The pioneering effort in applying the methods of stratigraphic correlation to Adirondack metasedimentary rocks was the careful work of Alling (1917) who was interested in stratigraphy as a guide to understanding the distribution of graphite deposits in the eastern and southern Adirondacks. Cannon (1937) mapped the Piseco Lake 15-minute quadrangle and focused particular attention on the "Piseco dome" (Fig. 16). More recently, the northern half of the Lake Pleasant 15-minute quadrangle was mapped by Bartholome (1956) as part of a study of the Speculator anorthosite sheet. B. Thomas (unpub. data) mapped the Harrisburg 15-minute quadrangle, and Nelson (1968) mapped the Ohio 15-minute quadrangle.

McLelland (1969; McLelland and others, 1978) has established a stratigraphic sequence for the southern Adirondacks (Fig. 17). McLelland and Isachsen (1980) presented formal descriptions of this sequence with type localities and present thicknesses that do not account for tectonic thickening and thinning. As suggested in the previous section, it is possible that some of the stratigraphic units in the southern Adirondacks are repeated around postulated, but not proven, F_1 isoclinal folds, which predate the Canada Lake nappe–Little Moose Mountain syncline. Such repetitions would reduce the number of formations and thickness of the section. Since neither mappable early isoclinal folds (F_1) nor on-the-ground connection of the potentially correlative units have been clearly identified at this time, the stratigraphy is presented as a continuous sequence from the structurally lowest Piseco Group to the structurally highest and probably youngest stratigraphic units (Fig. 17).

Basal Quartzo-Feldspathic Gneiss (of Piseco Group) (>1,000 m)

This formation was defined by McLelland and Isachsen (1980) and is the only unit in the Piseco Group in the southern Adirondacks. It consists primarily of equigranular to inequigranular, well-foliated, pink, gray, or pale-green gneisses that are typically charnockitic to granitic, but locally mangeritic. On fresh surfaces, such as roadcuts, all of these rock types are dark, olive drab. Interlayered with these rocks are very subordinate amphibolite, pyroxenite, and garnetiferous quartzite.

Charnockitic rocks predominate in the basal gneiss. As elsewhere in the Adirondacks, the occurrence of orthopyroxene is sporadic, and locally granitic gneiss may appear in an otherwise charnockitic terrane. The most common characteristic of quartzo-feldspathic gneisses in this unit is the presence of microperthite, which generally accounts for 50% to 60% of the mode, the remainder consisting of quartz (20% to 30%), sodic plagioclase (5% to 10%), biotite, hornblende, and pyroxene (5% to 15%). The inequigranular facies of the basal gneisses contains 5- to 10-cm-long megacrysts of microperthite which are tabular to augen-shaped and generally aligned parallel to the plane of foliation.

Lake George Group

Sacandaga Formation (200–1,200 m). The Sacandaga Formation, named by McLelland and Isachsen (1980), is the basal formation of the Lake George Group and a distinctive marker horizon in the southern Adirondacks. The lower two-thirds of the unit consists of well-layered, flaggy, sillimanite-garnet-quartz-feldspar leucogneisses (commonly containing 40% to 50% quartz), interlayered with sillimanite-garnet-biotite gneisses, quartzites, and minor calc-silicates and marble. Individual layers in the Sacandaga Formation average about 30 cm thick. Garnets, where present, are typically idioblastic and grow across foliation planes. In the leucogneisses, they locally form rims around sillimanite. The upper one-third of the Sacandaga Formation contains layered, two-pyroxene, mafic granulites, which are overlain in the uppermost 30 to 50 m by flaggy leucogneisses identical to those at the base of the formation.

Lake Durant Formation (200–1,000 m). Near its base, the Lake Durant Formation consists of pink and light-green granitic gneisses interlayered with thin calc-silicate rocks, amphibolites, and, rarely, marbles. In the upper three-quarters of the formation, interlayers are absent, and granitic gneisses dominate. In some places, these are inequigranular and massive, but generally they are 0.5 to 1 m thick units of tan to pink, equigranular granitic gneisses that display subtle layering defined by color variations. Locally, amphibolite enhances the layering. Although the thickness of the Lake Durant Formation is varied, the sequence of rocks as described above is consistent across the entire region.

Tomany Mountain Formation (300–1,000 m). The Tomany Mountain Formation consists of sillimanite-garnet-biotite-quartz-oligoclase gneiss (kinzigite), quartzite, sillimanite-garnet-quartz-feldspar leucogneisses, and minor amphibolite. The leucogneiss and kinzigite are characterized by an abundance of coarse-grained granitic veins and pods, suggesting that these rocks have undergone extensive anatexis. In some places the granitic layers are so large (2 m thick) that they appear as quartzo-feldspathic layers in the stratigraphic section, but close examination generally reveals crosscutting relationships.

Rooster Hill Megacrystic Gneiss (0–3,000 m). The dominant rock type in the Rooster Hill is a megacrystic gneiss consisting of 2- to 50-cm-long augen of microperthite in a groundmass of quartz, oligoclase, hornblende, biotite, orthopyroxene, garnet, and metallic oxide. The megacrysts almost invariably lie in the plane of foliation, but in a few places exhibit poor to almost random orientation. In such localities, the megacrysts approach a euhedral shape. Near its borders, the Rooster Hill Formation becomes increasingly equigranular.

Rare, discontinuous layers of garnetiferous paragneiss occur locally in the Rooster Hill Megacrystic Gneiss. In general, however, the megacrystic gneiss does not display compositional layering. In this respect, as well as in texture, the Rooster Hill Megacrystic Gneiss resembles a plutonic igneous rock, specifically, a quartz monzonite porphyry. However, neither it nor any of the numerous other megacrystic gneisses at various stratigraphic horizons are anywhere seen to exhibit crosscutting contacts. Such gneisses have been mapped over large areas and are found to maintain consistent stratigraphic positions as conformable sheetlike units. If the Rooster Hill Megacrystic Gneiss and similar megacrystic gneisses in the region have an intrusive origin, they must either have been intruded as very extensive sills or sheets subparallel to layering, or they have been tectonically swept into pseudoconformity during deformation, without a trace of any earlier crosscutting history.

An intrusive, igneous origin for the megacrystic gneisses is favored by the presence of shredded paragneiss inclusions and biotite-rich clots that appear to be reacted xenoliths. Zircon population studies by Eckelmann (1978) also suggest that the megacrystic gneisses have an intrusive, igneous origin. Whether or not the Rooster Hill Megacrystic Gneiss be metaplutonic, metasedimentary, or metavolcanic, its geometry is indistinguishable from that of definite stratigraphic units in the region. For this reason, it is here treated as a stratigraphic unit.

Peck Lake Formation (200–3,000 m). Garnet-sillimanite-biotite gneiss (kinzigite) is predominant in the Peck Lake Formation, which is, thus, lithically similar to the Tomany Mountain Formation. Intimately associated with the kinzigite gneisses are pods and layers of white, garnetiferous granite, which commonly exhibit crosscutting contacts and appear to be anatectites. In addition to the kinzigite-anatectite assemblage, the Peck Lake Formation contains subordinate garnetiferous leucogneiss, minor quartzite, and rare amphibolite.

The Bernhardt Mountain Member of the Peck Lake Formation is a thick megacrystic quartzofeldspathic gneiss that occurs near the middle of the formation and is lithically identical to the Rooster Hill megacrystic gneiss. Near contacts with the Bernardt Mountain megacrystic gneiss, there is a porphyroblastic, well-layered, biotite-quartz-oligoclase gneiss that resembles kinzigite except for the absence of sillimanite and garnet.

Green Lake Formation (100–500 m). The Green Lake Formation is characterized by excellent layering and great lithic variety. It consists mainly of sillimanite-garnet-quartz-feldspar leucogneisses, but is interlayered with an abundance of quartzite, amphibolite, and pyroxene granulite, as well as minor amounts of kinzigite. The Royal Mountain Member of the Green Lake Formation consists of pyroxene-quartz-plagioclase gneiss. The Royal Mountain occurs in several distinct bodies in the formation, not everywhere in the same stratigraphic position. The plagioclase composition is close to An_{40}, and the overall composition facies is that of quartz diorite. In many places the gneisses of the Royal Mountain are interlayered with 10- to 20-cm-thick layers of amphibolite or pyroxene granulite that are continuous over tens of meters. The widespread occurrence of such layering is suggestive of original stratification in the rock and suggests a metavolcanic or metasedimentary origin. However, subangular blocks of amphibolite, in places athwart the foliation, resemble xenoliths rather than tectonically dismembered layers, and suggest that the pyroxene-quartz-plagioclase gneiss crystallized from a melt. An origin by anatexis appears unlikely, because in dry pyroxene-quartz-plagioclase assemblages, the required temperatures would be on the order of 1,000 °C (Green, 1972), much in excess of the 700 to 800 °C temperatures estimated for granulite-facies metamorphism in the Adirondacks (Bohlen and Essene, 1977; Whitney, 1978).

Canada Lake Charnockitic Gneiss (500–800 m). The Canada Lake Charnockitic Gneiss is typical of Adirondack charnockitic gneisses. It is characterized by approximately 55% gray-green mesoperthite, 30% quartz, 10% sodic plagioclase, and minor amounts of mafic minerals. Orthopyroxene occurs only sporadically and is commonly partially altered to amphibole. The charnockitic gneisses in this formation are generally medium grained and equigranular. On fresh surfaces, they are characteristically dark green or olive drab, but on weathered surfaces, light tan to pinkish.

As is common with quartzo-feldspathic gneisses, the Canada Lake Charnockitic Gneiss is well foliated but shows little compositional layering. Where layering does occur, it is the result of thin (2- to 5- cm) layers of amphibolite or pyroxene granulite. Undisputed paragneiss layers are not present in this unit. As with the other quartzo-feldspathic units, no conclusive evidence pertaining to origin exists. However, a metavolcanic origin is a reasonable possibility in view of the fact that the unit is stratiform and can be traced continuously in the same stratigraphic position for more than 100 km.

Irving Pond Formation (>800–1,000 m). The Irving Pond Formation is the uppermost stratigraphic unit that has been recognized in the southern and central Adirondacks. This formation consists almost entirely of quartzite and garnetiferous, feldspathic quartzite. Layers of vitreous quartzite, 1 to 2 m thick, are commonly interlayered with similar thicknesses of slightly impure, garnetiferous feldspathic quartzite. Small quantities of granitic gneiss and charnockite are present locally. Pyroxene-plagioclase lenses, some appearing to be boudin, occur throughout the section. In the region east of Sacandaga Reservoir (Figs. 14, 15), a 10- to 20-m-thick layer of marble and associated calc-silicate rock occurs in the formation.

Correlation of the Southern and Central Adirondack Sequences

The correlation of units of the southern Adirondacks with those of the central Adirondacks is shown in Figure 17. The central Adirondack sequence was first established by deWaard (see Walton and deWaard, 1963) and subsequently extended eastward by Geraghty (1978). In the following, we briefly describe each of the central Adirondack units and discuss their correlation with the southern Adirondack sequence.

Basal Quartzo-Feldspathic Gneiss (1,000 m). This unit is the only formation distinguished in the Piseco Group and was designated the Basement Complex by Walton and deWaard (1963). Most of the Piseco Group in the central Adirondacks consists of charnockitic and granitic gneisses of the Pharaoh Mountain Gneiss.

Cedar River Formation (100–300 m). The Cedar River Formation (lower marble of Walton and deWaard, 1963) consists of medium- to coarse-grained, calcite marble with subordinate lenses of amphibolite, quartzite, and sillimanitic garnet-biotite gneiss and is correlated with the Sacandaga Formation (Fig. 17).

The Cedar River Formation has been mapped south and east along the Snowy Mountain dome from its type locality by Douglas and others (1976) (Figs. 14, 16), around the north end of the Snowy Mountain dome by Geraghty (1978), and south into the Humphrey Mountain syncline by Lettney (1969). These studies bring the Cedar River Formation into the general vicinity of the southern Adirondacks where Glennie (1973) and McLelland and others (1978) correlated it with similar units across the axis of the Glens Falls syncline between the Snowy Mountain and Piseco anticlines (Fig. 16). Mapping by McLelland and others (1978) extended the formation along the northern flank of the Piseco anticline. To the southeast, the marble-rich Cedar River Formation grades laterally into the Sacandaga Formation (Fig. 14). The details of this transition are described below.

Beginning at Oxbow Lake, 0.5 km northeast of Piseco Lake, in the Lake Pleasant 15-minute quadrangle (Figs. 14, 15, 16), the Cedar River Formation is traced almost continuously southeast along the northern flank of the Piseco anticline. Along this traverse, garnetiferous leucogneiss and quartzite increase at the expense of marble and calc-silicates. On the south side of Hamilton Lake (Lake Pleasant 15-minute quadrangle), quartzite and garnetiferous leucogneiss are predominant. Continuing to the southeast, calc-silicate, marble, and amphibolite are increasingly rare, until one passes into the typical Sacandaga Formation "just" north of Blackbridge, 3 km southwest of the town of Wells in the Lake Pleasant 15-minute quadrangle (Figs. 14, 15). Small-scale interbedding of lithologies is particularly well developed here. From this point on to the southeast, the rocks are clearly those of the Sacandaga Formation with local pods of calc-silicate and marble, as in the long roadcut along Route 30 at Pumpkin Hollow (Lake Pleasant 15-minute quadrangle).

We interpret this transition as resulting from an original sedimentary facies change now represented by the metamorphic rocks described above. The transition provides the evidence for the stratigraphic equivalence of the Sacandaga and Cedar River Formations (Fig. 17). Nowhere in the traverse described in the preceding paragraph have any features been recognized which suggest that the transition is due to tectonic interleaving.

Lake Durant Formation (200–1,000 m). DeWaard (1964) described the sequence in this formation as follows: "The section of diverse, layered metamorphic rocks includes pink and greenish leucocratic gneisses with thin metabasic layers, marble, and calc-silicate rocks." The entire thickness of the formation is not exposed at the type locality.

The Lake Durant Formation is the most lithically consistent unit in the central and southern Adirondacks. Furthermore, it is the only unit that can be mapped continuously and without lithologic change from the north limb of the Little Moose Mountain syncline, through the core of the Glens Falls syncline, and around the nose of the Piseco anticline into the southern Adirondacks proper (Fig. 14). This areal continuity and lithic consistency make the Lake Durant Formation a critical horizon for correlating stratigraphy of the southern and central Adirondacks and for unraveling the geologic structure.

Blue Mountain Lake Formation (Upper Marble) (300–1,300 m). The type locality is at Blue Mountain Lake in the Central Adirondacks where coarse-grained calcite marble, with subordinate lenses and layers of amphibolite, is exposed on islands in the lake. Dirk deWaard (1970, personal commun.) suggested that we use this as the type locality. Interlayered quartzite, gneiss, amphibolite, and calc-silicate rock make up a distinctive, mappable unit within this marble-rich formation, and a coarse-grained, garnetiferous, K-feldspar gneiss occurs locally near the base.

As the Blue Mountain Lake Formation is traced southward around the eastern nose of the Glens Falls syncline, it undergoes a facies change and grades into garnetiferous, biotite-rich gneisses (kinzigites) of the Tomany Mountain Formation (Figs. 14, 15, 16). The change is initiated by a gradual decrease in the percentage of marble. In the vicinity of Hamilton Mountain (Lake Pleasant 15-minute quadrangle), quartzites and garnetiferous amphibolites predominate in the sequence. Farther to the east on Dunham Mountain, whose summit lies 3 km west of the town of Wells (Figs. 14, 16), kinzigites are present in the section. Toward the east, the kinzigites increase over a short distance at the expense of all other rock types. Calc-silicates and marbles decrease in abundance eastward, and at the eastern base of Mount Dunham they occur only in local pods.

Toward the east, the Blue Mountain Lake Formation passes directly into the kinzigite-rich Tomany Mountain Formation, the transition being complete at the summit of Moose Mountain, 2.5 km south-southeast of the town of Wells in the Lake Pleasant 15-minute quadrangle (Figs. 14, 16).

A similar facies change occurs in this unit in the eastern Adirondacks, from marble-rich Blue Mountain Lake Formation to the aluminous Springhill Pond Formation (Fig. 11). As in the case of the Sacandaga Formation and the Cedar River Formation, we interpret these transitions to be the result of original sedimentary facies changes.

Little Moose Mountain Formation (1,000–1,500 m). The Little Moose Mountain Formation consists of an interlayered sequence of equigranular and inequigranular microperthite gneisses (including charnockitic gneisses), sillimanite-garnet-biotite-K-feldspar-quartz-plagioclase gneiss, and subordinate amphibolite and marble.

The inequigranular facies of the Little Moose Mountain Formation is well developed in both its eastern and western map areas in the Glens Falls syncline (Figs. 14, 15, 16). From here, it is correlated with the Rooster Hill Megacrystic Gneiss across the axis of the Piseco anticline. The correlation is based on similarity of lithology and stratigraphic position.

METAMORPHOSED IGNEOUS ROCKS AND PEGMATITES

Meta-Anorthosite

Several large domical masses of the metamorphosed anorthosite, totaling about 2,750 km² in map area, occur in the Adirondacks (Fig. 1). The largest body is the Marcy massif. Marcy-type anorthosite is plagioclase rich and consists of large blue-gray andesine megacrysts and a white plagioclase-rich matrix, whereas Whiteface-type anorthosite is mainly white and black–striped gabbroic anorthosite gneiss (Buddington, 1939). Although these rocks are often loosely referred to as "anorthosite" in the literature, they are clearly metamorphic, locally with striking granulite-facies coronites and in many places with gneissic fabric.

Although the bulk of these masses is true meta-anorthosite, metagabbroic anorthosite and metanoritic anorthosite also occur in the series, as do metanorite, metagabbro, a pyroxene-oxide rock with minor plagioclase, and ores of titaniferous magnetite and ilmenite. There is universal agreement among workers in the region that the anorthosite bodies are metamorphosed igneous plutons.

For a summary of geologic, geochronologic, mineralogical, and geophysical studies of the Adirondack meta-anorthositic rocks, the reader is referred to a symposium volume on the origin of anorthosite and related rocks (Isachsen, ed., 1969a) and a symposium summary paper by Isachsen (1969b). The following sections will focus only on new lines of evidence bearing on anorthosite petrogenesis that have evolved since the anorthosite symposium volume was published.

The Anorthosite–Mangerite–Charnockite Problem.
The main (Marcy) meta-anorthosite massif shown in Figure 1 crosscuts a variety of stratigraphic units including metasedimentary rocks and gneisses of granitic and syenitic composition (Fisher and others, 1971). The smaller massifs to the south similarly cut several east-west-trending map units (Figs. 15, 16). Where the meta-anorthosite is in contact with quartzo-feldspathic units, a mantle of mangerite gneiss grading to charnockite gneiss occurs.

A major problem that was discussed, but left unresolved at the 1968 anorthosite symposium, was the relationship between meta-anorthosite and the mantling mangerite-charnockite gneisses (e.g., Isachsen, 1969a; Buddington, 1972, 1976). The mangerite-charnockite gneisses in this association have been variously interpreted as comagmatic with anorthosite, as younger intrusives, and as contact anatectic products (Isachsen, 1969b).

The contact anatectic model best fits structural-stratigraphic relationships in the southern and central Adirondacks (Figs. 14, 15). The spatial relationships between these rock types have been mapped in the Snowy Mountain dome by deWaard and Romey (1963) who show that the zonal sequence from the meta-anorthositic core rocks outward is as follows: mangeritic gneiss

with andesine augen (xenocrysts), charnockitic gneiss with few andesine augen, equigranular charnockite. We interpret these compositional and textural relationships to be the result of localized contact anatexis caused by intrusion of anorthosite into charnockite of the Pharaoh Mountain Gneiss such as is exposed in the core of the Wakely Mountain nappe. The reasons are as follows:

1. The charnockitic Pharaoh Mountain Gneiss that cores the Wakely Mountain nappe is also present above the anorthosite in the core of the Snowy Mountain dome (Fig. 16) as well as in the Gore Mountain and Oregon domes (Fig. 16). Thus, the anorthosite bodies appear to be local intrusions in the Pharaoh Mountain Gneiss with textural-compositional transition zones that intervene between the anorthosite and Pharaoh Mountain Gneiss.

2. Not only are anorthosite massifs in contact with mangerite-charnockite, but they also intrude the Lake Durant Formation and marble of the Paradox Lake Formation along tens of kilometers of their borders (Isachsen and Moxham, 1969).

3. The intrusion of anorthosite into charnockite of the Pharaoh Mountain Gneiss would make local contact anatexis inevitable, considering the high P-T conditions of emplacement (about 8 kbar and 1,200 to 1,300 °C; Ashwal, 1978). At these depths, the country rocks themselves would have been at temperatures of 500 to 700 °C about 1.0 b.y. ago (e.g., Hargraves, 1976; Tarling, 1978). The anatectic melts so generated would expectably reintrude the more refractory anorthosite after it congealed, thus accounting for the andesine xenoliths and xenocrysts found in mangerite-charnockite bordering anorthosite bodies.

4. Field evidence for an anatectic origin for such bordering mangerite-charnockite rocks can be seen in the southern Adirondacks. Examples are well exposed in a series of roadcuts in the Lake Durant Formation along Route 10 over a distance of 5 km between Kennels Pond and Shaker Place (Piseco 15-minute quadrangle). Locally, the section is markedly swollen due to voluminous intrusion of anorthositic gabbros and related rocks. A few small pods of meta-anorthosite are also visible here. There is abundant evidence in this outcrop that quartzofeldspathic rocks in the Lake Durant Formation underwent widespread anatexis and even total melting. For example, it is possible to trace granitic and pegmatitic dikes and veins from positions within the meta-anorthositic gabbro into the enclosing parent rock, as they merge with normal Lake Durant gneiss. Similarly, otherwise comformable layers in the Lake Durant Formation may abruptly cut across enveloping mafic-rich layers and, thus, appear to be intrusive. In general, quartzofeldspathic gneisses exhibit flowage folding suggestive of igneous, or partly igneous, behavior.

5. Davis (1969) and Buddington (1939, 1972) presented bulk chemical data as well as mineralogical and field evidence that cast doubt upon the comagmatic origin of the anorthosites and the mangerite-charnockite rocks. These doubts have been magnified by rare-earth studies of associated mangerite-charnockite and anorthosite which demonstrate that the anorthosite possesses the expected positive Eu anomaly, but that the associated mangerite charnockite rocks lack the reciprocal negative

Eu anomaly and may even show a slight positive Eu anomaly (Simmons and Hanson, 1978; Seifert and others, 1977). This provides a strong argument against comagmatic origin.

6. Recently, Ashwal (1978) and Emslie (1978) argued cogently for a contact anatectic origin for rocks of the mangerite-charnockite suite. Emslie placed the site of anatexis in the deep crust, well below the current levels of exposure of the two suites. Ashwal was less specific, but implied in situ anatexis at current levels of exposure.

Relationship of Meta–Anorthosite to Piseco Group and Overlying Lake George Group. Walton and deWaard (1963) proposed that rocks of the anorthosite-charnockite suite and minor paragneisses (Eagle Lake Gneiss) constituted a pre-Grenvillian basement complex, which had been intruded, deformed, metamorphosed, and eroded prior to deposition of the marble-rich Paradox Lake Formation. Prior to the introduction of this new model of Adirondack geological history involving an anorthosite-cored basement complex and supracrustal sequence (Walton and deWaard, 1963), the prevailing interpretation was that the anorthositic rocks intruded an older terrane of stratified metamorphic rocks (Cushing, 1902; Miller, 1918; Buddington, 1939, 1969).

Field relations indicate that the meta-anorthosite suite intrudes the Lake George Group cover rocks that unconformably overlie the Piseco Group basement, and this is in agreement with the interpretations of the early workers cited above. The discordant relation of meta-anorthosite with the surrounding rocks was recognized by Walton in the eastern Adirondacks (Walton and deWaard, 1963) and is related to the intrusion of the meta-anorthosite at or near the contact between the Pharaoh Mountain Gneiss and the Paradox Lake Formation.

Strong evidence supporting the younger age of anorthosite intrusion also comes from field observations that anorthositic rocks intrude members of the "supracrustal" sequence of the Lake George Group in the southern Adirondacks (Husch and others, 1975; Isachsen and others, 1975). Ten sills or sheets of meta-anorthosite, meta-anorthositic gabbro and metagabbroic anorthosite intrude the stratigraphic section (Fig. 16). The 200- to 300-m-thick sheet of meta-anorthosite and metagabbroic anorthosite on Speculator Mountain (east of Piseco Lake on Fig. 14), for example, intrudes the Little Moose Mountain Formation. In several places, the Cedar River Formation also shows direct effects of intrusion by anorthosite. The best example, at map scale, can be seen along the northern margin of the Oregon dome meta-anorthosite massif where the Cedar River Formation is partly enclosed by meta-anorthosite (Figs. 14, 15, 16). Field relations here show no cataclasis or repetition of layering such as would be expected if the Cedar River Formation were faulted or folded into the meta-anorthosite. On the contrary, meta-anorthosite crosscuts the ~100-m thick portion of the Cedar River Formation along the northwestern base of Eleventh Mountain in the Thirteenth Lake 15-minute quadrangle. The discontinuous outcrop pattern of the Cedar River Formation along the southern margin of the meta-anorthosite in the Oregon dome

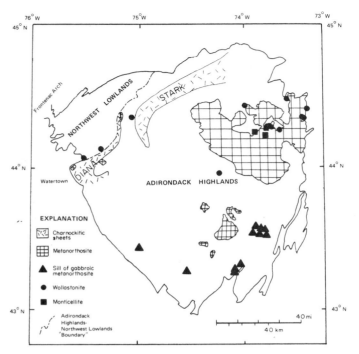

Figure 18. Wollastonite deposits and monticellite localities in the Adirondacks showing their proximity to meta-anorthosite bodies or charnockitic sheets.

(Figs. 14, 15) is thought to be due to intrusion by meta-anorthosite. An excellent example of such dismemberment may be seen in a roadcut where Route 8-30 intersects the southwestern tip of the Oregon dome meta-anorthosite, in the northeastern corner of the Lake Pleasant 15-minute quadrangle (McLelland and others, 1978). Here xenoliths of calc-silicate rock are present in metagabbroic anorthosite and meta-anorthositic ferrodiorite. This locality is only several hundred meters from the outer margin of the meta-anorthosite that is here in direct contact with the Lake Durant Formation. We interpret these xenoliths to represent relict portions of the Cedar River Formation.

The intrusive nature of the meta-anorthosite is also suggested by the distribution of wollastonite in the Adirondacks (Fig. 18). Localization of wollastonite near the Diana Complex, Stark Complex, and Marcy massif suggests a causal relationship. Where wollastonite occurs several kilometers from mapped contacts of plutons, plutonic rocks may be nearby at depth as suggested by negative aeromagnetic anomalies (Isachsen and others, 1979). We suggest that a causal relationship may be meaningful inasmuch as calcite and quartz coexist as a stable assemblage throughout most of the Adirondacks. The local occurrence of wollastonite near igneous plutons indicates that its presence may be due to heat derived from the intrusive or a reduction in the activity of CO_2 related to intrusive activity or metamorphism unassociated with igneous activity. The occurrence of monticellite in a calc-silicate assemblage in contact with the Marcy massif (Fig. 18) similarly permits an interpretation of contact metamorphism or low activity of CO_2 related to the anorthosite (Tracy and others, 1978).

Thus, we conclude that the meta-anorthosite was intruded over large areas of the Adirondacks either well within the basement or at, or near, the unconformity between the basement and cover rocks, and that charnockitic gneiss is a common anatectic product in the emplacement process. In addition, there are smaller meta-anorthosite suite plutons that were emplaced in the cover rocks, such as those present in the Lake Durant Formation, and these too caused local anatexis producing segregations and veins of granite and pegmatite.

Timing of Intrusion of Anorthosite Suite. The anorthosite suite apparently intruded early during F_2 folding, as indicated by relations of minor structural features to the plutons. Rotated, angular xenoliths of gneiss within the meta-anorthosite possess a strong F_1 foliation that predates intrusion. Both the meta-anorthosite and meta–country rocks have a later F_2 foliation. Contaminated meta-anorthosite and metagabbroic anorthosite are both folded by, and crosscut, F_2 minor folds immediately south of Shaker Place on Route 10 in the Piseco Lake 15-minute quadrangle. Therefore, the anorthosite intruded early during F_2 folding under synorogenic conditions. These conclusions are at variance with the widespread evidence cited by Emslie (1978) for the prevalence of anorogenic conditions during anorthosite intrusion.

Layered Charnockitic Complexes

The Diana, Stark, Tupper, Loon Lake, and Saranac complexes of the western Adirondacks (Fig. 3) are each interpreted as intrusive sheets of differentiated igneous rocks that have an average composition of quartz syenite (Buddington, 1939, 1948). These complexes have been folded and metamorphosed, and some of the complexes may connect in the third dimension (Buddington, 1939; Brock, 1980). The Diana Complex (Fig. 3) is thought to form the northwest (overturned) limb of a syncline that plunges northeastward (Buddington, 1939) but also contains a major anticline within it (Hargraves, 1969). The thickness of the Diana Complex reaches nearly 5,700 m southeast of Harrisville and thins to the northeast (Fig. 3). The Stark Complex forms the core of an isoclinal anticline, the Tupper Complex forms the core of an upright anticline, and the Loon Lake and Saranac sheets are exposed in domes.

Of the several layered complexes in the Adirondacks, the Diana has the thickest exposed section and shows the greatest compositional diversification. It grades from (1) a basal chilled (?) facies of augite quartz syenite to (2) ferroaugite syenitic gneiss containing lenses of shonkinite and feldspathic pyroxenite gneiss rich in ferroaugite, zircon, ilmenite, and magnetite, through (3) pyroxene-hornblende-quartz syenite gneiss and (4) hornblende-quartz syenite gneiss, into (5) hornblende (ferrohastingsite) granitic gneiss. Hypersthene is a minor constituent and confined to the more mafic facies. The gradation from pyroxenic to hornblendic facies is accompanied by a change in color from green to pink,

suggestive of increasing degree of oxidation in the later stages of differentiation. The ratio $FeO/FeO+MgO$ in pyroxenes and amphiboles also increases systematically upward in the sheet, consistent with the interpretation that the complex formed by fractional crystallization. Independent evidence of magmatic origin includes the presence of intrusive breccias (Buddington, 1939; Wiener, 1977) containing calc-silicate and other xenoliths and intrusive sheets of quartz syenite in metasedimentary rocks (Wiener, 1977).

It is uncertain whether or not the charnockitic layered intrusive complexes are melted equivalents of the charnockitic Pharaoh Mountain Gneiss formed by in situ contact anatexis associated with anorthosite intrusion. The layered intrusive complexes appear to occupy a stratigraphic position similar to that of the Pharaoh Mountain Gneiss (Fig. 3). Also, charnockitic rocks adjacent to anorthosite plutons have clearly undergone contact anatexis associated with heat from the plutons. Furthermore, isolated exposures of meta-anorthosite in the cores of layered complexes, as well as gravity minima, suggest presence of anorthosite in the subsurface beneath the layered complexes. Also, anorthosite and charnockite both intruded during the second phase of folding. If entire layered complexes were derived in situ by contact anatexis of Pharaoh Mountain Gneiss by anorthosite, it would explain the young 1.1-b.y. "metamorphic" ages of layered complexes and Pharaoh Mountain Gneiss basement. It remains uncertain, however, whether or not these complexes are related to the anorthosite bodies in time and space.

Hornblende Granite Gneiss

Pink to gray, medium-grained hornblende granitic gneisses with sparse biotite intrude the Diana, Stark, and other layered charnockitic complexes in the Adirondack Highlands (Fig. 3) (Buddington, 1939; Buddington and Leonard, 1962). These gneisses are mostly equigranular, in contrast to megacrystic hornblende granite in the Northwest Lowlands.

Pink to gray, well-foliated hornblende granite gneiss intrudes various metasedimentary rocks of the Northwest Lowlands, but is not shown on Figure 3 for simplicity and to emphasize the stratigraphic units. These gneisses contain megacrysts of K-feldspar in many places and have been called the Hermon granite gneiss (Buddington, 1929, 1939) and the Porphyroblastic Gneiss (Foose and Brown, 1976). The Hermon-type megacrystic granite gneiss in the Northwest Lowlands is biotite rich in the east and hornblende rich in the west (Foose and Brown, 1976). Much of the Poplar Hill Gneiss contains sheets of Hermon-type granite gneiss, which have been interpreted as stratigraphic units (Foose and Brown, 1976). Although these sheets are grossly concordant, studies in the Harrisville area indicate that they are discordant at outcrop scale and also that they are present in all stratigraphic units (Wiener, 1981), as is true for the whole Northwest Lowlands (compare Fig. 3 and Fisher and others, 1971). Furthermore, the megacrysts are most likely phenocrysts, as shown by their rectangular, random habit in places. Therefore, we interpret the Hermon-type hornblende granitic gneiss to be a metamorphosed porphyritic granite intrusive. It is suggested that the Hermon hornblende granite gneiss of the Northwest Lowlands (Buddington, 1929, 1939) represents the same magma suite as the "younger hornblende granite gneiss" of the Adirondack Highlands (Buddington and Leonard, 1962). Both suites are similar in their composition, appearance, and in their geologic relationship to the stratigraphy and intrusive rock suites.

Metagabbro and Amphibolite

In contrast to the silica-saturated metagabbro and metanorite in the border facies of the meta-anorthosite referred to earlier, there are younger olivine-bearing metagabbro and metanorite bodies that intrude nearly all other rock units (Fisher and others, 1971) and have undergone high-grade metamorphism. Most of these undersaturated rocks, hereafter lumped as metagabbros, are elliptical or lenticular in plan and have been interpreted as sheets, lenses, and torpedo-shaped bodies (e.g., Buddington, 1939, p. 222) (Figs. 3, 14). They also occur as dikes and locally as laccolithic masses (Alling, 1919, p. 64; Kemp and Alling, 1925, p. 26; Jaffe, 1946; deWaard, 1961). Two are located on Figure 14.

Some of the larger Adirondack gabbroic sheets are magmatically differentiated. Buddington (1963, p. 1173) described one near Santa Clara that passes from metamorphosed troctolite to gabbro gneiss over a distance of about 100 m. Other large bodies, such as those near Jay Mountain, Texas Ridge, and Tahawus Club in the eastern Adirondacks, often show cumulus layering with olivine gabbro, gabbro, troctolite, anorthosite, and ultramafic layers ranging in thickness from a few centimeters to several meters (Whitney, 1972, and written commun.).

The major primary minerals of the olivine metagabbros are cumulus plagioclase and olivine, and various amounts of intercumulus clinopyroxene. Troctolitic varieties without clinopyroxene are also common. Primary intercumulus orthopyroxene is present in places, and ilmenite is a ubiquitous minor constituent. Iron-magnesium ratios in the mafic minerals have a wide range; for example, olivines from Fo_7 to Fo_{18} have been observed, although most values lie between Fo_{70} and Fo_{50}. Original plagioclase may have been as calcic as An_{60}, but actual plagioclase, even where retaining its original igneous form, is commonly in the range of An_{40}-An_{30}, as a result of subsolidus reactions with olivine (McLelland and Whitney, 1980b). Metamorphic reactions between primary olivine and plagioclase have produced coronas of pyroxene and garnet between them, and the plagioclase has become heavily clouded with tiny (<10 μm) green spinels (Whitney and McLelland 1973; McLelland and Whitney 1980a, 1980b). Coronas of biotite, hornblende, and garnet have formed between ilmenite and plagioclase, giving the rocks a characteristic spotted appearance.

Where the gabbros are more strongly deformed, olivine is absent (due to metamorphic reaction), plagioclase is clear, and the rock passes into pyroxene-garnet granulite or garnet amphibo-

lite. In such rocks from 20% to 60% of the primary plagioclase has reacted with primary mafic minerals to produce hornblende, garnet, and slightly less calcic pyroxene; variations in the hornblende composition correlate directly with the bulk chemical composition of the rock (Buddington, 1952).

Discounting the addition of small amounts of water in many places, the metamorphic reconstitution of gabbroic rocks in the Central Highlands has been essentially an anhydrous isochemical process (Buddington, 1939, p. 258–282; 1952, p. 70–79; 1963). This has doubtless contributed to the common preservation of relict texture and minerals. Completely reconstituted gabbro also occurs as sharply crosscutting dikes of orthoamphibolite (e.g., Dietrich, 1957). In addition, Buddington (1939, p. 56–62) and Engel and Engel (1962) presented field and chemical evidence which suggests that thin sheets (30 cm to 3 m) and lenses of amphibolite interlayered in biotite-quartz-oligoclase paragneiss in the Northwest Lowlands probably originated by simple recrystallization of basaltic sills or flows. The flow evidence includes the presence of sharp rather than gradational contacts, consistently finer grained margins (probable chill zones), and a rare example of major discordance. Also, great consistency in mode of occurrence and bulk composition over a distance of some 80 m was demonstrated by Engel and Engel, (1962). The orthoamphibolites consist mainly of hornblende and andesine, and may contain augite, diopside, biotite, chlorite, scapolite, quartz, epidote, and garnet, as well as accessory calcite, apatite, magnetite, ilmenite, quartz, potash feldspar, and pyrite. Leake (1963) plotted analytical data of Engel and Engel (1962) on variation diagrams and found compositional trends to be very similar to those for the Karroo diabases of South Africa. From this he concluded that they had an igneous rather than sedimentary or metasomatic origin. An igneous parent is also consistent with cobalt contents of the amphibolites in question inasmuch as they are persistently very near the average value of 48 ppm found by Carr and Turekian (1961) that characterize basaltic rock. Furthermore, an igneous origin for these amphibolites is consistent with their lack of stratigraphic continuity and their position in a number of different stratigraphic positions.

In areas where relatively large bodies of olivine metagabbro intrude quartzo-feldspathic gneiss, the gneisses show evidence of partial melting in the form of irregular granitic and pegmatitic dikes that cut across the metagabbro but can be traced into the quartzo-feldspathic country rock where they merge imperceptibly with it. An important feature associated with this "contact anatexis" is an increase in the quantity of hornblende, biotite, and/or pyroxene in the quartzo-feldspathic rocks near the metagabbros; these mafic minerals tend to occur in pods that, in places, represent disaggregated and reacted xenoliths of metagabbro. This seemingly ambiguous relationship of intrusion by a gabbroic melt into felsic host rocks, and subsequent reintrusion of consolidated gabbro by anatectically generated granite, is an example of the "Sederholm effect," so named by Eskola (1961) after Sederholm (1926) who first explained the phenomenon.

In summary, most chemically analyzed amphibolites in the

Adirondacks have the composition of, and probably originated as, olivine-basalt or olivine gabbro. Poldervaart (1953) noted in a summary paper that amphibolites of this composition are the dominant type in metamorphic terranes.

Stratigraphically continuous thin layers of amphibolite, however, such as are commonly interlayered in carbonates, and granitic and charnockitic gneisses in the Adirondacks and elsewhere could have a metasedimentary origin. Orville (1969) suggested that reaction between carbonate-rich and carbonate-free metapelitic rocks under open-system conditions can explain such thinly layered amphibolites in granitoid rocks. By the proposed mechanism, layers in the parent rock that were originally richer in Ca-feldspar would become further enriched as K transferred out of them and into layers poorer in Ca-feldspar. A reciprocal transfer of Na+ ions would accompany this migration. Such an exchange of alkalies via an interstitial vapor phase was demonstrated experimentally by Vidale (1969) and could greatly accentuate subtle original compositional differences to produce the interlayering present.

Metadiabase Dikes

Narrow dikes or sheets of hypersthene metadiabase with chilled borders are scattered throughout the mylonitized Diana Complex (Buddington, 1939, p. 133). The dikes strike N to N30°W, at marked angles to the generally northeast-trending Diana Complex and northeast-trending F_3 axial surfaces. The dikes may be related to F_4 folds. When the Diana Complex strikes northwest near its southern extremity, the metadiabase occurs as sheets that are aligned parallel to the regional foliation. The interior portions of the dikes are generally porphyritic and show some alignment of the plagioclase phenocrysts. The metadiabase is mylonitized to the same degree as the enclosing rock and has an overprinted metamorphic foliation. Primary essential minerals are labradorite, hypersthene, and augite. Metamorphic minerals are reddish-brown biotite, green hornblende, and scapolite.

Metadiabase dikes, commonly metamorphosed to pyroxene granulites, are described also from other parts of the northwestern Adirondack Highlands. All these rocks contain plagioclase, augite, hypersthene, magnetite, and ilmenite, and most contain hornblende, garnet, and/or biotite as well (Buddington and Leonard, 1962, p. 59.)

These metadiabase dikes, with both metamorphic foliation and chill borders, may indicate a complex geological history, namely, two separate episodes of ductile deformation and metamorphism (Buddington, 1963, p. 1164). Alternatively, we suggest that the dikes intruded during a late stage of the Grenville orogeny, after third-phase folding, when the region had undergone uplift and had cooled to garnet amphibolite facies temperatures. The chill borders would form due to rapid crystallization of basaltic magma at about 1,000 °C chilled against wall rocks having a temperature of about 550 °C. Metamorphic recrystallization of plagioclase-augite-hypersthene basalt (norite) to hornblende-garnet-biotite-amphibolite occurred at ambient midcrustal amphibolite facies conditions.

Pegmatite Dikes

Both deformed and undeformed pegmatites occur in the Adirondacks. The former have very wide distribution and occur in most rock types (e.g., Buddington, 1939, p. 161–166).

The larger Adirondack pegmatites, those that have been commercially exploited for feldspar, are undeformed. They occur near the southern and eastern perimeter of the Adirondacks and have been systematically studied by Tan (1966). Compositionally, they are relatively rich in rare earths and virtually free of lithium and fluorine. Most of these pegmatites show several or all of the following textural zones: I, border zone with granitoid texture; II, wall zone with graphic texture; III, outer intermediate zone with graphic or pegmatoid texture; and V, quartz core with massive texture.

On the basis of mineral assemblage, Tan (1966) identified the following types: (1) hornblende-biotite pegmatite, (2) biotite pegmatite, (3) biotite-muscovite pegmatite, and (4) muscovite-beryl pegmatite. The temperature estimates suggest a thermal gradient of roughly 50 °C per zone from border to core. This lowering of temperature of crystallization is accompanied by a decrease in (1) the anorthite content of plagioclase, (2) the magnesium content of biotite, and (3) the manganese content of garnet.

The above observations clearly indicate that the major undeformed pegmatites of the Adirondacks are of magmatic origin. From the relative crystallization temperatures of the various mineral assemblages, their observed sequential field relations, and the variation in bulk chemical composition from quartz-monzonitic to granitic, Tan (1966, p. 127) suggested that the assemblages represent a differentiation sequence as follows:

Titanium-rich magma (700 °C)
1. Hornblende-biotite pegmatite (700 °C?)
2. Biotite-muscovite pegmatite (640–700 °C)
3. Biotite-pegmatite (590–650 °C)
4. Complex pegmatite (muscovite-beryl) (525–600 °C)

These crystallization temperatures when viewed in the context of geographical distribution of pegmatite types indicate a decreasing thermal gradient from the east-central Adirondacks to the southern Adirondacks at the time of pegmatite emplacement. The southeasternmost pegmatite is classified by Tan (1966) as transitional between types 1 and 2. Tan's proposed thermal gradient is in close agreement with that determined from feldspar and oxide thermometry by Bohlen and others (1980), and suggests that the undeformed pegmatites were introduced during the thermal peak of the Grenville orogenic event, but after deformation had ceased. The inferred date of this emplacement based on a Rb/Sr age on muscovite from biotite-muscovite pegmatite at Batchlerville in the southern Adirondacks is 930 ± 40 m.y., using = 47 × 10^{-11} yr^{-1} (B. Giletti, unpub. on specimen no. 2112 in New York State Museum Collection).

STRUCTURAL GEOLOGY OF THE NORTHWESTERN ADIRONDACKS

Four major phases of deformation affected the metamorphosed sedimentary and volcanic rocks of the Northwest Lowlands and western Adirondack Highlands (deLorraine, 1979; Wiener, 1981), whereas only the last three phases of deformation affected the metamorphosed plutonic rocks in this area (Wiener, 1981). In addition to the four major phases of deformation there is local evidence for a fifth minor phase of deformation. The structure of the Northwest Lowlands is better understood than that of the western Adirondack Highlands. Much detailed mapping remains to be done in both terranes to bear out the stratigraphic and structural hypotheses presented in this paper.

The first-phase folds are isoclinal northwest-directed(?) nappes, associated with peak Grenville metamorphism and formation of regional foliation and lineation. The second-phase folds are also isoclinal and are southeast-directed nappes. Widespread intrusive activity, high grade metamorphism, and mylonitization along the Adirondack Highlands–Northwest Lowlands "boundary" (Wiener, 1977, 1981; Geraghty and Isachsen, 1981) accompanied the second phase of isoclinal folding. A third phase of folding resulted in formation of open to tight, upright to southeast-overturned folds with northeast-trending axial surfaces (Fig. 19), which are the most obvious folds in the northwestern Adirondacks. The fourth phase of folding resulted in open, upright, northwest-trending folds and dome and basin patterns at intersections with third-phase folds. A fifth minor phase of folding resulted in open, upright, discontinuous folds with north northeast-trending axial surfaces only locally developed in the northwestern Adirondacks but more prevalent in the central Adirondacks. The relative age of the fourth- and fifth-phase folds is not certain.

In the following discussion, stratigraphic anticline refers to first-phase folds with older strata in the core, whereas stratigraphic syncline refers to first-phase folds with younger rocks in the core.

First-Phase Nappes

The earliest major and minor folds recognized in the northwestern Adirondacks are isoclinal folds in bedding characterized by axial plane foliation coincident with the regional foliation in metasedimentary and metavolcanic rocks. Also, regional lineations in these rocks are parallel to, and associated with, first-phase fold axes. Thus, the orientation of regional foliation and lineation is parallel to first-phase axial planes and fold axes, respectively. Peak metamorphism accompanied (and followed) the first phase of folding, as indicated by parallelism of minerals to first-phase axial plane foliation and axial lineation. First-phase fold axes generally plunge down the dip of the axial planes, indicating that many first-phase folds have a reclined profile. First-phase folding as defined in this study predates the F_1 folds of

Foose and Carl (1977) and McLelland and Isachsen (1980) Table 2).

The initial geometry and transport of the first-phase folds is difficult to establish owing to subsequent intense deformation and flow. The northwest-directed thrust in the Balmat no. 4 mine is a lower-limb fault on the first-phase Fowler syncline mapped by deLorraine (1979). There also is a northwest-directed early thrust in the Bigelow area (Foose, 1974) (Fig. 3). The regional configuration of first-phase isoclinal folds in the Northwest Lowlands further suggests that these nappes were northwest-directed (Fig. 20-1) with northwest-closing stratigraphic anticlines such as the Rock Island stratigraphic anticline and syncline (Figs. 19, 20-3). Regional mineral lineations are parallel to northwest- and southeast-plunging first-phase fold axes. These fold axes lie at a small angle to the suggested northwest transport direction, and if these angular relations have not been subsequently distorted, they imply considerable differential flow and varied plunges of first-phase fold axes.

The first-phase Power Line stratigraphic syncline, Graham stratigraphic anticline, and Graham stratigraphic syncline are refolded by second-phase isoclinal folds in the Harrisville area (Wiener, 1983) (Fig. 21b). Here, backfolding and rotation of first-phase folds by second-phase folds distorted the original orientation of first-phase folds so that anticlines close to the southeast (Wiener, 1981, 1983, Fig. 24). In the Moss Ridge area, a slight modification of Foose's (1974) map suggests the presence of two first-phase folds. A biotite-garnet-sillimanite gneiss layer within the leucogranitic gneiss may be an infold of Baldface Hill Gneiss into the Alexandria Bay Gneiss. If this is so, it indicates the presence of the first-phase Grove Road stratigraphic anticline and Grove Road stratigraphic syncline, which are refolded by the second-phase Moss Ridge anticline and Moss Ridge syncline (Figure 21a). The postulated Grove Road folds may be the same folds as the Rock Island folds, refolded by second- and third-phase folds.

Second-Phase Nappes

The dominant isoclinal folds in the northwest Adirondacks are second-phase nappes which, as defined in this study, are equivalent to F_1 folds of Foose and Carl (1977) and McLelland and Isachsen (1980) (Table 2). The second-phase folds are southeast-directed nappes with southeast-closing anticlines (Fig. 20-2), such as at Moss Ridge and Baldface Hill (Fig. 21). The southeast-directed thrust near Beaver Creek is a lower-limb fault on a second-phase nappe mapped by Brown (Foose and Brown, 1976) (Fig. 3).

Major and minor second-phase folds display folded regional foliation and lineation, which developed during the first phase of folding. Both axial plane foliation and slip cleavage and intersection lineation parallel to fold axes show varying degrees of development in second-phase folds, the strongest development being in thinly layered gneisses (Wiener, 1981).

Except where in the crests or troughs of later folds, axial

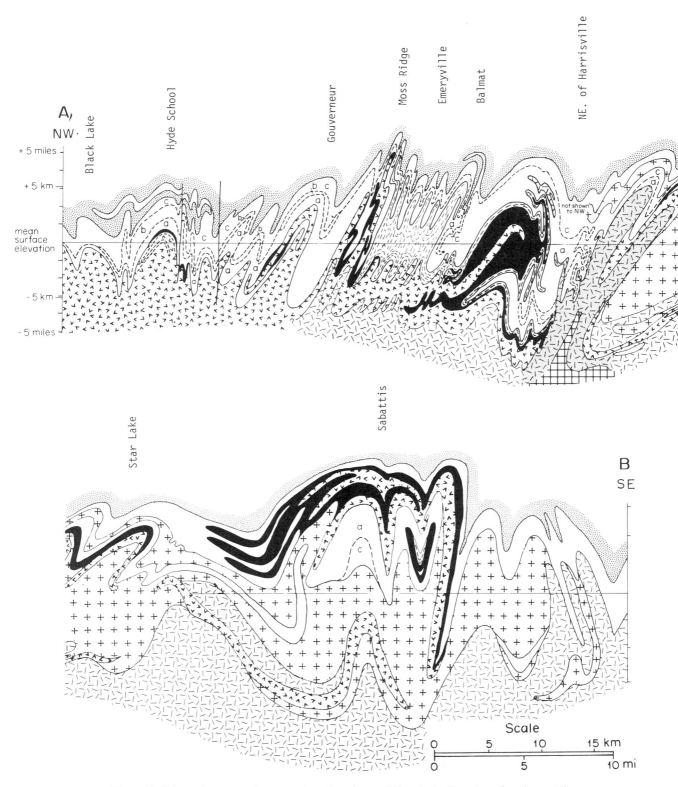

Figure 19. Schematic structural cross-section of northwest Adirondacks. Location of section on Figure 3. Symbols from Figure 3.

TABLE 2. CORRELATION OF AGES OF MAJOR FOLD PHASES

This study (whole Adirondacks)	McLelland and Isachsen (1980) (southern-central Adirondacks)	Foose and Carl (1977) (NW Adirondacks)	deLorraine (1979) (Balmat)
F_1-earliest folds in NW Adirondacks (Wiener, 1981); possible isoclinal fold in Lake Durant Fm	"pre-F_1 foliation forming event"		F_1-Fowler syncline, NW-directed fold-thrust nappe
F_2-major, recognized nappes directed to SE in NW Adirondacks, to S in S Adirondacks	F_1-Wakely Mountain nappe-Little Moose Mtn-Canada Lake syncline, directed to S. Amplitude >70 km	F_1-Moss Ridge isoclinal folds	F_2-isoclinal, SE-overturned folds, directed to SE
F_3-major N70°W to E-W to N45°E open to tight folds in Adirondacks	F_2-N80°W-trending, open upright folds. Piseco anticline, Glens Falls syncline, and others	F_2-N45°E-trending, open to tight folds, upright or overturned to SE; "Gouverneur anticline" and others. Responsible for dome elongation	F_3-open, discontinuous NNE-trending folds
F_4-NW-trending open folds; produce dome and basin patterns where interfere with F_3, NE-trending folds in NW Adirondacks	F_3-only a few folds present; open, discontinuous, NW-trending	F_3-open, NW-trending folds, which interfere with F_2 folds to produce dome-basin patterns	F_4-open, NW-trending folds
F_5-NNE-trending open folds; produce dome and basin patterns with WNW-trending F_3 in S.-central Adirondacks	F_4-open, discontinuous folds; produce dome-basin patterns where interfere with F_2 folds		

planes of second-phase folds generally strike northeast and dip parallel to the foliation, to the northwest or southeast (Figs. 19, 22). Fold axes generally plunge moderately to the north, northeast, or southwest. Movement line orientations for second-phase folds in the Harrisville-Pitcairn area indicate north-northwest over south-southeast motion about a north-northwest-plunging movement line (Wiener, 1983). The regional pattern of fold closures (Fig. 19), thrust faults, and movement line determinations suggests that most of the second-phase folds formed by northwest over southeast movement associated with development of southeast-directed nappes (Fig. 20-2), which locally developed lower-limb thrusts.

Intrusion and Mylonitization of Igneous Rocks during Second-Phase Deformation and the Adirondack High-lands–Northwest Lowlands Boundary. Extensive igneous activity accompanied the second phase of deformation, involving intrusion of gabbro, quartz syenite, and related rocks of the Diana, Tupper, and Stark complexes, and the hornblende granite (Fig. 3). These igneous rock suites are both folded by second phase folds and cut across second phase folds (Wiener, 1983) (Figs. 20-2, 22) indicating that intrusion occurred during the second deformational phase.

The Diana Complex of the Adirondack Highlands (Figs. 3, 20-2) underwent mylonitization along its northwestern margin during the second phase of regional deformation, based on the fact that mylonitic foliation in the Diana Complex is parallel to axial planes of major and minor second-phase folds and is folded by third-phase folds. The second regional phase of deformation was the initial phase of deformation of the Diana Complex and other igneous bodies in the northwestern Adirondacks (Wiener, 1983). Isoclinal folding of the Diana Complex into nappes (Fig. 20-2) produced a mylonitic fabric owing to lithic contrast between dry, massive, felsic plutonic rocks of the Diana Complex and various layered metasedimentary rocks of the Northwest Lowlands (Wiener, 1983). The mechanism of mylonitization is envisioned to be flattening and simple shear associated with isoclinal folding along this zone of lithic contrast (Wiener, 1981, 1983). Thus, intrusion and mylonitization of igneous rocks occurred during second-phase deformation.

Northeast-Trending Third-Phase Folds

Third-phase folds are the dominant fold set in the northwestern Adirondacks (Fig. 22) and, for that matter, the entire Adirondacks (Fig. 14). The third-phase folds of this study are equivalent to the F_2 folds of Foose and Carl (1977) and McLel-

METAMORPHISM AND INTRUSIVE ACTIVITY IN THE ADIRONDACKS

Fold characteristics	Metamorphism and associated anatexis	Intrusive activity	Absolute age (b.y.)
F_1-Axial plane regional foliation; mineral lineation parallel fold axes. Transport to northwest. Major folds difficult to recognize.	Peak amphibolite to granulite facies metamorphism of metasedimentary and metavolcanic rocks. Extensive partial melting of paragneisses resulting in granitic melts		1.10-1.02 peak metamorphism (Silver, 1969) (F_1-F_3)
F_2-Folded regional foliation and lineation in hinge regions; well-developed axial plane foliation	Sillimanite-grade in NW Adirondacks orthopyroxene and sillimanite in Highlands. Anatexis in Highlands only. Metamorphism of igneous rocks (gabbro 750-850°C, 7-8 kbar) (McLelland and Whitney, 1980a)	Major intrusive activity. Intrusion of anorthosite. Intrusion and subsequent F folding and mylonitization of charnockite sheets. Later intrusion and folding of hornblende granite and metagabbro	Anorthosite-1.19 (Ashwal, 1980); 1.13 (Silver, 1969). Charnockite-1.13 (Silver, 1969); 1.09 (Hills and Gast, 1964). Hornblende granite-1.05 (Foose and others, 1981)
F_3-Major open to tight folds in Adirondacks, largely responsible for aeromagnetic pattern. Axial plane slip or fracture cleavage and intersection lineations	Orthopyroxene, garnet, clinopyroxene in south-central Adirondacks. Sillimanite in NW Adirondacks.		1.10-1.02 (Silver, 1969) peak metamorphism (F_1-F_3)
F_4-Local axial plane slip or fracture cleavage, intersection lineations mostly present in NW Adirondacks	Retrograde metamorphism of sillimanite to muscovite (Foose, 1974), and diopside and forsterite to serpentine (Wiener, 1981). Garnet-amphibolite facies metamorphism of diabase dikes	(?)Intrusion and metamorphism of diabase dikes, which display chilled borders	
F_5-Local slip or fracture cleavage; mostly present in Adirondack Highlands	Retrograde metamorphism to muscovite in axial plane (600-650°C, 3-5 kbar) (McLelland and Isachsen 1980)	(?)Intrusion of muscovite pegmatite (Fisher and others, 1971)	930 ± 40 m.y. (Rb/Sr) (Fisher and others, 1971). Intrusion of pegmatite

Note: F_1, F_2, etc. indicate relative age of major folds.

land and Isachsen (1980) (Table 2). In the Northwest Lowlands, the third-phase folds are open to tight, upright to overturned folds with northeast-trending axial surfaces (Figs. 20-3, 22). Domes and basins are elongate to the northeast, parallel to these folds (Fig. 22). In the western and central Adirondacks, the third-phase folds swing in a smooth, continuous fashion to an easterly trend (Fig. 22). The principal folds shown in the cross section of the northwestern Adirondacks (Fig. 19) are the third-phase folds. These folds refold second- and first-phase isoclinal folds near Moss Ridge and Baldface Hill (Fig. 21) and elsewhere in the Northwest Lowlands (Fig. 20-3).

Minor third-phase folds display open to moderately tight profiles and may have axial plane slip cleavage or fracture cleavage. Mineral lineation is developed parallel to fold axes, and minerals are parallel to axial planes in many places.

Axial planes of third-phase folds strike north-northeast to northeast. From northwest to southeast in the Northwest Lowlands, the dip of axial planes changes from southeast to vertical to northwest (Fig. 19). Fold axes of third-phase folds generally plunge north, northeast, or southwest. Interference of third-phase folds with second-phase folds generally produces hook patterns owing to the small angle between second- and third-phase fold axes (Fig. 22), but heart patterns are also locally present.

Northwest-Trending Fourth-Phase Folds

Major and minor fourth-phase folds are mostly open, upright folds with northwest-trending axial surfaces, though locally these folds are tight and overturned. Fourth-phase folds in this study are equivalent to F_3 folds of Foose and Carl (1977) and McLelland and Isachsen (1980) (Table 2). These open folds are well developed in the northwestern Adirondacks (Fig. 22) but only locally present in the southern and central Adirondacks McLelland and Isachsen, 1980) (Fig. 16).

Axial planes of fourth-phase folds strike northwest (Fig. 22) and generally dip steeply. Where third-phase folds are overturned to the southeast, fourth-phase fold axes plunge northwest because both limbs of third-phase folds dip northwest, whereas in areas where third-phase folds are upright, fourth-phase fold axes plunge northwest and southeast on the respective northwest- and southeast-dipping limbs of third-phase folds. Mineral lineation is locally developed parallel to fold axes, and axial plane slip cleavage or fracture cleavage are also present in places. Locally, retrograde muscovite (after sillimanite) or serpentine (after diopside or forsterite) is present along the axial plane (Foose, 1974; Wiener, 1981).

Interference of fourth-phase major and minor folds with

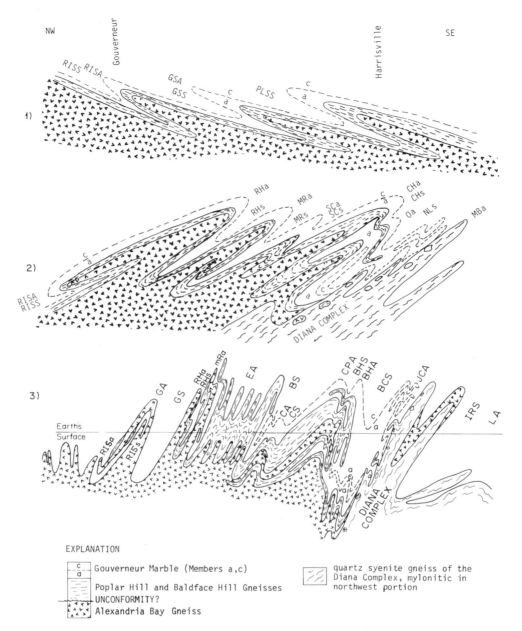

Figure 20. Schematic cross-sections depicting structural evolution of Northwest Lowlands. (1) First-phase nappes: RISS = Rock Island stratigraphic syncline; RISA = Rock Island stratigraphic anticline; PLSS = Power Line stratigraphic syncline; GSA = Graham stratigraphic anticline; GSS = Graham stratigraphic syncline. (2) Second-phase nappes: MBa = Middle Branch anticline; NLs = New Line syncline; Oa = Oswegatchie anticline; CHs = Cooper Hill syncline; CHa = Cooper Hill anticline; SCs = Sawyer Creek syncline; SCa = Sawyer Creek anticline; MRs = Moss Ridge syncline; MRa = Moss Ridge anticline; RHs = Reservoir Hill syncline; RHa = Reservoir Hill anticline. Note truncation of early phase folds and the types of xenoliths compared to the adjacent stratigraphic unit along northwest border of Diana Complex; also note that the mylonitic foliation in the Diana Complex is parallel to second-phase axial planes. (3) Third-phase folds: LA = Lowville anticline; IRS = Indian River syncline; JCA = Jenny Creek anticline; BCS = Big Creek syncline; BHA = Baldface Hill anticline; BHS = Baldface Hill syncline; CPA = Clark Pond anticline; CS = California syncline; CA = California anticline; BS = Balmat syncline; EA = Emeryville anticlinorium; GS = Gouverneur syncline; GA = Gouverneur anticline. Note folded mylonitic foliation in Diana Complex.

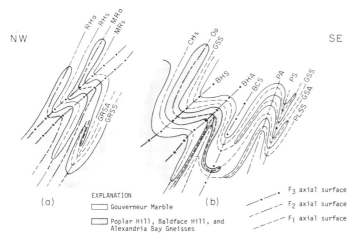

(a)

(b)

EXPLANATION
☐ Gouverneur Marble
☐ Poplar Hill, Baldface Hill, and
   Alexandria Bay Gneisses

—•— F₃ axial surface
— — F₂ axial surface
— — — F₁ axial surface

Figure 21. Schematic unscaled structural cross-sections depicting first three phases of northeast-trending folding. (a) Moss Ridge area, modified slightly from Foose (1974); RHa = Reservoir Hill anticline; RHs = Reservoir Hill syncline; MRa = Moss Ridge anticline; MRs = Moss Ridge syncline; GRSA = Grove Road stratigraphic anticline; GRSS = Grove Road stratigraphic syncline. (b) Baldface Hill–Harrisville area from Wiener (1981): CHs = Cooper Hill syncline; Oa = Oswegatchie anticline; GSS = Graham stratigraphic syncline; GRSS = Grove Road stratigraphic syncline; GSA = Graham stratigraphic anticline; BHS = Baldface Hill syncline; BHA = Baldface Hill anticline; BCS = Big Creek syncline; PA = Pitcairn anticline; PS = Pitcairn syncline; PLSS = Power Line stratigraphic syncline.

third-phase folds results in dome, saddle, and basin patterns (Fig. 22). The domes, saddles, and basins are generally elongate to the northeast, parallel to the higher amplitude third-phase fold set (Fig. 22).

North-Trending Fifth-Phase Folds

In the northwestern Adirondacks the fifth-phase folds are only locally developed. These generally open, upright folds have north-northwest to north-northeast-striking, steeply dipping, axial surfaces, which are discontinuous. Locally, these folds are tight and overturned. The structural basin south of Cranberry Lake is the result of interference of an open, fifth-phase, north-trending syncline and an east-northeast-trending third-phase syncline (Leavell, 1977) (Fig. 22). Other fifth-phase folds include the north-trending folds in the Pleasant Lake Gneiss southeast of the Stark Complex (Fig. 22). These folds are more prevalent to the east in the northern, central, and southern Adirondacks where they interfere with third-phase folds to produce dome and basin patterns (Fig. 16) (McLelland and Isachsen, 1980). However, fifth-phase folds are apparently only locally developed in the Northwest Lowlands, as in the areas northwest of the Clark Pond anticline and near Rossie (Fig. 22). In the Northwest Lowlands, the fifth-phase axial surfaces and hinge lines lie at a small angle to the third-phase folds, so that their distinction from third-phase folds is difficult, and it is possible that some folds identified as third-phase folds are actually fifth-phase folds.

STRUCTURAL GEOLOGY OF THE SOUTHERN AND CENTRAL ADIRONDACKS

Four unusually large fold sets are recognized in the southern and central Adirondacks (Fig. 16). These are designated F_2 through F_5 in order to correlate with second-through fifth-phase folds in the northwestern Adirondacks. Previously, these folds were referred to as pre-F_1 and F_1 to F_4 (McLelland and Isachsen, 1980) (Table 2). The absolute time intervals associated with phases of the deformational sequence are uncertain at the present. It is possible that the F_1, F_2, and F_3 (?) fold episodes are parts of a synchronous tectonic continuum within which only the relative order of folding is constant. Thus, F_1 folds might be developing in one area while F_2 folds were being formed in adjacent terrane. In contrast to this, we consider the F_4 and F_5 folds, and possibly F_3 folds as well, to have occurred distinctly later.

Possible F_1 Folds and F_1 and F_2 Foliations

Although no evidence for an F_1 folding episode is reflected directly in the regional map pattern (Figs. 14, 15), F_1 minor folds have been recognized locally, and there are possible stratigraphic repetitions in the central Adirondacks suggestive of larger F_1 nappes (see Stratigraphy). In the southern Adirondacks, however, stratigraphic repetitions that would accompany major F_1 regional folds have not yet been identified in mapping. It is possible that such a fold might be of dimensions exceeding the map area and thus remain undetected. It is also possible that the stratigraphy of the southern Adirondacks (Fig. 17) contains repetitions owing to F_1 isoclinal folding but that these have not yet been recognized.

To understand foliations related to F_1 folds, it is necessary to briefly describe their relation to F_2 folds. The major F_2 folds of the region are extremely large (Fig. 16) and are isoclinal. Associated with minor, tight to isoclinal F_2 folds is a strong axial plane foliation defined by mineral platelets that are suggestive of extensive flattening. For some time it was believed that this foliation was the oldest in the area. However, careful mapping in the hinge region of major F_2 folds has demonstrated that F_2 folds of all scales rotate an earlier F_1 foliation. This earlier F_1 foliation is itself defined by parallel platelets of quartz and feldspar and by oriented metamorphic minerals such as biotite and hornblende. Flattened garnets also occur within the early foliation planes.

Except in the noses of F_2 folds, it is difficult to distinguish between the F_1 and F_2 foliations. On the basis of examination of these fold hinges, the F_1 foliation is very intense, and it may account for most of the regional foliation, as in the northwestern Adirondacks. Intersection of the F_1 foliations with later foliations and with compositional layering results in strong intersection lineation, pencil gneisses, and rodding along the hinges of folds.

Major F_2 Folds

These are the earliest definitely recognizable map-scale folds in the area. Competent layers such as quartzite and charnockite

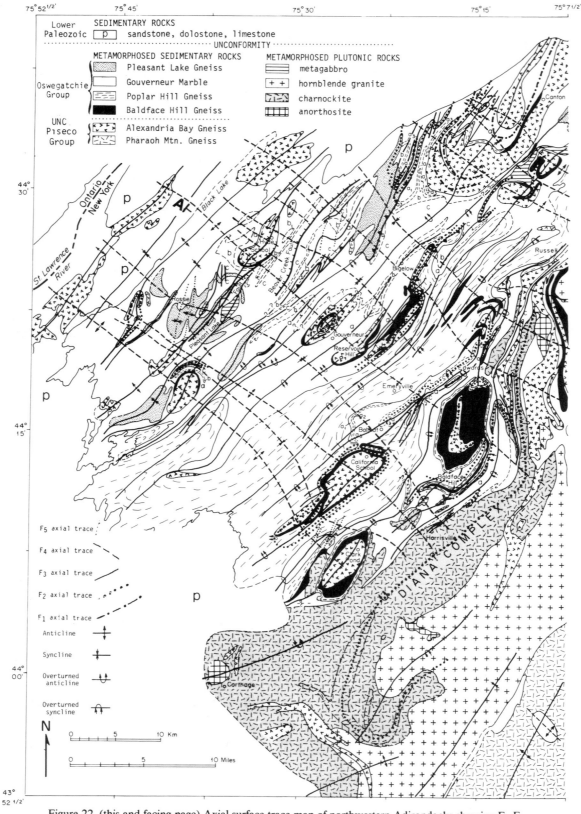

Figure 22. (this and facing page) Axial surface trace map of northwestern Adirondacks showing F_1-F_4 major phases of deformation and minor F_5 phase of deformation.

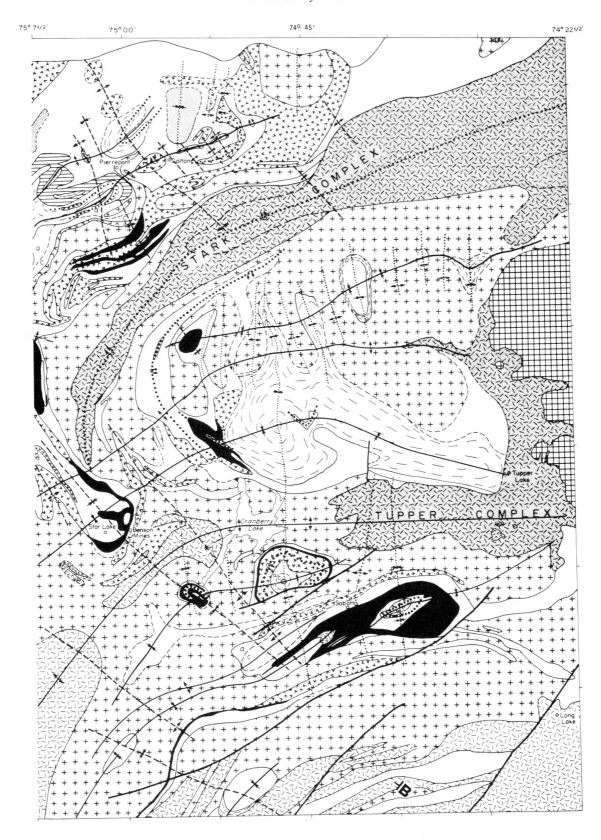

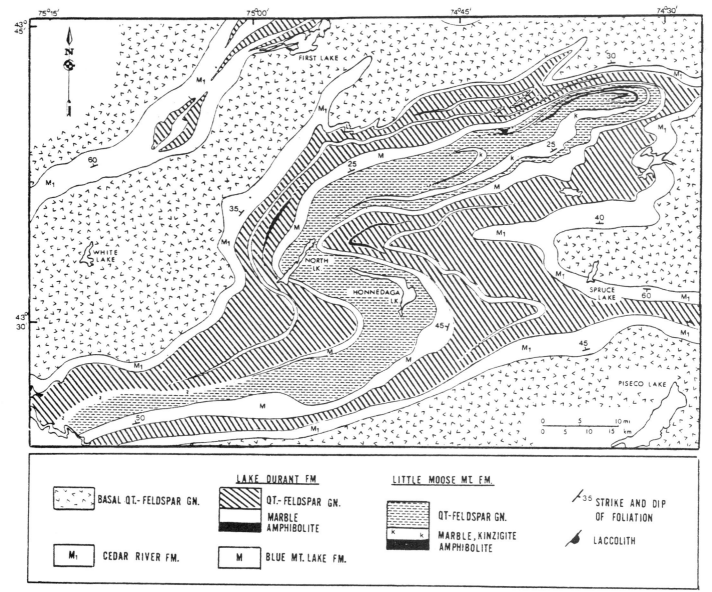

Figure 23. Geology of the west-central Adirondacks showing the F_2 isoclinal Little Moose Mountain syncline refolded by west-plunging F_3 open folds. The Little Moose Mountain syncline is cored by the Little Moose Mountain Formation. The Wakely Mountain nappe to the north and the Piseco anticline to the south are both cored by basal quartzo-feldspathic gneiss. Near the hinge of Little Moose Mountain syncline, an interpreted laccolith of metagabbro indicates the direction of younging. The geologic mapping in the northeastern quarter of the area was done by deWaard (1962). From McLelland and Isachen (1980).

tend to maintain constant thickness perpendicular to layering, whereas marbles, kinzigites, and some leucogneisses tend toward constancy of thickness parallel to F_2 axial planes.

The first major F_2 folds to be mapped in detail in the central or southern Adirondacks region were the Little Moose Mountain syncline and the Wakely Mountain nappe (deWaard, 1962, 1964). The axial plane of the Little Moose Mountain syncline (Fig. 16) dips gently to the north beneath the Wakely Mountain nappe, and the axes of both folds plunge 5° to 10° westward. The

axial trace of the Little Moose Mountain syncline can be followed northeastward along the Cedar River Flow for about 15 km, and deWaard (1964) mapped the Wakely Mountain nappe to the southwest as far as Pico Mountain, southeast of Old Forge, a total distance of more than 50 km. The axial traces of the Little Moose Mountain syncline–Wakely Mountain nappe continue another 50 km southwestward (Douglas and others, 1976) before disappearing beneath the Paleozoic cover (Fig. 23).

McLelland (1969) demonstrated the existence of a recum-

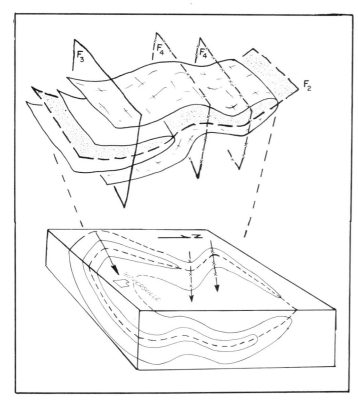

Figure 24. Block diagram depicting the manner in which folding of the Canada Lake nappe affects the trajectory of its axial trace. Modified from McLelland and Isachsen (1980).

bent isoclinal fold of major dimensions in the southern Adirondacks. The axial trace of the Canada Lake syncline extends about 100 km from east of Little Falls to north of Saratoga Springs (Fig. 16). This stratigraphic syncline is antiformal with an axis trending approximately east-west and plunging generally 10° to 15° eastward. The nappe has been refolded so that its axial trace follows a zigzag pattern between Canada Lake and Sacandaga Reservoir (Fig. 24). Further refolding, as well as faulting, accounts for the re-emergence of the nappe core (Irving Pond Formation) east of the reservoir. As will be indicated in the next section, the Canada Lake nappe is an overturned portion of a much larger stratigraphic syncline.

Major F_3 Folds

As shown in Figure 16, the principal members of the F_3 fold set are, from south to north, the Gloversville syncline, the Piseco anticline, the Glens Falls syncline, and the Spruce Lake anticline.

The F_3 folds are upright and relatively open. Their axes trend about east-west and appear to be coaxial with F_2 folds, at least in the southern Adirondacks. Horizontal dimensions of the F_3 folds are similar to those of F_2, although the amplitudes of F_2 folds are considerably larger. There is a pronounced linear concentration of minerals parallel to F_3 fold axes, owing to the

intersection of F_1, F_2, and, locally, F_3 foliations (McLelland and Isachsen, (1980).

F_4 Folds

F_4 folds are open, upright, and have a northwesterly trend. They are not widely developed in the southern or central Adirondacks, and only five major folds have been tentatively identified (Fig. 16). Locally, F_4 and F_3 are subparallel, and it is possible that some folds assigned to the F_4 generation are actually F_3 folds.

The F_4 folds south of the Piseco anticline display an axial plane foliation defined by oriented platelets of minerals, notably, quartz and feldspar. F_4 related lineations consist, in part, of preferred orientation of hornblende and biotite.

F_5 Folds

F_5 folds trend north-northeast, are open upright, and decrease in intensity and abundance from east to west (Fig. 16). In the northeastern Adirondacks, they are tight, and even isoclinal (P. R. Whitney, 1978, oral commun.). In the central and southern Adirondacks, F_5-related foliations are weak and are not generally associated with metamorphic recrystallization except in the eastern Adirondacks where, in places, flakes of muscovite occur parallel to the F_5 axial planes.

Throughout the area, F_5 folds are responsible for excellent examples of fold interference patterns (Fig. 16). For example, the conspicuous F_5 anticline that defines the long axis of the Snowy Mountain dome intersects several F_3 folds and results in alternate structural culminations and depressions. The most notable of these is the Piseco dome where third-phase folds and lineations plunge westward to the west of the F_5 axial trace and eastward east of the axial trace.

North of Sacandaga Reservoir (Figs. 14, 15), the outcrop patterns of the Basal Quartzo-feldspathic Gneiss and the Sacandaga Formation outline a series of structural depressions and culminations along the Piseco anticline. As in the Piseco dome, these are caused by interference between the F_3 Piseco anticline and the F_5 synclines and anticlines. Linear features paralleling the F_3 axis generally plunge toward the depressions and away from culminations.

An important F_5 syncline and anticline pair pass through the main body of Sacandaga Reservoir and its eastern arm, respectively (Figs. 16, 24). Interference of these folds with the Canada Lake nappe results in the southward swing of its axis and the reemergence of the core rocks of the nappe east of the reservoir (Fig. 16). Unfortunately, most of these F_5 folds lie beneath either water or Paleozoic rocks. However, the outcrop pattern along the shores of the eastern arm of Sacandaga Reservoir (Figs. 14, 15) indicates that the F_5 axis plunges south at about 10° to 15° and that the fold tightens in that direction. The north-northeast-trending normal fault along the long, straight, eastern arm of the Sacandaga Reservoir must offset the axial trace of the Canada

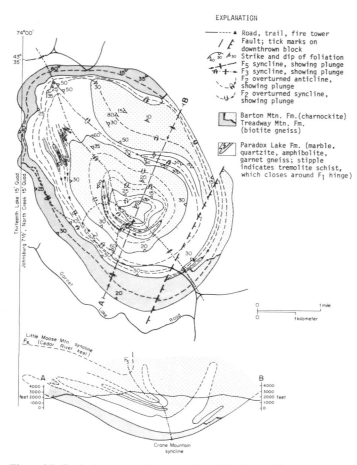

Figure 25. Geologic map and cross-section of the Crane Mountain area.

Lake nappe. The vertical displacement along this fault is at least 450 m, west side down (Isachsen and McKendree, 1977).

Another example of F_5-related interference is at Crane Mountain, which is a structural basin (Fig. 16) that results from the interference of an upright to overturned F_3 syncline, the Crane Mountain syncline, and an open F_5 syncline (Fig. 25). The map pattern is interpreted to be due to interference of three major fold sets (F_2, F_3, F_5) in a stratigraphy composed of two formations. The Paradox Lake Formation is overlain by the Lake Durant Formation, which grades along strike by original sedimentary facies changes into the Treadway Mountain Formation (Fig. 25). Map areas of Paradox Lake Marble between the map areas of Treadway Mountain–Barton Mountain Formations are related to F_2 isoclinal anticlines cored by Paradox Lake Marble (Fig. 25). There are four such anticlines, all of which are interpreted to close toward the northeast along the southwest side of Crane Mountain (Fig. 25). These F_2 isoclinal folds were subsequently refolded by an F_3 syncline that trends northwestward and then a northeast-trending F_5 syncline. The interference of the F_3 and F_5 synclines has resulted in the present basinal map pattern of the previously (F_2) isoclinally folded contact between the Paradox Lake Formation and the Barton Mountain–Treadway

Mountain Formations (Fig. 25). Evidence for an F_1 isoclinal fold is present at the end of the tremolite schist member of the Paradox Lake Formation (stipple on Fig. 25). The axial surface of this F_1 fold is refolded by F_2 isoclinal folds in the tremolite schist (Fig. 25).

REGIONAL SYNTHESIS AND STRUCTURAL FRAMEWORK

The sequence of rocks north of the Piseco anticline may conveniently be compared with the sequence to the south, along the line of section labeled A-A' on Figures 14, 15, and 16. Along the northern section of A-A', from Piseco Lake to the core of the Little Moose Mountain syncline, the sequence is the normal one shown in Figure 17 repeated by several F_3 folds. From Piseco Lake to the core of the Canada Lake syncline, more units are exposed than to the north, but the sequence is the mirror image of that to the north.

When combined with the pattern of F_3 folds shown in Figure 16 and 23, the symmetry reflected by the stratigraphy on either side of the Piseco anticline leads to the conclusion that the Little Moose Mountain syncline and the Canada Lake syncline are most likely different portions of the same fold. The fact that one portion of this fold closes downward as a syncline and the other closes upward as an anticline is due to their relative positions on the north and south limbs, respectively, of the Piseco anticline (Fig. 26). This major structure, here referred to as the Canada Lake–Little Moose Mountain syncline, has an amplitude exceeding 70 km (Fig. 26), and its axial surface has been traced for more than 100 km (Fig. 16).

Reconnaissance mapping strongly suggests that the F_2 folds underlying the Pine Ridge and Crane Mountain structural basins are on the lower, right-side-up, autochthonous limb of the Canada Lake–Little Moose Mountain syncline. The Paradox Lake Formation probably roots westward in the F_2 Little Moose Mountain syncline, as suggested by asymmetry of F_2 isoclinal folds on Crane Mountain (Fig. 25). Recent field work tentatively suggests that within 1 km east of Crane Mountain the axial trace

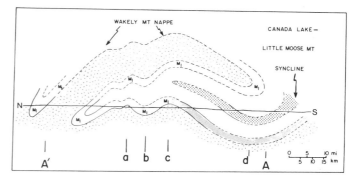

Figure 26. Generalized cross-section along A-A' of Figures 14, 15, and 16, showing isoclinal F_2 folds and the following open F_3 folds: a, Spruce Lake anticline; b, Glens Falls syncline; c, Piseco anticline; d, Gloversville syncline. Several map units have been omitted for clarity. Patterned rock unit symbols as in Figure 15. From McLelland and Isachsen (1980).

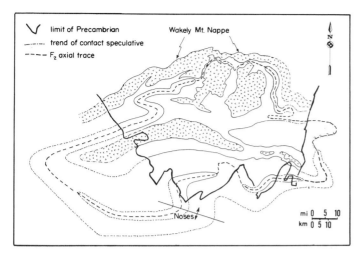

Figure 27. Generalized map showing the known axial trace of the Canada Lake–Little Moose Mountain syncline and its projection beneath Paleozoic cover. Heavy solid line marks the boundary between Proterozoic and Paleozoic rocks. Modified from McLelland and Isachsen (1980).

of the Little Moose Mountain syncline is rotated by an east-west F_3 anticline. From here, it may continue eastward toward Lake George (Fig. 16) or pass southward into the northern limb of the Glens Falls syncline. From here it may extend westward almost to the town of Wells where it may pass around the hinge line of the Glens Falls syncline and continue eastward and along the southern limb of this fold until passing beneath the Paleozoic cover to the east (Fig. 27). Regardless of its exact location, the axial trace of the Little Moose Mountain syncline must traverse the entire central Adirondacks, inasmuch as its continuation, the Canada Lake syncline, traverses all of the southern Adirondacks.

In order to visualize the spatial configuration of the axial surface of the Canada Lake–Little Moose Mountain syncline, we must recall that the axial trace must eventually form a closed loop on the Earth's surface, inasmuch as it has been deformed by at least three subsequent phases of plunging, noncylindrical folds. It is clear from the map pattern that this does not happen within the limits of Precambrian exposure. In Figure 27, we have schematically indicated the manner in which such closure might occur beneath Paleozoic cover. The western closure is suggested by aeromagnetic data (Zeitz and Gilbert, 1981) that show the magnetic signature of the basal quartzo-feldspathic gneisses of the Piseco anticline extending westward beneath the Paleozoic cover for some 50 km. The necessary F_3 anticline south and west of the map area is strongly supported by isolated outcrops of Proterozoic rocks between the southern limit of Figures 14 and 15 in the Noses area on the Mohawk River (Fisher and others, 1971).

The structural framework shown in Figures 16, 26, and 27 indicates the regional dimensions of the Canada Lake–Little Moose Mountain nappe. An overlying companion fold of similar dimensions, the Wakely Mountain nappe (Fig. 27), has been largely removed by erosion. The area as a whole is a regional

dome (Fig. 27) through which a window has been provided by subsequent erosion. A great deal of detailed work remains to be done in this region, but the broad structural framework is reasonably well outlined on the basis of present knowledge.

Axial Surface Trace Map of Adirondacks

A major conclusion of this research is that the outcrop patterns of major stratigraphic units in the Adirondacks are the result of interference between five fold sets, not all of which are recognized in all sections of the Adirondacks. The success of this model in the northwestern, southern, and central Adirondacks (Figs. 16, 22) suggests that the same interference patterns may extend throughout the rest of the Adirondacks. We have applied this interpretation based on evaluation of orientations of minor structural features and map patterns as shown on the Geologic Map of New York (Fisher and others, 1971), supplemented mainly with aeromagnetic data. The resulting analysis, shown in Figure 28, is intended only as a hypothetical working model deserving of further field investigation. In many instances, sets of several parallel anticlines and synclines are grouped together into a single axial trend corresponding to the strongest member of the set.

An area critical to the comparison and correlation of folds between the Adirondack Highlands (Fig. 16) and Northwest Lowlands (Fig. 22) is between the Wakely Mountain nappe and the Diana Complex of Buddington (1939, 1948) (shown on Fig. 22). This area is the least known geologically in the Adirondacks, but work has been undertaken in this region over the past 5 yr by students at the University of Massachusetts (Leavell, 1977; Potter, 1981, and in prep.; P. Fallon, in prep.). We have tentatively deduced the existence of a very large F_3 syncline north of the Arab Mountain (Fig. 28) on the basis of information from the State geologic map, aeromagnetic maps (Balsley and Bromery, 1965a, 1965b; Balsley and Buddington, 1960), and geologic reports (Miller, 1909; Buddington and Leonard, 1962). A large F_3 syncline (Potter, 1981; Buddington and Leonard, 1962) is south of the Arab Mountain anticline ("AMA" on Fig. 28) (Buddington and Leonard, 1962). This fold is extended southwestward and eastward on the basis of reconnaissance mapping, the state geologic map, aeromagnetic data (Balsley and Bromery, 1965a, 1965b), and prior geologic mapping (Miller, 1909).

Interference patterns of F_3 and F_5 folds appear well developed in and around the Marcy meta-anorthosite massif (north part of Fig. 28). For example, the structural basins north of the massif are interpreted as results of intersecting F_3 and F_5 synclines. Between and on either side of these synclines are F_5 anticlines of meta-anorthosite. Scattered mappable xenoliths of metasedimentary and mixed rocks occur in the east-central part of the meta-anorthosite massif (H. Jaffe and E. Jaffe, mapping in prep.), with metagabbroic border facies at their base (Fig. 28). These roof pendants are tentatively interpreted as downfolded remnants of the roof of the intrusive, such as would be preserved at the intersection of F_3 and F_5 synclines. This extrapolation is

Figure 28. Suggested framework of major ductile deformation in the Adirondacks. F_1 folds not shown due to scale of map. White areas within the Adirondack perimeter represent a variety of rock types, including quartzo-feldspathic gneisses. Black patches in the main meta-anorthosite body are xenoliths and mixed rocks (roof pendants) suggestive of downfolds. Crosses locate sills of gabbroic anorthosite. SLR = St. Lawrence River; LC = Lake Champlain; LG = Lake George; AMA = Arab Mountain anticline. Note swing of F_3 folds from N70°W in the south to northeast-trending in the northwest Adirondacks. Also note concentration of F_4 folds in northwest and F_5 folds in eastern and central Adirondacks, both of which produce dome and basin patterns where they interfere with F_3 folds. Geologic base map generalized and modified from Fisher and others (1971).

based partly on Davis's (1971) observation that in the northwestern part of the massif the gabbroic anorthosite border facies occurs in downfolds.

The anorthosite domes could alternatively be explained as anorogenic diapirs. The reason for preferring a multiple fold interpretation of their structural pattern is that the domes and basins are aligned with regional folds (Fig. 28) rather than arranged in a random pattern.

Nature and Correlation of Major Folds in Adirondacks

In the southern and central Adirondacks, four regional phases of folding have been recognized, designated F_2 to F_5 in this paper and F_1 to F_4 in a previous paper by McLelland and Isachsen (1980) (Table 2). Also, a "pre-F_1" foliation-forming event was recognized by McLelland and Isachsen (1980), herein correlated with F_1 in this study (Table 2). In the northwestern Adirondacks, four regional phases of deformation have been recognized (Brocoum, 1971; deLorraine, 1979; Wiener, 1977, 1981). F_1 to F_4 major folds recognized by deLorraine (1979), Brocoum (1971), and Wiener (1981) correlate with F_1 to F_4 folds as defined in this study. F_1 to F_3 folds recognized by Foose (1974) and Foose and Carl (1977) correlate with F_2 to F_4 folds in this study (Table 2). Much detailed work needs to be done to bear out these interpretations. Specific problems include the nature of F_1 folds in the central-southern Adirondacks, the distinction of F_1 and F_2 isoclinal folds in the northwest Adirondacks, and movement-sense of the F_1 and F_2 folds. With these qualifications, we attempt in the following section to present a unified theory of folding in the Adirondacks.

First-phase isoclinal major and minor folds in the Northwest Lowlands (Wiener, 1977, 1981) are characterized by axial plane foliation, which is the regional foliation. The Fowler syncline, exposed in the no. 4 mine at Balmat, New York, is an early isoclinal fold/thrust nappe with a lower limb fault and northwest-directed motion (deLorraine, 1979) (Table 2). Northwest-directed nappes exposed in the Gouverneur and Canton domes (Rock Island stratigraphic anticline and syncline) are interpreted as first-phase folds (Fig. 22). In the Harrisville area, first-phase nappes were backfolded and rotated by later folding (Wiener, 1983). First-phase folds have not been definitely recognized in the southern and central Adirondacks, but pre-F_2 foliation and possible stratigraphic repetition suggest their presence. Four subsequent phases of folding (F_2–F_5) have made recognition of first-phase folds difficult in the southern and central Adirondacks and, indeed, throughout the Adirondacks.

Second-phase folds are the major nappes recognized in the Adirondacks (Fig. 28). These folds display folded regional foliation and lineation that were generated in the first phase of deformation and generally have a well-developed axial plane foliation. In the Northwest Lowlands second-phase isoclinal folds are refolded by third-phase folds resulting in fishhook-shaped interference patterns (on both maps and cross section) such as in the Moss Ridge anticline (Foose, 1974) and Oswegatchie anticline

(Wiener, 1981, 1983) (Fig. 21B). Southeast-directed lower limb thrusts on second-phase nappes are developed near Beaver Creek (Fig. 22) (Foose and Brown, 1976). Movement line determinations in the Harrisville area and the closures of stratigraphic anticlines also indicate that the second-phase nappes were southeast directed. In the southern and central Adirondacks, the Wakely Mountain nappe and the Canada Lake–Little Moose Mountain syncline are second-phase folds and have amplitudes in excess of 70 km (Fig. 26) (McLelland and Isachsen, 1980) (Table 2). These isoclinal folds comprise a south-directed nappe system. The pattern of second-phase nappes in the Adirondacks suggests regional southeast- to south-directed movement from the Northwest Lowlands to the southern Adirondacks, respectively. This movement line is approximately perpendicular to the axial traces of third-phase folds.

Third-phase, open to tight, upright to overturned folds are the dominant structures in the Adirondacks (Figs. 16, 28). In the southern Adirondacks, these folds include the N70°W to east-west-trending Piseco anticline, Glens Falls syncline, and other folds (Fig. 16). Axial plane cleavage is locally well developed and results in intersection lineations. In the northwestern Adirondacks, the Gouverneur anticline and other third-phase folds are parallel to dome elongation, and third-phase axial traces change orientation smoothly from easterly to ~N45°E in this area (Fig. 22). Third-phase folds sweep in a broad arc from ~N70°W in the southern Adirondacks, to N45°E in the Northwest Lowlands (Fig. 28). The aeromagnetic pattern of the Adirondacks (Zietz and Gilbert, 1981) is largely a result of this folded configuration of differing rock types.

The fourth-phase folds have northwest-striking, steeply dipping axial planes (Fig. 28). These open folds are most prominent in the northwestern Adirondacks (Fig. 22), where they produce domes, saddles, and basins at intersections with third-phase folds.

The fifth-phase folds have north-northeast-striking, steeply dipping, discontinuous axial surfaces (Fig. 28). These open to locally tight folds produce dome and basin patterns at intersections with third-phase folds in the southern and central Adirondacks, but they are only locally present in the Northwest Lowlands (Fig. 28), where they trend north-northwest to north-northeast. It should be mentioned that the relative age of the fourth- and fifth-phase folds is not entirely certain, and the order could be reversed.

Nature and Timing of Metamorphisms with Reference to Folding in the Adirondacks

On the basis of uranium-lead isotope systems in a variety of Adirondack rock types, Silver (1969) concluded that Adirondack granulite-facies metamorphism occurred between 1,100 and 1,020 m.y. ago. Conditions of metamorphism were from ~600 °C, 6 kbar in the Northwest Lowlands to ~800 °C, 8 kbar in the central Adirondack Highlands (Essene and others, 1977). Field observations in the southern and northwestern Adirondacks indicate that high-grade metamorphism was essentially synchronous with the first three phases of folding, although anatexis ended after first-phase folding in the Northwest Adirondacks (Table 2). The fourth and fifth phases of deformation clearly postdate high-grade metamorphism.

The regional metamorphic foliation in the Adirondacks is parallel to axial planes of first-phase isoclinal folds, and regional lineation is parallel to the fold axes (Wiener, 1983) (Table 2). Anatectic layers of (garnet) granulite gneiss in paragneisses are parallel to and cut across first-phase axial planes and indicate that peak Grenville metamorphism and partial melting were associated with and continued after the first phase of deformation. The following observations relate to the second and third phases of deformation.

1. Tabular and elongate minerals are aligned, respectively, with the axial planes and axes of second- and third-phase folds (Foose, 1974; McLelland and Isachsen, 1980; Wiener, 1983). Most commonly, this alignment is shown by hornblende and/or biotite, but fold-associated lineations are also produced by silli-manite needles, blades of quartz, and by garnet, and aligned aggregates of orthopyroxene and clinopyroxene as well. Further-more, high-grade metamorphic minerals such as garnet, ortho-pyroxene, and clinopyroxene in the southern Adirondacks and sillimanite in the Northwest Lowlands are locally flattened in the plane of foliation of second- and third-phase folds, indicating that they were in equilibrium with the regional granulite facies met-amorphism that began during first-phase deformation. Thus, high-grade metamorphism lasted through the first three phases of deformation.

2. Anatectic patches, veins, and lenses of microperthite and quartz (or two feldspars and quartz) are deformed by second-phase minor folds in the southern Adirondacks and crosscut them in other places, indicating that anatexis was contemporaneous with second-phase folding. The relationship between anatexis and third-phase folds is uncertain. In the Northwest Lowlands, ana-texis has *not* been identified associated with either the second- or third-phase folding, indicating a lower metamorphic grade at this time.

Locally in the southern Adirondacks, and especially within leucogneisses, some euhedral garnets clearly postdate second- and third-phase folds in that they have grown across well-developed planes of foliation. These garnets commonly appear to form coro-nal rims around corroded sillimanite, suggesting that the garnets formed according to the generalized reaction:

(Fe, Mg, Ti) biotite + sillimanite + 2 quartz
= K-feldspar + garnet + Ti-rich-biotite
+ (Fe, Ti)-oxide + H_2O

This reaction is believed to occur at granulite facies grade (Kretz, 1961; Mueller and Saxena, 1977). In some places, the associated (Fe, Ti)-oxides exhibit delicate, skeletal shapes that are undeformed, indicating that granulite facies metamorphism lo-cally outlasted second- and third-phase folding. The mineral rela-tionships described above are found in leucogneisses that crop out

along New York Routes 8 and 10, about 2 km southeast of Speculator (Lake Pleasant quadrangle). These leucogneisses are part of the Blue Mountain Lake Formation near its contact with the Lake Durant Formation. Furthermore, garnet coronas in metamorphosed olivine gabbros formed during the second phase of deformation, since the gabbros intruded between the first and second phase of folding. Detailed studies of the coronal reactions indicate T = 750 to 850 °C and P = 7 to 8 kbar (McLelland and Whitney, 1980a).

In the Northwest Lowlands, euhedral sillimanite locally grew across axial planes of second-phase folds (Wiener, 1981), and the assemblage quartz-sillimanite-orthoclase was stable dur-ing third-phase folding in much of the region (Foose, 1974).

The isotherms of Bohlen and Essene (1977) and the isograds of deWaard (1969, 1971) show no relationship to F_1, F_2, or F_3 fold patterns, providing further evidence that the Grenville ther-mal peak outlasted the first three phases of folding.

The fourth and fifth phases of deformation appear to post-date granulite facies metamorphism (Table 2). The open, northwest-striking, fourth-phase folds are best developed in the Northwest Lowlands (Figs. 22, 28), whereas the open, north-northeast-striking fifth-phase folds are best developed in the Adi-rondack Highlands (Figs. 16, 28). In the Northwest Lowlands, retrograde muscovite after sillimanite is oriented parallel to northwest-striking fourth-phase axial planes in biotite-garnet-sillimanite gneisses (Foose, 1974), and retrograde serpentine after diopside and forsterite is also parallel to these axial planes (Wiener, 1981). In the Adirondack Highlands, retrograde mus-covite is oriented parallel to north-northeast-striking, fifth-phase axial planes in certain quartzites and feldspathic quartzites (McLelland and Isachsen, 1980). The coexistence of muscovite, feldspar, and quartz in these rocks indicates that the fourth- and fifth-phase folds formed at P-T conditions below the breakdown of muscovite. These P-T, aH_2O conditions may further be con-strained by noting that no F_4- or F_5-related anatexis occurs in these quartzo-feldspathic rocks. Based on the experimental results of Evans (1965) and Kerrick (1972) for the quartz-muscovite reaction referred to above and the results of Luth and others (1964) for the solidus of Ab + Or + quartz + H_2O, the F_4 and F_5 fold set may have formed at maximum P-T conditions of ~600 to 650 °C and 3 to 5 kbar pressure, considerably less than 600 to 800 °C and 6 to 8 kbar of peak Grenville metamorphism asso-ciated with the first three phases of folding (Table 2B). The fourth- and fifth-phase folding may, therefore, represent events distinctly younger than the events that produced the first three phases of folding. A 930 ± 40-m.y.-Rb/Sr mineral age for mus-covite from an undeformed pegmatite dike from the eastern Adi-rondacks (see Fisher and others, 1971) may serve to date this metamorphic event (Table 2). The erosion rate required to reduce the pressure due to overburden from 8 kbar to 4 kbar (12 km) during a period of 170 m.y. (1,100 to 930 m.y. ago) gives a value of 0.07 mm/yr. This falls within the range of estimates for current erosion rates (Bloom, 1978), suggesting that the date of pegmatite injection is geologically reasonable for the time of fourth- and fifth-phase folding. We suggest that the fourth- and fifth-phase folding are essentially synchronous responses to crustal uplift and buckling associated with late, cooling stages of the Grenville orogenic cycle.

Timing of Intrusion and Mylonitization of Igneous Rocks with Reference to Folding in the Adirondacks

No major intrusive events are associated with the first phase of deformation, but localized partial melting did occur as previously described.

Major intrusive activity in the Adirondacks accompanied the development of the second-phase folds (Table 2). The major intrusive rock suites truncate these isoclinal folds in some places, whereas elsewhere they are involved in the isoclinal folding. Thus, they are interpreted to have been intruded during the second phase of deformation (Wiener, 1977, 1983; McLelland and Isachsen, 1980) (Table 2). The oldest intrusive rock suite is interpreted to be the massif-type meta-anorthosite, with associated metagabbro and oxide-rich differentiate (Buddington, 1939; Buddington and Leonard, 1962; McLelland and Isachsen, 1980). Along its borders the anorthosite was intruded by the mangerite-charnockite suite (Buddington, 1939), which was probably produced by localized melting of quartzo-feldspathic rocks by nearby ~1,100 °C anorthosite magma. The charnockitic-mangeritic suite includes layered igneous complexes in the western Adirondacks, such as the Diana, Stark, Tupper, and Saranac Complexes (Buddington, 1939). The Diana Complex was emplaced early in the second phase of deformation, and later it underwent mylonitization and folding along its border during the same phase of regional deformation (Wiener, 1981, 1983). This resulted in development of penetrative foliation and grain-size reduction along the Carthage-Colton zone of mylonitization (Geraghty and Isachsen, 1981) (Figs. 2, 28).

Hornblende granite and then metagabbro intruded near the end of the second phase of deformation. The Hermon-type hornblende granite gneiss (Fig. 3) is folded by second- and third-phase folds in the Northwest Lowlands (Foose, 1974; Wiener, 1981). In the Adirondack Highlands, hornblende granite gneiss cuts mylonitic charnockitic gneisses of the Diana Complex (Buddington, 1939; Buddington and Leonard, 1962). The lack of mylonitic foliation in the hornblende granite suggests that it intruded after mylonitization. Relations with folds indicate intrusion near the end of or after the second phase of deformation (Wiener, 1983). Olivine metagabbro is apparently the youngest intrusive. However, the heat from the gabbroic magma apparently caused local anatexis along its margins, which produced granites from adjacent country rocks, which then locally intruded the metagabbro (McLelland and Isachsen, 1980). This effect may account for assignment of an old relative age for the metagabbro (Buddington and Leonard, 1962; Wiener, 1981). There may even be more than one suite of olivine metagabbro and hornblende granite. However, second-phase isoclinal folds in the Geers Corners meta-

gabbro near Harrisville indicate intrusion during the second deformational phase (Wiener, 1983). Metamorphism of the olivine gabbro in the Adirondack Highlands during the second deformational phase resulted in formation of garnet coronas at $T \cong$ 750 to 850 °C, $P \cong$ 7 to 8 kbar (McLelland and Whitney, 1980). In summary, anorthosite, charnockite, olivine gabbro, and possibly hornblende granite intruded and were later isoclinally folded during the second deformational phase. Mylonitization of some of the charnockite suites occurred during this isoclinal folding.

Regional Setting and Suggested Tectonic Evolution of the Adirondacks

Grenville metamorphism in the Adirondack Highlands occurred during the period 1,100 to 1,020 m.y. ago (Silver, 1969) at temperatures in the range 650 to 800 °C, and pressures on the order of 7 to 8 kbar (deWaard, 1969; Bohlen and Essene, 1977; McLelland and Whitney, 1977; Whitney, 1978; Stoddard, 1976). Tracy and others (1978) proposed a somewhat higher pressure range of 8 to 10 kbar for a portion of the Adirondacks. Such pressures correspond to a depth of metamorphism of at least 25 to 30 km. Inasmuch as the present Moho discontinuity lies 35 to 38 km beneath the current level of erosion in the Adirondacks (Katz, 1955), it appears that during the 1.0 to 1.1-b.y.-old Grenville metamorphism the continental crust within the region was on the order of 60 km thick. The Adirondacks, therefore, represent an erosional window into a tectonic setting that may provide an ancient analogue for deep-seated processes in present-day regions of double crustal thickness, such as the Tibetan Plateau (Dewey and Burke, 1973), the Turkish-Iranian Plateau (Sengor and Kidd, 1979), and the Altiplano. In addition, the evolution of the Adirondacks must ultimately be related to that of the much more extensive, largely unmapped, Grenville province of which it is a part (Baer and others, 1974).

The most unusual structural feature of the Adirondacks is the existence of extremely large folds reflecting pervasive ductile deformation. In map pattern these are best represented by the F_2 and F_3 fold sets, although F_1 folds are present but not everywhere recognized. As seen in Figure 28, the axial traces of F_2 and F_3 folds exhibit a regular pattern when viewed on a regional scale. F_2 nappes in the southern Adirondacks have a southward vergence, and those in the Northwest Lowlands verge to southeast. F_3 folds form a smooth arcing pattern, and their axial traces are approximately perpendicular to F_2 movement lines. On a broader scale, the fold trends of the Grenville province (Fig. 29) give the appearance of irregular flow. Two explanations of this pattern are (1) that apparently unaligned fold pattern of the Grenville province may result from multiple fold interference of linear fold axial traces formed in a compressional regime, or (2) that the fold pattern in the Grenville province largely represents an original lack of pronounced F_1 to F_3 parallelism (Figs. 28, 29). We suggest that the regular pattern of F_1, F_2, F_3, F_4, and F_5 folds in the Adirondacks (Fig. 28) prevails throughout the Grenville province and that multiple fold interference ac-

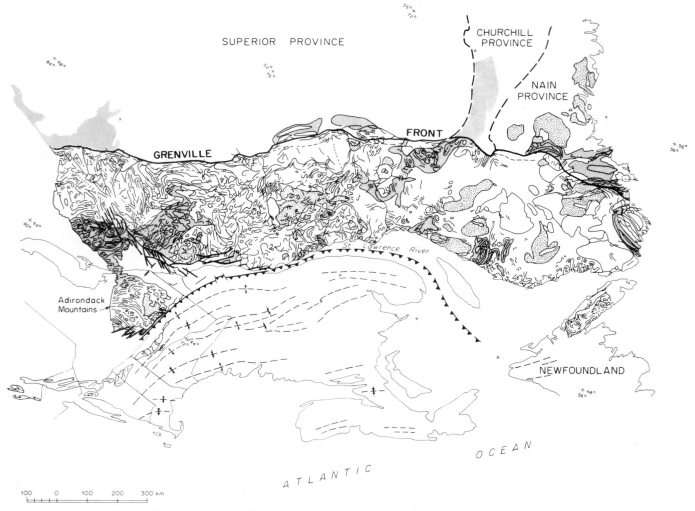

Figure 29. Map showing detailed foliation patterns in the Grenville province (west of thrust fault) and the structural pattern of the Appalachian foldbelt shown in more generalized fashion to the east. Anorthositic massifs are shown by stipple patterns, and dominantly metasedimentary terranes by gray screen pattern. Grenville terranes in the Appalachian foldbelt are enclosed by solid line, and structural trends are shown within them. Note trends of foliation in reactivated Grenvillian basement of Appalachians to the east.

counts for rounded and northeast or northwest elongate map patterns (Fig. 29). If the second hypothesis is correct, the pattern of large, apparently unaligned isoclinal-recumbent folds may be best explained by vertical, gravity tectonics operating in the ductile lower crust during the Grenville orogeny (see also Wynne-Edwards, 1969). On a more local scale, gravity-driven tectonics were invoked by Martignole and Schrijver (1970) to explain the structural evolution of the region around the Morin anorthosite, and this may also apply to the Marcy massif.

Vertical gravity tectonics may have been initiated in the Adirondacks when the crust attained a double thickness. The ensuing departure from isostatic equilibrium would have led to differential vertical uplift in which 25 km of cover were removed. Within the gradually warming lower crust, contrasts in both ductility and specific gravity would result in differential rates of

vertical movement. Initial vertical movements of this sort could produce early flattening resulting in subhorizontal F_1 nappes and foliation. Continued displacement would lead to diapiric uplifts and eventual lateral spreading beneath the more rigid upper crust. Experimental support for such gravity-driven tectonics comes from scale-model centrifuge experiments of Ramberg (1967) and Talbot (1974, 1979). The 6:1 ratio of amplitude to wavelength of folds produced by Ramberg (1967, Fig. 100) is approximately the same as for the Canada Lake–Little Moose Mountain syncline.

The mechanism of crustal thickening in the Adirondacks is not well understood, but several plate-tectonics models have been considered by McLelland and Isachsen (1980). The model they prefer postulates sequential underthrusting of 300 to 400-km-wide segments of continental crust whose ductility increases with heating. However, no unequivocal examples of suture zones have yet been recognized within the region, and such models remain speculative.

A strong tendency toward axial trace parallelism exists within both the F_4 and F_5 fold sets and suggests that these formed in response to late pulses of horizontal compression. The similarity in trend of F_5 folds, the Grenville Front, and the abundant north-northeast-striking normal faults and fracture zones of the Adirondacks is noteworthy (Fig. 1). Although many of these faults offset Paleozoic sedimentary rocks, their history of motion appears to extend back into Proterozoic time (Isachsen, 1974). The present fracture system may be inherited from one which originated on a Tibetan-type plateau surface during the Grenvillian orogeny and was steadily propagated downward during gradual uplift and erosion (Isachsen, 1974).

CONCLUSIONS

The Adirondack Mountains comprise the eroded core of a slightly north-northeast-elongate dome located at the easternmost edge of the North American craton. Geomorphic, geodetic, and structural data suggest that the dome is a neotectonic feature.

Except for scattered postmetamorphic dikes, the core of the dome exposes 1.1- to 1.0-b.y.-old metamorphosed sedimentary and igneous rocks, with granulite facies rocks in the Adirondack Highlands and amphibolite to granulite facies rocks in the Northwest Lowlands.

The stratigraphy of the metamorphic rocks of the Adirondacks consists of basal granitic gneisses of the Piseco Group overlain unconformably by metasedimentary rocks of the Oswegatchie Group in the Lowlands and correlative Lake George Group in the central, eastern, and southern Highlands. In the Piseco Group, charnockitic and granitic gneisses of the Pharaoh Mountain Gneiss are overlain by pink leucogranitic gneisses of the Alexandria Bay and Brant Lake Gneiss.

The basal formation of the Oswegatchie Group is the Baldface Hill Gneiss consisting of garnet-sillimanite gneiss and quartzite. This unit may be a distinctive marker horizon for future mapping. The overlying Poplar Hill Gneiss consists mostly of biotite-quartz-plagioclase gneiss with granitic portions. The Baldface Hill and Poplar Hill Gneisses correlate with the Eagle Lake Gneiss of the Lake George Group. The overlying Gouverneur Marble consists mostly of marble, siliceous marble, calc-silicate rocks, and pyritic schists that have been subdivided into five members. Carbonate equivalents of the Gouverneur Marble to the east include the Cedar River, Blue Mountain Lake, and Paradox Lake formations. Pelitic and calcareous gneisses and quartzites are facies equivalents of these carbonate rocks in parts of the Highlands, including the Cranberry Lake, Springhill Pond, Tomany Mountain, and Sacandaga Formations. The overlying Pleasant Lake Gneiss consists largely of biotite gneiss. In this stratigraphic framework, the upper and lower marble are thought to be equivalent so that the Pleasant Lake Gneiss is the youngest exposed formation. This hypothesis deserves further investigation.

Suites of anorthosite, charnockite, hornblende granite, and gabbro successively intruded this sequence during metamorphism and second-phase deformation. Anorthosite suite rocks intruded mostly within the Pharaoh Mountain Gneiss basement, but also within the Lake George Group cover rocks. Mangerite-charnockite suites that mantle the anorthosite are attributed to in situ anatexis of charnockitic Pharaoh Mountain Gneiss. Layered charnockitic complexes may also have originated by in situ melting due to heat from anorthositic magma, but this remains to be demonstrated. Megacrystic hornblende granite gneiss ("Hermon granite") intruded many formations as grossly concordant porphyritic granite sheets that were later swept into subparallelism by isoclinal folding.

Five phases of folding affected the entire stratigraphic sequence, whereas the last four phases affected the intrusive rocks. First-phase folds are isoclinal nappes that produced axial plane regional foliation and lineation associated with metamorphism. These folds are largely unrecognized except locally by detailed mapping and their northwest movement sense is speculative. Second-phase folds are high-amplitude nappes directed to the south in the central-southern Adirondacks (Wakely Mountain nappe, etc.) and to the southeast in the northwest Adirondacks with local thrusting. Second-phase folds refold the regional foliation and lineation and have axial plane foliation. Intrusion, mylonitization, and metamorphism of igneous rocks occurred during second-phase deformation.

Third-phase folds are open to tight, upright to overturned folds that swing from N45°E to N70°W from northwest to south and produce the dominant "grain" of the Adirondacks. Fourth-phase open folds trend northwest, are prevalent in the Lowlands, and may be associated with metamorphosed diabases. Fifth-phase open folds trend north-northeast to north-northwest, are prevalent in the Highlands, and may be associated with pegmatites. Fourth- and fifth-phase folds produce dome-and-basin patterns where they interfere with third-phase folds.

Upper amphibolite to granulite facies metamorphism during the 1.1- to 1.02-b.y.-old Grenville orogenic cycle outlasted the first three folding phases. Retrograde metamorphism, dike intrusion, and formation of lineament shear zones accompanied the fourth and fifth phases of folding about 930 m.y. ago.

Compressive or gravity tectonics associated with segmental continental underthrusting and continental collision may explain the earlier peak metamorphism and deformation of the Grenville orogenic cycle. The later phases of folding, retrograde metamorphism, igneous activity, and shearing may be associated with uplift and cooling of the double thickened Grenvillian crust.

ACKNOWLEDGMENTS

We acknowledge the numerous field workers in the Adirondacks who provided much of the basic data for this synthesis, especially H. Alling, H. Bannerman, R. D. Bloomer, C. E. Brown, J. S. Brown, A. F. Buddington, J. Carl, W. deLorraine, A. E. and C. E. Engel, M. P. Foose, E. Geraghty, R. Guzowski, H. and E.

Jaffe, D. Leavell, P. Leitzke, J. Lewis, W. Postel, J. Prucha, B. Turner, M. Walton, and D. deWaard. Many of these workers provided seminal ideas which led toward the present synthesis. The manuscript benefited from critical review by R. Fakundiny, P. Whitney, J. Moore, and M. Bartholomew. Manuscript preparation was aided by the New York State Geological Survey, Exxon Production Research Company, and Nuclear Regulatory Commission Contract No. 04-76-291 to Y. W. Isachsen. To all of the above we express grateful appreciation.

REFERENCES CITED

Alling, H. L., 1917, The Adirondack graphite deposits: New York State Museum Bulletin no. 199, 150 p.
——1919, Some problems of the Adirondack pre-Cambrian: American Journal of Science, 4th ser., v. 48, p. 47–68.
——1927, Stratigraphy of the Grenville of the Eastern Adirondacks: Geological Society of America Bulletin, v. 38, p. 795–804.
Ashwal, L. D., 1978, Petrogenesis of massif-type anorthosites: Crystallization history and liquid line of descent of the Adirondack and Morin Complexes [Ph.D. thesis]: Princeton, New Jersey, Princeton University, 136 p.
——1980, Nd and Sr isotope geochronology of the Marcy Anorthosite massif, Adirondacks, New York: Geological Society of America Abstracts with Programs, v. 12, no. 7, p. 380.
Baer, A. J., and others, 1974, Grenville geology and plate tectonics: Geoscience Canada, v. 1, no. 3, p. 54–60.
Balk, R., 1932, Geology of the Newcomb quadrangle: New York State Museum Bulletin no. 290, 106 p.
——1944, Comments on some eastern Adirondack problems: Journal of Geology, v. 52, p. 289–318.
Balsley, J. R., and Bromery, R. W., 1965a, Aeromagnetic map of Number Four quadrangle, Herkimer and Lewis Counties, New York: U.S. Geological Survey Geophysical Investigations Map GP-502, scale 1:62,500.
——1965b, Aeromagnetic map of the Old Forge quadrangle and part of the West Canada Lake quadrangle, Herkimer and Hamilton Counties, New York: U.S. Geological Survey Geophysical Investigations Map GP-501, scale 1:62,500.
Balsley, J. R., and Buddington, A. F., 1960, Magnetic susceptibility anisotropy and fabric of some Adirondack granites and orthogneisses: American Journal of Science, v. 258A, p. 6–20.
Bannerman, H. M., 1972, Geologic map of the Richville-Bigelow area, St. Lawrence County, New York: U.S. Geological Survey Miscellaneous Geological Investigation Map I-664.
Barnett, S. G., and Isachsen, Y. W., 1980, The application of Lake Champlain water level studies to the investigation of Adirondack and Lake Champlain crustal movements: EOS (American Geophysical Union Transactions), v. 61, no. 17, p. 210.
Bartholome, P. M., 1956, Structural and petrological studies in Hamilton County, New York [Ph.D. thesis]: Princeton, New Jersey, Princeton University, 113 p.
Berry, R. H., 1961, Precambrian geology of the Putnam-Whitehall area, New York: [Ph.D. dissertation]: New Haven, Yale University.
Bickford, M. E., and Turner, B. B., 1971, Age and probable anatectic origin of the Brant Lake Gneiss, southeastern Adirondacks, New York: Geological Society of America Bulletin, v. 82, no. 8, p. 2333–2342.
Bloom, A. L., 1978, Geomorphology: A systematic analysis of late Cenozoic landforms: New Jersey, Prentice Hall, Inc., 497 p.
Bloomer, R. O., 1969, Garnets in various rock types in the Canton area in northern New York: New York Academy of Science Annals, v. 172, p. 31–48.
Bohlen, S. R., and Essene, E. J., 1977, Feldspar and oxide thermometry of granulites in the Adirondack highlands: Contributions to Mineralogy and Petrology, v. 62, no. 2, p. 153–169.
Bohlen, S. R., Essene, E. J., and Hoffman, K. S., 1980, Update on feldspar and oxide thermometry in the Adirondack Mountains, New York: Geological Society of America Bulletin, Part I, v. 91, no. 2, p. 110–113.
Brock, B. S., 1980, Stark Complex (Dexter Lake area): Petrology, chemistry, structure, and relation to other green rock complexes and layered gneisses, northern Adirondacks, New York: Geological Society of America Bulletin, Part II, v. 91, no. 2, p. 443–504.
Brocoum, S. J., 1971, Structural and metamorphic history of the major Precambrian gneiss belt in the Hailesboro-West Fowler-Balmat area, Adirondack Lowlands, New York [Ph.D. thesis]: Columbia University, 194 p.
Brown, C. E., 1969, New talc deposit in St. Lawrence County: U.S. Geological Survey Bulletin 1272-D, 13 p.
Brown, J. S., 1936, Structure and primary mineralization of the zinc mine at Balmat, New York: Economic Geology, v. 31, no. 3, p. 233–258.
Brown, J. S., and Engel, A.E.J., 1956, Revision of Grenville stratigraphy and structure in the Balmat-Edwards district, northwest Adirondacks, New York: Geological Society of America Bulletin, v. 67, p. 1599–1622.
Buddington, A. F., 1929, Granite phacoliths and their contact zones in the northwest Adirondacks: New York State Museum Bulletin no. 281, p. 51–110.
——1934, Geology and mineral resources of the Hammond, Antwerp and Lowville quadrangles: New York State Museum Bulletin no. 296, 251 p.
——1939, Adirondack igneous rocks and their metamorphism: Geological Society of America Memoir 7, 354 p.
——1948, Origin of granitic rocks of the Northwest Adirondacks: Geological Society of America Memoir 28, 139 p.
——1952, Chemical petrology of some metamorphosed Adirondack gabbroic, syenitic, and quartz syenitic rocks: American Journal of Science, Bowen Volume, p. 37–84.
——1963, Isograds and the role of H_2O in metamorphic facies of orthogneisses of the northwest Adirondack area, New York: Geological Society of America Bulletin, v. 74, no. 9, p. 1155–1182.
——1965, The origin of three garnet isograds in Adirondack gneisses: Mineralogical Magazine, v. 34, no 268, p. 71–81.
——1969, Adirondack anorthositic series, in Isachsen, Y. W., ed., Origin of anorthosite and related rocks: New York State Museum and Science Service Memoir 18, p. 215–232.
——1972, Differentiation trends and parental magmas for anorthosite and quartz mangerite series, Adirondacks, in Shagam, R., and others, eds., Studies in earth and space sciences: Geological Society of America Memoir 132, p. 477–488.
——1976, Anorthosite-bearing complexes: Classification and parental magmas, in Studies in Precambrian: Department of Geology, Bangalore University, Bangalore, p. 115–141.
Buddington, A. F., and Leonard, B. F., 1962, Regional geology of the St. Lawrence County magnetite district, northwest Adirondacks, New York: U.S. Geological Survey Professional Paper 376, 145 p.
Buddington, A. F., Jensen, M. L., and Mauger, R. L., 1969, Sulfur isotopes and origin of northwest Adirondack sulfide deposits, in Larsen, L., ed., Igneous and metamorphic geology: Geological Society of America Memoir 115, p. 423–451.
Cannon, R. S., 1937, Geology of the Piseco Lake quadrangle: New York State Museum Bulletin no. 312, 107 p.
Carl, J. D., 1978, Sampling of a migmatite: New chemical data and possible parent material for the major paragneiss, Northwest Adirondacks, New York: Geological Society of America Abstracts with Programs, Northeastern Section, v. 10, no. 2, p. 36.
Carl, J. D., and Van Diver, B. B., 1975, Precambrian Grenville alaskite bodies as ash-flow tuffs, northwest Adirondacks, New York: Geological Society of America Bulletin, v. 86, no. 12, p. 1691–1707.
Carr, M. H., and Turekian, K. K., 1961, The geochemistry of cobalt: Geochimica et Cosmochimica Acta, v. 23, no. 1-2, p. 9–60.
Cushing, H. P., 1902, Recent geologic work in Franklin and St. Lawrence Counties, New York: New York State Museum Annual Report, v. 54, p. 23–82.

Cushing, H. P., and Newland, D. H., 1925, Geology of the Gouverneur Quadrangle: New York State Museum Bulletin 259, 122 p.

Cushing, H. P., Fairchild, H. L., Ruedemann, R., and Smyth, C. H., Jr., 1910, Geology of the Thousand Islands Region: New York State Museum Bulletin 145, 194 p.

Davis, B.T.C., 1969, Anorthosite and quartz syenitic series of the St. Regis quadrangle, New York, *in* Isachsen, Y. W., ed., Origin of anorthosite and related rocks: New York State Museum and Science Service Memoir 18, p. 281–287.

—— 1971, Bedrock geology of the St. Regis quadrangle, New York: New York State Museum and Science Service Map and Chart Series no. 16, 34 p.

deLorraine, W., 1979, Geology of the Fowler orebody, Balmat #4 Mine, Northwest Adirondacks, N.Y. [M.S. thesis]: University of Massachusetts, 159 p.

deWaard, D., 1961, Tectonics of a metagabbro laccolith in the Adirondack Mountains, and its significance in determining top and bottom of a metamorphic series: Koninklijke Nederlandse Akademie Van Wetenschappen Proceedings, ser. B., v. 64, no. 3, p. 335–342.

—— 1962, Structural analysis of a Precambrian fold: The Little Moose Mountain syncline in the southwestern Adirondacks: Koninklijke Nederlandse Akademic Van Wetenschappen Proceedings, ser. B., v. 65, no. 5, p. 404, 417.

—— 1964, Notes on the geology of the south-central Adirondack Highlands: New York State Geological Association, 36th Annual Meeting, Guidebook, p. 3–24.

—— 1969, Facies series and P-T conditions of metamorphism in the Adirondack Mountains: Amsterdam, Koninklijke Nederlandse Akademie Van Wetenschappen Proceedings, ser. B., v. 72, no. 2, p. 127–131.

—— 1971, Threefold division of the granulite facies in the Adirondack Mountains: Kristlanlinikum, v. 7, p. 85–93.

deWaard, D., and Romey, W. D., 1963, Boundary relationships of the Snowy Mountain anorthosite in the Adirondack Mountains: Koninklijke Nederlandse Akademie Van Wetenschappen Proceedings, ser. B, v. 66, no. 5, p. 251–264.

Dewey, J. F., and Burke, K.C.A., 1973, Tibetan, Variscan and Precambrian basement reactivation: Products of a continental collision: Journal of Geology, v. 81, no. 6, p. 683–692.

Dietrich, R. V., 1957, Precambrian geology and mineral resources of the Brier Hill quadrangle, New York: New York State Museum Bulletin 354, 121 p.

Douglas, B., Muller, D., and McLelland, J., 1976, Structural geology of the southwestern Adirondacks, N.Y.: Geological Society of America Abstracts with Programs, v. 8, p. 164.

Eckelmann, F. D., 1976, Petrogenetic significance of zircon morphology data for granitic gneisses, associated migmatites, and alaskites of the Grenville lowlands, northwest Adirondacks: Geological Society of America Abstracts with Programs, v. 8, no. 2, p. 166.

—— 1978, Paragneiss-orthogneiss distinction among charnockites of the southernmost Adirondacks based on zircon morphology data: Geological Society of America Abstracts with Programs, v. 10, p. 41.

Emslie, R. F., 1978, Anorthosite massifs, rapakivi granites, and late Precambrian rifting of North America: Precambrian Research, v. 7, no. 1, p. 61–98.

Engel, A. E., and Engel, C. G., 1953a, Grenville Series in northwest Adirondack Mountains, New York. Part I: General features of the Grenville Series: Geological Society of America Bulletin, v. 64, p. 1013–1048.

—— 1953b, Grenville Series in northwest Adirondack Mountains, New York. Part 2: Origin and metamorphism of the major paragneiss: Geological Society of America Bulletin, v. 64, p. 1049–1098.

—— 1962, Progressive metamorphism of amphibolite, Northwest Adirondack Mountains, New York: Geological Society of America, Buddington Volume, p. 37–82.

—— 1963, Metasomatic origin of large parts of Adirondack phacoliths: Geological Society of America Bulletin v. 74, no. 3, p. 349–352.

Eskola, P., 1961, Granitization of quartzose rocks: Finland Commission Geologique Bulletin no. 196 (Geological Society of Finland, C.R. no. 33), p. 483–498.

Essene, E. J., Bohlen, S. R., and Valley, J. W., 1977, Regional metamorphism in the Adirondacks: Geological Society of America, Abstracts with Programs, Northeastern Section, v. 9, no. 3, p. 260–261.

Evans, B. W., 1965, Application of a reaction rate method to the breakdown equilibria of muscovite and muscovite plus quartz: American Journal of Science, v. 263, p. 647–667.

Fisher, D. W., Isachsen, Y. W., and Rickard, L. V., 1971, Geologic map of New York: New York State Museum and Science Service Map and Chart Series, no. 15 , scale 1:250,000.

Foose, M. P., 1974, The structure, stratigraphy and metamorphic history of the Bigelow area, Northwest Adirondacks, New York [Ph.D. thesis]: New Jersey, Princeton University.

Foose, M. P., 1981, Geology, geochemistry and regional resource implications of a stratabound sphalerite occurrence in the Northwest Adirondacks, New York: U.S. Geological Survey Bulletin 1519, 18 p.

Foose, M. P., and Brown, C. E., 1976, A preliminary synthesis of structural, stratigraphic, and magnetic data from part of the northwest Adirondacks, New York: U.S. Geological Survey Open-File Report 76-281, 21 p.

Foose, M. P., and Carl, J. D., 1977, Setting of alaskite bodies in the Northwest Adirondacks: Geology, v. 5, no. 2, p. 77–81.

Foose, M. T., Mose, D. G., Nagel, M. S., and Tunsory, A., 1981, The Rb-Sr ages and structural and stratigraphic relationships of Precambrian granitic rocks in the Northwest Adirondacks, N.Y.: Geological Society of America Abstracts with Programs, Northeastern Section, v. 13, no. 3, p. 133.

Geraghty, E. P., 1978, Structure, stratigraphy, and petrology of part of the Blue Mountain 15′ quadrangle, central Adirondack Mountains, New York [Ph.D. thesis]: Syracuse, New York, Syracuse University, 186 p.

—— 1979, Nappe formation, fold-interference patterns, and structural evolution of the Blue Mountain area, central Adirondack Mountains, New York: Geological Society of America Abstracts with Programs, v. 11, no. 1, p. 14.

Geraghty, E. P., and Isachsen, Y. W., 1981, Extent and character of the Carthage-Colton Mylonite Zone, Northwest Adirondacks, N.Y.: Nuclear Regulatory Commission, Final Report, NUREG/CR-1865.

Geraghty, E. P., deWaard, D., and Turner, B. B., 1975, Preliminary correlation of stratigraphy between the eastern and central Adirondack highlands portion of the Grenville Province: Geological Society of America Abstracts with Programs, v. 7, p. 63–64.

Gilluly, J., 1945, Geologic map of the Gouverneur talc district, New York: U.S. Geological Survey Preliminary Map 3-163 (also unpublished manuscript on open file at New York State Geological Survey).

Glennie, J. S., 1973, Stratigraphy, structure, and petrology of the Piseco Dome area, Piseco Lake 15′ quadrangle, southern Adirondack Mountains, New York [Ph.D. thesis]: Syracuse, New York, Syracuse University, 45 p.

Grant, N. K., Maher, T. M., and Lepaki, R. J., 1981, The age and origin of leucogneisses in the Adirondack Lowlands, New York: Geological Society of America Abstracts with Programs, Annual Meeting, v. 13, no. 7, p. 463.

Green, T. H., 1972, Crystallization of calc-alkaline andesite under controlled high-pressure conditions: Contributions to Mineralogy and Petrology, v. 34, p. 140–166.

Guzowski, R. V., 1979, Stratigraphy, structure, and petrology of the Precambrian rocks in the Black Lake region, northwest Adirondacks, New York [Ph.D. thesis]: Syracuse, New York, Syracuse University, 186 p.

Hargraves, R. B., 1969, A contribution to the geology of the Diana syenite gneiss complex, *in* Isachsen, Y. W., ed., Origin of anorthosite and related rock: New York State Museum and Science Service Memoir 18, p. 343–356.

—— 1976, Precambrian geologic history: Science, v. 193, p. 363–371.

Hills, F. A., 1964, The Precambrian geology of the Glens Falls and Fort Ann quadrangles, southeastern Adirondack Mountains, New York (1:24,000) [Ph.D. dissertation]: Yale University.

Hills, F. A., and Gast, P. W., 1964, Age of pyroxene-hornblende granite gneiss of the eastern Adirondacks by the rubidium-strontium whole-rock method: Geological Society of America Bulletin, v. 75, p. 759–766.

Husch, J., Kleinspahn, K., and McLelland, J., 1975, Anorthositic rocks in the Adirondacks: Basement or non-basement?: Geological Society of America Abstracts with Programs, v. 7, p. 78.

Isachsen, Y. W., 1953, Geology of the Rensselaer Falls quadrangle, St. Lawrence County, New York (1:12,000) [Ph.D. dissertation]: Cornell University.

—— 1969a, editor, Origin of anorthosite and related rocks: New York State Museum and Science Service Memoir 18, 466 p.

—— 1969b, Origin of anorthosite: A summarization, *in* Isachsen, Y. W., ed., Origin of anorthosite and related rocks: New York State Museum and Science Service Memoir 18, p. 435–445.

—— 1974, Fracture analysis of New York State using multi-stage remote sensor data and ground study; possible application to plate tectonic modeling: International Conference on the New Basement Tectonics, First Proceedings, Utah Geological Association Publication no. 5, p. 200–217.

—— 1975, Possible evidence for contemporary doming of the Adirondack Mountains, New York, and suggested implications for regional tectonics and seismicity: Tectonophysics, v. 29, p. 169–181.

Isachsen, Y. W., Brown, C. E., and Zietz, I., 1979, Preliminary interpretation of an aeromagnetic map of most of the Adirondack Mountains, New York: Geological Society of America Abstracts with Programs, v. 11, no. 1, p. 17.

Isachsen, Y. W., Geraghty, E. P., and Wright, S. F., 1978, Investigation of Holocene deformation in the Adirondack Mountains Dome: Geological Society of America Abstracts with Programs, v. 10, no. 2, p. 49.

Isachsen, Y. W., and McKendree, W. G., 1977, Preliminary brittle structures map of New York: New York State Museum Map and Chart Series, no. 31.

Isachsen, Y. W., McLelland, J. M., and Whitney, P. R., 1975, Anorthosite contact relationships in the Adirondacks and their implication for geologic history: Geological Society of America Abstracts with Programs, v. 7, no. 1, p. 78–79.

Isachsen, Y. W., and Moxham, R. L., 1969, Chemical variations in plagioclase megacrysts from two vertical sections in the main Adirondack metanorthosite massif, *in* Isachsen, Y. W., ed., Origin of anorthosite and related rocks: New York State Museum and Science Service Memoir 18, p. 225–265.

Jacobi, R. S., 1964, Some petrologic, stratigraphic, and structural relationships of the Border Zone of the Diana Complex near Harrisville, New York [Master's thesis]: Syracuse, New York, Syracuse University.

Jaffe, H. W., 1946, Postanorthosite gabbro near Avalanche Lake in Essex County, New York: Journal of Geology, v. 54, p. 105–116.

Katz, S., 1955, Seismic study of crustal structure in Pennsylvania and New York: Bulletin of Seismological Society of America, p. 303–325.

Kemp, J. F., and Alling, H. L., 1925, Geology of the Ausable quadrangle: New York State Museum Bulletin 261.

Kerrick, D. M., 1972, Experimental determination of muscovite and quartz stability: American Journal of Science, v. 272, p. 946–958.

Kreiger, M. H., 1937, Geology of the Thirteenth Lake quadrangle, New York: New York State Museum Bulletin no. 308, 124 p.

Kretz, R., 1961, Some applications of thermodynamics to coexisting minerals of variable compositions. Examples: orthopyroxene-clinopyroxene and orthopyroxene-garnet: Journal of Geology, v. 69, p. 361–387.

Lea, E. R., and Dill, D. B., 1968, Zinc deposits of the Balmat-Edwards district, New York, *in* Ridge, J. D., ed., Ore deposits of the United States, 1933–1967 (Graton-Sales Volume): New York, American Institute of Mining, Metallurgical and Petroleum Engineers, v. 1, chap. 2, p. 20–48.

Leake, B. E., 1963, Origin of amphibolites from northwest Adirondacks, New York: Geological Society of America Bulletin, v. 74, no. 9, p. 1193–1202.

Leavell, D., 1977, The stratigraphy and structure of the Indian Mountain region, Cranberry Lake, New York [M.S. Special Problem]: Amherst, University of Massachusetts, and New York State Geological Survey, Open-File Report.

Leitzke, P. A., 1974, Discontinuity in fold structures between the Adirondack Highlands and Lowlands: Geological Society of America Abstracts with Programs, Northeastern Section, v. 5, no. 1 p. 47.

Leonard, B. F., and Buddington, A. F., 1964, Ore deposits of the St. Lawrence County magnetite district, northwest Adirondacks, New York: United States Geological Survey Professional Paper 377, 259 p.

Lettney, C. D., 1969, The anorthosite-charnockite series of the Thirteenth lake dome, south-central Adirondacks, *in* Isachsen, Y. W., ed., Origin of anorthosite and related rocks: New York State Museum and Science Service

Memoir 18, p. 329–342.

Lewis, J. R., 1969, Structure and stratigraphy of the Rossie Complex, northwest Adirondacks, New York [Ph.D. thesis]: Syracuse, New York, Syracuse University, 141 p.

Luth, W. C., Jahns, R. H., and Tuttle, O. F., 1964, The granite system at pressures of 4 to 10 kilobars: Journal of Geophysical Research, v. 69, p. 759–773.

Maher, T. H., Lepak, R. J., and Grant, N. K., 1981, Rb-Sr ages, crustal prehistory and stratigraphic sequence: Leucogneisses and marbles of the Adirondack Lowlands, New York: Geological Society of America Abstracts with Programs, v. 13, no. 3, p. 144.

Martignole, J., and Schrijver, K., 1970, Tectonic setting and evolution of the Morin anorthosite, Grenville Province, Quebec: Bulletin of Geological Societies, Finland, v. 42, p. 165–209.

McConnell, C. L., 1965, Geology of the southwest quarter of the Bolton Landing quadrangle, New York State, 1:24,000: Unpublished map and report at New York State Museum and Science Service, Geological Survey.

McLelland, J., 1969, Geology of the southernmost Adirondacks: New England Intercollegiate Geological Conference, 41st Annual Meeting Guidebook, section 11, p. 1–34.

McLelland, J., and Isachsen, Y. W., 1980, Structural synthesis of the southern and central Adirondacks: A model for the Adirondacks as a whole and plate tectonics interpretations: Geological Society of America Bulletin, Part II, v. 91, p. 208–292.

McLelland, J. M., and Whitney, P. R., 1977, The origin of garnet in the anorthosite-charnockite suite of the Adirondacks: Contributions to Mineralogy and Petrology, v. 60, no. 2, p. 161–181.

—— 1980a, A generalized garnet-forming reaction for metaigneous rocks in the Adirondacks: Contributions to Mineralogy and Petrology, v. 72, p. 111–122.

—— 1980b, Compositional controls on spinel clouding and garnet formation in plagioclase of olivine metagabbros, Adirondack Mountains, New York: Contributions to Mineralogy and Petrology, v. 73, p. 243–251.

McLelland, J., Geraghty, E., and Boone, G., 1978, The structural framework and petrology of the southern Adirondacks: New York State Geological Association, 50th Annual Meeting Guidebook, p. 58–103.

Miller, W. J., 1909, Geology of the Remsen quadrangle: New York State Museum Bulletin 126, 51 p.

—— 1911, Geology of the Broadalbin quadrangles, Fulton-Saratoga Counties, New York: New York State Museum Bulletin 153, 65 p.

—— 1916, Geology of the Lake Pleasant quadrangle, Hamilton County, New York: New York State Museum Bulletin no. 182, 75 p.

—— 1918, Adirondacks anorthosite: Geological Society of America Bulletin, v. 29, p. 399–462.

—— 1920, Geology of the Gloversville quadrangle: New York State Museum and Science Service, open-file maps, scale 1:62,500.

—— 1923, Geology of the Luzerne quadrangle: New York State Museum Bulletin nos. 245–246, 66 p.

Moore, J. M., Jr., and Thompson, P. H., 1980, The Flinton Group: A late Precambrian metasedimentary succession in the Grenville Province of eastern Ontario: Canadian Journal of Earth Sciences: v. 17, p. 1685–1707.

Mueller, R. F., and Saxena, S. K., 1977, Chemical petrology: New York, Springer-Verlag, 398 p.

Nelson, A. E., 1968, Geology of the Ohio quadrangle: U.S. Geological Survey Bulletin 1251-F, p. F1–F46.

Nielsen, P. A., 1971, Metamorphism and metamorphic isograds in the northwest lowlands of the Adirondacks [M.A. thesis]: New York, State University of New York at Binghamton, 59 p.

Orville, P. M., 1969, A model for metamorphic differentiation origin of thinly layered amphibolites: American Journal of Science, v. 267, p. 64–86.

Poldervaart, A., 1953, Metamorphism of basaltic rocks—A review: Geological Society of America Bulletin 64, p. 259–274.

Postel, A. W., 1952, Geology of the Clinton County magnetite district, New York: U.S. Geological Survey Professional Paper 237, 88 p.

Potter, D. B., 1981, Stratigraphy and structural geology of the Loon Pond Syncline, central Adirondack Mountains, New York: Geological Society of

America Abstracts with Programs, Northeastern Section, v. 13, no. 3, p. 170.

Ramberg, H., 1967, Gravity, deformation, and the earth's crust as studied by centrifuged models: New York, Academic Press, 214 p.

Romey, W. D., Elberty, W. R., Jr., Jacobi, R. S., Christofferson, R., Schrier, T., and Tietbohl, D., 1980, A structural model for the northwestern Adirondacks based on leucogranitic gneisses near Canton and Pyrites, New York: Geological Society of America Bulletin, Part II, v. 91, no. 2, p. 505–588.

Sederholm, J. J., 1926, On migmatites and associated Precambrian rocks of southwestern Finland, II: The region around the Barosundsfjord W. of Helsingfors and neighboring areas: Bulletin de la commission Geologique de Finlande, v. 77, 143 p.

Seifert, K. E., and others, 1977, Rare earths in the Marcy and Morin anorthosite complexes: Canadian Journal of Earth Sciences, v. 14, p. 1033–1045.

Sengor, A.M.C., and Kidd, W.S.F., 1979, The post-collisional tectonics of the Turkish-Iranian Plateau, and a comparison with Tibet: Tectonophysics, v. 55, p. 361–377.

Silver, L. T., 1965, U-Pb isotopic data in zircons of the Grenville series of the Adirondack Mountains, New York: American Geophysical Union Transactions, v. 46, p. 164.

—— 1969, A geochronologic investigation of the anorthosite complex, Adirondack Mountains, New York, *in* Isachsen, Y. W., ed., Origin of Anorthosites and related rocks: New York State Museum and Science Service Memoir 18, p. 233–252.

Simmons, E. C., and Hanson, G. N., 1978, Geochemistry and origin of massif type anorthosites: Contributions to Mineralogy and Petrology, v. 66, p. 119–135.

Smyth, C. H., Jr., and Buddington, A. F., 1926, Geology of the Lake Bonaparte quadrangle: New York State Museum Bulletin no. 269, 106 p.

Spooner, C. M., and Fairbairn, H. W., 1970, Strontium 87/strontium 86 initial ratios in pyroxene granulite terrranes: Journal of Geophysical Research, v. 75, no. 32, p. 6706–6713.

Stoddard, E. F., 1976, Granulite facies metamorphism in the Colton–Rainbow Falls area, northwest Adirondacks, New York [Ph.D. thesis]: University of California, Los Angeles, 271 p.

Talbot, C. J., 1974, Fold nappes as asymmetric gneiss domes and ensialic orogeny: Tectonophysics, v. 24, p. 259–276.

—— 1979, Infrastructural migmatitic upwelling in east Greenland interpreted as thermal convective structures: Precambrian Research, v. 8, p. 77–93.

Tan, L. P., 1966, Major pegmatite deposits of New York State: New York State Museum and Science Service Bulletin no. 408, 138 p.

Tarling, D. H., ed., 1978, Evolution of the Earth's crust: New York, Academic Press, 443 p.

Tracy, R. J., Jaffe, H. W., and Robinson, P., 1978, Monticellite marble at Cascade Mountain, Adirondack Mountains, New York: American Mineralogist, v. 63, no. 9-10, p. 991–999.

Turner, B., 1979, Precambrian structure and stratigraphy of the southeastern Adirondack Uplands: New York State Geological Association Guidebook, v. 51, p. 426–446.

Tyler, R. D., 1979, Chloride metasomatism in the southern part of the Pierrepont quadrangle, northwest Adirondacks, New York [Ph.D. thesis]: Binghamton, State University of New York at Binghamton, 548 p.

Vidale, R., 1969, Metasomatism in a chemical gradient and the formation of calc-silicate bands: American Journal of Science, v. 267, p. 857–874.

Walton, M. S., 1961, Geologic maps of the eastern Adirondacks: New York State Museum and Science Service, open-file maps, scale 1:24,000.

Walton, M. S., and deWaard, D., 1963, Orogenic evolution of the Precambrian in the Adirondack highlands, a new synthesis: Koninklijke Nederlandse Akademie Van Wetenschappen Proceedings, ser. B, v. 66, no. 3, p. 98–106.

Walton, M. S., and Turner, B. B., 1963, Refolding of the Skiff Mountain tectonic unit attending emplacement of the Adirondack anorthosite: American Geophysical Union Transactions, v. 44, p. 106.

Whitney, P. R., 1972, A layered intrusion near Jay Mountain, New York: Geological Society of America Abstracts with Programs, v. 4, no. 1, p. 51.

—— 1978, The significance of garnet "isograds" in granulite facies rocks of the Adirondacks, *in* Fraser, J. A., and others, eds., Metamorphism in the Canadian Shield: Canada Geological Survey Paper 78-10, p. 357–365.

Whitney, P. R., and McLelland, J. M., 1973, Origin of coronas in metagabbros of the Adirondack Mts., New York: Contributions in Mineralogy and Petrology, v. 39, p. 81–98.

Wiener, R. W., 1977, Timing of emplacement and deformation of the Diana Complex along the Adirondack Highlands–Northwest Lowlands boundary: Geological Society of America Abstracts with Programs, v. 9, no. 3, p. 329–330.

—— 1979, Stratigraphic and structural relations along the Adirondack Highlands–Northwest Lowlands boundary: Geological Society of America Abstracts with Programs, Northeast Section, v. 11, no. 1, p. 59.

—— 1981, Stratigraphy, structural geology and petrology of bedrock along the Adirondack Highlands–Northwest Lowlands boundary near Harrisville, N.Y. [Ph.D thesis]: Amherst, University of Massachusetts, 195 p.

—— 1983, Adirondack Highlands–Northwest Lowlands "boundary": A multiply folded intrusive contact with fold-associated mylonitization: Geological Society of America Bulletin, v. 94, no. 9., p. 1081–1108.

Wynne-Edwards, H. R., 1967, Westport map-area, Ontario, with special emphasis on the Precambrian rocks: Geological Survey of Canada Memoir 346, 142 p.

—— 1969, Tectonic overprinting in the Grenville Province, *in* Wynne-Edwards, H. R., ed., Age relations in high-grade metamorphic terrains: Geological Association of Canada Special Paper 5, p. 5–16.

Zietz, I., and Gilbert, F. P., 1981, Aeromagnetic map of New York, Adirondack Mountains sheet: U.S. Geological Survey Geophysical Investigation Map GP-938 (sheet 5 of 5), scale 1:250,000.

MANUSCRIPT ACCEPTED BY THE SOCIETY AUGUST 2, 1983

Geological Society of America
Special Paper 194
1984

Geology and geochronology of Canada Hill granite and its bearing on the timing of Grenvillian events in the Hudson Highlands, New York

Henry L. Helenek
Department of Geological Sciences
Bradley University
Peoria, Illinois 61625

Douglas G. Mose
Department of Geology
George Mason University
Fairfax, Virginia 22030

ABSTRACT

High-grade Precambrian (Proterozoic Y) granulite gneisses, paragneiss and rusty gneisses (metasedimentary/metavolcanic, ?) in the Hudson Highlands of southeastern New York State are intruded by plutons of Storm King granite gneiss and Canada Hill granite. The metasedimentary/metavolcanic (?) gneisses were deformed multiply during the Grenvillian dynamothermal metamorphism in three phases of deformation (D_1, D_2, D_3). Plutons of Storm King granite gneiss contain S_1 planar and L_2 linear fabrics and were deformed in all phases of deformation with the metasedimentary/metavolcanic (?) gneisses. This indicates a pre-tectonic to early syntectonic age for intrusion of the Storm King granite gneiss. Plutons of Canada Hill granite cut D_1 and D_2 fabric elements in both Storm King granite gneiss and the metasedimentary/metavolcanic (?) gneisses. Canada Hill granite contains a weak S_3 planar fabric and was deformed during D_3. This indicates a late-tectonic age for the Canada Hill granite.

Plutons of Canada Hill granite are for the most part confined to and have migmatitic borders with paragneiss. A closed system metasomatic process is indicated for formation of the Canada Hill granite. Rather than a magmatic episode, formation of Canada Hill granite may represent a time of regional granitization brought about by redistribution of chemical components within the host paragneiss. This mechanism is supported by Rb-Sr systematics for the paragneiss and Canada Hill granite.

The age determinations for paragneiss (Rb-Sr whole-rock age = 1147 ± 43 m.y.) and Canada Hill granite (Rb-Sr whole-rock age = 913 ± 45 m.y.), when examined in conjunction with reported radiometric data for the Hudson Highlands, provide evidence for two distinct metamorphic-plutonic episodes within the Grenvillian event. The first episode occurred at about 1140 m.y. and is recorded in the age for paragneiss. This episode was characterized by Storm King plutonism, metamorphism of the Storm King granite and metasedimentary/metavolcanic (?) gneisses, and development of D_1 fabric elements. The Storm King granite gneiss intruded as sheets and sills prior to and during the early stages of D_1. The second episode at 1000 m.y. was characterized by hornblende granulite metamorphism, plutonism, Canada Hill granitization, and the development of coaxial D_2 and D_3 fabric elements.

INTRODUCTION

The Hudson Highlands are an east-west trending range of low-lying, rugged hills (maximum elevation 500 m) in southeastern New York State about 70 km north of New York City. The Highlands form the northern part of the Reading Prong. The Reading Prong is a narrow belt of crystalline rock extending from western Connecticut, through New York and New Jersey, and terminates in Pennsylvania near the town of Reading (Fig. 1). Crystalline rocks of the Reading Prong are for the most part Precambrian in age. They represent the most extensive and continuous exposure of Proterozoic rocks in the New England physiographic province.

The oldest Precambrian rocks (Proterozoic Y) in the Hudson Highlands are granulite gneisses, paragneiss and rusty gneisses (Fig. 2). These rocks were subjected to hornblende granulite facies metamorphism and deformed during the Grenvillian dynamothermal event (Dodd, 1965; Dallmeyer and Dodd, 1971).

The granulite gneisses (Yqp) consist of various quartz-plagioclase gneisses dominated by hypersthene-bearing assemblages. The quartz-plagioclase gneisses contain subordinate amphibolite, pyroxene-hornblende-plagioclase gneiss, quartz-plagioclase leucogneiss, calc-silicate, and minor quartzitic and calcareous metasediments. Various protoliths were proposed for the granulite gneisses. Hotz (1953) and Sims (1953) considered the quartz-plagioclase gneisses as early syntectonic dioritic intrusions. More recent workers interpreted the sequence as volcanic-volcaniclastic in origin (Dodd, 1965; Offield, 1967; Baker and Buddington, 1970; Jaffe and Jaffe, 1973).

Paragneiss (Ypn) contains subordinate amphibolite, rusty weathering paragneiss, quartzite, and minor calcareous and ferruginous metasediments. The paragneiss originally was interpreted as plutonic (Berkey and Rice, 1919) and subsequently as metasedimentary (Dodd, 1965) in origin.

The rusty gneisses (Yrg) are a heterogeneous group of rocks consisting of paragneiss, amphibolite, sillimanitic schist, quartzite, pyroxene gneisses, biotite-hornblende gneiss, quartz-plagioclase gneiss, calc-silicate and marble. Weathering imparts a distinctive rusty brown color to several of these rocks in the field. Most workers (Berkey and Rice, 1919; Lowe, 1950, 1958; Dodd, 1965) considered the pelitic and calcareous members of the rusty gneisses as metasedimentary in origin. Dodd (1965) assigned a metavolcanic origin to amphibolite, biotite-hornblende gneiss and quartz-plagioclase gneiss.

The granulite gneisses, paragneiss and rusty gneisses are intruded by Canada Hill granite (Ych) and Storm King granite (Ysk). Berkey and Rice (1919) initially described these rocks and applied the terms Canada Hill and Storm King to designate major episodes of plutonism in the Hudson Highlands. Since first proposed by Berkey and Rice (1919), the validity and temporal relationship of these plutonic episodes have been the subject of considerable discussion.

The type Canada Hill granite of Berkey and Rice (1919, p. 52) included at least two petrographically distinct rocks: (1)

strongly foliated, medium-grained garnet-biotite granite gneiss (gneissic Canada Hill granite), and (2) weakly foliated to massive, coarser grained biotite granite (massive Canada Hill granite). Massive Canada Hill was interpreted as a pegmatitic facies of gneissic Canada Hill granite even though petrologic kinship was not demonstrated. Canada Hill granite containing the rusty gneisses was considered as mixed Grenville-Canada Hill granite. Mixed Grenville-Canada Hill granite with the two varieties of Canada Hill granite were mapped collectively as Canada Hill granite. Storm King granite intruded gneissic Canada Hill granite and mixed Grenville-Canada Hill granite. Structural relationships between massive Canada Hill granite and Storm King granite were not established clearly. Berkey and Rice (1919) and Colony (1921) interpreted a penetrative fabric present in both granites as primary foliation resulting from magmatic flow. Canada Hill granite was assigned the older age.

Lowe (1950, 1958) considered the gneissic Canada Hill granite of Berkey and Rice as typical Canada Hill granite. He proposed a metasomatic origin for the rock. Canada Hill granite and metasomatized Grenvillian metasediments were indistinguishable from one another and grouped together as the Highlands complex. Foliation in the Canada Hill granite was a relict planar fabric inherited from metasomatized Grenvillian metasediments. Linear and planar fabric elements in the Storm King granite were primary structures formed during magmatic flow. Lowe concluded the Storm King granite was a post-kinematic plutonic intrusion younger than the Canada Hill granitization phase. U/Pb zircon ages of 1170 m.y. for the Canada Hill granite and 1060 m.y. for the Storm King granite (Tilton and others, 1960) supported the temporal relationship indicated for these rocks from field studies.

Dodd (1965) pointed out that the age for Canada Hill granite (Tilton and others, 1960) was obtained from a hypersthene-quartz-plagioclase gneiss and not Canada Hill granite. This age represented the time of recrystallization of the gneiss during Grenvillian dynamothermal metamorphism. The lack of evidence supporting a distinct Canada Hill plutonic episode prompted Dodd to abandon use of the term. Dodd also presented evidence indicating that linear and planar structures in the Storm King granite were secondary fabric elements. He interpreted the Storm King granite as syntectonic with the Grenvillian dynamothermal metamorphism. Offield (1967) questioned the validity of the Canada Hill granite as a field unit and recommended redefinition of the lithology if the term was to be used. Helenek (1971) considered the Canada Hill granite a valid field unit and proposed the granite formed by partial anatexis of paragneiss. Jaffe and Jaffe (1973) considered the Storm King granite as part of the pre-Grenvillian stratigraphic sequence. The gneissic fabric of the granite was interpreted as secondary foliation. A metavolcanic origin was assigned to the Storm King granite.

Clearly, considerable difference of opinion exists regarding the validity of the terms Canada Hill granite and Storm King granite as designating major plutonic episodes in the Hudson Highlands. The crux of the problem centers on the Canada Hill

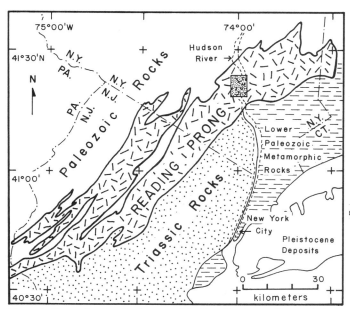

Figure 1. Generalized geologic map of the Reading Prong (Precambrian) and surrounding rocks in New York and New Jersey. The West Point-Bear Mountain region of the Hudson Highlands in New York State is outlined and stippled. A geologic map of this area is given in Fig. 2.

granite. The definition for Canada Hill granite proposed by Berkey and Rice (1919) included rocks having diverse fabric elements and mineralogical compositions. This definition for Canada Hill granite must be regarded as ambiguous. Various petrogenetic schemes are possible depending upon which rock is selected as typical Canada Hill granite.

PLAN OF INVESTIGATION

A detailed field mapping and Rb/Sr isotopic study was carried out in the West Point-Bear Mountain area of the Hudson Highlands (Fig. 1) west of the Hudson River (West Point, New York, and Peekskill, New York, 7½′ quadrangles). Here, plutons of Canada Hill granite and Storm King granite intrude a series of metasedimentary/metavolcanic rocks. The purpose of the investigation was to determine the temporal relationships between successive plutonic episodes accompanying Grenvillian dynamothermal metamorphism. Specifically, this required (1) definition and description of the Canada Hill granite and Storm King granite, (2) determination of the structural relationships between the plutonic units and metasedimentary/metavolcanic gneisses, and (3) determination of Rb/Sr whole rock ages for Canada Hill granite and paragneiss. The West Point-Bear Mountain region was considered critical for several reasons: (1) the area was situated within the West Point, New York, 15′ quadrangle originally mapped by Berkey and Rice (1919), (2) type localities for Canada Hill granite and Storm King granite were located within 4 km of the study area, (3) field relationships between Canada Hill granite, Storm King granite and metasedimentary/metavolcanic rocks could be established readily, and (4) the area lay adjacent to

the Popolopen Lake, New York, 7½′ quadrangle recently mapped by Dodd (1965). This provided for comparison of plutonic rock units with the type localities described by Berkey and Rice (1919) and permitted correlation of field units with those mapped by Dodd (1965) in the adjacent Popolopen Lake quadrangle.

DESCRIPTION AND DEFINITION OF ROCK UNITS

Storm King granite gneiss

Storm King granite gneiss is the prominent ridge-forming lithology in the Hudson Highlands. Characteristically, it forms large domical outcrops which weather to a light pinkish tan color. On fresh exposure, it is medium gray to dark gray in color. Storm King granite gneiss is found in the West Point pluton and the Bear Mountain pluton (Ysk, Fig. 2).

Storm King granite gneiss is a medium- to coarse-grained (3.0 mm to 1.0 cm), leucocratic (color index, 3 to 14; average color index, 7), foliated and lineated, equigranular hypidioblastic rock composed of quartz, alkali-feldspar, plagioclase feldspar and hornblende. Biotite and garnet are varietal minerals. Opaques, sphene, apatite and zircon are accessory minerals. Clinopyroxene is rare. Inequigranular very coarse-grained to pegmatitic, weakly layered to massive patches and lenses of hornblende granite are found within the granite gneiss.

The mineralogical composition of Storm King granite gneiss (appendix 1) shows variability in the alkali-feldspar: plagioclase feldspar ratio and the composition of free plagioclase feldspar in the rock. Two facies of granite gneiss result from variations in the alkali-feldspar: plagioclase feldspar ratio, hornblende granite gneiss and hornblende alkali-granite gneiss. Hornblende granite gneiss is composed of two separate feldspars, slightly perthitic microcline and plagioclase feldspar (An_{22-26}, calcic oligoclase). Both feldspars are present in approximately equal amounts. Microcline mesoperthite is the dominant feldspar in hornblende alkali-granite gneiss. Plagioclase feldspar is subordinate and albitic (An_4) in composition. The Bear Mountain pluton consists of hornblende alkali-granite gneiss. Hornblende granite gneiss is found in the West Point pluton.

Quartz is commonly strained and contains inclusions of xenoblastic microcline, hornblende, plagioclase feldspar and opaque oxide. Slightly perthitic microcline, microcline-microperthite and microcline mesoperthite are common alkali-feldspars. Microcline and microcline-microperthite contain inclusions of quartz, hornblende and plagioclase feldspar. Microcline mesoperthite typically is free of inclusions. Plagioclase feldspar is poorly twinned and locally myrmekitic. Where a minor constituent, it is interstitial to larger grains of quartz and mesoperthite. Hornblende (X = pale brown; Y = moderate greenish-brown, dark green; Z = dark greenish-brown) forms hypidioblastic grains altered to opaque oxides and hematite along cleavages and grain boundaries. Partial alteration to chlorite and biotite is rare. Biotite (X = light tan; Y = Z = reddish-brown) and

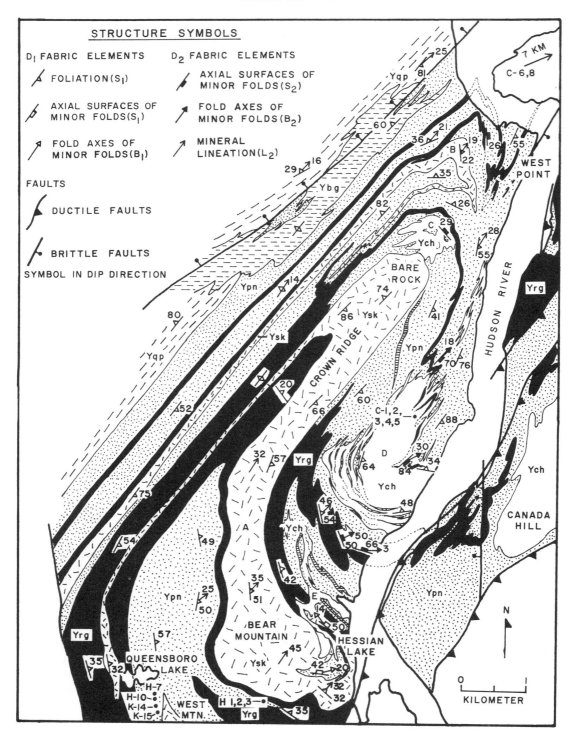

Figure 2. Geologic map of the West Point-Bear Mountain region. Yqp, granulite gneisses; Ybg, biotite granite gneiss; Ypn, paragneiss; Yrg, rusty gneisses; Ysk, Storm King granite gneiss; Ych, Canada Hill granite. Plutons of Storm King granite gneiss are designated: A, Bear Mountain pluton; B, West Point pluton. Plutons of Canada Hill granite are designated: C, Stoney Lonesome pluton; D, Crystal Lake pluton; E, Brooks Lake pluton. Sample locations for Canada Hill granite (C series) and paragneiss (H and K series) used in isotopic studies are indicated. Geologic data west of Bear Mountain taken from Dodd (1965). Data east of the Hudson River modified from Ratcliffe (unpublished data).

brownish-red garnet are found in the border facies of the granite gneiss.

Storm King granite gneiss has strong penetrative linear and planar fabrics. In the interior of plutons, the linear fabric dominates to the complete exclusion of the planar fabric. The strong foliated fabric of the granite gneiss results from compositional layering and preferred dimensional and crystallographic orientation of minerals. Compositional layering is formed by thin (1 mm to 3 mm), discontinuous, alternating quartzo-feldspathic layers. The planar structure is enhanced by weathering where quartzose layers stand out boldly against feldspathic layers. A second foliation, parallel to quartzo-feldspathic layering, results from the preferred planar orientation of hornblende. The hornblende foliation accentuates the quartzo-feldspathic layering and gives the granite gneiss a strongly foliated appearance in weathered outcrop. A biotite schistocity accompanies the hornblende foliation in marginal portions of Storm King plutons. The schistocity results from preferred crystallographic orientation of biotite crystals and parallels the quartzo-feldspathic and hornblende foliations. Linear structure in the granite gneiss is due to preferred orientation of pencil-sized rods of hornblende crystals.

Representative modes for Storm King granite gneiss are listed in appendix 1. A chemical analyses for sample 5-78 is given in appendix 3.

Canada Hill granite

Canada Hill granite is found in the Stoney Lonesome pluton, the Crystal Lake pluton and the Brooks Lake pluton (Ych, Fig. 2). Typically it forms large white domical exposures.

Canada Hill granite is a leucocratic (color index, 0 to 9; average color index, 4), grayish-white, equigranular rock composed of quartz, white and gray mottled alkali-feldspar, white plagioclase feldspar and biotite. Locally the granite contains numerous lavender or reddish-brown garnets. Grain-size is variable from medium-grained (ranging from 0.3 to 2.0 cm) to very coarse-grained and pegmatitic (ranging from 2.0 to 30.0 cm). Strained quartz, slightly perthitic microcline (about 5% albite lamellae) and plagioclase feldspar form equant anhedral grains giving the rock a xenomorphic-granular texture. Poorly to well-twinned plagioclase feldspar (An_{7-24}) is zoned, myrmekitic and altered partially to sericite and epidote. Equant, poikilitic garnet is altered partially along fractures, and reddish-brown biotite is altered partially to aggregates of chlorite, opaques, muscovite and epidote. Sphene and zircon are rare accessory minerals; sillimanite, graphite, fluorite and tourmaline are varietal minerals. Late mineralization resulted in partial alteration of biotite, garnet and plagioclase feldspar, and the formation of calcite, muscovite and epidote in pervasive, widely spaced microfractures.

Canada Hill granite is massive but locally has a weak planar fabric. Planar structure in the granite is centimeter to decimeter scale quartzo-feldspathic compositional layering. Alternating layers consist of varying proportions of quartz, alkali-feldspar and plagioclase feldspar. Planar structure barely is perceptible due to the low color index of the granite and the lack of significant color contrasts between adjacent quartzo-feldspathic layers. Very coarse-grained and pegmatitic Canada Hill granite is exclusively massive. Representative modes for Canada Hill granite are listed in appendix 2. A chemical analysis of sample C-5 is given in appendix 3.

This rock corresponds to the faintly foliated to massive, coarse-grained biotite granite described by Berkey and Rice (1919) at King's quarry, south of Garrison. The rock also was recognized by Dodd (1965) in the Popolopen Lake quadrangle as gray leucogranite and pegmatite. The Canada Hill granite as defined above is a distinct, mappable field unit. The use of the term Canada Hill granite should be retained.

Paragneiss

Paragneiss is the major metasedimentary rock in the West Point-Bear Mountain region. The paragneiss (Ypn, Fig. 2) contains minor amphibolite, rusty weathering paragneiss, quartzite and minor calcareous and ferruginous metasediments.

Paragneiss is a medium-grained, gray to bluish-gray, equigranular rock with pronounced compositional layering and biotite schistocity. Outcrops of paragneiss have a slabby appearance. Alternating quartzo-feldspathic bands (about 3 mm thick) and thin layers of biotite (about 0.5 mm) impart a strong foliation to the rock. Microfolds, especially within schistose layers, impart a strong linear fabric to the rock. Locally linear structure is accentuated by fibroblastic sillimanite.

Paragneiss consists of strained xenoblastic quartz, nonperthitic or slightly perthitic microcline, plagioclase feldspar and biotite. Plagioclase feldspar (An_{12-24}, oligoclase) is poorly to well-twinned, myrmekitic and altered partially to white mica and less commonly to epidote. Biotite (X-straw yellow; Y = Z = dark reddish-brown) is altered partially to chlorite, epidote, white mica and opaque oxide. Zircon, apatite and opaque oxide are accessory minerals. Sillimanite and garnet are varietal minerals. Graphite, cordierite and opaque sulfides are sparse and local in occurrence. Representative modal analyses of paragneiss are listed in appendix 4.

Paragneiss is migmatitic along contacts with plutons of Canada Hill granite. Migmatitic paragneiss is a coarsely banded, light gray rock (Fig. 3). It consists of alternating decimeter-scale layers of leucosome containing Canada Hill granite and paragneiss melanosome. Canada Hill granite within leucosome is a medium-grained (2 mm to 6 mm), white rock with a xenomorphic-granular texture. Plagioclase feldspar is oligoclase (An_{16-22}). Melanosome is a medium-grained (0.5 mm to 2 mm), coarsely foliated, dark, inequigranular rock with a lepidoblastic texture. Melanosome is composed of quartz, plagioclase feldspar, biotite, garnet and sillimanite. The proportion of these minerals varies from sample to sample. Poorly to well-twinned plagioclase feldspar (An_{16-22}) is myrmekitic and partially altered to sericite and epidote. Biotite (X = straw yellow; Y = Z = dark reddish-brown) is altered partially to chlorite, epidote and opaques. Garnet is unal-

Figure 3. Migmatite developed along the contact between the Crystal Lake pluton and paragneiss. Canada Hill granite forms light colored leucosome. Refractory melanosome forms dark layers. Canada Hill granite cuts across prominent S_1 compositional layering in melanosome. Coin is 2.4 cm in diameter.

tered and contains irregular, non-oriented inclusions of apatite, quartz, biotite and opaque minerals. Non-perthitic microcline and zircon are common accessory minerals. Representative modal analyses of migmatite are given in appendix 5.

Berkey and Rice (1919) and Lowe (1950, 1958) correlated the paragneiss with gneissic Canada Hill granite at King's quarry, near Garrison. Recent workers (Hotz, 1953; Sims, 1953; Dodd, 1965; Baker and Buddington, 1970; Jaffe and Jaffe, 1973) have presented convincing evidence supporting a metasedimentary parentage for the paragneiss. Also, fabric elements in the paragneiss are identical with those in the associated rusty metasedimentary/metavolcanic gneisses. It is clear that paragneiss and the rusty gneisses are genetically related and reflect a similar tectonic history. It is recommended that the paragneiss be deleted from the original definition of Canada Hill granite.

Rusty Gneisses

The rusty gneisses consist of paragneiss, amphibolite, sillimanitic schist, quartzite, pyroxene gneiss, biotite-hornblende gneiss, quartz-plagioclase gneiss, calc-silicate and marble (Yrg, Fig. 2). A characteristic rusty color is imparted to these rocks in the field due to weathering of sulfides in several lithologies (paragneiss, sillimanitic schist, quartzite, pyroxene gneiss). The rusty gneisses stand out clearly where interlayered with paragneiss.

The rusty gneisses have planar and linear fabrics similar to those found in paragneiss. Compositional layering results from varying proportions of essential minerals in all lithologies. Biotite schistocity is developed especially in paragneiss, sillimanite schist, biotite-hornblende gneiss, and some quartzite and calc-silicate gneisses. Linear structure results from alignment of micro-

fold axes in schistose layers, rodding of hornblende crystals and preferred dimensional orientation of sillimanite. Modal analyses of several lithologies in the rusty gneisses are given in appendix 6.

Berkey and Rice (1919) considered the rusty gneisses as Grenvillian metasediments. The rusty gneisses and paragneiss were mapped as mixed Grenville-Canada Hill granite.

STRUCTURAL GEOLOGY

Fold Geometry

Rocks of the West Point-Bear Mountain area were deformed multiply during the Grenvillian dynamothermal event (Dodd, 1965; Helenek, 1966; Dallmeyer, 1972). Three generations of folds occur in the rocks (Fig. 4). The earliest folds (F_1) are appressed, intrafolial, isoclinal folds. Immediately east of Bear Mountain at Hessian Lake, Storm King granite gneiss is folded with paragneiss and rusty gneisses into minor F_1 folds. Lithologic layering (S_o) in paragneiss and rusty gneisses is discordant against Storm King granite gneiss. Hornblende foliation and biotite schistocity in the granite gneiss and compositional layering and biotite schistocity in the paragneiss and rusty gneisses parallel axial surfaces of F_1 folds. Planar structure in both Storm King granite gneiss and the host gneisses is continuous across lithologic contacts in hinges of tight F_1 folds. Along the Crown Ridge and Bare Rock, quartzo-feldspathic layering parallels hornblende foliation (S_1) in the Storm King granite gneiss. At Queensboro Lake, planar structure in the Storm King granite gneiss is continuous across lithologic contacts in the hinge of an F_1 fold. The prominent planar structure in the Storm King granite gneiss and metasedimentary/metavolcanic rocks is an axial plane foliation (S_1)

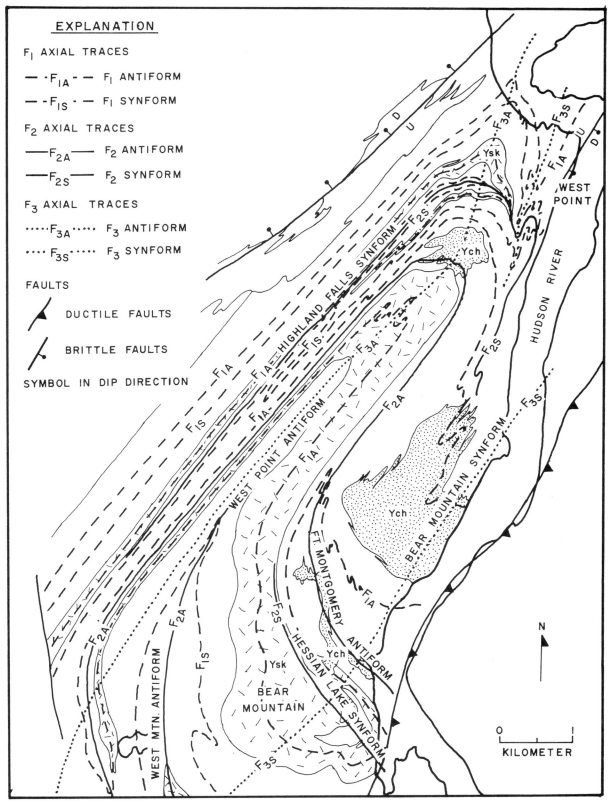

Figure 4. Axial traces of major F_1, F_2 and F_3 folds in the West Point-Bear Mountain region. Ysk, plutons of Storm King granite gneiss; Ych, plutons of Canada Hill granite.

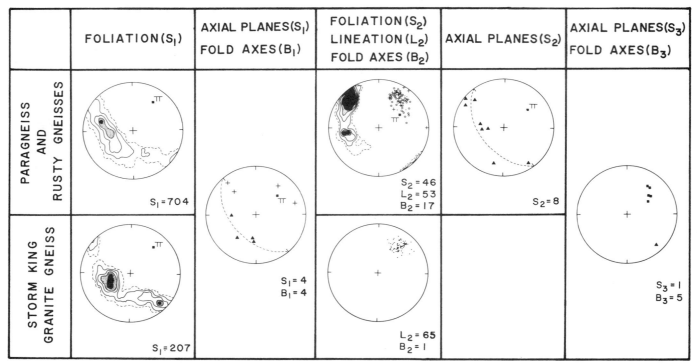

Figure 5. Lower hemisphere equal-area projections showing orientation of D_1, D_2 and D_3 fabric elements in Storm King granite gneiss, paragneiss and rusty gneisses. Solid triangles, poles to axial surfaces of minor folds; crosses, fold axes of minor folds; solid squares, fold axes of regional folds; circles, microfold axes in schistose layers; dots, hornblende lineation. Contours for S_1 and S_2 foliation; 10, 7.5, 5, 2.5, 1%/1% area. Number of fabric elements measured is indicated to lower right.

which parallels lithologic layering (S_o) in the limbs of F_1 folds and transects lithologic layering (S_o) in the hinges of these folds. Linear structure related to F_1 folding did not develop or was eradicated during subsequent recrystallization. Coincidence of fabric elements in Storm King granite gneiss and the metasedimentary/metavolcanic rocks (Fig. 5) indicates that the Storm King granite gneiss was deformed during F_1-folding.

A second phase of intense deformation resulted in reclined, isoclinal folds (F_2). The West Mountain antiform, Hessian Lake synform, Ft. Montgomery antiform and Highland Falls synform (Fig. 4) were formed during this deformation. S_1 planar structure in Storm King granite gneiss, paragneiss and rusty gneisses was refolded and reoriented during D_2. S_o and S_1 planar fabric elements were transposed locally into the axial surfaces of F_2 folds (Fig. 5). Penetrative mineral lineation (hornblende, sillimanite) and axes of microfolds in schistose layers in Storm King granite gneiss, paragneiss and rusty gneisses parallel fold axes of minor and regional F_2 folds (Fig. 5).

A final phase of open, upright, isoclinal folds (F_3) refolded F_1 and F_2 fabric elements (Fig. 5 and 6). This resulted in the West Point antiform and Bear Mountain synform. A weak hornblende foliation in the Storm King granite gneiss at Bare Rock parallels F_3 axial surfaces. The F_3 fold axis is oriented N51°E at 52°N and is approximately coaxial with D_2 fold axes (B_2) and mineral lineations (L_2) (Fig. 5).

Structural Relationships between Storm King Granite Gneiss and Paragneiss/Rusty Gneisses

Plutons of Storm King granite gneiss are discordant with lithologic layering (S_o) in paragneiss and rusty gneisses. Discordance occurs in the Bear Mountain pluton immediately south and east of Hessian Lake and in the crest of the West Point antiform. Dodd (1965, p. 18) reported lithologic discordance between the Storm King granite gneiss and rusty gneisses at West Mountain in the Popolopen Lake quadrangle. The West Point pluton is discordant with paragneiss and rusty gneisses along the northwest limb of the West Point antiform.

Although there is lithologic discordance in these instances, plutons of Storm King granite gneiss have gross structural concordance with the metasedimentary/metavolcanic gneisses. Planar structure in both rocks is axial planar to highly appressed F_1 folds. F_1 folds involve both the granite gneiss and metasedimentary/metavolcanic rocks. Linear fabric elements in both rocks parallel fold axes of isoclinal F_2 folds (Fig. 5). D_1 and D_2 fabric elements in both rocks are refolded by F_3 folds. Inclusions of rusty gneisses within the Storm King granite gneiss also have fabric elements concordant with the host granite gneiss. Field relationships and fabric data indicate that the Storm King granite gneiss was involved in all phases of folding which deformed the host gneisses. This implies that Storm King granite gneiss was in

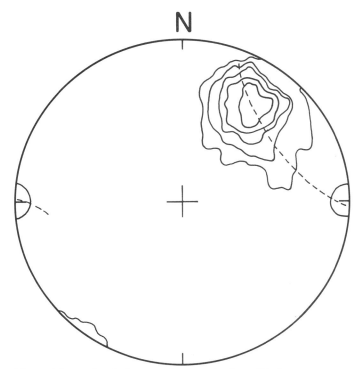

Figure 6. Lower hemisphere equal-area projection of D_2 linear structures (L_2) from Storm King granite gneiss, paragneiss and rusty gneisses. Linear structures plotted are microfold axes in schistose layers, hornblende lineation, and fold axes of minor folds. D_2 linear structures are refolded by F_3 and show dispersion along a small circle. Contours: 20, 15, 10, 5, 0.7%/1% area; N = 136.

situ prior to the onset of or early in Grenvillian deformation. Linear and planar structure within the Storm King granite gneiss are secondary fabric elements formed during reconstitution of the rock during Grenvillian dynamothermal metamorphism.

Structural Relationships between Canada Hill Granite and Storm King Granite Gneiss and Paragneiss

Canada Hill granite within the Stoney Lonesome, Crystal Lake and Brooks Lake plutons is a very-coarse grained to pegmatitic, massive rock. The granite contains inclusions of paragneiss and rusty gneisses. Contacts between Canada Hill plutons and host gneisses are gradational. Paragneiss grades into migmatitic paragneiss. Migmatitic paragneiss in turn grades into Canada Hill granite with abundant inclusions of paragneiss and rusty gneisses. Massive Canada Hill granite free of inclusions is found only in the core of the Crystal Lake pluton. Canada Hill granite in all plutons cuts S_1 and S_2 planar structures (foliation, schistocity, axial surfaces of minor folds) and L_2 linear structures (hornblende, sillimanite, microfold axes) in paragneiss and rusty gneisses. Along the southern margin of the Brooks Lake pluton near Hessian Lake, inclusions of Storm King granite gneiss are incorporated into the Canada Hill granite (Fig. 7). S_1 planar and L_2 linear structures (hornblende) in the Storm King granite gneiss also are cut by the Canada Hill granite. Locally a faint foliation parallel to the axial surface of F_3 folds occurs in Canada Hill granite.

Geological relationships clearly show that Canada Hill granite postdates D_1 and D_2 and is younger than both Storm King granite gneiss and the metasedimentary/metavolcanic rocks. Foliation related to D_3 and the location of the granite in hinges of

Figure 7. Canada Hill granite (Ych) truncating S_1 planar structure (trend indicated by lines) in Storm King granite gneiss (Ysk). Faint foliation diagonally from upper left to lower right of photograph is S_3 foliation in the Canada Hill granite. Along the southern contact of the Brooks Lake pluton northeast of Hessian Lake. Coin is 2.4 cm in diameter.

D_3 folds indicates the Canada Hill granite was syntectonic to and deformed during the D_3 phase of Grenvillian deformation.

MINERAL PARAGENESES AND METAMORPHISM

Mineral assemblages in metasedimentary/metavolcanic gneisses, Storm King granite gneiss and Canada Hill granite are listed in appendix 7. Dodd (1965), Dallmeyer and Dodd (1971), and Jaffe and Jaffe (1973) assigned mineral assemblages in pelitic and basic metasedimentary/metavolcanic rocks to the hornblende granulite facies of regional metamorphism. Considerable agreement exists for the magnitude of intensive parameters during Grenvillian metamorphism. Maximum temperatures of $750° \pm 50°C$ were estimated based on mineral assemblages found within pelitic and basic rocks (Dodd, 1965; Jaffe and Jaffe, 1973). Dallmeyer and Dodd (1971) proposed a slightly lower thermal maximum of $725° \pm 25°C$. Estimates of pressure based upon stability of cordierite-bearing assemblages in paragneiss indicate maximum load pressure of 4 ± 1 kb with $P_{H_2O} < P_L$ (Dallmeyer and Dodd, 1971; Jaffe and Jaffe, 1973; Hall and others, 1975).

Mineral assemblages in pelitic and basic metasedimentary/metavolcanic rocks have penetrative linear and planar fabric elements formed during F_1 and F_2 folding. This indicates that maximum conditions of metamorphism were attained during recrystallization of these rocks during D_1 and D_2. The absence of fabric elements related to F_3 folds suggests that recrystallization of the gneisses in the West Point-Bear Mountain area effectively terminated prior to the onset of D_3.

Mineral assemblages in the Storm King granite gneiss are compatible with either metamorphic recrystallization under hornblende granulite conditions or magmatic precipitation. The development of a tectonite fabric in the granite gneiss suggests recrystallization of these minerals in response to deformational stresses accompanying dynamothermal metamorphism. Also contacts of plutons lack both chilled facies in the granite and contact metamorphic effects in host gneisses. This evidence indicates metamorphic recrystallization of the Storm King granite gneiss at elevated temperatures and pressures, and precludes a post-tectonic age for intrusion of the granite gneiss.

Mineral assemblages in Canada Hill granite also are compatible with metamorphic recrystallization or magmatic precipitation. Penetrative D_1 and D_2 fabric elements do not occur in the Canada Hill granite since intrusion of the granite postdated D_2. Some recrystallization of the granite is indicated by the development of faint S_3 foliation.

EVALUATION OF THE STORM KING AND CANADA HILL PLUTONIC EPISODES

Storm King Plutonic Episode

A plutonic origin was proposed for Storm King granite gneiss based upon chemical composition, fabric and contact relationships with host gneisses. Lowe (1950) and Dodd (1965) pre-

sented petrologic evidence indicating the Storm King granite gneiss had a minimum melt composition in the system quartz-albite-orthoclase-water. These data support an igneous origin for the Storm King granite gneiss but do not require a plutonic origin. Jaffe and Jaffe (1973) pointed out that meta-rhyolites also have compositions approximating minimum melt compositions in the granite system.

Lowe (1950) interpreted linear fabric in the Bear Mountain pluton as primary structure related to passive, post-kinematic intrusion of Storm King granite gneiss. Based on evidence presented above, linear and planar fabric elements in the Storm King granite gneiss are secondary structures formed during deformation of the rock. Because initial emplacement of the granite gneiss pre-dates deformation, arguments based on secondary structures contribute little to establishing a plutonic/volcanic origin for the granite gneiss.

Jaffe and Jaffe (1973) noted concordance between the Storm King granite gneiss and host gneisses nine kilometers west of Bear Mountain in the Monroe quadrangle. Although the granite gneiss is concordant at map scale, Storm King granite gneiss commonly is transgressive in detail within the Popolopen Lake quadrangle (Dodd, 1965). Lithologic discordance between plutons of Storm King granite gneiss and host paragneiss is also evident in the West Point-Bear Mountain region. Lithologic discordance combined with a eutectic composition supports a magmatic plutonic origin for Storm King granite gneiss.

Pre-kinematic and syn-kinematic, subconcordant, sheet-like, granitoid intrusions were described in the Fiskenaesset region of Greenland (Myers, 1976) and in the Svecofennian belt of Norway (Hietanen, 1975). Such granitoid intrusions form an exceptionally large proportion of continental crust exposed in these regions (Myers, 1976). It is concluded that the Storm King granite gneiss represents a period of intrusion of granitoid sheets and sills into a Proterozoic metasedimentary/metavolcanic terrane. Intrusion was contemporaneous with and somewhat younger than formation of the metasedimentary/metavolcanic protoliths and pre-tectonic and early syntectonic to the Grenvillian dynamothermal metamorphism.

Canada Hill Plutonic Episode

Canada Hill granite is a late-kinematic rock emplaced during D_3. The granite shows considerable variation in the proportion of essential minerals. The composition deviates from minimum melt compositions in the quarternary system quartz-albite-orthoclase-water and is enriched in orthoclase component (Fig. 8). Crystal-melt equilibrium is not indicated by these data. Plutons of Canada Hill granite are for the most part confined to paragneiss and rusty gneisses. Contacts with host rocks are migmatitic. The intimate association of Canada Hill granite and the major paragneiss indicates a fundamental petrogenetic relationship between the two rocks.

Several mechanisms of granitization have been proposed based upon study of migmatites: (1) igneous injection of magma,

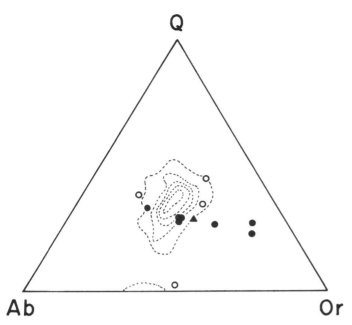

Figure 8. Mineralogical composition of Canada Hill granite from the interior of plutons (dots) and migmatitic leucosome (circles) plotted in terms of modal quartz, plagioclase feldspar and potash feldspar onto the Q-Ab-Or plane of the quarternary system Ab-Or-Q-H_2O. Potash feldspar has been corrected for 5% perthitic albite. Average composition for Canada Hill granite is indicated by solid triangle. Dashed contours represent plutonic rocks that carry 80 percent or more normative albite, orthoclase and quartz (Tuttle and Bowen, 1958).

(2) anatexis, (3) external metasomatism (metasomatism by external fluids or aqueous vapors), and (4) internal metasomatism (metamorphic segregation) (White, 1966; Yardley, 1978). Granitization by injection of magma or external metasomatism is an open system process. The bulk chemical composition of the granitized system is changed through introduction of components. Anatexis and internal metasomatism are closed system mechanisms in which bulk chemical composition remains unchanged (Yardley, 1978). All mechanisms were invoked to explain the formation of migmatites. Brown (1967) favored an external metasomatic origin for the Loch Coire migmatite complex of Scotland. White (1966), Hedge (1972), and Yardley (1978) preferred metamorphic segregation (internal metasomatism) to explain migmatite formation. Misch (1968) argued that both metasomatism and metamorphic segregation formed the Skagit migmatites. Olsen (1977) preferred a model involving anatectically induced metamorphic segregation. On the basis of field and experimental data, v. Platen (1965), Mehnert (1968), Winkler (1974), and Ashworth (1976) interpreted granitization as proceeding by total or partial anatexis of paragneiss.

Yardley (1978) presented several criteria for distinguishing different mechanisms of granitization accompanying the migmatization process. These criteria were applied to the Canada Hill granite-paragneiss migmatites. The ratio of refractory to nonrefractory minerals in paragneiss is about 1:1. Leucosome-melano-

some pairs from individual migmatites were averaged in this ratio to estimate the mineralogical composition of homogenized migmatite (appendix 8). This computation must be regarded as an approximation since the exact amount of leucosome and melanosome is difficult to determine. The mineralogical composition of homogenized migmatite approximates that of the paragneiss. The similarity in composition between homogenized migmatite and paragneiss implies a closed system mechanism for migmatization (Yardley, 1978). Such mechanisms include anatexis and internal metasomatism.

Plagioclase feldspar from migmatitic leucosome-melanosome pairs have identical compositions (appendix 5). Mineral assemblages in the granite and paragneiss are similar. Only the proportion of essential minerals varies. The migmatites consist of extensive, closely formed leucosomes which show no rotation of enclosed host rock. Coarse-grained to pegmatitic granite follows the foliation planes within the host paragneiss. When the criteria of Yardley (1978) are applied to the Canada Hill-paragneiss migmatites, the major process indicated is a closed system internal metasomatic process. Rather than a magmatic episode, formation of Canada Hill granite may represent a period of regional granitization brought about by a process of internal metasomatism. These data support the recent proposal of Ratcliffe (personal communication, 1981) that a metasomatic process was important in forming the Canada Hill granite. Given the P_L, T, and P_{H_2O} conditions of metamorphism, it is difficult to disregard the role partial anatexis may have played in contributing to the granitization process.

ISOTOPE STUDIES

A Rb-Sr radiometric study was undertaken to determine time and isotopic relationships between the paragneiss and the Canada Hill granite. Samples of non-migmatitic paragneiss were collected over a relatively small area south of the Bear Mountain pluton (H-1, H-2, H-3, H-7, H-10, K-14 and K-15 on Fig. 2). Samples of Canada Hill granite were taken from the Crystal Lake pluton (C-1, C-2, C-3, C-4, C-5 on Fig. 2) and the main mass of Canada Hill granite east of the Hudson River (C-6, C-8). Sample locations are given in appendix 9.

Samples weighing about 5 to 15 kg. were collected, crushed and individually split into 10 g portions which were powdered. The powders were split into 0.2 to 0.3 g portions for isotopic analysis. Each analysis was done using ^{84}Sr and ^{87}Rb spikes, ultrapure HF, $HClO_4$ and HCl, and Teflon beakers. The solutions were passed through pyres cation-exchange resin columns to obtain Rb and Sr fractions.

The mass spectrometric analyses were done at Florida State University in 1975-76 using a 12-inch radius of curvature, 60° sector, single focusing mass spectrometer with triple-filament source, Faraday cup collector, vibrating reed electrometer and an expanded scale stripchart recorder. The Eimer and Amend Standard $SrCO_3$ was analyzed 22 times at Florida State University between 1973-76 and yielded an average $^{87}Sr/^{86}Sr$ ratio of

0.7980 ± 0.0002 (1σ) when $^{86}Sr/^{88}Sr$ is normalized to 0.1194 so no correction of the strontium isotopic data from the samples was made to correct for machine fractionation. The ^{87}Rb blanks averaged 1.3 ng, and the ^{86}Sr blanks averaged 4.1 ng; compared to the Rb and Sr concentrations in the samples analyzed, these blank values are not significant.

Analyses of National Bureau of Standards standard K-feldspar (NBS-70a feldspar) that were performed over the course of this study are in close agreement with those reported by Compston and others (1969) and by DeLaeter and Abercrombie (1970). This indicates no major systematic errors in the isotope tracer calibrations.

All the Sr isotopic compositions were calculated from analyses of sample Sr and spike Sr mixtures. The $^{85}Rb/^{87}Rb$ ratio used was 2.593; the decay constant used for ^{87}Rb is 1.42×10^{-11} yr^{-1} (Steiger and Jager, 1977).

The Rb-Sr age and initial $^{87}Sr/^{86}Sr$ ratio on the isochron diagrams were calculated using the regression treatment described by York (1966). The one-standard-deviation experimental error in $^{87}Rb/^{86}Sr$ used to calculate the age and initial $^{87}Sr/^{86}Sr$ ratio was derived from an examination of duplicate analyses done over the past six years, and these estimates include splitting errors. The one-standard-deviation experimental error in $^{87}Rb/^{86}Sr$ was calculated to be 2 percent, and the one-standard-deviation experimental error in $^{87}Sr/^{86}Sr$ was calculated to be 0.05 percent. These error estimates, without the error induced by sample splitting, are estimated to be 1 percent for $^{87}Rb/^{86}Sr$ and 0.02 percent for $^{87}Sr/^{86}Sr$. The isochron diagrams show the magnitude of experimental error as an "error box" whose dimensions are two standard deviations (4% by 0.10%). The errors assigned to the age and initial $^{87}Sr/^{86}Sr$ ratio are given at the 68 percent confidence level. Isotopic data are presented in appendix 10; sample locations for the Canada Hill granite and the host paragneiss are shown in Fig. 2 and appendix 9.

Age Determination on Paragneiss and Canada Hill granite

Samples of non-migmatitic paragneiss sampled over a distance of about 2 km yielded a Rb-Sr isochron age of 1147 ± 43 m.y. and a $^{87}Sr/^{86}Sr$ ratio at that time of 0.7046 ± 0.0014 (Fig. 9). Samples of Canada Hill granite yielded a Rb-Sr whole-rock age of 913 ± 45 m.y. with an initial $^{87}Sr/^{86}Sr$ ratio of 0.7186 ± 0.0017 (Fig. 10). A plot of the Canada Hill granite isotopic analyses (Fig. 10) shows that most of the analyses fall slightly above or slightly below the best-fit regression line. We believe that the small nonlinearity is due to initial strontium heterogeneity of the samples. Although inaccurate, we believe this age of about 915 m.y. to be a good approximation of the time of granitization.

DISCUSSION OF AGE DETERMINATIONS

The age determinations reported in this paper, when examined in conjunction with radiometric ages from other studies in

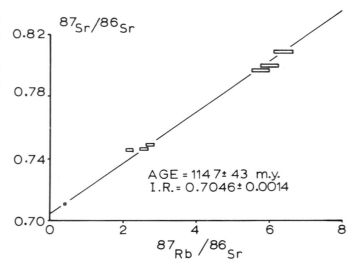

Figure 9. Rb-Sr whole-rock isochron obtained from samples of the paragneiss (1147 ± 43 m.y.)

this area, provide strong evidence for the existence of two distinct metamorphic-plutonic events that have chronological similarities to the younger but better known Taconic orogeny (about 470 m.y. ago) and the Acadian orogeny (about 350 m.y. ago) in the Appalachians. Evidence for the first event is the paragneiss discussed in this paper which yields an age of about 1145 m.y. This is essentially the same, within the estimated analytical errors in age determinations, as the reported ages for metavolcanic rocks in this area (Rb-Sr age = 1139 ± 25 m.y., recalculated using 1.42×10^{-11} yrs^{-1} decay constant from data reported by Murray and others, 1977; U-Pb concordia diagram upper intercept age of about 1170 m.y. reported by Alienikoff and others, 1982; $^{207}Pb/^{206}Pb$ age of about 1150 m.y., recalculated using new U-Pb decay constants (Steiger and Jager, 1977) from data reported by Tilton and others, 1960).

These Rb-Sr ages are interpreted to record a time of high grade metamorphism in this area. This conclusion is similar to that reached in the nearby Manhattan prong (Mose, 1982). The

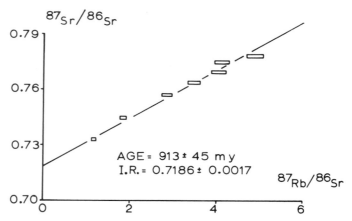

Figure 10. Rb-Sr whole-rock isochron obtained from samples of the Canada Hill granite (913 ± 45 m.y.).

prograde metamorphism is thought to have obliterated the depositional ages of these metasediments and metavolcanics in the Hudson Highlands and reset the Rb-Sr chronometer due to Sr isotopic homogenization during metamorphism. Subsequent metamorphic events are thought not to reset the Rb-Sr age because the rocks were depleted in water which facilitated the isotopic homogenization during the first event.

The magma which produced the Storm King granite gneiss intruded the metasediments and metavolcanics in this area. Plutons of Storm King granite gneiss yield a U-Pb concordia diagram upper intercept of about 1118 ± 55 m.y. (Alienikoff and others, 1982) with a lower intercept age which corresponds to Devonian time. A single zircon extract from the Storm King granite gneiss yields U-Pb data (Tilton and others, 1960) that fall on a chord with an upper intercept of about 1120 m.y. (taken from Aleinikoff and others, 1982) and a lower intercept of about 940 m.y. Although the meaning of lower intercept ages on U-Pb concordia diagrams is not well known, it may be important to note that Devonian time corresponds to an episode of low-grade regional metamorphism in this area, and that 940 m.y. is approximately the age of the youngest granite formed in Grenvillian time in this area. Further studies may someday improve our understanding of the true meaning of intercept ages.

In any case, it can be seen that Rb-Sr whole rock ages and U-Pb upper intercept ages for metasediments, metavolcanics and Storm King granite gneiss yield about the same age which, when taken together, is an age of about 1140 ± 25 m.y. That these rock units should be considered together is supported by the fact that all have D_1, D_2 and D_3 fabric elements. In any case, this age of about 1140 m.y. is interpreted to represent the first major metamorphic-plutonic event in the Grenville terrane.

Other rock units in this area were not affected by the D_1 event. For example, the Canopus pluton which yields an age of 1068 ± 45 m.y. (age recalculated using 1.42×10^{-11} yr^{-1} decay constant, from data in Ratcliffe and others, 1972) and the Canada Hill granite which has an age of 913 ± 45 m.y. contain only D_3 structures. Since D_2 and D_3 have almost parallel fabric elements, both of which are oblique to those of D_1, we favor grouping D_2 and D_3 into a single deformational episode. The ages mentioned above suggest that the maximum age for D_3 is about 960 m.y. (maximum age for the Canada Hill granite) and the minimum age for D_2 is 1023 m.y. (minimum age for Canopus pluton). Since D_2 and D_3 are coaxial we suspect they developed closely spaced in time at about 1000 m.y. On this basis, we propose that the Grenvillian rocks of the Hudson Highlands contain a record of two separate events. The first is the 1140 m.y. event characterized by metamorphism, plutonism and the development of D_1 fabric elements. The second is the approximately 1000 m.y. event characterized by metamorphism, plutonism, granitization and the development of D_2 and D_3 fabric elements. Although this is the first documented evidence for two separate events in the Grenvillian of the Appalachians, a similar conclusion has been reached about Grenvillian rocks in Canada (Baer, 1981).

Rb-Sr systematics on the paragneiss and Canada Hill granite also support a polymetamorphic-polydeformational history for the Reading Prong. High initial $^{87}Sr/^{86}Sr$ ratios are thought to be characteristic of rocks formed by closed system reconstitution (partial melting or internal metasomatism) of material (such as the paragneiss) with high average $^{87}Sr/^{86}Sr$ ratios at the time of reconstitution (Faure and Powell, 1972). The Rb-Sr study of the paragneiss shows that it had an average $^{87}Sr/^{86}Sr$ ratio of 0.7046 at about 1150 m.y. ago. By 915 m.y. ago, its average $^{87}Sr/^{86}Sr$ ratio would have risen to about 0.717 (average Rb/Sr ratio of the paragneiss is 3.8). A closed system internal metasomatic process involving redistribution of chemical components at this time within the paragneiss to form the Canada Hill granite would be expected to produce a rock with an initial $^{87}Sr/^{86}Sr$ ratio of about 0.717. Considering the approximate nature of our estimate of the average $^{87}Sr/^{86}Sr$ ratio for the paragneiss (7 samples), the calculated initial ratio of 0.717 for the Canada Hill granite and its determined ratio of $0.7186 \pm .0017$ are remarkably similar.

It is interesting to compare various time-temperature estimates for this area. As discussed earlier, the Canada Hill granite formed at 913 ± 45 m.y. at a temperature of about $725 \pm 25°C$ (thermal maximum during D_3). Dallmeyer and others (1975) and Dallmeyer and Sutter (1976) presented $^{40}Ar/^{39}Ar$ data indicating argon retention in cooling hornblende occurred at 900 ± 30 m.y. (blocking temperatures estimated to be $500°$-$550°C$) and in cooling biotite at 760 ± 60 m.y. (blocking temperatures estimated to be $300°$-$350°C$). Based on this evidence, these workers estimated an uplift rate of 0.03 to 0.07 mm/yr for the cooling Grenville terrane. Although the Rb-Sr age for Canada Hill granite (913 ± 45 m.y.) and the $^{40}Ar/^{39}Ar$ retention age for hornblende (900 ± 30 m.y.) may appear similar, the uncertainty of the ages and temperatures can easily yield a difference of at least 90 m.y. over which the terrane cooled about $150°$ to $250°C$. If the cooling took 90 m.y. and was associated with regional uplift, and if the geothermal gradient is assumed to have been about $35°C/km$, the cooling event would correspond to an uplift rate of about 0.05 to 0.10 mm/yr. This estimate is similar to that of Dallmeyer and others (1975).

PETROGENETIC SCHEME FOR PROTEROZOIC ROCKS OF THE HUDSON HIGHLANDS

A summary of Grenvillian petrogenetic, tectonic and metamorphic events in the West Point-Bear Mountain region of the Hudson Highlands is presented in appendix 11.

Storm King granite gneiss and Canada Hill granite are distinctive rock types and valid field mapping units. It is recommended the names be retained. The Storm King granite gneiss intruded as granitoid sheets and sills into metasedimentary and metavolcanic rocks. Structural data indicate a pre-tectonic to early syntectonic age for plutons of Storm King granite gneiss. Canada Hill granite represents a period of late-tectonic granitization. The terms Storm King and Canada Hill should be retained to designate major episodes of plutonism (sensu lato) in the Hudson Highlands. These plutonic episodes may correspond to two temporally separated phases of Grenvillian dynamothermal metamorphism.

Evidence supporting two distinct episodes of Grenvillian orogenic activity comes from several sources. Offield (1967) proposed a polymetamorphic history based on stratigraphic relationships in the Reading Prong. In the West Point-Bear Mountain region, the oblique orientation of D_1 fabric elements to the almost coaxial D_2 and D_3 fabric elements supports two separate deformational events. Finally, two orogenic episodes are indicated by U-Pb mineral and Rb-Sr whole rock isotopic data.

The chronology of these events is 1140 m.y. for the older event (Rb-Sr age on paragneiss, this report; U-Pb data from Tilton and others, 1960; U-Pb data from Aleinikoff and others, 1982) and 1000 m.y. for the younger event (Rb-Sr age on Canada Hill granite, this report; Rb-Sr age on the Canopus pluton, Ratcliffe and others, 1972; U-Pb data from Aleinikoff and others, 1982; U-Pb data from Klemic and others, 1959). These events are separated by about 150 m.y. If Phanerozoic orogenic episodes serve as analogues, this interval is equivalent to the total time span for the combined Taconic and Acadian events in the Appalachians. A polymetamorphic-polydeformational history for the Hudson Highlands is more consistent with the data than a single event spanning 150 m.y. over which the Grenville terrane was subjected continuously to the elevated temperatures and pressures of dynamothermal metamorphism.

ACKNOWLEDGEMENTS

We would like to thank R. T. Dodd, F. D. Eckelmann, L. M. Hall, Howard and Elizabeth Jaffe, K. E. Lowe, D. P. Murray, N. M. Ratcliffe and S. Schaffel for numerous discussions concerned with various aspects of Highlands geology. Comments from reviewers improved the paper considerably. Field work was supported by grants from the New York State Geological Survey, Brown University and Bradley University.

APPENDIX 1. MODAL ANALYSES OF STORM KING GRANITE GNEISS

	Bear Mountain pluton				West Point Pluton			
	51-64	53-64	5-78	13-78	44-64	57-66	46-64	54-66
Quartz	28.5	39.3	41.3	26.1	35.0	28.6	36.1	28.4
Microcline	--	--	--	1.2	21.0	--	21.2	6.4
Microperthite	--	--	1.5	1.4	11.6	28.9	8.8	25.2
Mesoperthite	66.9	48.8	47.8	59.9	--	--	--	--
Plagioclase	1.2	6.5	3.2	2.8	25.1	28.5	23.1	33.2
An-content	n.d.	n.d.	4	n.d.	26	23	24	22
Hornblende	2.6	4.6	4.9	6.1	4.0	12.1	5.4	2.6
Biotite	--	tr.	--	0.1	0.4	--	4.4	--
Clinopyroxene	--	--	--	--	--	--	--	1.4
Opaques	0.1	0.3	1.2	1.2	2.5	1.5	0.7	1.5
Apatite	tr.	tr.	0.2	0.4	0.1	0.3	tr.	0.3
Zircon	0.1	0.2	tr.	tr.	tr.	0.1	tr.	0.1
Alteration	0.8	0.3	tr.	0.7	0.4	0.2	0.3	0.9
	100.2	100.0	100.1	99.9	100.1	100.2	100.0	100.0

APPENDIX 2. MODAL ANALYSES OF CANADA HILL GRANITE

	C-1	C-2	C-3	C-4	C-5	C-6	C-8	Average*
Quartz	22.7	26.5	26.5	26.6	28.6	26.0	32.4	27.0 ± 2.9
Microcline	66.8	36.0	52.0	62.9	37.0	35.2	25.0	45.0 ± 15.7
Plagioclase	10.2	28.4	21.1	7.7	31.7	30.0	40.8	24.3 ± 12.0
An-content	24	20	24	22	20	16	7(?)	19 ± 6
Garnet	0.3	9.1	tr.	--	--	8.8	--	2.6 ± 4.3
Biotite	tr.	tr.	0.4	2.8	2.6	tr.	1.4	1.0 ± 1.2
Zircon	tr.	--	--	--	--	--	tr.	tr.
Sphene	--	--	--	--	tr.	tr.	--	tr.
Tourmaline	--	--	--	--	--	tr.	--	tr.
SECONDARY MINERALS								
White mica	tr.	tr.	tr.	tr.	tr.	tr.	0.3	tr.
Epidote	tr.	tr.	tr.	tr.	tr.	tr.	0.1	tr.
Chlorite	--	--	tr.	--	--	--	--	tr.
Calcite	tr.	tr.	--	tr.	tr.	--	tr.	tr.
Opaques	tr.	--	tr.	tr.	tr.	--	tr.	tr.
TOTAL	100.0	100.0	100.0	100.0	99.9	100.0	100.0	99.9

*Average values are calculated to one standard deviation from the mean value.

APPENDIX 3. CHEMICAL ANALYSES AND CIPW NORMS FOR STORM KING GRANITE GNEISS AND CANADA HILL GRANITE

	Storm King[1] granite gneiss* 5-78	Canada Hill[2] granite* C-5
SiO_2	75.5	74.3
TiO_2	0.20	0.23
Al_2O_3	11.4	14.0
Fe_2O_3	1.64	0.08
FeO	2.00	0.79
MnO	0.03	0.00
MgO	0.04	0.29

APPENDIX 3. CHEMICAL ANALYSES AND CIPW NORMS FOR STORM KING GRANITE GNEISS AND CANADA HILL GRANITE (continued)

	Storm King granite gneiss* 5-78	Canada Hill granite* C-5
CaO	0.79	1.08
Na_2O	3.03	3.15
K_2O	4.70	5.04
P_2O_5	0.05	0.12
H_2O	0.18	0.42
Total	99.56	99.50
CIPW Norms		
Q	37.20	33.71
Or	27.84	30.06
Ab	25.70	26.75
An	3.62	4.45
C	--	1.63
Di	0.25	--
Hy	1.95	1.62
Ilm	0.46	0.46
Mt	2.32	0.23
Ap	--	0.31
Total	99.34	99.22

1. Storm King granite gneiss sample locality: northeast of Bare Rock along Route 9W (Peekskill, N.Y. 7-1/2' quadrangle).
2. Canada Hill granite sample locality: along Route 9W at interchange due west of Crystal Lake (Peekskill, N.Y., 7-1/2' quadrangle).
* Analyst: Scott Argast, Analytical Laboratory, SUNY-Binghamton.

APPENDIX 4. MODAL ANALYSES OF PARAGNEISS

	1-66	45-64	49-64	58-66	6-66	5-66	Average*
Quartz	24.1	39.8	31.3	29.4	35.9	43.2	34.0 ± 7.0
Microcline	28.8	19.0	36.5	18.6	25.7	10.5	23.2 ± 9.1
Plagioclase	29.4	27.1	16.9	35.4	18.8	38.3	27.7 ± 8.6
An-content	24	24	16	24	19	12	20 ± 5
Biotite	16.7	8.8	9.6	12.0	19.4	8.0	12.4 ± 4.6
Garnet	0.2	3.7	4.7	3.1	tr.	--	2.0 ± 2.1
Apatite	0.8	0.2	0.5	0.5	0.3	--	0.4 ± 0.3
Zircon	0.1	0.1	0.1	0.1	tr.	tr.	0.1 ± 0.1
Sphene	--	--	tr.	0.1	--	--	tr.
Opaques	0.2	0.3	tr.	0.2	--	--	0.1 ± 0.1
Alteration	0.1	1.1	0.5	0.9	tr.	tr.	0.4 ± 0.5
TOTAL	100.4	100.1	100.1	100.3	100.1	100.0	100.3

*Average values are calculated to one standard deviation from the mean value.

APPENDIX 5. MODAL ANALYSES OF LEUCOSOME AND MELANOSOME
FROM CANADA HILL GRANITE-PARAGNEISS MIGMATITE

	LEUCOSOME				MELANOSOME		
	6-66	2-66	5-66	P/4/1	6-66	2-66	5-66
Quartz	37.8	2.5	43.3	33.6	43.5	58.1	48.8
Microcline	18.9	46.9	38.3	42.7	5.2	0.4	--
Plagioclase	40.6	42.1	14.8	20.9	15.3	21.4	11.3
An-content	16	22	16	16	16	22	16
Biotite	1.7	7.8	2.3	1.6	29.2	19.1	40.0
Garnet	--	--	--	tr.	6.8	--	--
Sillimanite	--	--	--	0.2	--	--	--
Apatite	0.6	tr.	--	--	tr.	1.0	--
Zircon	tr.	0.1	tr.	tr.	tr.	tr.	tr.
SECONDARY MINERALS							
White mica	tr.	tr.	tr.	0.3	--	tr.	tr.
Sericite	0.6	tr.	0.2	0.7	tr.	tr.	tr.
Epidote	tr.	0.4	tr.	tr.	--	tr.	tr.
Chlorite	tr.	0.2	0.8	--	--	tr.	--
Calcite	tr.	--	--	tr.	--	--	--
Opaques	tr.	tr.	0.2	tr.	--	tr.	tr.
TOTAL	100.2	100.0	99.9	100.0	100.0	100.0	100.1

APPENDIX 6. MODAL ANALYSES OF QUARTZITE, BIOTITE-HORNBLENDE GNEISS,
AMPHIBOLITE, AND CALC-SILICATE FROM THE RUSTY GNEISSES

	Rusty Quartzite	Biotite hornblende gneiss	Amphibolite	Pyroxene amphibolite	Rusty calc-silicate gneiss
	50-64	57-64	2-65	10-65	11-65
Quartz	72.2	17.4	--	--	--
Microcline	8.8	--	--	--	--
Microperthite	--	18.7	--	--	--
Plagioclase	9.0	43.0	26.4	40.1	32.3
An-content	n.d.	n.d.	46	45	n.d.
Hornblende	--	12.2	71.7	37.2	--
Clinopyroxene	--	--	1.6	18.4	28.7
Biotite	6.7	5.5	0.2	3.2	31.7
Garnet	1.4	--	--	--	--
Sillimanite	0.8	--	--	--	--
Zircon	0.1	tr.	--	tr.	tr.
Apatite	--	0.6	0.1	0.2	tr.
Sphene	tr.	--	--	--	--
Graphite	0.1	--	--	--	0.9
Opaques	0.2	2.3	0.1	0.5	3.4
Alteration	0.9	0.3	0.2	0.6	0.3
Total	100.2	100.0	100.3	100.2	100.0

APPENDIX 7. DIAGNOSTIC MINERAL ASSEMBLAGES IN STORM KING GRANITE GNEISS,
PARAGNEISS, RUSTY GNEISSES, AND CANADA HILL GRANITE*

Storm King granite gneiss

    quartz-microcline-plagioclase-hornblende garnet
    quartz-microcline-plagioclase-hornblende-clinopyroxene
    quartz-microcline-plagioclase-hornblende-biotite

Canada Hill granite

    quartz-microcline-plagioclase-garnet-biotite
    quartz-microcline-plagioclase-biotite
    quartz-microcline-plagioclase-sillimanite graphite

Paragneiss and Rusty Gneisses

1. Pelitic

    microcline-quartz-plagioclase-sillimanite-cordierite-biotite-garnet graphite
    microcline-quartz-plagioclase-biotite-garnet graphite
    microcline-quartz-plagioclase-biotite-garnet-sillimanite graphite
    microcline-quartz-plagioclase-biotite-sillimanite-cordierite
    quartz-plagioclase-sillimanite-cordierite-biotite-garnet graphite
    quartz-plagioclase-sillimanite-cordierite-garnet graphite
    quartz-plagioclase-garnet-biotite graphite

2. Quartzofeldspathic

    microcline-plagioclase-quartz-hornblende-biotite orthopyroxene

3. Basic

    plagioclase-hornblende-quartz-clinopyroxene biotite
    plagioclase-hornblende-clinopyroxene biotite
    plagioclase-hornblende-biotite
    plagioclase-hornblende-clinopyroxene-orthopyroxene biotite

4. Calc-Silicate

    plagioclase-clinopyroxene-biotite
    clinopyroxene-hornblende-plagioclase graphite
    clinopyroxene-quartz-microcline-hornblende-biotite

5. Ferruginous

    garnet-eulite-graphite quartz
    garnet-clinopyroxene-magnetite-graphite

*Note: Data from Dallmeyer and Dodd (1971), Helenek (1971), and Jaffe and Jaffe (1973).

APPENDIX 8. AVERAGE MINERALOGICAL COMPOSITION OF HOMOGENIZED MIGMATITE
COMPARED TO AVERAGE PARAGNEISS*

	6-66	2-66	5-66	Average Homogenized Migmatite	Average Paragneiss
Quartz	40.7	30.3	46.1	39.0	34.0 ± 7.0
Microcline	12.1	23.7	19.2	18.3	23.2 ± 9.1
Plagioclase	28.0	31.8	13.1	24.3	27.7 ± 8.6
An-content	16	22	16	18	20 ± 5
Biotite	15.5	13.5	21.2	16.7	12.4 ± 4.6
Garnet	3.4	--	--	1.1	2.0 ± 2.1
Apatite	0.3	0.5	--	0.3	0.4 ± 0.3
Zircon	tr.	0.5	tr.	0.2	0.1 ± 0.1
Total	100.0	100.3	99.6	99.9	99.8

*Note: Data from appendices 4 and 5. Leucosome and melanosome in migmatite are averaged in a ratio of 1:1.

APPENDIX 9. DESCRIPTION OF SAMPLES USED FOR ISOTOPIC ANALYSES

Canada Hill granite.—Samples of Canada Hill granite from the Crystal Lake pluton (C-1, C-2, C-3, C-4, C-5) were obtained from a series of large, freshly exposed (1969) outcrops at the intersection of Routes 218 and 9W, about 4.2 km north of the Bear Mountain Bridge traffic circle (Peekskill, New York, 7 1/2' quadrangle). Samples of Canada Hill granite from the east side of the Hudson River (C-6, C-8) were obtained from fresh exposures along Route 301. Sample C-8 is located 1.1 km southwest of Canopus Lake along Route 301 (Oscawana Lake, New York, 7 1/2' quadrangle). Sample C-6 is located 3.9 km southwest of Canopus Lake along Route 301 (West Point, New York, 7 1/2' quadrangle).

All samples contain quartz, slightly perthitic microcline, plagioclase feldspar, and biotite. Samples from the Crystal Lake pluton are coarser grained (0.3 mm to 3.0 cm) than samples of Canada Hill granite from the Hudson River (0.3 to 7.0 mm). Microcline is fresh and unaltered. Plagioclase feldspar is poorly to well twinned, occasionally myrmekitic and partially altered to sericite and epidote. Reddish-brown biotite is partially altered to chlorite, epidote, mica, and opaque oxides and rarely bleached to a pale green color. Samples C-1, C-2, C-3, and C-6 contain garnet in varying amounts. Accessory minerals include sphene, zircon, and tourmaline. Secondary calcite, epidote, and muscovite fill widely-spaced microfractures in all samples. Only sample C-8 is highly sheared. Remaining samples all lack cataclastic fabrics.

Paragneiss.—Paragneiss was sampled at two locations. Samples H-1, H-2, and H-3 were obtained south of Seven Lakes Drive approximately 1 km southeast of the intersection of Seven Lakes Drive with Route 6 (Popolopen Lake, New York, 7 1/2' quadrangle). Samples H-7, H-10, K-14, and K-15 were obtained at the east end of the parking area at Anthony Wayne Recreation Area along Palisades Interstate Parkway (Popolopen Lake, New York, 7 1/2' quadrangle).

All samples are medium-grained (about 1 mm) and foliated and consist of quartz, non-perthitic or slightly perthitic microcline, plagioclase feldspar, and biotite. Microcline is fresh and unaltered. Plagioclase feldspar is poorly to well twinned, myrmekitic, and partially altered to sericite and less commonly epidote. Biotite (X = straw yellow, Y = Z = dark reddish-brown) is partially altered to chlorite, epidote, white mica, and opaque oxides. Alteration, however, is minor. Garnet is present in samples H-10 and K-14. Zircon, apatite, and opaque oxides are accessory minerals. All samples lack cataclastic fabrics.

APPENDIX 10. ISOTOPIC ANALYSES OF CANADA HILL GRANITE AND HOST PARAGNEISS

Sample	$(^{87}Sr/^{86}Sr)_n$	^{87}Rb (ppm)	^{86}Sr (ppm)	$^{87}Sr/^{86}Sr$ (atomic ratio)
Canada Hill Granite:				
C-1	0.7749	56.86	13.60	4.133
C-2	0.7572	40.22	13.89	2.862
C-3	0.7639	47.33	13.42	3.487
C-4	0.7782	71.12	14.36	4.895
C-5	0.7696	48.58	11.85	4.053
C-6	0.7446	19.28	10.31	1.849
C-8	0.7328	27.29	22.97	1.175
Paragneiss:				
H-1	0.7963	56.83	9.74	5.770
H-2	0.7994	56.59	9.31	6.010
H-3	0.8085	59.15	9.15	6.393
H-7	0.7460	36.39	14.01	2.568
H-10	0.7486	34.85	12.63	2.729
K-14	0.7451	30.68	14.00	2.166
K-15	0.7108	14.04	33.26	0.417

APPENDIX 11. SUMMARY OF GRENVILLIAN PETROGENETIC, TECTONIC, AND METAMORPHIC EVENTS IN THE WEST POINT-BEAR MOUNTAIN REGION OF THE HUDSON HIGHLANDS, NEW YORK

Orogenic Episode	Deformational Event	Fold System	Type of Folds	Important Tectonic Features	Metamorphic Event	Important Crystalloblastic and Other Features	Plutonic Events	Age of Events
			Post-Grenvillian uplift of the Hudson Highlands			Retrogression of high grade mineral assemblages with cooling		Biotite blocking temperatures, 300°- 350°C (760 m.y.) Hornblende blocking temperatures, 500°- 550°C (900 m.y.)
Episode II	D_3	F_3	Open, upright, isoclinal folds with vertical axial surfaces; fold axes almost coaxial with F_2-folds; refolding of D_1 and D_2 fabric elements	Local axial plane foliation (S_3)	$M_{p\epsilon 2}$		Canada Hill granitization	Canada Hill granitization (913 m.y.)
							Emplacement of Canopus pluton	Crystallization of Canopus pluton (1068 m.y.)
	D_2	F_2	Regional isoclinal folds; refolding of D_1 fabric elements	Development of penetrative mineral lineation (L_2); prominent regional mineral lineation; local development of axial plane foliation (S_2); S_1 foliation preserved	Thermal maximum, 1000 m.y. $T = 725 \pm 25°C$, $P = 4 \pm 1$ kb.	Recrystallization and rotation of prismatic and platy minerals into F_2 axial planes and fold axes		
Episode I	D_1	F_1	Appressed, intrafolial, isoclinal folds; axial surfaces moderately to steeply inclined with variable orientation	Development of penetrative axial plane foliation (S_1) subparallel to primary lithologic layering (S_0); prominent regionally developed foliation in Precambrian rocks	Thermal maximum 1140 m.y. $M_{p\epsilon 1}$(?)	Recrystallization of metasedimentary and metavolcanic rocks, and Storm King granite	Emplacement of Storm King granite gneiss as sills and sheets to form the Bear Mtn. and West Point plutons	Recrystallization of paragneiss (1147 m.y.), leucogneiss (1169 m.y.) and Storm King granite gneiss (1118 m.y.)
Sequence of sedimentary and volcanic rocks; deposition and extrusion prior to 1170 my								

REFERENCES CITED

Aleinikoff, J. N., Grauch, R. I., Simmons, K. R., and Nutt, C. J., 1982, Chronology of metamorphic rocks associated with uranium occurrences, Hudson Highlands, New York-New Jersey (abs.): Geol. Soc. America Abs. with Programs 1982, v. 14, no. 1-2, p. 1.

Ashworth, J. R., 1976, Petrogenesis of migmatites in the Huntly-Portsoy area, north-east Scotland: Mineralog. Mag., v. 40, p. 661–682.

Baer, A. J., 1981, Two orogenies in the Grenville belt?: Nature, v. 290, p. 129–131.

Baker, D. R., and Buddington, A. F., 1970, Geology and magnetite deposits of the Franklin quadrangle and part of the Hamburg quadrangle, New Jersey: U.S. Geol. Survey Prof. Paper, 638, 73 p.

Berkey, C. P., and Rice, M., 1919, Geology of the West Point quadrangle, N.Y.: N.Y. State Mus. Bull., nos. 225-226, 152 p.

Brown, P. E., 1967, Major element composition of the Loch Coire migmatite complex, Sutherland, Scotland: Contr. Mineralogy Petrology, v. 14, p. 1–26.

Colony, R. J., 1921, The magnetite iron deposits of southeastern New York: N.Y. State Mus. Bull., nos. 249-250.

Compston, W., Chappell, B. W., Arriens, P. A., and Vernon, M. J., 1969, On the feasibility of NBS 70a K-feldspar as a Rb-Sr age reference sample: Geochimica et Cosmochimica Acta, v. 33, p. 753–757.

Dallmeyer, R. D., 1972, Precambrian structural history of the Hudson Highlands near Bear Mountain, New York: Geol. Soc. America Bull., v. 83, p. 895–904.

Dallmeyer, R. D., and Dodd, R. T., 1971, Distribution and significance of cordierite in paragneiss of the Hudson Highlands, southeastern New York: Contr. Mineralogy Petrology, v. 33, p. 289–308.

Dallmeyer, R. D., and Sutter, J. F., 1976, [40]Ar/[39]Ar incremental-release ages of biotite and hornblende from variably retrograded basement gneisses of the northeasternmost Reading Prong, New York: their bearing on Early Paleozoic metamorphic history: Amer. J. Sci., v. 276, p. 731–747.

Dallmeyer, R. D., Sutter, J. F., and Baker, D. J., 1975, Incremental [40]Ar/[39]Ar ages of biotite and hornblende from the northeastern Reading Prong: their bearing on Late Proterozoic thermal and tectonic history: Geol. Soc. America Bull., v. 86, p. 1435–1443.

DeLaeter, J. R., and Abercrombie, I. D., 1970, Mass spectrometric isotopic dilution analyses of rubidium and strontium in standard rocks: Earth and Planetary Science Letters, v. 9, p. 327–330.

Dodd, R. T., 1965, Precambrian geology of the Popolopen Lake quadrangle, southeastern New York: New York State Mus. and Sci. Serv., Map and Chart Ser., no. 6, 39 p.

Faure, G., and Powell, J. L., 1972, Strontium isotope geology: New York, Springer-Verlag, 188 p.

Hall, L. M., Helenek, H. L., Jackson, R. A., Caldwell, K. G., Mose D., and Murray, D. P., 1975, Some basement rocks from Bear Mountain to the Housatonic Highlands, in Ratcliffe, N.M., ed., New England Intercoll. Geol. Conf., 67th Ann. Mtg. Guidebook, New York, p. 1–29.

Hedge, C. E., 1972, Source of leucosomes of migmatites in the Front Range, Colorado: Geol. Soc. America Memoir, no. 135, p. 65–72.

Helenek, H. L., 1966, Stratigraphic and structural relationships in Precambrian gneisses of the Hudson Highlands, Bear Mountain, New York (abs.): Geol. Soc. America Abs. with Programs 1966, Northeastern Section, p. 24–25.

Helenek, H. L., 1971, An investigation of the origin, structure and metamorphic evolution of major rock units in the Hudson Highlands [Ph.D. thesis]: Providence Brown Univ., 244 p.

Hietanen, A., 1975, Generation of potassium-poor magmas in the northern Sierra Nevada and the Svecofennian of Finland: U.S. Geol. Survey Jour. Research, v. 3, p. 631–645.

Hotz, P. E., 1953, Magnetite deposits of the Sterling Lake, New York—Ringwood, New Jersey area: U.S. Geol. Survey Bull., 982-F, p. 153–244.

Jaffe, H. W., and Jaffe, E. B., 1973, Bedrock geology of the Monroe quadrangle, Orange County, New York: New York State Mus. and Sci. Serv., Map and Chart Ser., no. 20, 74 p.

Klemic, H., Eric, J. H., McNitt, J. R., and McKeown, F. A., 1959, Uranium in Phillips Mine-Camp Smith area, Putnam and Westchester Counties, New

York: U.S. Geol. Survey Bull., v. 1074-E, p. 165–197.

Lowe, K. E., 1950, Storm King granite at Bear Mountain, New York: Geol. Soc. America Bull., v. 61, p. 137–190.

Lowe, K. E., 1958, Pre-Cambrian and Paleozoic geology of the Hudson Highlands: New York State Geological Association, Guidebook, Peekskill, Trip D, p. 41–52.

Mehnert, K. R., 1968, Migmatites: New York, Elsevier, 393 p.

Misch, P., 1968, Plagioclase compositions and non-anatectic origin of migmatitic gneisses in northern Cascade Mountains of Washington state: Contr. Mineralogy Petrology, v. 17, p. 1–70.

Mose, D. G., 1982, 1,300 million-year-old rocks in the Appalachians: Geol. Soc. America Bull., v. 93, p. 391–399.

Murray, D. P., Mose, D. G., and Helenek, H. L., 1977, Chemical evolution of quartz-plagioclase gneisses in the Reading Prong, New York (abs.): Geol. Soc. America Abs. with Programs 1977, v. 9, no. 3, p. 303–304.

Myers, J. S., 1976, Granitoid sheets, thrusting, and Archean crustal thickening in West Greenland: Geology, v. 4, no. 5, p. 265–268.

Offield, T. W., 1967, Bedrock geology of the Goshen-Greenwood Lake area, N.Y.: New York State Mus. and Sci. Serv., Map and Chart Ser., no. 9, 78 p.

Olsen, S. N., 1977, Origin of the Baltimore Gneiss migmatites at Piney Creek, Maryland: Geol. Soc. America Bull., v. 88, p. 1089–1101.

Ratcliffe, N. M., Armstrong, R. L., Chai, B. H-T., and Senechal, R. G., 1972, K-Ar and Rb-Sr geochronology of the Canopus pluton, Hudson Highlands, New York: Geol. Soc. America Bull., v. 83, p. 523–530.

Sims, P. K., 1953, Geology of the Dover magnetite district, Morris County, New Jersey: U.S. Geol. Survey Bull., 982-G, p. 245–304.

Steiger, R. H., and Jager, E., 1977, Subcommission on geochronology: Convention on the use of decay constants in geo- and cosmochronology: Earth and Planetary Science Letters, v. 36, p. 359–362.

Tilton, G. R., Wetherill, G. W., Davis, G. L., and Bass, M. N., 1960, 1000-million year-old minerals from the eastern United States and Canada: Jour. Geophys. Research, v. 65, p. 4173–4179.

Tuttle, O. F., and Bowen, N. L., 1958, Origin of granite in the light of experimental studies in the system $NaAlSi_3O_8$-$KAlSi_3O_8$-SiO_2-H_2O: Geol. Soc. America Memoir, no. 74, 153 p.

v. Platen, H., 1965, Experimental anatexis and the genesis of migmatites, in Pitcher, W. S., and Flinn, G. W., Controls of metamorphism: New York, John Wiley and Sons, p. 203–218.

White, A. J.R., 1966, Genesis of migmatites from the Palmer region of South Australia: Chem. Geology: v. 1, p. 165–200.

Winkler, H.G.F., 1974, Petrogenesis of metamorphic rock (3rd ed.): New York, Springer-Verlag, 320 p.

Yardley, B.W.D., 1978, Genesis of the Skagit Gneiss migmatites, Washington, and the distinction between possible mechanisms of migmatization: Geol. Soc. America Bull., v. 89, p. 941–951.

York, D., 1966, Least-squares fitting of a straight line: Canadian Jour. Physics, v. 44, p. 1079–1089.

MANUSCRIPT ACCEPTED BY THE SOCIETY AUGUST 2, 1983

Printed in U.S.A.

Geological Society of America
Special Paper 194
1984

The Reading Prong of New Jersey and eastern Pennsylvania: An appraisal of rock relations and chemistry of a major Proterozoic terrane in the Appalachians

Avery Ala Drake, Jr.
U.S. Geological Survey
Reston, Virginia 22092

ABSTRACT

The Proterozoic Y terrane of the Reading Prong of eastern Pennsylvania and New Jersey consists of light-colored, sodic-rich rocks containing intercalated amphibolite, the Losee Metamorphic Suite, calcarerous and quartzofeldspathic metasedimentary rocks, the intrusive Hexenkopf Complex, the Byram Intrusive Suite, and quartz-poor monzonite, syenite, and related pyroxene granite.

The Losee consists of oligoclase-quartz gneiss and amphibolite that in places has been partly mobilized to form venite and albite oligoclase granite. Rocks of charnockitic affinity may be a partial melt of an amphbiolite-rich phase of the Losee. The Losee is thought to be metamorphosed quartz keratophyre and related sodic basalt. It is probably basement to the calcareous and quartzofeldspathic metasedimentary rocks.

The calcareous rocks are mostly marble, amphibolite, pyroxene gneiss, and epidote- and scapolite-bearing gneisses. They are interlayered with quartzofeldspathic gneiss of two general types: biotite-quartz-feldspar gneiss and potassic feldspar gneiss. At places, the potassic feldspar gneiss has melted and has formed small bodies and layers of potassium-rich granite. The quartzofeldspathic rocks are of continental margin type and are thought to be a clastic wedge containing layers and lenses of calcareous rocks. The source of the clastic material was probably a granitic terrane because of the large amount of potassic feldspar. Some volcanic material may be present in this sequence, but the evidence is equivocal.

The Hexenkopf Complex consists of severely altered mafic plutonic rock. It appears to lie beneath the Losee, and if so, is the oldest known rock in this part of the Reading Prong.

The Byram Intrusive Suite consists of hornblende granite and alaskite that form syntectonic and conformable sheets within the metamorphic rocks, as well as scattered small bodies of biotite granite that resulted from the granitization of biotite-quartz-feldspar gneiss. Byram leucosome forms arterites from both biotite-quartz feldspar gneiss and amphibolite. The Byram probably results from the anatectic melting of older rocks, but at the present level of erosion, there is no evidence that this has taken place. The Byram probably had its origin in the source terrane of the quartzofeldspathic gneiss.

The quartz-poor monzonite, syenite, and related pyroxene granite are not well understood. These rocks form syntectonic conformable sheets like the rocks of the Byram Intrusive Suite, with one exception, in which a sheet of quartz syenite appears to cut across the structure of hornblende granite. These rocks are also probably anatectic, perhaps originating at a lower level than the Byram. This place of origin is not certain,

75

however, because the relation of the quartz-poor rocks to the Byram is not really known.

The rocks in eastern Pennsylvania and New Jersey were metamorphosed in at least upper amphibolite facies, and most were probably metamorphosed in hornblende granulite facies. There is some evidence of polymetamorphism in northeasternmost New Jersey, but that concept needs further evaluation.

The metamorphic, intrusive, and deformational event in the Reading Prong can be dated at about 1 b.y. ago, so it is clearly Grenvillian. The rocks in the Reading Prong are very much like those of the Adirondacks and probably like those of the Honey Brook Upland. They have similarities to rocks of the Green Mountains, Berkshires, and basement massifs of western Connecticut. They are not at all like the rocks of Avondale-West Chester Massif, the Baltimore Gneiss, or the rocks of the northern and southern Blue Ridge.

In a few small areas, the Proterozoic Y rocks are overlain by a sequence of interlayered metasedimentary and metavolcanic rocks that are named the Chestnut Hill Formation. These rocks are at a lower metamorphic grade and are much less homogenized than the Proterozoic Y rocks and are thought to be of probable Proterozoic Z age.

INTRODUCTION

The Reading Prong is one of the major massifs of Proterozoic Y rocks in the Appalachian region of eastern North America (Fig. 1). It extends from Reading in eastern Pennsylvania to east of the Hudson River in New York. The Prong's eastern terminus is generally taken to mark the boundary between the central and northern Appalachians (Drake, 1980). The Proterozoic rocks of the Prong are bounded on the north by very lightly metamorphosed Cambrian and Ordovician sedimentary rocks of the Lehigh and Kittatinny valleys, and on the south by Mesozoic rocks of the Newark basin. The Proterozoic rocks were strongly involved with their Cambrian and Ordovician cover rocks in regional nappes during the Taconic orogeny. The nappes were later deformed by ramp and flat-type thrust faults and attendant folds during the Alleghanian orogeny (Drake, 1969, 1970, 1978, 1980; Drake and Lyttle, 1980; Faill and MacLachlan, 1980; MacLachlan, 1979a, b; MacLachlan and others, 1975). Also during the Alleghanian orogeny, Silurian and Devonian rocks were folded and faulted into the Prong to form the Green Pond syncline (GP on Fig. 2) (Bayley and others, 1914). Finally, the entire assemblage was cut by faults of Mesozoic and perhaps younger age (Bayley and others, 1914; Ratcliffe, 1980). Because these subjects have been covered extensively elsewhere, they will not be considered further here.

The Proterozoic rocks of the Reading Prong consist of intrusive plutonic rocks and attendant migmatites, quartzofeldspathic and calcareous metasedimentary rocks, sodium-rich rocks of problematic origin, hypersthene-bearing charnockitic rocks of equally problematic origin, and a minor proportion of distinctly younger metasedimentary and metavolcanic(?) rocks. Perhaps half this assemblage consists of plutonic intrusive and closely related rocks. Most of the intrusive rocks appear to belong to three suites (Drake, 1969; Drake and Lyttle, 1980): (1) hornblende granite and microperthite alaskite; (2) quartz-poor clinopyroxene-bearing rocks generally called syenite, and prob-

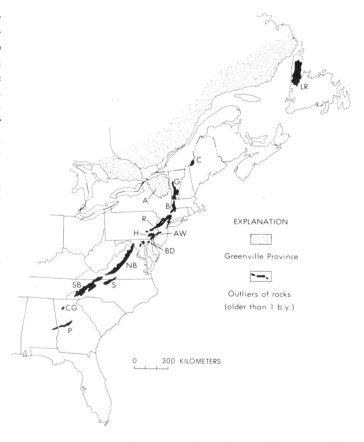

Figure 1. Distribution of rocks older than 1 b.y. years in eastern North America. Symbology: LR, Long Range; C, Chain Lakes Massif; A, Adirondack Mountains; G, Green Mountains; B, Berkshire Massif; R, Reading Prong; H, Honey Brook Upland; AW, Avondale-West Chester Massif; BD, Baltimore gneiss "domes"; NB, northern Blue Ridge; SB, southern Blue Ridge; S, Sauratown Mountains; CG, Corbin Granite; and P, Pine Mountain belt. Modified from Rankin (1976).

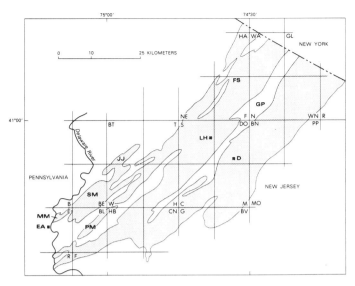

Figure 2. Distribution of Proterozoic Y rocks of the Reading Prong in New Jersey. Quadrangles (7½-minute) are: Riegelsville, R; Frenchtown, F; Easton, E; Bloomsbury, BL; High Bridge, HB; Califon, CN; Gladstone, G; Bernardsville, BV; Bangor, B; Belvidere, BE; Washington, W; Hackettstown, H; Chester, C; Mendham, M; Morrisville, MO; Blairstown, BT; Tranquility, T; Stanhope, S; Dover, DO; Boonton, BN; Pompton Plains, PP; Newton East, NE; Franklin, F; Newfoundland, N; Wanaque, WN; Ramsey, R; Hamburg, HA; Waywanda, WA; and Greenwood Lake, GL. Places mentioned in text: GP, Green Pond syncline; E, Easton; D, Dover; LH, Lake Hopatcong area; FS, Franklin Sterling area; JJ, Jenny Jump Mountain; SM, Scotts Mountain; MM, Marble Mountain; PM, Pohatcong Mountain.

ably related clinopyroxene-mesoperthite or microantiperthite granites (Baker and Buddington, 1970; Young, 1971, 1972, 1978); and (3) microantiperthite alaskite of complex origin.

Traditionally, the bulk of the rocks of the Prong were considered to be intrusive and were divided into Byram "granite gneiss" (feldspar dominantly potassic feldspar), Losee "diorite gneiss" (feldspar dominantly plagioclase), and Pochuck "gabbro gneiss" (mafic rocks), which were thought to grade into each other through intermediate forms (Spencer and others, 1908; Bayley, 1941). Metasedimentary units, however, were recognized, and include the Franklin Limestone (Spencer and others, 1908), the Pickering Gneiss near the Delaware River (Bayley, 1941), and the Moravian Heights Formation of Fraser (in Miller and others, 1939, 1941) in eastern Pennsylvania. Since the close of World War II, the rocks of the Prong have been the subject of many detailed investigations in eastern Pennsylvania, New Jersey, and immediately adjacent New York. The most important of these studies include those by Baker and Buddington (1970), Buckwalter (1959, 1962), Dodd (1965), Drake (1969), Hague and others (1956), Hotz (1952), Offield (1967), Sims (1958), Smith (1969), and Young (1971, 1972, 1978). Most of these workers abandoned the classic formations, which were found to contain rocks of different compositions and origins, in favor of more precise lithologic units. The rock units recognized in Pennsylvania and New Jersey also occur in New York, but only in

what is known as the western Hudson Highlands (Hall and others, 1975). This segment of the Reading Prong is separated by the Ramapo-Canopus fault zone from the eastern Hudson Highlands, a terrane of somewhat different rocks that have a different tectonic style and aeromagnetic signature (Hall and others, 1975). Because the rocks of the eastern Hudson Highlands differ from the rocks in the bulk of the Reading Prong, they are not considered here.

Since my Delaware Valley paper was published (Drake, 1969), I have been involved in or have supervised the detailed mapping of roughly an additional 500 km² of Reading Prong terrane and, more importantly, have obtained 51 new chemical analyses of Reading Prong rocks. Most of these analyses are of samples collected by R. I. Tilling and me as the beginning of a study of heat-producing elements in Reading Prong rocks and heat in the Prong. Unfortunately, this study has not progressed beyond the beginning stages because of other assignments. The bulk of the other samples were collected to support Kastelic's (1980) work in Warren County, New Jersey. The remainder were collected routinely to support my work in eastern Pennsylvania. The purpose of this paper is to attempt to summarize the character of the Proterozoic Y terrane of the Reading Prong. I believe that the effort is particularly important at this time, as the search for suspect terranes within the Appalachian orogen is on the upswing. Good data on the character of basement massifs are critical to such studies. In no sense can this paper be considered a study of igneous petrology as neither isotope, nor minor element work has been done. The chemical data presented here are used to characterize the rocks regionally and to help define what are thought to be major rock units of regional extent.

In this light, the name Byram is reintroduced and redefined as the Byram Intrusive Suite to include the hornblende granite and microperthite alaskite so common throughout the Reading Prong. The name Losee is reintroduced and redefined as the Losee Metamorphic Suite to include the oligoclase-quartz gneiss and albite-oligoclase granite, which are also common throughout the Prong. The rocks of the Losee appear to be unconformably beneath the metasedimentary gneisses and to be a basement to them. In addition, the new name Hexenkopf Complex is applied to a suite of unusual mafic rocks of probable oceanic origin, which lie beneath the Losee in eastern Pennsylvania, and the new name Chestnut Hill Formation, to a low-grade sequence of metasedimentary and metavolcanic(?) rocks along the Delaware River.

Chemical data are presented in the tables from southwest to northeast along the strike of the Prong. Modal data for the rocks, with the exception of amphibolite, pyroxene gneiss, and biotite-quartz-feldspar gneiss, are presented in triangular diagrams to avoid tables as far as possible.

I am indebted to John Aaron, Bob Davis, Bob Kastelic, Atilla Kilinc, Peter Lyttle, Karl Seifert, and Val Zadnik who worked with me at one time or another in the Reading Prong. Thanks are also due Lyttle and Eric Force, U.S. Geological Survey, for reviews of an earlier version of this paper. Paul Sims instructed me in the field study of Precambrian rocks, gave me advice in my beginning study of the Reading Prong, and quite recently discussed with me the possible origin of the Losee Metamorphic Suite. I am grateful for all these efforts.

METASEDIMENTARY ROCKS

All modern workers in the Reading Prong have recognized that the abundant and widely distributed metasedimentary rocks therein include both calcareous and quartzofeldspathic (aluminosiliceous) rocks and that these rocks are interlayered. These rocks would appear to be more abundant in easternmost Pennsylvania and New Jersey than in the southwestern part of the Prong in the Reading area. However, this may be more apparent than real, because of the extremely poor exposure in that area. Buckwalter (1959, 1962) was forced to map broad areas of mixed rock, as sequences of metasedimentary rocks could not be readily separated from granitic sheets. In other areas, strongly migmatized metasedimentary rocks are lumped with the intrusive rocks, giving a false impression of their relative abundance.

Calcareous Rocks

Rocks though to be metamorphosed calcareous sedimentary rocks include calcite and dolomite marble, pyroxene gneiss, pyroxenite (as used by Kastelic, 1980), skarn, quartz-epidote gneiss, epidote-scapolite-quartz gneiss, and lime-silicate gneiss. These calcareous rocks are interlayered with the quartzofeldspathic rocks and clearly pre-date the emplacement of the various intrusive sheets.

Marble. Calcite and lesser dolomite marble is a typical rock unit within the Proterozoic Y terrane of the Reading Prong, although quantitatively it is not especially important. It is most abundant in the Franklin-Sterling area, New Jersey (Fig. 2), where it is the host rock for the famous zinc deposits. Here, the marble was named the Franklin Limestone (Wolff and Brooks, 1898). Most subsequent workers mapped all the marble in the Reading Prong, and even elsewhere within the central Appalachians, as Franklin Limestone. I choose not to use Franklin as a stratigraphic name, as it has no real stratigraphic meaning.

Marble, or at least nonsilicated marble, is most abundant along the northwest margin of the Reading Prong. The prominent marble belt of the Franklin-Sterling area can be traced northwest along strike to Big Island, New York (Offield, 1967), and southwest to near Andover in the northwest corner of the Stanhope quadrangle (Fig. 2). Thick layers are also present along the northwest margin of the Prong in Jenny Jump, Scotts, and Marble mountains (Fig. 2). Along strike to the southwest in Pennsylvania are prominent layers of marble at Chestnut Hill and the Pine Top-Camelhump massif (Fig. 3).

Only small bodies of marble have been recognized elsewhere within the Reading Prong. Sims (1958) reported several small bodies within the Dover district (Fig. 2). Drake (1967b) mapped a small body of marble in Pohatcong Mountain in the

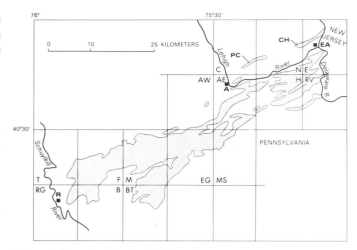

Figure 3. Map showing distribution of Proterozoic Y rocks of the Reading Prong in eastern Pennsylvania. Quadrangles (7½-minute) are: Easton, EQ; Nazareth, N; Catasauqua, C; Riegelsville, RV; Hellertown, H; Allentown East, AE; Allentown West, AW; Milford Square, MS; East Greenville, EG; Manatawny, M; Fleetwood, F; Temple, T; Boyertown, BT; Birdsboro, B; and Reading, RG. Towns: Easton, E; Allentown, A; and Reading, R. Places mentioned in text: CH, Chestnut Hill; PC, Pine Top-Camelhump massif; RH, Rattlesnake Hill.

Bloomsbury quadrangle (Fig. 2) and (1969) reported layers of nonsilicated marble associated with pyroxene gneiss in Rattlesnake Hill in the Riegelsville quadrangle (Fig. 3). The only other marble reported from the Prong is a small lens near the contact with rocks of the Newark basin in the northeastern part of the Boyertown quadrangle (Fig. 3) (Buckwalter, 1959).

Much of the marble is a coarse- to locally fine-grained calcite rock that contains fewer dolomitic layers. Most marble contains some disseminated graphite, and where silicated, a little chondrodite, scapolite, phlogopite, clinopyroxene, tremolite, serpentine, and sulfides. Locally, the marble contains lenses of quartz-rich gneiss, which is probably metasandstone.

In contrast to the above, is the largely dolomitic marble that crops out on Marble Mountain (Fig. 2) and Chestnut Hill (Fig. 3). At these places, the bulk of the unit has been altered to serpentine rock, talc rock, tremolite rock, or more rarely, skarn. These rock types are end members, and virtually any mineralogic combination can be found. Abundant quarries in this area formerly exploited these altered rocks for both facing stone and talc.

The marble is clearly interlayered with other metasedimentary rocks. In the Franklin-Sterling area, a lower marble layer of the Franklin Limestone overlies an interlayered sequence of potassic feldspar gneiss, biotite-quartz-plagioclase gneiss, and amphibolite; internally, the marble contains an interval of quartzofeldspathic and quartz-rich gneisses (Hague and others, 1956). The Franklin layer is overlain by a thick sequence of quartzofeldspathic gneiss and amphibolite, which is in turn overlain by a layer of the Wildcat Marble of Hague and others (1956), which lithologically cannot be distinguished from the Franklin layer. The Wildcat layer is overlain by a thick sequence of interlayered quartzofeldspathic gneiss and amphibolite (Hague and others, 1956).

To the southwest in western New Jersey and eastern Pennsylvania, the marble bodies are within sequences of potassic feldspar gneiss, biotite-quartz-plagioclase gneiss, and amphibolite. The small body in the Dover district (Fig. 2) is an anomaly, in that it is between a large body of charnockitic hypersthene-bearing gneiss and a sheet of albite-oligoclase granite (Sims, 1958). More typically, the bodies in Pohatcong Mountain (Fig. 2) and Rattlesnake Hill (Fig. 3) are associated with potassic feldspar gneiss and amphibolite, and the small body in the Boyertown quadrangle (Fig. 3) is in a mixed terrane of graphitic quartzofeldspathic gneiss and amphibolite (Buckwalter, 1959). Throughout the Prong, therefore, marble is interlayered with both other calcareous and quartzofeldspathic metasedimentary rocks.

Pyroxene Gneiss. Pyroxene gneiss, although not abundant, is present throughout the Reading Prong in New Jersey. In Pennsylvania it has only been recognized in Rattlesnake Hill (Fig. 3). The gneiss, therefore, is like most other calcareous rocks and is apparently quite sparse in the southwestern part of the Reading Prong.

Pyroxene gneiss is quite heterogeneous (Table 1). There appears to be three end-member types: clinopyroxene-plagioclase rocks, clinopyroxene-potassic feldspar rocks, and clinopyroxene-scapolite rocks. The gneiss is interlayered or closely associated with marble, other clearly calcareous rocks, and amphibolite. All modern workers in the Reading Prong, therefore, agree that the unit results from the metamorphism of carbonate-bearing metasedimentary rocks. However, metasomatism probably has had a hand in the formation of some pyroxene gneiss bodies.

Epidote-Scapolite-Quartz Gneiss. Epidote-scapolite-quartz gneiss is reported only from the Franklin and Hamburg quadrangles (Fig. 2) by Baker and Buddington (1970). This unit is somewhat of a catchall for calcium-rich gneisses characterized by epidote and(or) scapolite. Some varieties are so quartz-rich as to be epidote- and scapolite-bearing metaquartzites. The analyzed sample (Table 2, sample 6) is from a quartz-rich phase of the unit.

The epidote-scapolite-quartz gneiss forms two major layers in the Franklin and Hamburg quadrangles (Fig. 2) as well as layers within an important mixed gneiss unit in the Edison district (Baker and Buddington, 1970). Most of the rocks in this mixed

TABLE 2. CHEMICAL ANALYSES AND C.I.P.W. NORMS OF BIOTITE-QUARTZ-FELDSPAR AND EPIDOTE-SCAPOLITE-QUARTZ GNEISS

	1	2	3	4	5	6
	Chemical Analyses (Weight Percent)					
SiO_2	75.7	71.2	68.0	72.1	72.5	52.5
Al_2O_3	13.3	13.3	14.5	14.4	13.4	14.6
Fe_2O_3	0.51	3.3	3.8	0.8	2.8	10.6
FeO	0.40	1.8	2.1	3.8	1.4	6.7
MgO	0.30	0.4	0.91	1.1	0.08	1.7
CaO	1.3	1.0	1.4	1.2	0.82	4.9
Na_2O	4.8	4.2	4.7	2.8	3.4	3.4
K_2O	2.5	3.4	3.5	3.8	4.7	1.3
H_2O+	0.62	0.45	0.42	0.6	0.66	0.45
H_2O-	0.11	0.02	0.15	--	0.07	0.05
TiO_2	0.26	0.28	0.72	0.37	0.28	2.7
P_2O_5	0.07	0.04	0.22	0.08	0.04	1.1
MnO	0.01	0.05	0.04	0.15	0.05	0.6
CO_2	0.01	0.17	0.01	0.08	0.01	0.08
Total	99.9	99.6	100.5	101.3	100.2	100.7
	C.I.P.W. Norms					
Quartz	36.0	31.8	23.3	34.6	33.3	17.3
Orthoclase	15.0	20.0	20.6	22.2	27.8	7.8
Albite	40.4	35.6	39.8	23.6	28.8	28.8
Anorthite	5.6	5.0	6.1	5.0	3.9	17.0
Corundum	1.7	0.9	0.7	3.8	1.3	1.2
Hypersthene	0.8	1.0	2.3	8.6	0.2	4.3
Magnetite	0.5	4.9	4.6	1.2	3.5	15.3
Ilmenite	0.6	0.6	1.4	0.8	0.6	5.2
Hematite	0.2	--	0.6	--	0.5	--
Apatite	0.3	--	0.3	0.3	--	2.7
Total	101.1	99.8	99.7	100.1	99.9	99.6

1. Quartz-microcline-plagioclase-biotite gneiss from outcrops just east of unnumbered county road about 200 m (airline) southwest of Oley Furnace, Fleetwood 7 1/2-minute quadrangle, Pa. Rapid-rock analyses by Hezekiah Smith, U.S. Geological Survey.
2. Quartz-plagioclase-microcline-biotite gneiss from outcrops along Lower Belvidere Road on east bank of Delaware River about 0.5 km north of its intersection with Marble Mountain Road. Analyst same as 1.
3. Quartz-microcline-plagioclase-hornblende gneiss from outcrops on lowest north slope of Scotts Mountain about 0.25 km west of Hazen Cemetery. Rapid-rock analysis by K. Coates and H. Smith, U. S. Geological Survey.
4. Garnetiferous quartz-feldspar-biotite gneiss from outcrops along secondary road about 0.7 km north of west end of Splitrock Reservoir, Boonton 7 1/2-minute quadrangle, N.J. Analyst same as 1.
5. Biotite-quartz-feldspar gneiss from outcrops along road on west shore of Tamarack Lake about 0.5 km south of the village of Summit Lake, Franklin 7 1/2-minute quadrangle, N.J. Analyst same as 1.
6. Epidote-scarpolite-quartz gneiss from outcrops along road on west shore of Tamarack Lake about 1 km south of the village of Summit Lake, Franklin 7 1/2-minute quadrangle, N.J. Analyst same as 1.

TABLE 1. MODES (VOLUME PERCENT) OF PYROXENE GNEISS*

	1	2	3	4	5	6	7	8
Clinopyroxene	38.0	27.5	36.0	41.0	10.8	24.0	4.6	25.9
Hornblende	--	--	--	--	--	--	2.8	--	tr.
Plagioclase	55.5	--	57.0	--	0.4	69.7	57.7	--
K-Feldspar	--	67.5	2.0	--	--	--	1.4	32.8
Biotite	0.5	--	--	--	--	--	--	--
Sphene	2.0	0.5	2.0	6.5	--	0.9	1.2	1.3
Quartz	1.5	3.5	0.5	--	--	--	33.9	tr.
Epidote	1.5	0.5	1.0	--	--	--	--	--
Magnetite	0.5	1.5	0.5	--	1.3	1.3	1.0	--
Apatite	0.5	--	1.0	0.5	4.9	1.3	0.2	--
Scapolite	--	--	--	52.0	82.6	--	--	40.0

*Note: 1-2, from Drake (1969, p. 58); 3-4, from Sims (1958, p. 15); 5-7, from Baker and Buddington (1970, p. 7); 8, from Haque and others (1956, p. 445).

gneiss unit are potassic feldspar- and iron-rich. In the Edison district, the epidote-scapolite-quartz gneiss appears to grade into the potassic feldspar gneiss, suggesting a facies change within a sequence of bedded sedimentary rocks (Baker and Buddington, 1970).

Volumetrically, the epidote-scapolite-quartz gneiss is not important in the Reading Prong. Scientifically and economically, however, it is of great importance as it is associated with major magnetite deposits, and along with parts of the potassic feldspar gneiss, may represent metamorphosed iron-formation (note that sample 6, Table 2, contains more than 17 percent iron oxides). These rock types certainly must have been a major, perhaps the only, source of iron for many of the Highlands magnetite deposits.

Quartz-Epidote Gneiss. A layer of gneiss composed primarily of quartz, epidote and chlorite was mapped by Kastelic (1980) in the Washington quadrangle (Fig. 2). The unit contains varying but minor amounts of plagioclase and calcite. On weathered surfaces, it has a characteristic layering on the scale of a few millimeters that may be relict bedding. Like the epidote-scapolite gneiss, it is associated with a major body of potassic feldspar gneiss and probably has a facies relation with that rock type. The quartz-epidote gneiss is the host rock for the Pequest magnetite mine (Kastelic, 1980).

In retrospect, I am aware that I have seen this rock type southwest of the occurrence in the Washington quadrangle. Unfortunately, I included it in my unit of undivided metasedimentary and metavolcanic rock rather than in the potassic feldspar gneiss where it more logically belongs. I would suggest that calcareous quartz-bearing rocks may be more common in the Reading Prong than has been believed and that it might be useful to look for them in the areas of known magnetite deposits.

Skarn and Pyroxenite. Light- to dark-green, medium- to coarse-grained, massive equigranular rocks composed almost entirely of diopside, salite, or ferrosalite have been mapped at many places in the New Jersey Highlands segment of the Reading Prong (Drake, 1967b; Hotz, 1952; Sims, 1958). These rocks have been termed "pyroxene skarn" and are thought to result from the metasomatism of marble and other calcareous rocks by the addition of silica and iron and the removal of calcium and carbon dioxide (Hotz, 1952; Sims, 1958). The source of the metasomatic fluids is thought to be the magma or the Bryam Intrusive Suite. All mapped bodies are small, and most are related to magnetite deposits.

Similar but darker rocks are composed primarily of hornblende with lesser amounts of pyroxene, biotite, apatite, and calcite. In addition to composition, these rocks differ from pyroxene skarn in that they are layered and foliated. Sims (1958) believes that hornblende skarn represents the last stage of metasomatic alternation of calcareous rocks, but he considers that these rocks may be amphibolite altered by magnetite ore-forming fluids derived from Bryam magma.

Kastelic (1980) mapped rocks identical with those called "pyroxene skarn" in the Washington quadrangle (Fig. 2). How-

ever, he prefers the name "pyroxenite" (Table 3) for these rocks, because he does not believe that the rocks were formed by the contact metasomatism of marble or other calcareous rock, as he found little detailed chemical or mineralogical similarity between these pyroxene-rich rocks and undisputed unmetamorphosed pyrometasomatic rocks associated with young intrusives. He believes, therefore, that the pyroxenite formed from impure siliceous dolomite by dedolomitization reactions during regional metamorphism.

These complex rocks are not abundant in the Reading Prong, nor are they of any particular importance to the regional aspects of that Proterozoic Y terrane. Their origin is, however, of extreme importance to an understanding of the origin of the magnetite deposits of that terrane, and I hope that some geologist will be inspired to study these pyroxene- and hornblende-rich rocks in detail.

Lime-silicate gneiss. A small body of calcareous gneiss composed of hornblende, plagioclase, calcite, chlorite, epidote,

TABLE 3. CHEMICAL ANALYSES AND C.I.P.W. NORM OF PYROXENITE*

Chemical Analysis (Weight Percent)	
SiO_2	52.9
Al_2O_3	2.5
Fe_2O_3	3.8
FeO	7.8
MgO	11.3
CaO	19.9
Na_2O	1.1
K_2O	0.29
H_2O+	0.19
H_2O-	0.11
T_2O_2	0.05
P_2O_5	0.02
MnO	0.48
CO_2	0.02
Total	100.5

C.I.P.W. Norm	
Quartz	3.7
Orthoclase	1.7
Anorthite	1.0
Diopside	75.9
Hypersthene	4.8
Jadite	7.2
Magnetite	5.5
Calcite	0.1
Ilmenite	0.1
Water	0.3
Total	100.3

*Note: Sample from dump of Washington magnetite mine, Washington quadrangle, N.J. (from Kastelic, 1980).

and diopside was mapped by Buckwalter (1959) in the northeastern part of the Boyertown quadrangle (Fig. 3). This rock is directly on strike with a small body of marble (see above) and appears to grade into amphibolite. The entire calcareous assemblage is within a terrane of graphitic quartzofeldspathic gneiss. The rock relations in these southwesternmost exposures is exactly the same as in New Jersey, where there are far more calcareous rocks.

Quartzofeldspathic gneiss

Quartzofeldspathic gneiss has been mapped throughout the Reading Prong. In the southwest, such rocks are sparse, but Buckwalter (1959, 1962) recognized two varieties of such rocks which he called "graphitic gneiss" and "quartz-biotite-feldspar gneiss." To the east, in the Delaware Valley area, these rocks are much more abundant, particularly in Scotts and Pohatcong Mountains (Fig. 2). Here, Drake (1969) mapped biotite-quartz-plagioclase gneiss, potassic feldspar gneiss, and sillimanite-bearing gneiss. In the Dover district in New Jersey, Sims (1958) recognized biotite-quartz-plagioclase gneiss, garnetiferous biotite-quartz-feldspar gneiss, sillimanitic garnetiferous biotite-quartz-feldspar gneiss, and graphitic biotite-quartz-feldspar gneiss, although these rocks are not abundant. All these varieties were mapped as biotite-quartz-feldspar gneiss. In the Lake Hopatcong area, quartzofeldspathic rocks are sparse, but Young (1971) mapped a unit of biotite-feldspar-quartz gneiss, which, judging from his modes, contains both potassic feldspar and plagioclase dominant types. In the central and northeastern New Jersey Highlands, Smith (1969) mapped a unit of quartz-feldspar-biotite gneiss, although such rocks are uncommon. In the Franklin and Hamburg quadrangles (Fig. 2), Baker and Buddington (1970) mapped abundant quartz-potassium feldspar, biotite-quartz-feldspar, and quartz-microcline gneisses. Hague and others (1956) mapped a unit of microcline gneiss in that part of the northeastern New Jersey Highlands that they studied. These quartzofeldspathic rocks are much more abundant there than in other areas to the south in New Jersey. In nearby areas in New York, Hotz (1952) mapped sparse quantities of garnetiferous quartz-biotite gness; Offield mapped biotite-quartz-feldspar gneiss, quartz-microcline gneiss, and graphitic quartz-plagioclase gneiss; and finally, Dodd (1965) mapped rusty biotite-quartz-feldspar gneiss and gray biotite-quartz-feldspar gneiss. At one place or another, all these rocks contain some graphite, garnet, sillimanite, magnetite, and sulfide. Many of these rocks could be considered Pickering Gneiss as used by Bayley (1914, 1941). I choose not to use that formational name for the quartzofeldspathic gneiss because it is interlayered with marble that occurs in different stratigraphic positions (see above). Neither rock type has real stratigraphic meaning, and neither should be considered to constitute a formation.

It is my impression that the quartzofeldspathic gneiss consists of two rock types. One type contains substantially more plagioclase than potassic feldspar, although a few specimens

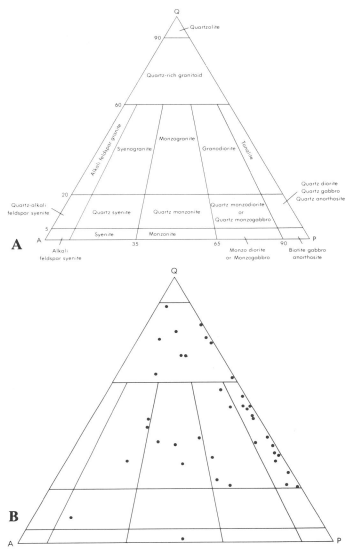

Figure 4. Q.A.P. (Streckeisen, 1976) plots. A. Q-A-P (quartz-alkali feldspar-plagioclase) plot showing rock names. B. Q-A-P plot of biotite-quartz-feldspar gneiss. Data from: Baker and Buddington (1970), Drake (1969), Hague and others (1956), Sims (1958), Smith (1969), and Young (1971).

contain roughly equal amounts of the two feldspars (Fig. 4). I call these rocks biotite-quartz-feldspar gneiss.

The other rock type contains potassic feldspar far in excess of plagioclase and does not show the variations in modal composition typical of the biotite-quartz-feldspar gneiss (Fig. 5). I call these rocks potassic feldspar gneiss. Each rock type appears to contain interlayers of the other, but I believe that they are valid units and so mapped them in the Delaware and Lehigh valleys (Drake, 1967a, b, and unpublished data; Drake and others, 1967, 1969). Such rock types were also mapped separately by Baker and Buddington (1970).

Biotite-Quartz-Feldspar Gneiss. Biotite-quartz-feldspar gneiss is a highly variable unit in both composition (Table 4) and

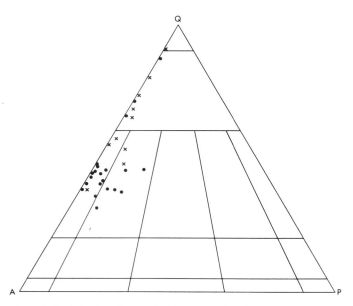

Figure 5. Q-A-P plot of potassic feldspar gneiss. Data from Baker and Buddington (1970), Drake (1969), Hague and others (1956), and Kastelic (1980). Sillimanite-bearing samples, X; analysis 4, Table 4.

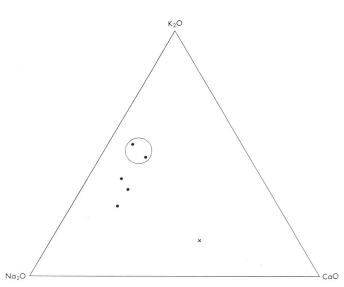

Figure 6. K_2O-Na_2O-CaO (weight percent) plot of biotite-quartz-feldspar and epidote-scapolite-quartz gneiss. Biotite-quartz-feldspar gneiss, (.); epidote-scapolite-quartz gneiss, (x).

texture but is characterized by conspicuous biotite and prominent compositional layering. It occurs throughout the Reading Prong of eastern Pennsylvania, New Jersey, and adjacent New York. How much of the graphitic gneiss mapped by Buckwalter (1959, 1962) in the southwestern part of the Prong is biotite-quartz-feldspar gneiss is uncertain. Buckwalter (1959) reported that graphitic gneiss in the Boyertown quadrangle (Fig. 3) contains 20-50 percent potassic feldspar and 10-35 percent plagioclase. Although his major quartzofeldspathic unit is called "graphitic gneiss," graphite is absent in most exposures (Buckwalter, 1959, 1962). This unit probably includes both biotite-quartz-feldspar and potassic feldspar gneiss as used herein. In any case, biotite-quartz-feldspar gneiss (analysis 1, Table 2) has been mapped in the Fleetwood quadrangle (Fig. 3).

The unit differs in composition from place to place and from layer to layer, reflecting differences in composition of original beds. At many places, the gneiss contains small lenticular bodies of amphibolite, pyroxene gneiss, skarn, or marble. Much of the biotite-quartz plagioclase gneiss contains layers or veins of alaskite or pegmatite along the original metasedimentary layering, or, much more rarely, in crosscutting veins. The leucosome in mineralogy and texture is identical with the alaskitic and pegmatitic rocks of the Byram Intrusive Suite. Contacts between the biotite-quartz-feldspar gneiss and rocks of the Byram Suite are rarely sharp, because the gneiss grades into granite by a gradual decrease in biotite and plagioclase and a corresponding increase in microperthite and in places, hornblende.

I believe that these migmatic rocks are arterites that result from the injection and permeation of Byram magma and magmatic fluids. In this, I follow the original work of Fenner (1914) and the more recent work of Sims (1958), Buckwalter (1959,

1962), and Baker and Buddington (1970). This interpretation is made because of the field evidence presented above. Although an anatectic origin for the leucosome of the migmatites is intellectually appealing, it is doubtful because of the heterogeneity of the gneiss, the homogeneity of the leucosome, the absence of granitic bodies or blotches within the gneiss, and the absence of enrichment of mafic minerals along the contacts of biotite gneiss and leucosome.

All modern workers have considered the biotite-quartz-feldspar gneiss, although migmatized, to be a metamorphosed sedimentary rock, most likely a graywacke containing aluminous and carbonaceous beds. Chemical analyses of rocks from the unit (Table 2 and Fig. 6) are not too different from those of some graywackes (Pettijohn and others, 1973, p. 200). The biotite-quartz-feldspar gneiss does differ from graywacke in the dominance of CaO over MgO, and in two samples, the dominance of K_2O over Na_2O. The chemistry of the gneiss, however, is much less like that of other sandstones reported by Pettijohn and others (1973). Perhaps the chemical differences result from modification during metamorphism and migmatization. In any case, graywacke-like sand is the most likely protolith for the unit. The sillimanite-bearing layers probably represent more aluminous interbeds and the graphite-bearing layers, carbonaceous interbeds.

Potassic Feldspar Gneiss. Potassic feldspar gneiss has been mapped in the Franklin and Hamburg quadrangles (Baker and Buddington, 1970), the Delaware Valley (Drake, 1967a, b; Drake and others, 1969), the Franklin-Sterling area (Hague and others, 1956), the Belvedere and Washington quadrangles (Kastelic, 1980) and adjoining New York (Offield, 1967). This rock type also probably constitutes at least part of the graphitic gneiss (see above) mapped in the southwestern part of the Reading

TABLE 4.   MODES (VOLUME PERCENT) OF BIOTITE-QUARTZ-FELDSPAR GNEISS

	1	2	3	4	5	6	7	8	9	10	11
Plagioclase	5.5	43.0	40.0	57.5	75.0	17.5	28.5	16.6	43.0	15.4	17.4
K-feldspar	3.0	1.5	--	--	--	25.0	6.0	11.2	1.5	11.9	2.0
Quartz	14.0	37.5	48.5	27.5	tr.	25.0	47.0	63.6	40.0	63.5	65.6
Biotite	21.5	6.5	10.0	11.5	19.0	1.5	6.0	2.0	5.0	8.4	7.6
Garnet	10.0	6.5	--	0.5	5.0	27.0	6.5	--	7.0	0.5	3.4
Sillimanite	--	--	--	--	--	2.0	--	--	--	--	3.4
Magnetite	0.5	2.5	0.5	2.5	--	--	--	--	3.5	--	0.5
Pyrite	0.5	0.5	--	--	tr.	1.0	tr.	1.1	--	--	--
Rutile	--	--	--	--	0.5	1.0	0.5	--	--	--	--
Zircon	--	--	--	--	tr.	tr.	--	--	--	tr.	--
Chlorite	--	0.5	0.5	0.5	--	--	--	--	--	--	--
Epidote	--	--	--	--	tr.	tr.	0.5	--	--	--	--
Apatite	--	--	--	--	tr.	--	0.5	1.1	--	tr.	--
Graphite	--	1.5	0.5	--	--	--	1.5	4.4	--	--	--
Muscovite	--	--	--	--	--	tr.	3.0	--	--	--	--

1. Garnetiferous biotite-quartz-feldspar gneiss from Marble Mountain, Bangor 7 1/2-minute quadrangle (Drake, 1969).
2. Garnetiferous biotite-quartz-feldspar gneiss from Chestnut Hill, Easton 7 1/2-minute quadrangle (Drake, 1969).
3. Biotite-quartz-plagioclase gneiss from Morgan Hill, Easton 7 1/2-minute quadrangle (Drake, 1969).
4. Biotite-quartz-plagioclase gneiss from Musconetcong Mountain, Frenchtown 7 1/2-minute quadrangle (Drake, 1969).
5. Garnetiferous biotite-plagioclase gneiss from Splitrock Resevoir area, Boonton 7 1/2-minute quadrangle (Sims, 1958).
6. Sillimanitic biotite-quartz-feldspar gneiss from White Meadow Lake area, Dover 7 1/2-minute quadrangle (Sims, 1958).
7. Graphitic biotite-quartz-feldspar gneiss from Cooks Lake area, Boonton 7 1/2-minute quadrangle (Sims, 1958).
8. Graphitic biotite-quartz-feldspar gneiss from near Russia, Franklin 7 1/2-minute quadrangle (Baker and Buddington, 1970).
9. Garnetiferous biotite-quartz-feldspar gneiss from near Newark Reservoir, Franklin 7 1/2-minute quadrangle (Baker and Buddington, 1970).
10. Garnetiferous biotite-quartz-feldspar gneiss from Lake Hopatcong area (Young, 1971).
11. Garnetiferous biotite-quartz-feldspar gneiss from Lake Hopatcong area (Young, 1971).

Prong (Buckwalter, 1959, 1962) and part of the quartzofeldspathic gneiss mapped by Hotz (1952) in New York and Young (1971) in the Lake Hopatcong area. The unit is characterized by its high content of potassic feldspar and quartz and by its paucity of plagioclase (Fig. 5). Much of the unit is quite heterogeneous and contains varying proportions of microcline, microperthite, and quartz; some layers are so quartz-rich that they approach quartzite in composition. Other parts of the unit are quite homogeneous, are granofels rather than gneiss, and are very granitic in appearance. Sillimanite, biotite, and almandine are sporadic, even in the most granitic-appearing phases. Magnetite is very abundant in many places, especially in the Edison area in the Franklin quadrangle and the Oxford area in the Washington quadrangle (Fig. 2), where the unit resembles metamorphosed siliceous iron-formation. In some places, the unit contains layers of biotite-quartz-feldspar gneiss, and in the Franklin-Hamburg area (Fig. 2), it has a facies relation with epidote-scapolite-quartz gneiss (see above).

Baker and Buddington (1970) recognized two types of potassic feldspar gneiss: quartz-potassium feldspar gneiss (heterogeneous) and quartz-microcline gneiss (homogeneous), which they mapped separately. I also recognize both end members, but where I have worked, these rocks appear to occur within the same outcrop belt.

The heterogeneous nature, generally quartz-rich composition, and interlayering and interlensing relation of much of the potassic feldspar gneiss with rocks of sedimentary affinity suggest a possible sedimentary origin. Chemically (Table 4), the unit is similar to some arkoses (Pettijohn and others, 1973, p. 179), although the potassic feldspar gneiss has higher K_2O-Na_2O ratios. The four available chemical analyses (Table 5) show that the potassic feldspar gneiss contains 37.9 to 63 percent normative feldspar; such high values are typical of arkosic rocks.

In the Delaware Valley (Drake, 1969), the more granitic-appearing parts of the potassic feldspar gneiss form small sheets and occur as veins, lenses, and blotches within rocks of apparent

TABLE 5.   CHEMICAL ANALYSES AND C.I.P.W. NORMS
            OF POTASSIC FELDSPAR GNEISS

	1	2	3	4
	Chemical Analysis (weight percent)			
SiO_2	73.3	76.6	79.5	71.0
Al_2O_3	11.6	12.1	9.3	13.0
Fe_2O_3	3.3	1.9	0.46	3.0
FeO	1.8	1.2	0.44	1.3
MgO	0.25	0.31	0.13	0.32
CaO	0.30	0.06	0.34	0.07
Na_2O	1.8	0.51	0.14	0.95
K_2O	6.2	5.5	7.5	9.3
H_2O+	0.45	1.1	0.60	0.74
H_2O-	0.07	0.04	0.07	0.04
TiO_2	0.24	0.19	0.17	0.33
P_2O_5	0.02	0.03	0.05	0.06
MnO	0.01	0.02	0.02	0.03
CO_2	0.02	0.01	0.02	--
Total	99.4	99.6	98.7	100.1
C.I.P.W. norms				
Quartz	38.6	52.7	50.3	29.3
Orthoclase	37.2	33.4	45.0	55.5
Albite	15.2	4.2	1.1	8.4
Anorthite	1.4	0.3	1.7	--
Corundum	1.4	5.3	0.4	1.0
Hypersthene	0.7	1.1	0.4	0.8
Magnetite	4.9	2.8	0.7	3.2
Ilmenite	0.5	0.5	0.3	0.6
Hematite	--	--	--	0.8
Total	99.9	100.3	99.9	99.7

1. From old roadcut on north side of U.S. Route 309 about 0.14 km west of interchange with Rock Road, Allentown East 7 1/2-minute quadrangle, Pa. Rapid-rock analyses by Hezekiah Smith, U. S. Geological Survey.
2. From large outcrops on west bank of Delaware River about 1.7 km northeast of Easton water filtration plant. Analyst same as 1.
3. From outcrops on north slope of Danville Mountain on Lake Just It Road, just north of boundary between Washington and Blairstown 7 1/2-minute quadrangle, N.J. Rapid-rock analyses by Z. A. Hamlin, U. S. Geological Survey.
4. From outcrops at S.E. abutment of dam, Intersol Reservoir, Scotts Mountain, Bloomsburg 7 1/2-minute quadrangle, N.J. From Drake (1969).

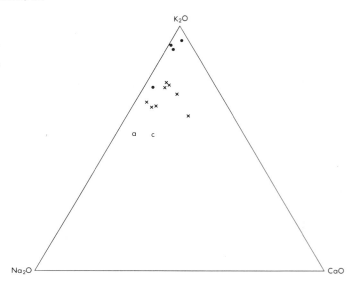

Figure 7. K_2O-N_2O-C_aO (weight percent) plot comparing the chemistry of potassic feldspar gneiss, (.), with that of microcline granite gneiss (x) from the Adirondack Mountains (Buddington, 1957; Buddington and Leonard, 1962), the average of 72 analyses of calc-alkali granites (c) (Nockolds, 1954), and average of 48 analyses of alkali granites (Nockolds, 1954), (a).

the quartz-microcline gneiss and supported this interpretation by pointing out that its rather bizarre chemistry is similar to that of potassium-rich Tertiary vitric tuff from the Great Plains (Swineford and others, 1955). This tuff contains 3.08 to 6.07 percent K_2O and from 0.93 to 3.03 percent Na_2O. The K_2O-Na_2O ratios of 14 samples of this tuff range from 3 to 7, whereas those of the potassic feldspar gneiss range from 3.4 to 54. The K_2O-Na_2O ratio of the most granitic-appearing sample (Table 5, analysis 1) is within or close to the range of ratios in the vitric tuffs. Although not suggested by Baker and Buddington (1970), it is possible that the potassic feldspar gneiss may result from the metamorphism of mixed arkosic sediment and potassium-rich tuff.

Buddington (1957) and Buddington and Leonard (1962) have described microcline granite gneiss from the Adirondacks that is quite similar to the potassic feldspar gneiss of the Reading Prong (Fig. 7). The potassic feldspar gneiss, however, is even more potassium-rich than the Adirondack rocks, which are believed to be, in part, metasomatized biotite-quartz-plagioclase gneiss and partly magmatic. In the Reading Prong, there is no evidence of potassium metasomatism in the biotite-quartz-feldspar gneiss, nor is there any evidence for the emplacement of bodies of potassium-rich granite in any unit other than the potassic feldspar gneiss. I believe, therefore, that the potassic feldspar gneiss results from the metamorphism of arkosic sediments, perhaps containing some siliceous iron-formation and potassium-rich tuff, which have undergone partial anatexis. If part of the graphitic gneiss at the southwestern end of the Reading Prong is potassic feldspar gneiss, the unit is also migmatized by alaskites of the Byram Intrusive Suite, and, as has been pointed out above, that area is a megamigmatite.

meta-sedimentary aspect. Field evidence and chemical similarity, and the fact that the granitic-appearing (analyses 1, Table 5) rocks are only slightly less siliceous than the layered rocks, suggest to me that the more granitic phases of the potassic feldspar gneiss were produced within the unit by partial anatexis forming a venitic migmatite. Baker and Buddington (1970) reported similar relations within their quartz potassium feldspar gneiss, which contains seams and layers of quartz-microcline gneiss that also occurs in mappable bodies. They inferred a magmatic origin for

A

B

Figure 8. Photographs of rocks of Losee Metamorphic Suite in Bloomsbury quadrangle, New Jersey. A, oligoclase-quartz gneiss interlayered with amphibolite. B, oligoclase-quartz gneiss containing a layer of amphibolite and veins of anatectic albite-oligoclase granite.

LOSEE METAMORPHIC SUITE (HERE REINTRODUCED, RENAMED, AND REDEFINED)

Rocks consisting largely of sodic plagioclase and quartz and characterized by an extremely low color index were named Losee Gneiss by Spencer and others (1908) for exposures near Losee Pond (now Beaver Lake) in the Franklin 7½-minute quadrangle, Sussex County, New Jersey. The unit was generally considered to be a metamorphosed intrusive diorite. Most modern workers found that as originally mapped (Spencer and others, 1908; Darton and others, 1908; Bayley and others, 1914; Bayley, 1941), the Losee includes quartzofeldspathic gneiss and rocks of charnockitic affinity as well as the typical sodic gneisses. For that reason, these workers abandoned the term Losee in favor of lithologic units. Losee as originally defined, however, is a distinctive rock that varies little, other than in the ratio of quartz to sodic plagioclase, amount of interlayered amphibolite, and in how granitic it appears. The unit is very leucocratic and rarely contains more than 5 percent of mafic minerals. Biotite is most common, but locally, the rock contains minor amounts of hornblende, magnetite, augite, or hypersthene. The mafic minerals are commonly altered to chlorite and epidote. Parts of the unit are well layered and foliated (Fig. 8), others are poorly foliated granofels, and still others are granitoid or pegmatitic in aspect. All these rock types have very high Na_2O-K_2O ratios, very high normative albite, and differ petrographically only in the relative amount of sodic plagioclase and quartz (Drake, 1969; Table 6; Fig. 9, Fig. 10); I believe that they constitute a petrochemical suite. For this reason,

I reintroduce, rename, and redefine the name Losee as the Losee Metamorphic Suite to include oligoclase-quartz gneiss (and similar units as mapped by other geologists), albite-oligoclase granite (a term long established in Reading Prong usage), and albite pegmatite.

Rocks of the Losee Metamorphic Suite are abundant throughout New Jersey and adjacent New York but have not been recognized west of the Allentown East quadrangle (Fig. 3) in Pennsylvania. Only Sims (1958) and I (1969) have mapped both gneissic and granitoid phases of this sodic suite. Hague and others (1956) and Hotz (1952), however, have presented evidence that the unit was mobile, at least in part, in their areas and consider the Losee to be an intrusive rock. Sims considers the gneissic phase of the unit to be a metamorphosed graywacke because of its marked layering. Young (1971) believes the unit to be a metamorphosed quartz keratophyre, and Dodd (1965) believed it to be a metamorphosed sodic volcanic rock. Other workers are noncommittal. I (1969) suggested that the gneissic part of the Losee was a quartz keratophyre or sodic tuff. I still prefer a quartz keratophyre origin, but it is virtually impossible to prove.

I believe that graywacke is an unlikely protolith for the unit. Rocks of the Losee are compared with the most sodium-rich graywacke I could find in the literature in Fig. 9. The graywackes approach the normative albite content of the Losee rocks, although none fall in their field. Graywacke, particularly very sodic graywacke, characteristically occurs in turbidite deposits typically containing at least as much shale as sandstone in the total sedi-

TABLE 6. CHEMICAL ANALYSES AND C.I.P.W. NORMS
OF ROCKS OF THE LOSEE METAMORPHIC SUITE

	1	2	3
	Chemical Analyses (weight percent)		
SiO_2	76.2	65.4	64.1
Al_2O_3	12.3	16.0	17.6
Fe_2O_3	2.8	3.6	1.3
FeO	1.2	1.8	2.1
MgO	0.06	1.3	1.2
CaO	0.04	2.0	5.8
Na_2O	6.5	7.1	4.8
K_2O_3	0.20	1.0	0.88
H_2O+	0.22	0.61	0.66
H_2O-	0.03	0.01	0.06
TiO_2	0.23	0.68	0.33
P_2O_5	0.04	0.07	0.23
MnO	--	0.01	0.05
CO_2	0.01	0.15	0.08
Total	99.8	99.7	99.2
	C.I.P.W. norms		
Quartz	38.0	14.6	19.8
Orthoclase	1.1	6.1	5.6
Albite	55.0	60.8	41.4
Anorthite	--	8.9	24.5
Corundum	1.5	0.1	--
Wollastonite	--	--	1.2
Diopside	--	0.2	1.4
Hypersthene	0.2	3.1	2.7
Magnetite	3.2	3.7	1.9
Ilmenite	0.5	1.4	0.6
Titanite	--	--	--
Hematite	0.6	1.1	--
Apatite	--	0.3	0.3
Total	100.1	100.3	99.4

1. Oligoclase-quartz gneiss from quarry just north of Reading Railroad tracks 0.45 km east of west border of Allentown East 7 1/2-minute quadrangle, Pa. Rapid-rock analysis by Hezekiah Smith, U. S. Geological Survey.
2. Oligoclase-quartz gneiss from outcrops along unnumbered county road about 0.2 km west of Meriden, Boonton 7 1/2-minute quadrangle, N.J. Analyst same as 1.
3. Quartz-oligoclase gneiss from outcrops along N.J. Route 23 about 0.5 km southwest of Beaver Lake village, Franklin 7 1/2-minute quadrangle, N.J. Sample is from type outcrop belt of the "Losee Diorite Gneiss" of Spencer and others (1908). Analyst same as 1.

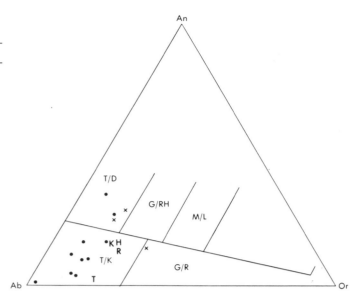

Figure 9. Normative feldspar plot (O'Connor, 1965) comparing rocks of the Losee Metamorphic Suite with soda-rich graywackes (K, "Kulm" graywacke; T, "Tanner" graywacke; R, Rensselaer graywacke; H, average of 17 analyses of "Harz" graywacke; and X, and charnockitic rocks. Plutonic/volcanic rock fields are: T/K, trondhjemite/quartz keratophyre; G/R, granite/rhyolite; T/D, tonalite/dacite; G/RH, granodiorite/rhyodacite; and M/L, quartz monzonite/quartz latite. Graywacke data from Pettijohn and others, 1973, and Blatt and others, 1972. Losee data from this study, Drake (1969), Bayley and others (1914), Sims (1958), and Young (1971). Data on rocks of charnockitic affinity from this study and Baker and Buddington (1970).

ment package. Even the thickest sequence of amalgamated graywacke beds I have seen, the proximal graywacke of the Proterozoic X Sabará Formation in the Quadrilátero Ferrífero of Brazil, contains about 15 percent shale (Drake and Morgan, 1980). I have never seen any report of a rock remotely resembling a metapelite associated with the Losee.

Hotz (1952) found that, in detail, contacts of Losee with metasedimentary rocks are discordant, although on a large scale, they are generally conformable. This gives some support for an intrusive origin for the Losee. The contact zones are, at most places, transitional, and the rock is migmatitic. At some places, the Losee contains folded inclusions of pyroxene amphibolite. Clearly, some of the unit in his area has been mobile.

Hague and others (1956) found that at most places in the Franklin-Sterling area the contacts of Losee with other rocks are conformable, although small-scale discordances are found in some places. Many contacts are gradational, but the unit contains blocks of amphibolite, some of which have remanent folds. Hague and others have interpreted the Losee as a phacolithic magmatic intrusion.

Sims (1958) found no direct evidence for an intrusive origin for his oligoclase-quartz-biotite gneiss (layered Losee). He did, however, map small bodies of albite-oligoclase granite, which he believed to be a remobilized phase of the gneiss.

In the Delaware Valley, I (1969) recognized both layered and foliated granofels as well as more massive granitic-appearing phases of the Losee. The granitic-appearing rocks were mapped as albite-oligoclase granite, following Sims (1958). These rocks contain the discontinuous layers and remanent folds recognized by Hotz (1952) and Hague and others (1956). This phase of the Losee shows evidence of mobility. Mapping of the layered and granitic phases of the sodic rock, in retrospect, was rather arbi-

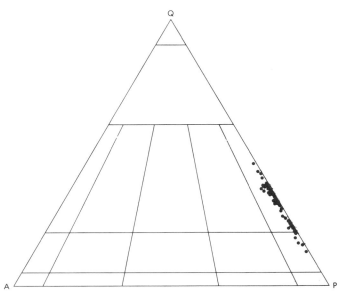

Figure 10. Q-A-P plot of rocks of the Losee Metamorphic Suite. Data from this study, Drake (1969), Kastelic (1980), Hague and others (1956), Sims (1958), Baker and Buddington (1970), and Young (1971).

Figure 11. Layered oligoclase-quartz gneiss grading into albite-oligoclase granite and albite pegmatite. Interstate 78 cut through Musconetcong Mountain, Bloomsbury quadrangle.

trary, as the phases grade imperceptibly into each other. Extremely poor exposure aided and abetted this arbitrary separation.

In any case, at least part of the Losee can be shown to have been mobile. If the Losee, or any part of it, is an intrusive rock, the bulk of it is trondhjemite, that is, a tonalite (Fig. 10) containing less than 10 percent mafic minerals and having oligoclase or andesine as the feldspar (Streckeisen, 1976). Only five of the available modes contain so little quartz as to be quartz diorite (Fig. 10). Chemically, the analyzed rocks of the Losee span the trondhjemite-tonalite fields (Fig. 9). Of the available analyses, the sample from the type body of Losee (Table 8, analysis 3) contains by far the most CaO. This sample and one from the Riegelsville quadrangle (Fig. 3) are the two tonalites of Fig. 9.

However, the bulk of the petrographic and chemical data presented above pertains to oligoclase-quartz gneiss (of my usage) and not to rocks having intrusive characteristics. No real mineralogic or chemical differences exist between the phases sampled, and these phases all imperceptibly grade into each other as well as into pegmatite phases, as seen in Fig. 11, where oligoclase-quartz gneiss (at the left of the hammer) fades into albite-oligoclase granite (at lower left and lower right parts of the photograph) and into albite pegmatite (at the top of the photograph). In these rocks, small, very sodic hornblendes in the oligoclase-quartz gneiss have been recrystallized into large crystals and clusters of crystals in the pegmatite.

I agree with Sims (1958) that the albite-oligoclase granite is an anatectic phase of the oligoclase-quartz gneiss, but differ in that I interpret the anatectite as having been produced essentially at the current level of erosion, rather than at a greater depth. Leucosome can be best seen in those parts of the Losee that contain abundant interlayered amphibolite, as in the Bloomsbury quadrangle, New Jersey (Fig. 8), and in the Franklin-Sterling area (Fig. 2). At these places, discontinuous layers, lenses, veins, and knots of leucosome are apparent in the amphibolite. A typical feature of the Losee containing interlayered amphibolite is the transition of amphibolite through lighter colored amphibolite to leucogneiss and finally to more granitic-appearing rock. Many geologists have described this feature as having the appearance of replacement. Petrographic study, however, discloses no replacement textures. Actually, the relation results from the production of an anatectic leucosome. Typically, the mafic layers have coarse rims of hornblende or biotite reflecting remanent surplus mafic minerals. This rim does not form when leucosome is added to a mafic unit.

If the argument is accepted that the granitic phase of the Losee results from the anatexis of oligocene-quartz gneiss, one is still faced with the problem of finding a protolith for that rock. I find the best candidate to be a volcanic rock in general and quartz keratophyre in particular. The field relations, the layered nature, the interlayered amphibolite, and the relative homogeneity support the volcanic origin, and the chemistry almost demands the quartz keratophyre interpretation. In this model, the Losee is a metamorphosed volcanic pile of quartz keratophyre and interlayered basalt or basaltic tuff. Conceivably, the pile could have contained some intrusions of trondhjemite. After its evolution as a metamorphic complex, the Losee was partially mobilized to form the albite-oligoclase granite phase.

The most important aspect of the Losee, however, is its

extremely sodic nature. These rocks clearly originated in an oceanic domain, unlike the other rocks of the Reading Prong. This oceanic nature would hold true even if the Losee were metagraywacke.

ROCKS OF UNCERTAIN ORIGIN

Most of the rocks in the Reading Prong have had complex histories, many of which can be worked out to a reasonable level of satisfaction. Some rocks, however, may have more than one protolith or may have origins so complex that they can only be interpreted by analogy. Two such units are treated herein: amphibolite and rocks of charnockitic affinity.

Amphibolite

Amphibolite as used here includes both amphibolite and pyroxene amphibolite. These rocks are composed largely of hornblende, andesine, clinopyroxene and, less commonly, smaller amounts of biotite and orthopyroxene. Typical modes are given in Table 7. Amphibolite, by one name or another, has been mapped throughout the Reading Prong. In the early days of geologic work in New Jersey, amphibolite was named Pochuck Gneiss (Wolff and Brooks, 1898) for Pochuck Mountain in the Hamburg quadrangle (Fig. 2). The Pochuck was thought to be a gabbro gneiss of intrusive origin. Bayley (1941), in fieldwork done in 1906, divided the Pochuck as originally defined into an older sequence of dark gneiss of metasedimentary origin and a younger sequence of intrusive origin. He assigned the metasedimentary sequence to his Pickering Gneiss (see above) and retained the name Pochuck for what he believed to be igneous

differentiates of Byram and Losee magma. As mapped, this unit included the Hexenkopf Complex as named here. Fraser (in Miller and others, 1939, 1941) retained the name Pochuck for the amphibolite in the eastern Pennsylvania Reading Prong. The Pochuck as mapped by Fraser also includes rocks of the Hexenkopf Complex named here. More recently, Buckwalter (1959) mapped amphibolites as Pochuck Gneiss in the Boyertown quadrangle (Fig. 3). Farther southwest, however, he (1962) mapped similar rocks as hornblende gneiss. All other modern workers, with the exception of Hague and others (1956), who used hornblende gneiss, have mapped these rocks as amphibolite and have abandoned the term Pochuck because the amphibolites probably have different protoliths and occur in different stratigraphic positions. The name Pochuck, thereby, has neither stratigraphic nor petrologic meaning.

Amphibolite (Table 8) is associated with all other rock units in the Reading Prong, and its origin (or origins) is (are) difficult to decipher because of the lack of primary features. Much of the amphibolite is layered, whereas some is essentially massive. Relations are further muddled by extensive migmatization. Much of the amphibolite, like the biotite-quartz-feldspar gneiss, contains layers and lenses of alaskite or pegmatite leucosome and is an arteritic migmatite. Migmatization has been especially severe in the southwestern part of the Prong where Buckwalter (1959, 1962) has mapped separate units on the basis of the amount of contained leucosome. Amphibolite was probably a widespread unit in this area because of the abundant amphibolite migmatite. Buckwalter (1959, 1962) believes that the bulk of the amphibolite in his area is of metasedimentary origin as it is interlayered and intergradational with quartzofeldspathic gneiss and marble that are obviously metasedimentary rocks. In the Boyertown

TABLE 7. MODES (VOLUME PERCENT) OF AMPHIBOLITE*

	1	2	3	4	5	6	7	8	9	10
Plagioclase	31.0	62.0	46.0	60.0	64.0	25.0	55.1	29.9	42.2	36.3
Hornblende	21.0	28.0	28.0	28.0	8.0	55.4	40.3	58.7	41.2	45.1
Clinopyroxene	15.5	4.0	9.5	2.0	11.0	10.4	--	--	7.5	--
Orthopyroxene	--	1.0	4.0	--	12.0	3.0	--	--	--	--
Biotite	19.0	--	11.0	1.0	--	--	--	0.6	tr.	5.9
Magnetite	1.5	4.0	0.5	4.0	4.0	3.2	1.0	1.9	5.8	2.1
Epidote	4.2	1.5	0.5	2.0	--	--	2.9	--	--	--
Apatite	tr.	1.0	0.5	1.0	1.0	tr.	0.7	0.5	tr.	tr.
Chlorite	tr.	1.5	--	1.0	--	--	--	--	--	2.0
Sphene	tr.	--	--	--	--	--	--	--	3.3	tr.
Calcite	1.0	1.0	--	1.0	--	--	--	--	--	--
Scapolite	5.7	--	--	--	--	--	--	--	--	--
Quartz	--	--	--	--	--	3.1	--	8.5	--	8.6

*Note: 1-3, from Drake (1969, p. 57); 4-5, from Sims (1958, p. 16); 6-7, from Baker and Buddington (1970, p. 19); 8, from Hague and others (1956, p. 445); 9-10, from Kastelic (1980, p. 132). The chemistry of samples 5 and 7 are reported in analyses 5 and 3 in Table 8.

TABLE 8. CHEMICAL ANALYSES (WEIGHT PERCENT) OF AMPHIBOLITE

	1	2	3	4	5	6	7	8	9	10	11
SiO_2	45.7	47.1	47.1	46.7	53.2	48.4	51.3	56.8	68.3	43.98	42.80
Al_2O_3	15.2	16.0	15.6	13.5	14.4	16.7	12.3	15.3	11.6	12.01	12.03
Fe_2O_3	2.2	0.70	2.9	7.6	2.9	7.3	3.8	3.2	2.4	6.60	6.56
FeO	11.6	11.6	8.9	6.8	8.4	3.6	6.0	3.7	2.6	12.20	13.25
MgO	6.5	5.6	8.1	9.2	5.6	6.6	7.6	4.8	2.9	5.46	4.67
CaO	9.6	10.0	9.4	8.0	7.7	8.8	13.0	4.8	4.6	11.99	11.76
Na_2O	2.7	2.9	2.6	3.1	3.4	3.3	1.7	3.3	2.3	2.93	2.91
K_2O	1.7	0.51	1.7	0.9	1.6	1.8	2.6	5.03	2.8	1.10	1.13
H_2O+	1.1	0.77	0.94	1.3	1.5	1.9	1.4	1.3	1.3	1.33	1.13
H_2O-	0.06	0.01	0.06	0.04	0.13	0.13	0.17	0.19	0.20		
TiO_2	1.5	3.5	1.3	1.4	1.5	1.0	0.46	0.86	0.35	2.25	2.24
P_2O_5	0.64	1.3	0.22	0.38	0.27	0.28	0.25	0.18	0.09	0.28	0.17
MnO	0.24	0.23	0.14	0.18	0.24	0.14	0.25	0.16	0.10	0.05	0.05
CO_2	0.08	0.04	0.08	0.03	0.06	0.07	0.44	0.28	0.99	0.18	0.20
Total	98.8	100.3	99.0	99.1	100.9	100.0	101.3	99.9	100.5	100.36	98.9

1. Amphibolite migmatite paleosome from outcrops along Antietam Creek Road about 300 m south of Antietam Reservoir, Birdsboro 7 1/2-minute quadrangle, Pa. Rapid-rock analysis by Hezekiah Smith, U. S. Geological Survey.
2. Amphibolite from outcrops along Antietam Creek Road directly opposite Antietam Reservoir, Birdsboro 7 1/2-minute quadrangle, Pa. Map unit contains about 20-30 percent interlayered alaskite. Analyst same as 1.
3. Small unmapped amphibolite body along unnumbered county road that traverses Musconetcong Mountain south from Finesville, N.J. Outcrop is 3.2 airline km southeast of Finesville and about 100 m south of Byram sample 20 (see Table 9), Riegelsville 7 1/2-minute quadrangle. Analyst same as 1.
4. From outcrops along Belvidere Road on east bank of Delaware River about 1.2 km south of intersection with Marble Mountain Road, Easton 7 1/2-minute quadrangle, N.J.-Pa. Rock erroneously shown on Easton quadrangle map (Drake, 1967a) as potassic feldspar gneiss. Analyst same as 1.
5. Black amphibolite from Oxford Stone quarry about 1.4 airline km S.20 E. from Bridgeville, Belvidere 7 1/2-minute quadrangle, N.J. Rapid-rock analysis by K. Coates and H. Smith, U. S. Geological Survey.
6. Fine-grained, grayish-green amphibolite from Lommasons Glen, Belvidere 7 1/2-minute quadrangle, N.J. Analyst same as 5.
7. Fine-grained, gray amphibolite from outcrops about 0.7 km S.40 E. from Lommasons Glen, Belvidere 7 1/2-minute quadrangle, N.J. Sample taken about 4 m from alaskitic pegmatite sheet. Analyst same as 5.
8. Fine-grained, gray amphibolite from same location as sample 7, but taken 1 m from alaskitic pegmatite sheet. Analyst same as 5.
9. Fine-grained, gray amphibolite paleozome from migmatite from same location as sample 7 but taken at contact with alaskitic pegmatite sheet. Analyst same as 5.
10. From Pardee Mine, Newfoundland 7 1/2-minute quadrangle, N.J. (Bayley and others, 1914).
11. From Greenwood Lake 7 1/2-minute quadrangle, N.J. (Darton and others, 1908)

quadrangle, Buckwalter (1959) thought that some amphibolite might be metagabbro because it was only weakly foliated. In a later report, Buckwalter (1962) found that his metagabbro was gradational into typical well-foliated amphibolite and that it might well be of metasedimentary origin as well.

In the Delaware Valley (Drake, 1969), amphibolite is both associated and interlayered with calcareous and quartzofeldspathic metasedimentary rocks as well as with oligoclase-quartz gneiss of the Losee Metamorphic Suite. Small to moderate-sized bodies occur within rocks of the Byram Intrusive Suite, and migmatite terranes have been mapped in many places. Much of this amphibolite is thought to be of metasedimentary origin because it is interlayered with metasedimentary rocks and some contains relict calcite (Table 7). Some of these rocks could be metamorphosed flows or mafic tuffs, but direct evidence is lacking, and such rocks are not suggested by rock distribution or map pattern.

Some of the amphibolite in the Delaware Valley, however, is thought to be metavolcanic or metavolcaniclastic. This amphibolite (mode 5, Table 7, and analysis 5, Table 8) is interlayered with oligoclase-quartz gneiss of the Losee Metamorphic Suite. This relation is considered at greater length under the discussion of that unit.

Analysis 3 (Table 8) is from a small body within hornblende granite of the Byram Intrusive Suite, and analysis 4 is from a layer within a body of potassic feldspar gneiss. Both these samples are higher in MgO than are those from the southwest end of the Prong. Analysis 4 is particularly high, probably reflecting the dolomitic nature of the marble (see above) and impure marble of the Marble Mountain-Chestnut Hill area.

Farther northeast in the Belvidere and Washington quadrangles, New Jersey (Fig. 2), Kastelic (1980) found layers and lenses of amphibolite within biotite-quartz-feldspar gneiss, oligoclase-quartz gneiss of the Losee Metamorphic Suite, and the intrusive rocks of the Byram Intrusive Suite. All these bodies were too small to map at a scale of 1:24,000, except for a layer along the northwest edge of Scotts Mountain in the Belvidere quadrangle. Arteritic migmatites have formed in this area, but not in bodies large enough to map.

Analysis 5 (Table 8) is of amphibolite interlayered with oligoclase-quartz gneiss of the Losee Metamorphic Suite, an extremely sodium-rich unit (see above). This sample is more sodic and siliceous and contains less MgO and CaO than the other unaltered amphibolites from which analyses are available. Unfortunately, analyses of other amphibolites interlayered with oligiolase-quartz gneiss are not available at this time. In any case, this amphibolite is thought to have a volcanic or volcaniclastic origin.

Analysis 6 (Table 8) is from an unmapped body of amphibolite within a layer of biotite-quartz-feldspar gneiss in close juxtiposition to a sheet of microantiperthite alaskite. The relatively high content of Na_2O and K_2O probably results from alteration related to the emplacement of this alaskite sheet.

Analyses 7, 8 and 9 (Table 8) are of samples collected by Kastelic (1980) from an amphibolite layer near a small intrusive sheet of alaskitic pegmatite; they demonstrate the alteration of the unit by the permeation of granitic material. Sample 7 was collected 4 m from the contact, sample 8 was collected 1 m from the contact, and sample 9 was collected at the contact. SiO_2 increases as the contact is approached, total Fe decreases, MgO decreases, CaO decreases, Na_2O increases, K_2O increases, TiO_2 decreases, P_2O_5 decreases, and MnO decreases. Permeation processes as shown above have probably been important in the formation of the arteritic migmatite terranes of the Reading Prong.

Farther northeast in the Dover district, Sims (1958) found that amphibolite was widely distributed both as inclusions within intrusive rocks and as discrete layers within and associated with metasedimentary rocks. Most of these bodies were too small to map conveniently at a scale of 1:31,680. Arteritic migmatites are present in the area but are not mapped separately. Sims (1958) believes that the field evidence in the area suggests that most of the amphibolites there result from the metamorphism of calcareous or magnesium sedimentary rocks but that some might be metamorphosed mafic igneous rock.

Analysis 8 (Table 8) is of an amphibolite body that is the host for a magnetite deposit in the Dover district. This amphibolite is interlayered with altered calcareous rocks as well as biotite-quartz-feldspar gneiss and is similar to other amphibolites thought to be metasedimentary rocks. An interesting feature of this mine is that the ore contains a layer of magnetite-chalcedony rock (Bayley, 1910). Does this represent metamorphosed cherty iron-formation?

North of the Dover district, in the Lake Hopatcong area (Fig. 2), Young (1971) mapped amphibolite bodies within large sheets of both hornblende granite and clinopyroxene granite and noted that amphibolite is interlayered with oligoclase-quartz gneiss. He interpreted these rocks as a metamorphosed equivalent of an originally volcanic association.

In the Franklin-Sterling area (Fig. 2) to the northeast, four types of amphibolite, mapped as hornblende gneiss, have been recognized (Hague and others, 1956). One type (mode 8, Table 7) is characterized by alternating light and dark layers, contains thin layers of biotite gneiss, and appears to grade into biotite gneiss. This variety of amphibolite contains features that have been considered to be possible pillows (Hague and others, 1956). A second type of amphibolite differs little from the above, other than in fabric. A third type of amphibolite contains lenses of marble and lenses and layers of pyroxene gneiss. This amphibolite is identical with the second type of amphibolite described above except for the intercalated calcareous rocks. A fourth type of amphibolite is an augen gneiss and has discordant contacts with quartzofeldspathic gneiss. As usual with this rock type, a protolith is difficult to determine. Some would appear to be the result of metamorphism of calcareous sedimentary rocks. The pillow structures, if real, suggest that some of the rock results from the metamorphism of basaltic flows, yet the bulk of that rock type is identical with rocks of apparent metasedimentary origin. Perhaps, as suggested by Hague and others (1956) this amphibolite originated by the interfingering of calcareous sediments and basaltic flows. The augen gneiss is probably metaintrusive because of its discordant contacts.

In the Franklin and Hamburg quadrangles, which include the Franklin-Sterling area discussed above, Baker and Buddington (1970) found small and varying amounts of amphibolite within all metasedimentary units as well as within the intrusive rocks. They recognize the amphibolite types of Hague and others (1956) and equivocated as to their origins. They have, however, presented excellent evidence that some amphibolite, at least, is metagabbro. This amphibolite contains differentiated anorthositic layers, both the gabbro and anorthosite having primary interlocking textures.

In nearby New York, Hotz (1952) found no evidence for a metaigneous origin for the amphibolites he studied and favored a metasedimentary protolith. Offield (1967) found little evidence in his area upon which to base an origin for the amphibolite, but he pointed out that the amphibolite is interlayered with other rock units, which suggest a metasedimentary origin. Analysis 11 (Table 8) is from his area and differs little from analysis 10, which, because of the very low SiO_2 content, is thought to be of a metasedimentary rock. Finally, Dodd (1965) favors a metavolcanic origin for the bulk of the amphibolite in his area but believes that part might represent sills rather than flows. The bulk of the amphibolite in his area occurs with quartz-plagioclase (oligoclase-quartz) gneiss and rocks of charnockitic affinity.

In an attempt to determine the origin of the amphibolites in New Jersey, Kastelic (1980) studied the sparse available minor-element chemistry for these rocks in the Reading Prong. In these data, he found a negative correlation between Niggli Mg and Cr and Ni, high contents of Ba, and La-Ce ratios greater than 0.4. If the concepts of van de Kamp (1968) are used, this suggests a sedimentary parentage for the bulk of the amphibolite.

To sum up, evidence is abundant that a great deal of the amphibolite in the Reading Prong results from the metamorphism of sedimentary rocks and that at least some probably results from the metamorphism of mafic intrusive rocks. The amphibolites within the Losee Metamorphic Suite are thought to be metamorphosed volcanic rocks. Other amphibolites may be metavolcanic rocks, but the evidence is equivocal.

Rocks of charnockitic affinity

Even more enigmatic than the Losee Metamorphic Suite are the hypersthene-bearing rocks of charnockitic affinity. These rocks, composed of oligoclase or andesine, quartz, hypersthene, and lesser and varying amounts of augite, hornblende, and biotite, are similar in all respects, including the characteristic greasy luster, to charnockites from all over the world. They are quite abundant in the Dover district, New Jersey (Sims, 1958), and to the north in the Lake Hopatcong (Young, 1961) and Franklin-Sterling (Baker and Buddington, 1970) areas. Smaller bodies were mapped in the Delaware Valley (Drake, 1967a, b; Drake and others, 1967, 1969) and at the southwestern end of the

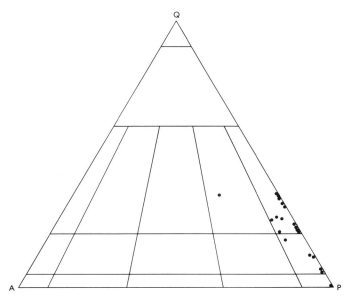

Figure 12. Q-A-P plot of rocks of charnockitic affinity. Data from Drake (1969), Baker and Buddington (1970), Sims (1958), and Young (1971).

Reading Prong (Buckwalter, 1959, 1962). In adjacent New York, abundant hypersthene-bearing rocks were mapped by Hotz (1952), Dodd (1965), and Offield (1967).

The charnockitic rocks are granitic-appearing, but at most places are characterized by conspicuous alterating light and dark layers as well as by discontinuous bodies of amphibolite and pyroxene amphibolite. The origin of rocks of this type is a difficult problem throughout the world, and no unanimity of opinion exists on the origin of these rocks in the Reading Prong. Petrographically, the bulk of the rocks are tonalites, some are quartz diorites, and one sample each of granodiorite and diorite were counted (Fig. 12). In charnockite terminology, most of these rocks are enderbites. They differ only slightly from rocks of the Losee Metamorphic Suite (Fig. 10) by containing more mafic minerals and, in some specimens, more potassic feldspar.

The charnockitic rocks (Table 9) are also chemically similar to rocks of the Losee Metamorphic Suite (Table 8), differing primarily in their higher content of normative orthoclase, anorthite, and pyroxene. Two of the charnockitic rocks are tonalites, and one barely falls in the granite field (Fig. 9).

Spatially, the charnockitic rocks contain amphibolite layers and are closely related to rocks of the Losee Metamorphic Suite. In the southwestern part of the Reading Prong (Buckwalter, 1959, 1962), the charnockitic rocks appear to grade into amphibolite. The charnockitic rocks form remanent lenses within granite of the Byram Intrusive Suite and have been migmatized by that unit. In the Delaware Valley (Drake, 1967a, b; Drake and others, 1967, 1969), the bulk of the charnockitic rocks form remanent lenses within rocks of the Byram Intrusive Suite, but they are spatially associated with amphibolite and with rocks of the Losee Metamorphic Suite and with venitic migmatites of the Losee. On Musconetcong Mountain in the Bloomsbury quadran-

TABLE 9. CHEMICAL ANALYSES AND C.I.P.W.
NORMS OF GNEISSES OF CHARNOCKITIC AFFINITY

	1	2	3
	Chemical Analyses (weight percent)		
SiO_2	62.7	68.3	67.69
Al_2O_3	16.8	15.5	15.99
Fe_2O_3	1.5	1.0	0.64
FeO	3.5	3.3	2.42
MgO	2.4	1.7	1.16
CaO	4.1	3.8	2.88
Na_2O	5.0	3.9	4.64
K_2O	1.7	1.9	2.86
H_2O+	0.92	0.55	0.35
H_2O-	0.08	0.07	0.14
TiO_2	0.48	0.56	0.48
P_2O_5	0.50	0.22	0.16
MnO	0.06	0.09	0.05
CO_2	0.01	0.08	0.04
Total	99.8	100.97	99.50

C.I.P.W. norms

Quartz	14.0	26.0	16.86
Orthoclase	10.0	11.1	22.80
Albite	43.0	33.0	39.30
Anorthite	18.1	18.1	10.84
Corundum	0.2	0.4	--
Wollastonite	--	--	--
Diopside	--	--	2.29
Hypersthene	10.6	8.7	4.94
Magnetite	2.1	1.4	0.93
Ilmenite	0.9	1.1	0.91
Apatite	1.3	0.3	0.34
Total	100.2	100.1	99.21

1. Quartz-andesine-hornblende-hypersthene-augite gneiss from outcrops along Pennsylvania Route 73 about 1.5 km northwest of Shanesville, Boyertown 7 1/2-minute quadrangle, Pa. Rapid-rock analysis by Hezekiah Smith, U. S. Geological Survey.
2. Quartz-andesine-hypersthene gneiss from outcrop 0.2 km north of Randolph Township Consolidated School, Mendham 7 1/2-minute guardrangle, N.J. Analyst same as 1.
3. Hypersthene-quartz-oligoclase gneiss from just north of the old road about 0.8 km northwest of confluence of Pacock Brook and Canistear Reservoir, Newfoundland 7 1/2-minute quadrangle (Baker and Buddington, 1970, p. 8).

gle (Fig. 2; Drake, 1967), charnockitic rocks appear to form a border phase of a venitic migmatite zone within a major body of the Losee.

In the Dover district (Sims, 1958), the charnockitic rocks always contain intercalated amphibolite. When not in contact with rocks of the Byram Intrusive Suite, they are adjacent to Losee rocks in all places but one, where they abut biotite-quartz-feldspar gneiss. Sims (1958) mentioned no interlayering of the units along that contact. The charnockitic rocks are migmatized by alaskite and pegmatite of the Byram.

In the Lake Hopatcong area (Fig. 2), Young (1971) found charnockitic rocks associated with rocks of the Losee Metamorphic Suite. In northern New Jersey and adjacent New York, Hotz (1952) and Offield (1967) found that charnockitic rocks are so closely related to the rocks that are here called Losee, that the two groups were mapped together as one unit. Elsewhere in southeastern New York, Dodd (1965) recognized the close relation of charnockitic rocks to Losee rocks, but mapped the two groups as separate units.

Contrary to the findings of the above geologists, Baker and Buddington (1970) found charnockitic rocks interlayered with obvious metasedimentary quartzofeldspathic gneiss at one place in the Franklin-Sterling area. In addition, they found that the charnockitic rocks in that area contain disseminated graphite, an observation not made by any other Reading Prong geologist. In the Roseland area of Virginia, graphite occurs in rocks of probable igneous origin that are interlayered with metasedimentary rocks, showing that carbon does diffuse (Eric Force, written commun., 1981). Graphite, however, is a known constituent of igneous rocks and meteorites.

Sims (1958) believes that the charnockitic rocks of the Dover district (Fig. 2) are quartz diorite and related diorite and are intrusive, but pointed out that there is no evidence that they did not form by high-grade metamorphism and local anatexis. Hotz (1952) believes that the charnockitic rocks are intrusive, as does Buckwalter (1959, 1962). Buckwalter, however, believes that these rocks are a badly contaminated phase of what here is called the Byram Intrusive Suite.

Baker and Buddington (1970), on the other hand, believe that the charnockitic rocks are metasedimentary because of their contained disseminated graphite and their interlayering at one locality with garnetiferous metasedimentary rocks. They recognize, however, the difficulties of such an origin, in particular, in finding a suitable protolith.

A third origin is preferred by Offield (1967), Dodd (1965), and Young (1971). These geologists believe that the charnockitic rocks, like those of the Losee Metamorphic Suite, result from the metamorphism of sodic volcanic rocks, both quartz keratophyre and basalt. Although I called these rocks charnockitic quartz diorite, I (1969) believe, as do they, that the charnockitic rocks result from the high-grade metamorphism of quartz keratophyre and basalt; however, I believe that partial anatexis was also involved and that these rocks represent an end member of the venitic migmatites produced within the Losee Metamorphic

Suite. I have refrained from including the charnockitic rocks in the Losee, because of the relations described by Baker and Buddington (1970) in the Franklin-Sterling area (Fig. 2). Perhaps there are charnockitic rocks of more than one origin or perhaps the anatectic charnockitic sheets locally picked some carbon from the metasedimentary pile. Hopefully, future work will solve this problem.

INTRUSIVE ROCKS

Intrusive rocks of four general types constitute about half the Proterozoic Y rocks of the Reading Prong. The most important and widespread of these rocks is a suite of microperthite-bearing hornblende granite, alaskite, and attendant biotite granite and pegmatite that crops out throughout the Prong. This suite is here named the Byram Intrusive Suite.

Sparse, small bodies of alaskite characterized by microantiperthite crop out at several places in eastern Pennsylvania and western New Jersey. These rocks have certain similarities to those of the Byram Intrusive Suite. These microantiperthite bodies, however, are thought to have a more complex origin and therefore are excluded from that intrusive suite.

A suite of generally quartz-poor, clinopyroxene-bearing monzonitic to syenitic rocks is abundant in north-central and northeastern New Jersey, northwest of the Green Pond syncline (Fig. 2). Mesoperthite- and clinopyroxene-bearing granites are also fairly common in this area and are thought to be comagmatic with the quartz-poor rocks.

A small area of eastern Pennsylvania is underlain by extremely altered and silicified mafic rocks that in this paper are named the Hexenkopf Complex. These rocks were probably mafic diorite and perhaps pyroxenite prior to their alteration. The Hexenkopf Complex clearly predates the emplacement of the Byram Intrusive Suite.

Byram Intrusive Suite (here reintroduced, renamed, and redefined)

Granitoid rocks consisting of quartz, potassic feldspar, oligoclase, and containing varying amounts of, and as much as ten percent mafic minerals crop out throughout the Reading Prong. In the early days of geologic work in New Jersey such rocks were named Bryam Gneiss for typical exposures in Byram Township, Stanhope 7½-minute quadrangle, Sussex County, New Jersey (Spencer and others, 1908). Detailed mapping since World War II has shown that the Byram as mapped by early workers (Spencer and others, 1908; Darton and others, 1908; Bayley and others, 1914; Bayley, 1941; and Fraser *in* Miller and others, 1939, 1941) also contains an intrusive sequence of clinopyroxene-mesoperthite monzonite and syenite and related clinopyroxene granite, as well as microantiperthite granite, all of which are thought to have a different origin. For this reason, the term Byram was abandoned by most modern workers.

The modern work, however, clearly shows that granitoid

rocks containing microperthite or, in some places microcline, have close petrographic and petrochemical similarities throughout the Reading Prong. For this reason, the name Byram is here reintroduced, renamed, and redefined as the Byram Intrusive Suite to include the comagmatic microperthite- or microcline-bearing hornblende granite, alaskite, lesser biotite granite, and related pegmatite of the Proterozoic Y terrane of eastern Pennsylvania, New Jersey, and adjacent New York.

The rocks of the Byram Intrusive Suite are abundant throughout the Reading Prong where they occur as regionally conformable sheets and pods and as refolded bodies that in the past were defined as phacoliths. Although regionally conformable, these granitic bodies locally cut across the structure of the older rocks. The Byram rocks are gneissoid, that is, they have a primary flow layering and lineation. Many of the rocks have a secondary metamorphic foliation, and in some places they have been gneissified. In the gneiss phases, the original microperthite in the rock has typically unmixed into microcline and free oligoclase. All modern workers agree that rocks of the Byram Intrusive Suite have all the features usually ascribed to syntectonic granites. Pegmatites thought to be units of the Byram are late stage features, and in some places markedly crosscut their country rocks.

Petrographically, the Byram consists of five different rock types: hornblende granite (and granite gneiss), alaskite, biotite granite, pegmatite, and sparse amounts of hornblende-quartz syenite. the syenite is included in the Byram Intrusive Suite because it is identical with hornblende granite in other than quartz content and forms a lens on the crest of an antiform within a large sheet of hornblende granite. Hornblende granite differs from alaskite primarily in its content of mafic minerals. Biotite granite is a hybrid rock formed by the assimilation and modification of biotite-quartz-feldspar gneiss by alaskite (Dodd, 1965; Drake, 1969; Sims, 1958). Fig. 13 gives available modes for rocks of the Byram Intrusive Suite. The alaskites in general contain a higher percentage of alkali feldspar than do the hornblende granites. Unfortunately, no data are available for the southwestern end of the Prong, as Buckwalter (1959, 1962) published no modes. The center of gravity of the plot lies to the left of the syenogranite-monzogranite boundary. A few rocks plot as alkali feldspar granite and quartz monzonite. The syenitic rocks are quite rich in alkali feldspar. A comparison of these modal data with those of the biotite-quartz-feldspar gneiss (Fig. 4) shows that although a few of the metasedimentary rocks are within the Byram field, the bulk of the metasedimentary rocks are much richer in plagioclase and(or) quartz than are those of the Byram. A comparison of Byram rocks with rocks of the potassic feldspar gneiss (Fig. 5) shows that the bulk of the potassic feldspar gneiss is far richer in alkali feldspar and quartz than are rocks of the Byram, although a few samples of potassic feldspar gneiss are within the Byram field.

Table 10 gives new chemical analyses of rocks of the Byram Intrusive Suite. These samples were collected over a strike length of about 145 km from the Fleetwood quadrangle (Fig. 3) in the southwest to the Franklin quadrangle (Fig. 2) in the northeast.

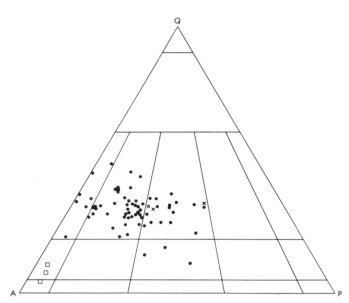

Figure 13. Q-A-P plot of rocks of the Byram Intrusive Suite: hornblende granite and alaskite, (.); biotite granite, (x); hornblende-quartz syenite, (□). Data source: Drake (1969); Hague and others (1956); Young (1971); Smith (1969); Baker and Buddington (1970); Sims (1958); and Kastelic (1980).

The chemistry of these samples is remarkably similar. In general, the alaskites are somewhat more siliceous and have slightly higher K_2-Na_2O ratios than do the hornblende granites. Analysis 17, of biotite granite, reflects the hybrid nature of this rock. The chemistry of sample 19 suggests that it also is a hybrid rock. Field relations support this interpretation, as that sample was collected from a small sheet of hornblende granite adjacent to a unit of mixed metasedimentary gneisses.

Some chemical and normative data from Table 10 are plotted on Fig. 14. These data are compared with equivalent data from the Storm King Granite (Lowe, 1950). The Storm King data plot with the Byram data, confirming the correlation previously made by Drake (1969). This shows that granitoid rocks of similar chemistry can be traced from the southwest end of the Reading Prong to at least the Hudson River in New York, a distance of about 205 km. The hornblende-quartz syenite, the pegmatite, and Lowes' Storm King Granite all lie on the Byram differentiation trend as shown by Fig. 14B, whereas the rocks thought to be metamorphosed sedimentary rocks do not. Most of the rocks of the Byram are less potassic than average alkali or calc-alkali granites (Fig. 14D). A comparison of Fig. 6 with Fig. 14D shows that two of the five samples of biotite-quartz-feldspar gneiss have chemistry similar to that of the Byram rocks.

All modern authors have presented evidence that much of the Byram Intrusive Suite originated from the crystallization of a melt or some mobile material, although most recognize that some of the rocks result from migmatization, replacement, or assimilation of country rocks. Evidence to support this concept includes: relative uniform chemistry of the rocks (Table 10); euhedral and elongate character of the contained zircon, and direct observation

of field relations, that is, local crosscutting relations and the formation of migmatites containing alaskite leucosome in both biotite-quartz-feldspar gneiss and amphibolite. I believe that a fair amount of the hornblende granite in the area in which I have worked results from the almost complete assimilation of amphibolite by alaskite. Drake and others (1961) described a stream traverse in the Frenchtown quadrangle (Fig. 2) in which hornblende granite was traced into hornblende granite containing ghostlike schlieren of amphibolite and finally into a rock containing about equal proportions of leucosome and amphibolite.

Buckwalter (1962) has described similar relations in the Birdsboro quadrangle (Fig. 3). He described chemical data from samples of amphibolite paleosome, alaskite leucosome, and mixed material at the contact of the two rock types demonstrating chemical gradients in the intervening rock so that SiO_2, Na_2O, K_2O increase and Al_2O, FeO, Fe_2O_3, CaO, MgO, and TiO_2 decrease. A similar relation was described above in the description of amphibolite. The mixing of amphibolitic and granitic rocks cannot be measured as few mixed or contaminated-appearing rocks have been sampled for analysis; good typical granitic rock was always desired. An exception is biotite granite. Sample 19 (Table 10) is clearly a hybrid rock. Sample 18 is also suspect because of its relatively low SiO_2 content and relatively high CaO content.

The origin of the granitic material of the Byram Intrusive Suite is unclear. Sims (1958) is quite precise in believing that rocks of the suite originated predominantly from the differentiation of a magma of granitic composition, most of the magma consolidating to form hornblende granite and the more mobile felsic differentiates forming alaskite. Hotz (1952), Dodd (1965), Hague and others (1956), Baker and Buddington (1970), and Young (1971) favored a magmatic origin, the source of the magma not being specified. Offield (1967) appeared to favor a magma of anatectic origin, perhaps produced near the present level of erosion and Kastelic (1980) favors an anatectic magma produced from the potassic feldspar gneiss.

Using the gahnite-franklinite geothermometer, temperatures in the Reading Prong have been at least as high as 760 °C (Carvalho and Sclar, 1979); which would permit the melting of rocks of the Byram Intrusive Suite in a hydrous environment. I, therefore, also favor an anatectic origin for the Byram, but doubt that the potassic feldspar gneiss is the source, although direct observation shows that some melt formed in that unit (see above). The potassic feldspar gneiss, however, seems too deficient in Na_2O to be a potential source for rocks of the Byram. Some melt also formed in rocks of the Losee Suite, but these rocks are far too Na_2O-rich to be a source for the Byram. Abundant migmatites have developed within both amphibolite and biotite-quartz-feldsdpar gneiss. Byram-type rocks could hardly have been produced from amphibolites, but some phases of the biotite-quartz-feldspar have an appropriate composition (see above). The bulk of this unit, however, would appear to be too heterogeneous to produce the fairly homogeneous rocks of the Byram Intrusive Suite. Much of the biotite-quartz-feldspar gneiss is migmatitic, but

TABLE 10. CHEMICAL ANALYSES AND C.I.P.W. NORMS OF ROCKS OF THE BYRAM INTRUSIVE SUITE

Chemical analyses (weight percent)

	1	2	3	4	5	6	7	8	9	10	11	12	13	14	15	16	17	18	19	20	21
SiO_2	72.1	72.1	71.2	72.1	76.2	72.0	73.2	70.3	73.7	75.1	73.5	61.8	73.2	71.2	74.2	75.7	80.4	68.3	63.3	71.2	70.8
Al_2O_3	12.5	14.9	13.3	13.3	12.2	13.2	13.1	12.7	13.4	13.0	14.1	15.7	12.2	13.3	13.6	13.8	9.3	15.1	17.0	14.6	11.8
Fe_2O_3	1.8	0.10	3.5	1.1	0.83	1.9	1.1	3.8	0.5	0.48	0.02	1.4	1.7	1.5	0.32	0.10	0.3	1.4	2.0	1.1	2.6
FeO	2.6	1.2	1.9	1.9	0.60	0.44	1.0	2.3	1.1	0.56	0.28	3.5	1.3	1.6	0.04	0.96	1.4	3.8	2.2	2.4	3.2
MgO	0.24	0.32	0.10	0.30	0.20	0.60	0.20	0.40	0.43	0.40	0.14	0.38	0.18	0.56	0.06	0.30	0.76	0.45	1.5	0.21	0.20
CaO	1.3	1.7	0.70	1.0	1.5	1.2	0.80	0.30	1.7	1.3	0.15	4.9	1.9	1.0	0.19	1.2	2.3	2.5	4.2	1.3	1.3
Na_2O	3.3	3.0	3.7	3.8	3.4	3.4	3.2	2.4	3.1	3.0	3.3	5.1	4.1	3.6	2.8	3.4	1.2	3.0	5.0	3.2	1.2
K_2O	4.2	5.6	5.0	4.6	3.8	5.6	6.0	6.1	3.9	4.1	7.3	5.3	4.1	5.5	7.2	4.8	3.0	4.9	2.3	5.8	8.1
H_2O+	0.59	0.63	0.53	0.68	0.40	0.07	0.66	1.2	0.73	0.63	0.39	0.26	0.28	0.58	0.40	0.43	0.49	0.68	0.75	0.6	0.55
H_2O-	0.06	0.07	0.04	0.04	0.05	--	--	--	0.05	0.07	0.04	0.20	0.11	0.16	0.11	0.01	--	0.01	0.05	0.03	0.04
TiO_2	0.37	0.21	0.51	0.31	0.19	0.24	0.47	0.45	0.18	0.11	0.01	0.68	0.32	0.31	0.01	0.10	0.24	0.56	0.31	0.32	0.30
P_2O_5	0.11	0.08	0.06	0.09	0.04	0.14	0.11	0.09	0.07	0.04	0.02	0.12	0.03	0.08	0.02	0.06	0.12	0.26	0.39	0.06	0.03
MnO	0.09	0.03	0.08	0.04	0.04	0.06	0.06	0.06	0.02	0.02	--	0.16	0.03	0.05	0.02	0.02	0.03	0.07	0.07	0.03	0.07
CO_2	0.02	0.07	0.03	0.02	0.08	0.24	0.05	0.05	0.22	0.01	0.07	0.08	0.07	0.01	0.02	0.08	0.12	0.08	0.08	0.08	0.45
Total	99.3	100.0	100.6	99.3	99.5	99.1	99.95	100.2	99.1	98.8	99.3	99.6	99.5	99.45	99.0	101.0	99.7	101.1	99.2	100.9	100.6

C.I.P.W. Norms

	1	2	3	4	5	6	7	8	9	10	11	12	13	14	15	16	17	18	19	20	21
Quartz	31.2	29.2	29.2	29.0	39.1	28.3	30.4	31.4	36.5	38.8	26.0	3.1	31.2	23.8	30.7	33.8	56.4	24.2	15.2	26.8	29.6
Orthoclase	25.6	33.4	29.5	27.8	22.2	33.9	36.1	36.4	23.9	25.0	43.9	31.7	24.5	33.4	43.4	28.4	18.4	28.2	13.3	34.5	47.8
Albite	28.8	25.2	31.4	33.0	28.8	28.8	27.8	20.4	27.2	26.2	27.8	44.0	34.6	35.5	23.6	28.8	10.0	25.2	43.0	27.2	10.0
Anorthite	6.4	7.5	2.8	5.0	7.2	4.2	3.1	0.6	7.5	6.4	0.8	3.9	3.1	0.8	1.1	5.0	10.6	10.8	17.8	5.6	3.1
Corundum	0.1	0.6	0.6	--	--	--	--	--	1.1	1.1	0.5	--	--	--	0.8	1.0	0.3	--	--	1.0	--
Wollastonite	--	--	--	--	--	--	--	--	--	--	--	4.2	--	--	--	--	--	--	--	--	--
Diopside	--	--	--	--	--	1.9	0.6	--	--	--	--	6.3	2.0	2.8	--	--	--	--	--	--	2.4
Hypersthene	3.2	2.5	0.2	2.8	0.6	--	--	3.2	2.2	1.5	0.9	6.3	1.4	0.8	0.1	2.4	4.5	6.0	5.4	3.4	2.5
Magnetite	2.6	0.2	4.6	1.6	1.2	0.9	1.6	5.6	0.7	0.7	--	2.1	2.6	2.1	--	0.2	0.5	2.1	3.0	1.6	3.7
Ilmenite	0.8	0.5	0.9	0.6	0.5	0.5	0.9	0.9	0.3	0.2	--	1.4	0.6	0.6	--	0.2	0.5	1.2	0.6	0.6	0.6
Apatite	--	--	0.3	0.3	--	0.3	0.3	0.3	--	--	--	0.3	0.1	0.3	--	0.2	0.3	0.7	1.0	--	--
Hematite	--	0.3	0.2	--	--	1.3	--	1.3	0.3	--	--	--	--	--	0.3	0.3	--	--	--	0.3	--
Total	98.7	100.0	99.7	100.2	99.6	100.1	100.8	100.9	100.3	99.9	99.9	103.3	100.1	100.1	100.0	100.1	101.5	99.9	99.8	101.0	99.8

1. Hornblende granite from outcrops just to east of Pennsylvania Route 662 about 1.3 km south of Pricetown, Fleetwood 7 1/2-minute quadrangle, Pa. Rapid-rock analyses by Hezekiah Smith, U. S. Geological Survey.

2. Hornblende granite from outcrops along Antietam Creek Road about 0.3 km south of Antietam Reservoir, Birdsboro 7 1/2-minute quadrangle, Pa. Same analyst as 1.

3. Hornblende granite from outcrops along unnumbered County road about 0.85 airline km from Hill Church, Manatawny 7 1/2-minute quadrangle, Pa. Same analyst as 1.

4. Hornblende granite from outcrops along unnumbered County road about 0.35 airline km from Rittenhouse Gap, Manatawny 7 1/2-minute quadrangle, Pa. Same analyst as 1.

5. Microperthite alaskite from outcrops under powerlines on south slope of South Mountain about 1.2 km south of Stones Throw Road, Allentown East 7 1/2-minute quadrangle, Pa. Analyst same as 1.

6. Hornblende granite from outcrops along South Mountain Drive about 0.3 km west of its intersection with Hayes Street, South Mountain, Hellertown 7 1/2-minute quadrangle, Pa. Rapid-rock analyses by P. Elmore, L. Artis, J. Glen, G. Chloe, S. Butts. Kelsey, and K. Smith, U. S. Geological Survey.

7. Microperthite alaskite from outcrops 0.1 km south of Switchback Road about 0.4 km east of its intersection with U.S. Route 46, Hellertown 7 1/2-minute quadrangle, Pa. Analyst same as 6.

8. Magnetite granite from outcrops on north slope of mountain 0.3 km north of Switchback Road about 1 km west of its intersection with Campbell Pond Road, Hellertown 7 1/2-minute quadrangle, Pa. Analyst same as 6.

9. Sheared microperthite alaskite from small quarry on east bank of Delaware River about 1.2 km south of Riegelsville, N.J.-Pa. Riegelsville 7 1/2-minute quadrangle. Rock erroneously mapped as albite-oligoclase granite by Drake and others (1967). Analyst same as 6.

10. Microperthite, alaskite from stream exposure just north of unnumbered County road that traverses Musconetcong Mountain south from Finesville, N.J. Outcrop is 3 airline km southeast of Finesville, Riegelsville 7 1/2-minute quadrangle. Analyst same as 1.

11. Granite pegmatite from large outcrops along west bank of Delaware River about 0.45 km northeast of Easton water filtration plant, Easton 7 1/2-minute quadrangle, N.J.-Pa. Analyst same as 1.

12. Hornblende-quartz syenite from outcrops along county road on south slope of Oxford Mountain about 0.45 km southeast of its intersection with N.J. Route 31, Washington 7 1/2-minute quadrangle, N.J. Rapid-rock analyses by K. Coates and H. Smith, U. S. Geological Survey.

13. Microperthite alaskite from outcrops to west of unimproved road in Warren County House Mountain about 1.1 km south of Warren County Welfare House, Washington 7 1/2-minute quadrangle, N.J. Analyst same as 12.

14. Microperthite alaskite from outcrops on north side of U.S. Route 46, about 0.8 km southwest of Townsbury, Washington 7 1/2-minute quadrangle, N.J. Analyst same as 12.

15. Granite pegmatite from Lommasons Glen, Belvidere 7 1/2-minute quadrangle, N.J. Analyst same as 12.

16. Strongly foliated microperthite alaskite from outcrops along unnumbered County road about 0.45 km northwest of Hibernia, Boonton 7 1/2-minute quadrangle, N.J. Analyst same as 12.

17. Biotite granite from outcrops along road parallel to southwest end of Splitrock Reservoir about 0.25 km northwest of southwest corner of reservoir. Boonton 7 1/2-minute quadrangle, N.J. Sample is from rock body between outcrop belts of biotite-quartz-plagioclase gneiss and hornblende granite. Same analyst as 1.

18. Hornblende granite from outcrops on Lake Denmark Road directly opposite southeast corner of small pond south of Lake Denmark, Dover 7 1/2-minute quadrangle, N.J. Analyst same as 1.

19. Hornblende granite from outcrops along unnumbered County road about 0.8 km N. 33° W of the crest of Sterling Hill, Franklin 7 1/2-minute quadrangle, N.J. Rock probably is granitized by meta-sedimentary gneiss. Analyst same as 1.

20. Microperthite alaskite from outcrops along Greenpond Road about 0.2 km northeast of Lake Telemark, Boonton 7 1/2-minute quadrangle, N.J. This locality is within a body of albite-oligo-clase granite. Analyst same as 1.

21. Microcline-oligoclase alaskite from outcrops within a body of rock mapped as hypersthene-quartz-oligoclase gneiss along road about 0.5 km north of Y-intersection at Stockholm, Franklin 7 1/2-minute quadrangle, N.J. Analyst same as 1.

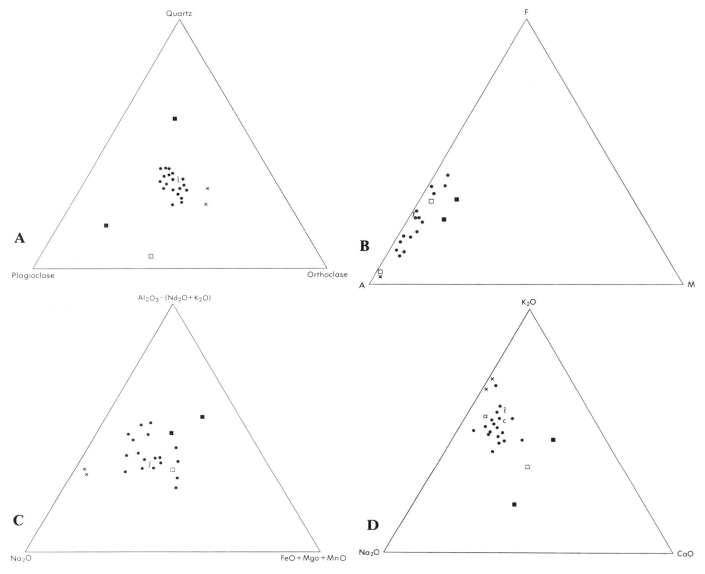

Figure 14. Plots of bulk chemistry (weight percent) of rocks of Byram Intrusive Suite. A. Quartz-orthoclase-plagiocase plot: alaskite and hornblende granite (.); pegmatite (x); hornblende-quartz syenite (□); granitized metasedimentary rocks (■); Storm King Granite as used by Lowe (1950) (ℓ). B. A-F-M plot: symbology same as A. C. Modified A-K-F plot: symbology same as A. D. Na_2O-K_2O-CaO plot comparing Byram rocks with average alkali (a), and calc-alkali (c), granites (Nockolds, 1954): (.), symbology same as A.

the leucosome is believed to have originated elsewhere because of its homogeneity as opposed to the heterogeneity of the biotite-quartz-feldspar gneiss. In addition, no small- to moderate-sized granitic bodies representing possible melt have been recognized within biotite-quartz-feldspar gneiss terranes of easternmost Pennsylvania and New Jersey.

In the southwestern part of the Reading Prong, rocks of the Byram Intrusive Suite appear to be much more abundant than elsewhere. Metasedimentary rocks occur only in strongly mig-matized lenses. This area is probably the deepest exposed level of

the Reading Prong, an idea supported by independent structural information. If rocks of the Byram result from the anatectic melting of quartzofeldspathic rocks at present erosion levels, this process would probably have taken place in this southwestern area. Unfortunately, the terrane is very poorly exposed, and relations are virtually impossible to observe.

The source of the Byram, then, is moot. These rocks do not appear to have originated at the present erosion level, with the possible exception of those of the southwestern Reading Prong. This leaves an unknown source at greater depth, perhaps quartzo-

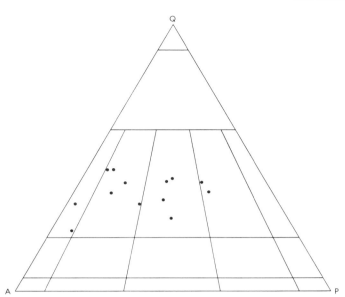

Figure 15. Q-A-P plot of microantiperthite alaskite. Data from Drake (1969, unpublished) and Kastelic (1980).

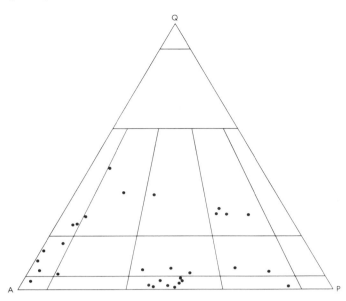

Figure 16. Q-A-P plot of monzonite, syenite, and related pyroxene granite. Data from Baker and Buddington (1970), Young (1971, 1972, 1978), and Hague and others (1956).

feldspathic gneiss, as described above, or an earlier granitic terrane. Isotope and minor-element studies would help solve this problem.

Microantiperthite alaskite

Microantiperthite alaskite is a minor rock type in the Reading Prong, and so far as I know, has only been mapped by Drake (1969) and Kastelic (1980). Sims (1958) recognized the rock type in the Dover district but did not map it separately. Compositionally, this unit ranges from granodiorite through monzogranite and syenogranite to alkali feldspar granite. Antiperthite is considered an alkali feldspar for the plot of Fig. 15.

Microantiperthite alaskite is difficult to separate from alaskite of the Byram Intrusive Suite without thin-section study. Mafic minerals including hornblende, biotite, and augite are sparse, shredded, strongly altered to chlorite, and occur in clots rather than in layers or streaks. Quartz characteristically is strongly lenticulated. Patch and plume antiperthites are the most common, but string types also occur.

Microantiperthite rocks seem to have formed where sodic rocks of the Losee Metamorphic Suite are juxtaposed with rocks of the Byram Intrusive Suite and all microantiperthite alaskite contains layers and veins of microperthite alaskite. I believe that the rock owes its origin to an interaction of Losee sodic rocks with magma or fluids of the Byram. A Losee protolith is supported by the one available chemical analysis (Drake, 1969), which shows a Na_2O-K_2O ratio of 2.

Sims (1958) visualized an origin much like that described above, but Buddington (*in* Sims, 1958) reversed the concept, stating that the albite-oligoclase granite resulted from the albitization of microantiperthite alaskite. In addition, some of the pyrox-

ene granite associated with the quartz-poor intrusive suite (see below) is a microantiperthite rock. It would appear that microantiperthitic rocks may have had more than one origin. For that reason, I have not assigned them to either the Losee, the Byram, or to quartz-poor suites.

Monzonite, Syenite, and related Pyroxene granite

Quartz-poor rocks generally described as syenites or quartz syenites (Baker and Buddington, 1970; Young, 1971, 1972, 1978), and what is thought to be related pyroxene granite, are abundant in northern New Jersey northwest of the Green Pond Syncline between the New York state line and the Stanhope and Tranquility quadrangles. The quartz-poor rocks range in composition from quartz monzodiorite to quartz-alkali feldspar syenite and alkali feldspar syenite, the bulk being monzonite and quartz monzonite (Fig. 16). The pyroxene granites range from granodiorite to alkali feldspar granite. Chemically analyzed rocks (Table 11) reported herein include pyroxene trondhjemite, pyroxene granite, pyroxene alaskite, pyroxene monzonite, and pyroxene-quartz monzonite. I have never worked with these rocks, but judging from Tilling's and my sampling, the rock units must be much more heterogeneous than one would judge from Baker and Buddington (1970), as rocks sampled from syenite units are granite and vice versa. Granitic and quartz-poor phases of the pyroxene-bearing sequence may be more closely intercalated than one would judge from the literature. In any case, there can be little doubt that the pyroxene granite is related to the quartz-poor rocks.

Other than in their paucity of quartz and contained pyroxene, these rocks differ from those of the Byram Intrusive Suite by having mesoperthite (largely antiperthitic) as their primary feld-

TABLE 11. CHEMICAL ANALYSES AND C.I.P.W. NORMS OF
MONZONITE, SYENITE, AND RELATED PYROXENE GRANITE

	1	2	3	4	5	6
	Chemical Analyses (weight percent)					
SiO_2	69.0	59.2	59.9	62.7	70.0	74.6
Al_2O_3	12.7	16.3	16.6	16.8	12.5	12.3
Fe_2O_3	5.7	2.3	2.4	2.7	2.5	2.4
FeO	3.1	6.3	5.8	2.8	3.8	1.4
MgO	0.39	0.39	0.4	0.3	0.24	0.1
CaO	1.7	3.3	3.4	2.8	2.2	0.72
Na_2O	4.2	4.7	4.8	4.8	3.6	3.4
K_2O	2.8	5.4	5.2	5.5	4.4	4.5
H_2O+	0.6	0.68	0.62	0.51	0.51	0.51
H_2O-	0.04	0.04	0.08	0.04	0.06	0.04
TiO_2	0.71	0.98	0.91	0.62	0.56	0.23
P_2O_5	0.14	0.29	0.27	0.17	0.08	0.03
MnO	0.11	0.22	0.20	0.13	0.14	0.02
CO_2	0.03	0.02	0.07	0.04	0.12	0.12
Total	101.2	100.1	100.6	99.9	100.7	100.3
C.I.P.W. Norms						
Quartz	29.5	2.5	3.8	8.2	26.8	36.0
Orthoclase	16.7	31.7	30.6	32.8	26.1	26.7
Albite	35.6	39.8	40.4	40.4	30.4	28.8
Anorthite	7.2	7.8	5.8	8.3	5.0	3.6
Corundum	--	--	--	--	--	0.5
Wollastonite	0.1	0.5	1.5	0.1	0.7	--
Diopside	0.2	4.6	4.6	3.8	2.8	--
Hypersthene	0.9	7.1	6.5	1.2	2.3	0.3
Magnetite	8.1	3.2	3.5	3.9	3.7	3.5
Ilmenite	1.4	2.0	1.7	1.2	1.1	0.5
Hematite	0.2	--	--	--	--	--
Apatite	0.3	0.7	0.7	0.3	0.3	--
Total	100.2	99.9	99.1	100.2	100.2	99.9

1. Pyroxene trondhjemite, mapped as pyroxene granite, from
outcrops along unnumbered County road about 0.9 km southeast
of Edison, Franklin 7 1/2-minute quadrangle, N.J. Rapid-rock
analyses by Hezekiah Smith, U. S. Geological Survey.
2. Pyroxene monzonite, mapped as granite, from massive
outcrops along N.J. Route 15 about 0.6 km north of south
boundary of Franklin 7 1/2-minute quadrangle, N.J. Analyst
same as 1.
3. More mafic-appearing pyroxene monzonite, mapped as granite,
from same locations as 2. Analyst same as 1.
4. Pyroxene-quartz monzonite, mapped as syenite gneiss, from
outcrops at Blue Herron Camp on north shore of Blue Herron
Lake, Franklin 7 1/2-minute quadrangle, N.J. Analyst same as
1.
5. Pyroxene granite, mapped as pyroxene syenite, from outcrops along N.J. Route 15 about 0.15 km north of south
boundary of Franklin 7 1/2-minute quadrangle, N.J. Analyst
same as 1.
6. Pyroxene alaskite from outcrops along unnumbered county
road just north of south boundary and 2.85 km east of
southwest corner of Franklin 7 1/2-minute quadrangle, N.J.
Analyst same as 1.

spar and in their high Fe-Mg ratios (Baker and Buddington, 1970; Young, 1971, 1972, 1978; Table 11). In addition, rocks of the quartz-poor sequence are more femic (Fig. 17A) and contain more CaO (Fig. 17B) than rocks of the Byram. The apparent differentiation trend (Fig. 17C) is the same as that of the Byram Intrusive Suite.

The monzonites, syenites, and related pyroxene granite in the Franklin and Hamburg quadrangles (Baker and Buddington, 1970), with one exception, are considered to be intrusive sheets that have a primary flow foliation. One body of pyroxene syenite wraps around the hinge of a major antiform and is considered to be an orthogneiss having a secondary foliation (Baker and Buddington, 1970). I could see little difference in the supposed gneissoid rocks and the orthogneiss. Baker and Buddington (1970) do not present data on the relative ages of the monzonite, syenite, and related pyroxene granite and the hornblende granite and alaskite of the Byram Intrusive Suite: in fact, they totally ignore the issue. The two suites are in contact only in the hinge zone of the major Beaver Lake antiform where, hornblende granite forms an upper border to the syenite gneiss and the alaskite fills the arch of the plunging hinge (Baker and Buddington, 1970, pl. 1). These relations could easily be interpreted to be the result of a body of syenite grading upward into more quartz-rich rocks.

In the Lake Hopatcong area (Fig. 2), Young (1971) mapped two sheets of well-foliated syenite gneiss without specifying the origin of the foliation. In addition, he (1971, 1972) mapped a sheet of quartz syenite that transects the fold structure in surrounding hornblende granite of the Byram Intrusive Suite. This quartz syenite lacks both foliation and lineation and from field relations appears to be intrusive into the hornblende granite. To my knowledge, this is the only place in the Reading Prong that the relation between any syenitic rock and rocks of the Byram Intrusive Suite has been described. Young (1972) believes that the quartz syenite probably was derived from anatexis of a pre-existing syenite gneiss. If this is correct, then quartz-poor rocks were emplaced during two different magmatic episodes. This I doubt, unless the conformable sheets of monzonite, syenite, and pyroxene granite are comagmatic with the Byram Intrusive Suite.

The relation of the monzonites, syenites, and related pyroxene granite to the rocks of the Byram Intrusive Suite, in both time and space, is perhaps the most befuddling current problem in Reading Prong geology. Young (1978) has pointed out the regional distribution of the two rock types and has carefully considered all pertinent evidence. He favors a hypothesis involving partial melting at successively deeper crustal levels to produce the different rocks and postulates a higher geothermal gradient for the area containing the quartz-poor rocks. I can do little more than repeat the need for further detailed field, geochronologic, and isotope studies.

HEXENKOPF COMPLEX (HERE NAMED)

A body of unusual altered mafic rock underlies an area of roughly 15 km[2] in the extreme northwest corner of the Riegelsville quadrangle (Drake and others, 1967), the extreme southwest corner of the Easton quadrangle (Drake, 1967a), the extreme southeast corner of the Nazareth quadrangle (Aaron, 1975), and the northeast corner of the Hellertown quadrangle (Drake, unpublished data, see Fig. 3). Although extremely heterogeneous, this body consists of three general rock types: hornblende-augite-quartz-andesine gneiss, epidote-augite-hornblende-plagioclase

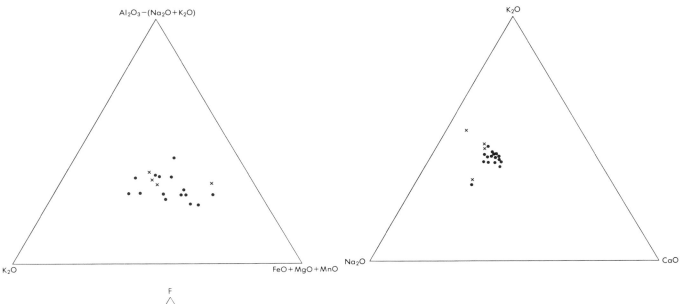

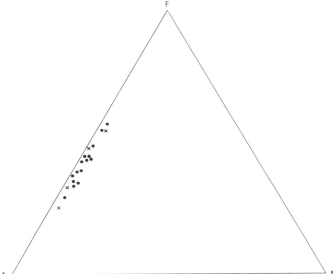

Figure 17. Plots of bulk chemistry (weight percent) of monzonite, syenite (.), and pyroxene granite (x). Data from this study, Baker and Buddington (1970), and Young (1971, 1972, 1978). A. Modified A-K-F plot. B. Na_2O,-K_2O-CaO plot. C. A-F-M plot.

gneiss, and quartz-garnet-augite granofels. These rocks had been described earlier (Drake, 1969), but neither the dimensions nor the importance of the body was known at that time. This body of extremely altered mafic rocks is here named the Hexenkopf Complex for exposures on and near Hexenkopf Hill in the Riegelsville quadrangle, Pennsylvania (Fig. 3).

On the south, the Hexenkopf Complex is in thrust (both Taconic [Musconetcong] and Alleghanian [Raubsville] Age) contact with lower Paleozoic rocks of the Lehigh Valley sequence (Fig. 18). To the north, it is in thrust contact (Alleghanian age) with alaskite of the Byram Intrusive Suite and biotite-quartz-feldspar gneiss. To the east and west, it is overlain by metasedimentary rocks, and at one place in the west, it is intruded by alaskite of the Byram Intrusive Suite. The west-central part of the complex is overlain by rocks of the Losee Metamorphic Suite.

About 60 percent of the complex consists of hornblende-augite-andesine-quartz gneiss and lesser granofels, which contains varying amounts of epidote, biotite, sphene, garnet, apatite, magnetite, pyrite, chlorite, and zircon. This rock is strongly sericitized, chloritized, and silicified. In many places, it is strongly veined by albite pegmatite. At places, the rock also contains veins of microperthite alaskite. Sample 3 (Table 12) is fairly typical of this rock type, if that is possible. This rock type probably results from the metamorphism and alteration of mafic diorite and(or) gabbro. Sample 2 (Table 12) reflects the extreme alteration that some of this rock has undergone.

A rock composed principally of epidote, augite, and hornblende constitutes about 25 percent of the Hexenkopf Complex. It is a very heavy, massive, dense, grayish-olive, well-foliated gneiss that contains sparse amounts of biotite, sphene, and magnetite, as well as widely varying amounts of quartz. This rock appears to be a metamorphosed and altered pyroxenite. Analysis 1 (Table 12) is of a typical, again if that is possible, sample.

Quartz-garnet-augite granofels constitutes perhaps 15 per-

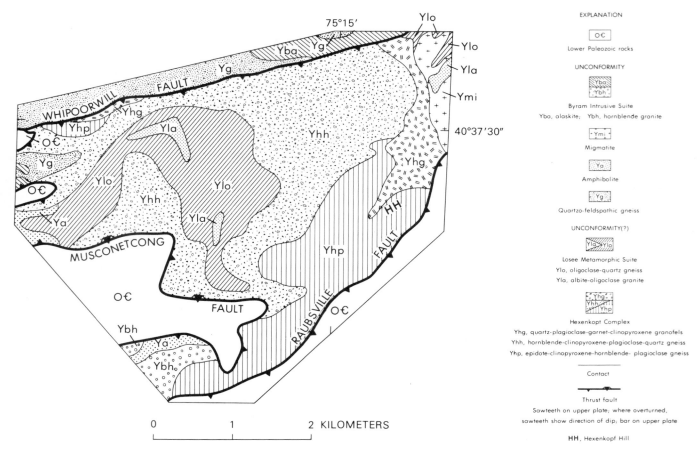

Figure 18. Geologic map of the Hexenkopf Complex.

cent of the complex. This unit is medium fine to medium grained, equigranular, and highly siliceous. In addition to the essential minerals it contains minor amounts of saussuritized plagioclase and chlorite. I (Drake, 1969) formerly considered this unit to be a hornfels related to the emplacement of the Hexenkopf plutonic rocks. However, it is now clear that neither the quartzofeldspathic nor Losee rocks that surround the Hexenkopf Complex could possibly have served as a protolith for the quartz-garnet-clinopyroxene granofels, no matter what had happened to them; in fact, all these rocks overlie the complex. My best guess is that the granofels is a metamorphosed and altered impure chert, which lay atop an igneous body that ranged in composition from pyroxenite to mafic diorite. I consider the granofels to be a meta-sedimentary part of the complex, which would appear to constitute an oceanic rock suite. The complex clearly predates all the other Proterozoic Y rocks.

STRATIGRAPHY OF THE LAYERED ROCKS

Hague and others (1956) proposed a stratigraphic sequence for the layered rocks in the Franklin-Sterling area based on their careful detailed work in this relatively well-exposed terrane. They considered the oldest metasedimentary rocks to be a sequence of interlayered amphibolite, potassic feldspar gneiss, biotite-quartz-feldspar gneiss containing garnet and graphitic-rich intervals, pyroxene gneiss, and local quartzitic layers. These rocks constitute the "Hamburg Mountain Gneiss" series and are overlain by the marble layer of the Franklin Limestone that contains an interval, the median gneiss, of interlayered biotite-quartz-feldspar gneiss, potassic feldspar gneiss, amphibolite, and quartzite. The marble layer is overlain by another clastic sequence, the "Cork Hill Gneiss" zone, of interlayered gneisses similar to the "Hamburg Mountain Gneiss" series. These rocks are overlain by the "Wildcat Marble" layer, which in turn is overlain by another thick interlayered sequence of metasedimentary gneisses called the "Pochuck Mountain Gneiss" series. Hague and others (1956) estimated that this total sequence is probably more than 3,000 m thick. They placed the base of this section at its contact with rocks of the Losee Metamorphic Suite, which they believed to be intrusive.

Baker and Buddington (1970) recognized a similar stratigraphy in the same general area. They also placed the base of their sequence at its contact with rocks of the Losee Metamorphic Suite. No other workers in New Jersey or Pennsylvania have

TABLE 12.   CHEMICAL ANALYSES AND C.I.P.W. NORMS OF
ROCKS OF THE HEXENKOPF COMPLEX

	1	2	3
		Chemical Analyses (weight percent)	
SiO_2	55.8	73.9	57.5
Al_2O_3	9.6	7.4	10.5
Fe_2O_3	6.0	2.9	3.8
FeO	0.24	2.4	5.0
MgO	8.8	2.3	6.5
CaO	17.0	7.5	9.7
Na_2O	0.20	0.7	2.0
K_2O	0.04	1.0	2.1
H_2O+	0.91	0.73	1.2
H_2O-	--	--	--
TiO_2	0.82	0.04	0.97
P_2O_5	0.25	0.22	0.31
MnO	0.12	0.03	0.17
CO_2	0.05	0.08	0.14
Total	99.9	99.2	99.9
	C.I.P.W. Norms		
Quartz	18.8	51.8	13.7
Orthoclase	--	6.1	12.2
Albite	1.6	5.8	16.8
Anorthite	25.6	14.5	13.9
Corundum	--	--	--
Wollastonite	23.4	6.2	5.3
Diopside	--	6.4	12.0
Hypersthene	22.2	4.4	12.9
Magnetite	--	4.4	5.6
Ilmenite	0.6	--	2.0
Titanite	1.2	--	--
Hematite	6.1	--	--
Apatite	0.7	0.3	0.7
Total	100.2	99.9	101.2

1. Hornblende-clinopyroxene-epidote gneiss from outcrops 0.1 km north of Applebutter Road about 0.65 km east of pipeline crossing, Hellertown 7 1/2-minute quadrangle, Pa. Rapid-rock analyses by P. Elmore, L. Artis, J. Glenn, G. Chloe, S. Botts, J. Kelsey, and K. Smith, U. S. Geological Survey.
2. Hornblende-clinopyroxene gneiss from outcrops on ridge about 0.2 km east of Buttermilk Road about 0.6 km north of its intersection with Gaffney Hill Road, Hellertown 7 1/2-minute quadrangle, Pa. Analyst same as 1.
3. Hornblende-clinopyroxene gneiss from outcrops on Hexenkopf Hill along Hexenkopf Road about 0.75 km south of its intersection with Tumble Creek Road, Easton 7 1/2-minute quadrangle, Pa. Analyst same as 1.

attempted to define a stratigraphy, probably because of the difficulties imposed by an abundance of intrusive rocks and a paucity of exposure. I know that this was true in my area.

In New York, immediately adjacent to New Jersey, Offield (1967) recognized a stratigraphy similar to that recognized by Hague and others (1956) and by Baker and Buddington (1970). He also found that these rocks overlie rocks of the Losee Metamorphic Suite and, with imaginative insight, he interpreted the Losee rocks, as well as the rocks of charnockitic affinity, as basement to the calcareous and quartzofeldspathic metasedimentary rocks. This interpretation was based in part on the stratigraphic relations described above, in part on what are believed to be

metamorphic discontinuities, and in part on the recognition of a possible unconformity between the calcareous and quartzofeldspathic gneisses and the underlying rocks of charnockitic affinity in southeastern New York. The evidence for the unconformity is the apparent truncation of structural trends in the charnockitic rocks by the physically overlying metasedimentary sequence, and the lack of stratigraphic repetition in the charnockitic rocks on opposite sides of metasedimentary rock-filled synforms (Offield, 1967).

A possible unconformity also exists in the Hellertown quadrangle (Fig. 3), where metasedimentary gneisses appear to be across contacts between rocks of the Losee Metamorphic Suite and those of the Hexenkopf Complex, as well as being directly on Hexenkopf rocks (Fig. 18). Both of the above observations are only interpretations of equivocal evidence. Their importance, however, cannot be discounted.

Major masses of Losee and charnockitic rocks crop out in antiforms, but as Offield (1967) has correctly pointed out, they also crop out in the cores of synforms. For the proposed stratigraphic relations to hold, some rock must be structurally overturned. There is abundant evidence for overturning during Proterozoic Y deformation in New Jersey and eastern Pennsylvania. The bulk of the folds of map scale in the area are isoclinal and are strongly overturned to the northwest. These structures (F_2) fold an earlier foliation and plunge parallel with the regional lineation. Isoclinal mesoscopic folds (F_1) predate the overturned folds and, together with the distribution of rock types, strongly suggest earlier regional isoclinal folds. I have never been able to really define such folds. Hall and others (1975), however, described such folds in the western segment of the Hudson Highlands in New York. At least two subsequent (F_3 and F_4) generations of folds have been recognized at places in New Jersey and eastern Pennsylvania. One wonders whether major regional Proterozoic Y nappes like those in the Adirondacks (McLelland, 1979) are not present in the Reading Prong, almost totally masked by the severe later deformations? One also wonders whether the sequence gneiss, marble, gneiss, marble, gneiss in the Franklin-Sterling area does not result from the refolding of a major early isoclinal fold? In any case, structural geology does not contradict the interpretation that the Losee Metamorphic Suite is basement to the metasedimentary gneisses.

DEPOSITIONAL ENVIRONMENTS OF THE LAYERED ROCKS

Although the Losee and the calcareous and quartzofeldspathic gneisses are quite distinct, determination of their paleoenvironments is very difficult because of metamorphism and deformation.

Blatt and others (1972) have developed a chemical end-member diagram (Fig. 19) in an attempt to relate the chemical composition of sandstones to their tectonic setting. The diagram makes a fairly effective separation, with some overlap, between "eugeosynclinal" sandstone, sandstone deposited in deep

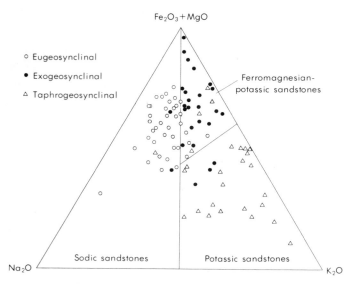

Figure 19. Diagram relating chemical composition (weight percent) of sandstones to tectonic setting. After Blatt and others (1972).

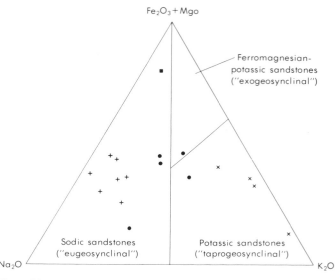

Figure 20. Diagram relating the chemical composition of Reading Prong layered rocks to tectonic setting. Symbols: biotite-quartz-feldspar gneiss (.); potassic feldspar gneiss (x); epidote-scapolite-quartz gneiss (■); quartz-oligoclase gneiss (Losee Metamorphic Suite) (+). Data from this study and Drake (1969).

fault basins on a craton, and sandstones of clastic wedges that spread out over the edges of cratons from sources in peripheral folded mountains.

Available chemical data for layered rocks of the Reading Prong are plotted (Fig. 20) on the discrimination diagram of Blatt and others (1972). Rocks of the Losee Metamorphic Suite plot well within the "eugeosynclinal" field and below or to the left of virtually all the sodic sandstones shown on Fig. 20. This plot supports the belief that these rocks are quartz keratophyre, not sodic graywacke. Most of the quartzofeldspathic rocks plot, not too surprisingly, in the potassic sandstone field; all but one are in the zone of overlap of "exogeosynclinal" sandstones. I believe that these rocks must represent a clastic wedge on the edge of a craton, because what is known of the regional geology seems to preclude the existence of a deep fault trough within a craton. The interlayered marble and other calcareous rocks support the above interpretation. We know that iron was redistributed within the quartzofeldspathic gneiss during regional metamorphism, and it may well have been subtracted from the analyzed rocks.

The above data support a model of a clastic wedge of quartzofeldspathic rock containing layers and lenses of limestone and other impure calcareous rocks deposited on a basement of very sodic oceanic rocks. A granitic terrane was probably the source for the quartzofeldspathic rocks because of the relative abundance of potassium-rich rocks. Some volcanic or volcaniclastic material may have been added during sedimentation, but evidence is equivocal.

If the above interpretation is correct, it supports the idea that the Losee is basement to the metasedimentary gneisses. The two sequences represent different regimes, and I believe that a better case can be made for the Losee to be basement to the metasedimentary gneisses than for the reverse to be true.

ISOTOPIC AND RELATIVE AGES OF THE INTRUSIVE ROCKS

Only two ages have been reported for rocks of the Reading Prong in New Jersey and Pennsylvania. Long and others (1959) obtained a Pb-Pb age of 883 m.y. on uraninite from marble near Phillipsburg, New Jersey (Easton quadrangle), and an age of 833 m.y. on monazite from rocks of the Losee Metamorphic Suite from near Chester, New Jersey (Chester quadrangle). Several rocks have been dated from the Hudson Highlands of New York. These dates have been summarized by Rankin and others (1983); those that can be related to rocks described here follow. Hornblende granite and late tectonic alaskite of probable Byram-type were dated by Rb-Sr techniques as having ages of 1,106 ±90 m.y. and 996 +94 m.y., respectively, and Lowe's (1950) Storm King Granite, almost certainly a member of the Byram Intrusive Suite (see above), was dated by U-Pb techniques as having an age of 1,149 ±45 m.y. A partial melt of paragneiss of "Canada Hill" type was dated by Rb-Sr techniques as having an age of 913 ±90 m.y. The "Canada Hill Granite" is a rock very much like the Losee Metamorphic Suite (Drake, 1969); a partial melt, then, would correspond to the albite-oligoclase granite phase of the Losee. The dates from the Hudson Highlands seem reasonable for a Grenvillian terrane, whereas the few from New Jersey appear to be too young.

Rocks of the Hexenkopf Complex are proabably the oldest rocks in the Reading Prong of eastern Pennsylvania because they appear to lie beneath rocks of the Losee Metamorphic Suite, which is believed (see above) to be beneath the other layered rocks. The Byram Intrusive Suite is known to intrude all Proterozoic rocks except the quartz-poor sequence of monzonite, syenite, and related pyroxene granite. At the only place that these intrusive rocks are in contact, rocks of the Byram are located above an orthogneiss of the quartz-poor sequence in a major antiform. At another place, a sheet of quartz syenite cuts cross structures in hornblende granite of the Byram. The relation of the the two intrusive sequences simply cannot be established at this time.

The albite-oligoclase granite has been shown to be an anatectic melt of the Losee Metamorphic Suite, which I believe resulted from the thermal event connected with the emplacement of the Byram Intrusive Suite. The microantiperthite alaskite probably results from the potassium metasomatism of albite-oligoclase granite. The Byram, albite-oligoclase granite, and microantiperthite alaskite probably are about the same age. The ages of the monzonite, syenite, and related pyroxene granite are not known, but it is difficult for me to visualize a second major period of plutonism, whether pre- or post-Byram.

METAMORPHIC CHARACTER

The critical minerals in the quartzofeldspathic gneiss throughout the Reading Prong in eastern Pennsylvania and western New Jersey, quartz-microcline-oligoclase-sillimanite-biotite-almandine, indicate a metamorphic grade of at least high amphibolite facies. Mafic rocks contain the critical minerals plagioclase-diopside-hypersthene-hornblende-biotite, implying hornblende granulite facies metamorphism. These rocks are interlayered with one another, so it is likely that the entire area has undergone hornblende granulite metamorphism. Young (1971) has presented data to show that the minimum temperature of metamorphism was 700°C in the Lake Hopatcong area, and Carvalho and Sclar (1979) have shown that metamorphic temperatures in the Franklin-Sterling area were at least 760°C. Partial melting of some rocks is consistent with such temperatures.

Offield (1967) presented a more complex metamorphic history for the rocks along the New York-New Jersey State line. There, the rocks southeast of the Green Pond syncline contain unaltered hypersthene and rutile; plagioclase is only slightly antiperthitic; the potassic feldspar is orthoclase or microperthite, microcline being absent except in intrusive rocks; and rocks of appropriate composition contain primary hornblende. These rocks were probably metamorphosed in the high hornblende or perhaps pyroxene granulite facies. No worker in New Jersey has discussed the metamorphism of this belt of rocks.

Offield (1967) believes that the rocks northwest of the Green Pond syncline and southeast of the outlier of lower Paleozoic rocks in the Waywanda and Hamburg quadrangles (Fig. 2) are at somewhat lower metamorphic grade because hypersthene is sparse and is commonly altered, plagioclase is antiperthitic or mesoperthitic, both microperthite and microcline are prominent, sphene not rutile is the prominent titanium mineral, and rocks of appropriate composition contain primary hornblende. The altered hypersthene and other features suggest a retrograde event in

this belt of rocks. These observations agree with those I have made in rocks along strike in the Delaware Valley.

Rocks, except for the Losee Metamorphic Suite, northwest of the lower Paleozoic outlier in the Waywanda and Hamburg quadrangles are thought by Offield (1967) to be in upper amphibolite facies, because hypersthene is present only in Losee rocks, nonperthitic microcline is the dominant potassic feldspar and plagioclase is nonantiperthitic oligoclase, and granitic rocks are sparse. To the southwest, I cannot recognize this metamorphic belt. The rocks in Jenny Jump Mountain (Fig. 2), the most northwesterly area I have studied, contain both microperthite and antiperthitic plagioclase, and granitic intrusions are abundant. These rocks appear to be in the lower hornblende granulite facies, as are the other rocks in the Delaware Valley and adjacent New Jersey and Pennsylvania.

Offield (1967) has also presented evidence suggesting that the rocks between the Green Pond syncline and the lower Paleozoic outlier are polymetamorphic. He made a particular point of the fact that the epidote-bearing gneisses are associated with and apparently mantle charnockitic rocks in the Franklin and Hamburg quadrangles (Baker and Buddington, 1970). He felt that anomalous association of rocks in different metamorphic facies could best be explained by superposed metamorphism. No one else has recognized this relation and its importance to the regional geology. I doubt this interpretation, however, because the calcareous rocks in this area contain abundant and varying amounts of scapolite as well as epidote, minerals that are alteration products of other minerals. I believe that these minerals result from metasomatic alterations by granitic fluids as they do in the marble. This is not to deny that there has been retrogressive metamorphism in this belt of rock, because evidence, such as altered hypersthene, suggests retrogression. There is, therefore, no evidence to refute the concept that the Losee Metamorphic Suite and the rocks of charnockitic affinity constitute an older basement upon which the calcareous and quartzofeldspathic gneisses were deposited.

CHESTNUT HILL FORMATION (HERE NAMED)

In contrast to the gneisses and granitoid rocks at high metamorphic grade, discussed above, is a poorly exposed sequence of arkose, ferruginous quartzite, quartzite conglomerate, metarhyolite, and metasaprolite along the north border of the Reading Prong on Marble Mountain (Fig. 2) and Chestnut Hill (Fig. 3). These rocks are best exposed on both banks of the Delaware River in the Easton quadrangle (Drake, 1967a). They differ from the other metasedimentary and metavolcanic rocks in that bedding can be recognized, they have not been so homogenized by metamorphism as to lose their clastic parenthood, and they are at a much lower metamorphic rank.

These rocks were included in the Pickering Gneiss by Bayley (1914, 1941) and in the Moravian Heights Formation by Fraser (*in* Miller and others, 1939). Both those units, however, contain metasedimentary rocks of at least sillimanite grade. Therefore, my

A

B

Figure 21. Chestnut Hill Formation: A. Photograph showing rock fragments in arkosic sandstone. White block to left of center is pegmatite. Coin for scale is 25 mm in diameter. B. Photomicrograph showing rock fragments in arkosic sandstone. Field of view about 9 mm.

colleagues and I (Drake, 1967a; Drake and others, 1969; Aaron, 1975) mapped the low-grade sequence as undivided metasedimentary and metavolcanic rocks, and I (1969) assigned them to an informal sequence of younger metasedimentary and metavolcanic rocks. At that time, unfortunately, I erroneously included some magnesium-rich rocks as well as some quartz-epidote gneiss (see above) and fine-grained amphibolite, which I now know to be related to marble and the quartzofeldspathic gneiss.

I herein assign the arkose, ferruginous quartzite, quartzite conglomerate, metarhyolite, and metasaprolite that crop out on Marble Mountain in Warren County, New Jersey, and Chestnut Hill in Northampton County, Pennsylvania, to the Chestnut Hill Formation for exposures on that hill along the west bank of the Delaware River in the Easton quadrangle in east-central Pennsylvania. Other typical exposures occur on Marble Mountain across the Delaware River in New Jersey. This formation is not to be confused with the Chestnut Hill schists and gneisses of Hall (1881), an abandoned unit in the Philadelphia area that consists of a mixture of schist of the Wissahickon Formation and granitized schist of the Wissachickon.

The arkose of the Chestnut Hill Formation is a purple, fine- to medium-grained massive granular rock of distinct sedimentary appearance. It is characterized by pink clastic feldspar grains, poorly defined bedding, and abundant rock fragments, including pegmatite (Fig. 21A). Thin-section study shows that it consists primarily of quartz and potassic feldspar grains and rock fragments (Fig. 21B).

Ferruginous quartzite is typically a gray, hard, brittle, foliated to nonfoliated rock that has a splintery fracture. Quartz and hematite are essentially the only minerals in the rock, and

hematite-rich parts are a dusky red-purple. The hematite appears to have been added to the rock after consolidation and deformation.

The quartzite conglomerate differs from the ferruginous quartzite by containing pebbles of vein quartz, quartzite, and dark greenish-gray metamorphic rock. The clasts are as large as 3 cm in diameter and most are flattened and sheared. The conglomerate contains hematite in the same manner as the quartzite.

Nearly massive, very dusky red, very fine-grained rock looks like a metamorphosed rhyolite. The rock consists primarily of a mass of chlorite, clay mineral-mica aggregate, and very fine-grained quartz. Hematite occurs as discrete grains within the aggregate of flaky minerals. The rock has a layered or vein-like nature that suggests a flow structure.

Purple slaty rocks are closely associated with rocks that might be metarhyolites. Although these rocks were originally thought to be just strongly cleaved metarhyolite, chemical analyses (Drake, 1969) show that they contain more than 20 percent Al_2O_3, an unlikely composition for a rhyolite. The rocks are very much like volcanic rock-derived metasaprolite from the Blue Ridge of Virginia (Drake, 1969); a similar origin is suggested.

All these rocks have been sheared and probably have been closely folded. They have not, however, experienced the metamorphic and tectonic history of the Proterozoic Y rocks of the Reading Prong. They surely must be younger, and for that reason, I provisionally consider them to be of Proterozoic Z age.

Bayley (1914) described rocks that he correlated with those herein called the Chestnut Hill Formation. At two of his localities, Pottersville Falls in the Gladstone quadrangle and near Andover in the Stanhope quadrangle (Fig. 2), the rocks are clearly

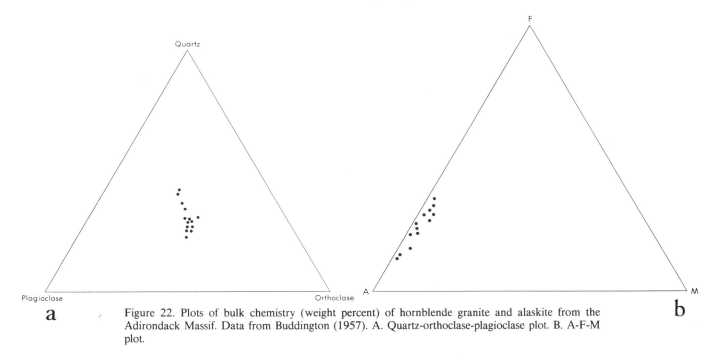

Figure 22. Plots of bulk chemistry (weight percent) of hornblende granite and alaskite from the Adirondack Massif. Data from Buddington (1957). A. Quartz-orthoclase-plagioclase plot. B. A-F-M plot.

mylonite, not metasedimentary or metavolcanic rocks. I have not seen Bayley's other localities.

I believe that there are small exposures of Chestnut Hill Formation in both the Hellertown and Allentown East quadrangles (Fig. 3). These rocks must be reexamined, however, before a stratigraphic assignment can be made. In any case, it appears that at least minor amounts of post-Grenvillian and pre-Lower Cambrian rocks are present in the Reading Prong.

COMPARISON WITH OTHER BASEMENT MASSIFS

The Adirondack Massif, needless to say, is the classic area of Grenvillian rocks in the eastern United States and is the standard with which the rocks of the other basement massifs are compared. The older Precambrian rocks of the basement massifs of the Appalachians, from the Reading Prong north, are generally considered to be more like those of the Adirondacks than they are like the rocks of the massifs to the south. The rocks of the Adirondacks and the other more important basement massifs are briefly compared below with those from the Reading Prong.

Adirondack Massif

The Proterozoic Y rocks of the Reading Prong have long been known to closely resemble those of the Adirondack massif to the north; the base of this massif is a sequence of charnockitic quartzofeldspathic gneiss overlain by several sequences of stratified quartzite, marble, calc-silicate gneiss, quartzofeldspathic gneiss, and metaigneous rocks of the anorthosite-charnockite suite as well as intrusive granite, pegmatite, and metagabbro (Rankin and others, 1983). There is no unanimity of opinion

on the relation of these rocks to one another, but their similarity to most of the rocks described here is obvious. A particular similarity is found in the basal sequence of rocks of charnockitic affinity that is overlain by a sequence of calcareous and quartzofeldspathic metasedimentary rocks that are intruded by sheets of granite and related alaskite. The individual rock units of the Adirondack massif have almost a one-to-one correlation with those of the Reading Prong, particularly the biotite-quartz-feldspar and potassic feldspar gneisses. Granitic-appearing rocks occur within the quartzofeldspathic gneisses in both the Adirondacks and the Reading Prong (see above). Their chemistry is compared in Fig. 7. Sheets of hornblende granite and alaskite are particularly characteristic of both massifs. Comparison of Fig. 14 with Fig. 22 shows that these rocks from the Adirondacks are very similar chemically to those of the Byram Intrusive Suite. Most of the Adirondack hornblende granites and alaskites are, however, somewhat richer in normative orthoclase and poorer in normative quartz than are similar rocks from the Byram Intrusive Suite (Fig. 14A, Fig. 20A). The differentiation trends of the Adirondack and Byram rocks (Fig. 14B, Fig. 20B) are remarkably similar. In addition, the alaskites from the Adirondacks contain fluorite (Buddington, 1957), as do similar rocks in the Franklin and Hamburg quadrangles (Baker and Buddington, 1970). Surprisingly, no other Reading Prong worker has reported fluorite in the Byram alaskites. Buddington obviously would look for it, and several Reading Prong geologists are Buddington's students. Perhaps fluorite only occurs in the Franklin and Hamburg quadrangles. In any case, the rocks of the Adirondacks and Reading Prong, both metasedimentary and intrusive, are remarkably similar in aspect, contain magnetite deposits, and have similar petrologic problems, such as the origin of the potassium-rich, granitic-appearing rocks in quartzofeldspathic gneiss and origin of quartz-poor syenite and monzonite.

Eastern Hudson Highlands

As was pointed out in the introduction, the rocks of the eastern Hudston Highlands differ from those of the rest of the Reading Prong. These rocks consist mostly of biotite granite gneiss and amphibolite of uncertain origin (metavolcanic?) and layered metasedimentary biotite-hornblende-quartz-feldspar gneiss (Hall and others, 1975). Minor amounts of intrusive hornblende granite gneiss and other metasedimentary gneisses are also present. Chemical and modal data are not available for these rocks. Although different from the rocks of the bulk of the Reading Prong, they are similar to the Fordham Gneiss in the New York City area and to rocks of the small basement massifs south of the Berkshire Massif in western Connecticut (Hall and others) 1975). The biotite granitic gneiss has been dated by Rb-Sr whole-rock techniques as having an age of 1,250-1,300 m.y. (Hall and others, 1975). If these dates are correct, the rocks are somewhat older than those in the rest of the Reading Prong.

Berkshire Massif, Housatonic Highlands, and other small western Connecticut massifs

Rocks of the Berkshire Massif, of the Housatonic Highlands and of other small basement massifs in western Connecticut are reported to be lithologically similar (Hall and others, 1975; Rankin and others, 1983). Most work has been done in the Berkshires, so those rocks will be briefly described. The bulk of the rocks there are graphitic and nongraphitic quartzofeldspathic gneiss; included are lesser felsic and mafic metavolcanic rocks and a distinctive calc-silicate unit. Other quartzofeldspathic units appear to contain calcareous layers or lenses (Ratcliffe and Zartman, 1976).

The metasedimentary and metavolcanic rocks are intruded by the Tyringham Gneiss, which forms in broadly concordant sheets, much like the intrusive rocks of the Reading Prong. The Tyringham ranges from granite to granodiorite in composition. The Tyringham yields slightly discordant U-Pb ages, which suggest an emplacement about 1,050 m.y. ago (Ratcliffe and Zartman, 1976).

The metasedimentary and metavolcanic sequences in the Berkshires, and presumably in the massifs of western Connecticut are similar to those in the Reading Prong. Not enough data are available to make any correlation of the Tyringham Gneiss with the rocks of the Byram Intrusive Suite. The major differences between the Berkshire rocks and those of the Reading Prong are the general paucity of intrusive rocks and the lack of rocks similar to those of the Losee Metamorphic Suite and to the rocks of charnockitic affinity. Interestingly, the Proterozoic Y rocks in the Berkshire Massif are uncomfortably overlain by the Dalton Formation of Proterozoic Z and Early Cambrian age that contains meta-arkose and quartz-pebble conglomerate (Ratcliffe and

Zartman, 1978) reminiscent of the Chestnut Hill Formation of the Reading Prong. The Dalton perportedly grades into the clean Cheshire Quartzite (Ratcliffe and Zartman, 1978). In the Reading Prong, however, there is no evidence of a gradational relation between the Chestnut Hill and the Lower Cambrian Hardyston Quartzite.

Green Mountain Massif

The Proterozoic Y rocks of the Green Mountains have not been studied in any great detail. They are collectively known as the Mount Holly Complex and comprise a sequence of biotite gneiss, locally muscovitic, which is massive and granitoid in some places and layered in others (Rankin and others, 1983). Less abundant rock types include amphibolite, hornblende gneiss, mica schist, quartzite, calc-silicate, granofels, marble, and numerous intrusive bodies of pegmatite and foliated granitoid rocks. These rocks are certainly Grenville-type, but too few data are available for any direct comparison with rocks of the Reading Prong.

Honey Brook Upland

The Honey Brook Upland is underlain by very poorly exposed, undated, but presumably Proterozoic Y rocks that lie unconformably beneath the Proterozoic Z Chickies Quartzite. Four major types of rock crop out (more properly, fail to crop out) in the upland: granulite gneiss, anorthosite and related rocks, layered gneiss, and Pickering Gneiss (Drake and others, 1978). The granulite gneiss consists of hypersthene-hornblende-mesoperthite-quartz-andesine gneiss that has all the features of rocks of the charnockite series. Plagioclase-rich rocks ranging from anorthosite to diorite constitute the anorthsite suite in the Honey Brook Upland area, which is intimately associated with the charnockitic granulite gneiss. Layered gneiss has light and dark phases, which are closely interlayered. The leucocratic gneiss consists of microcline microperthite-quartz gneiss and biotite-oligoclase-microperthite-quartz gneiss. The dark gneiss is hornblende-andesine amphibolite. I believe that these light and dark gneisses are metamorphosed felsic and mafic volcanic or volcaniclastic rocks. The Pickering Gneiss is a metasedimentary sequence consisting of graphitic quartz-plagioclase gneiss and microperthite-quartz gneiss that in places contains hypersthene. The unit contains layers of calc-silicate gneiss and marble. These rocks are very Grenvillian, much like the rocks in the Reading Prong to the north. If the layered gneiss were sodium-rich, the sequence charnockite, Losee, quartzofeldspathic gneiss would be duplicated. The Honey Brook area also appears to lack intrusive rocks of Byram aspect. The rocks in the Honey Brook Upland are the southernmost that closely resemble those of the Reading Prong.

Avondale-West Chester Massif

The rocks in the Avondale-West Chester Massif are very

high grade and consist primarily of poorly foliated hypersthene-salite-microantiperthite granofels and potassic feldspar-quartz granofels that contains minor amounts of garnet, clinopyroxene, biotite, and hornblende (Bascom and Stose, 1932; Wagner and Crawford, 1975; Drake and others, 1978). These rocks were called Baltimore Gneiss by Bascom and Stose (1932) and have as little similarity to that unit as they do to the rocks of the Reading Prong. The rocks of the Avondale-West Chester massif have yielded zircon concordia intercept ages of 980 and 1,060 m.y. (Grauert, 1974; Grauert and others, 1973).

Baltimore Gneiss domes

The Baltimore Gneiss domes, actually refolded recumbent folds, are cored by quartzofeldspathic gneiss of granodioritic to granitic composition (Crowley, 1976). This quartzofeldspathic gneiss has layered and augen phases and contains some amphibolite. These rocks are collectively known as Baltimore Gneiss. At places, the gneiss has melted, forming venites and granitoid bodies that cut across the layering of the gneiss. Crowley (1976) believes, and I concur, that the Baltimore Gneiss probably is a metamorphosed pile of largely felsic volcaniclastic rocks. The Baltimore Gneiss has yielded ages in the range of 1 to 1.2 b.y. (Rankin and others, 1983). In spite of the presence of amphibolites, the Baltimore Gneiss is not at all like the rocks of the Reading Prong.

Northern Blue Ridge

The northern Blue Ridge of Virginia and Maryland has a core of Proterozoic Y rocks that collectively have been called the "Virginia Blue Ridge Complex." Most of these rocks have been described as metamorphosed intrusive rocks because little paragneiss has been identified. Work in recent years (Herz, 1968; Bartholomew, 1977; Eric Force, written commun., 1981; Norman Herz, written commun., 1981) has discovered a terrane in central Virginia that, although not similar to that of the Reading Prong, bears a close resemblance to the Grenville terranes of the Adirondacks and Quebec. These rocks of pre-Grenvillian to Grenvillian age have a relative age sequence of: (1) granulite-facies metasedimentary rocks and mangerites, (2) anorthosite and leucocratic granite, and (3) ferrodiorite and charnockite. The intrusive rocks form as broadly concordant sheets within the metasedimentary rocks.

Southern Blue Ridge.
Both orthogneiss and paragneiss crop out in the southern Blue Ridge (Rankin and others, 1983). Although detailed descriptions of this terrane are sparse, orthogneiss appears to be much more prevalent than paragneiss. Abundant age determinations clearly confirm the Grenvillian age of these rocks. At the present state of knowledge, however, these rocks do not appear to particularly resemble the rock of the Reading Prong.

CONCLUSIONS

The oldest Proterozoic Y rocks, except for those of the Hexenkopf Complex, of the Reading Prong in eastern Pennsylvania and New Jersey are thought, but have not been proved, to be those of the sodic Losee Metamorphic Suite. These rocks are most probably metamorphosed quartz keratophyre and lesser interlayered spilitic basalt, which are now respectively oligoclase-quartz gneiss and amphibolite. Some of the oligoclase-quartz gneiss may be metamorphosed trondhjemite. It is also possible, but highly unlikely, that some of the oligoclose-quartz gneiss is metamorphosed sodic graywacke. Rocks of charnockitic affinity are also highly sodic and probably are partially melted, amphibolite-rich phases of the Losee. The chemistry of these rocks demands an oceanic affinity.

The Hexenkopf Complex consists of severely altered mafic rocks that crop out over only a few square kilometers in eastern Pennsylvania. The complex appears to be overlain by rocks both of the Losee Metamorphic Suite and quartzofeldspathic gneiss, which suggests an unconformity.

The Losee is thought, but has not been proved, to be the basement upon which a thick sequence of calcareous and quartzofeldspathic sedimentary rocks was deposited. There is some evidence for an unconformity between the calcareous and quartzofeldspathic rocks and the underlying Losee. The calcareous and quartzofeldspathic rocks are now marble, amphibolite, pyroxene gneiss, epidote- and scapolite-bearing gneisses, biotite-quartz-feldspar gneiss, and potassic feldspar gneiss.

After the emplacement of these rocks, sheets of hornblende granite and alaskite of the Byram Intrusive Suite were intruded into the older rocks during an intense period of regional metamorphism and deformation about 1 b.y. ago. The emplacement of the intrusive sheets was accompanied by the formation of migmatites and, in places, the total transformation of metasedimentary rocks into igneous-appearing rocks in much the manner described long ago by Fenner (1914).

Large sheets of quartz-poor monzonite and syenite and related pyroxene granite were also syntectonically emplaced in the older rocks in parts of north-central New Jersey. The timing of the emplacement and the relation of the quartz-poor rocks to those of the Byram Intrusive Suite remains moot.

Anatectic melt and migmatites formed within the Losee Metamorphic Suite and, to a lesser extent, in the potassic feldspar gneiss during the regional metamorphic event. The melt within the Losee crystallized in small to moderately large bodies of albite-oligoclase granite. The melt within the potassic feldspar gneiss crystallized into unmapped bodies of alkali feldspar granite and quartz-rich granitoid. Small bodies of microantiperthite alaskite were formed by K_2O metasomatism where rocks of the Losee and Byram were juxtaposed during the thermal event.

The metamorphic history of the Reading Prong is not clear at this time. All the Proterozoic Y rocks were metamorphosed in at least upper amphibolite facies and most in hornblende granulite facies. There is some evidence of polymetamorphism along the

New York-New Jersey state line. This concept, however, needs more study.

After the Proterozoic Y events, arkosic and quartzose sedimentary rocks and some metavolcanic(?) rocks of the Chestnut Hill Formation were deposited. These rocks are only sparsely present along the northwest edge of the Reading Prong and are probably of Proterozoic Z age.

To establish a paleoenvironment for the Proterozoic Y rocks of the Reading Prong is difficult. The two layered sequences, the Losee and the calcareous and quartzofeldspathic gneisses are quite different and distinct. The first has an oceanic affinity, and the second has the character of a continental-margin sequence. The available data support a model of a clastic wedge of quartzofeldspathic rock containing layers and lenses of limestone and other impure calcareous rocks deposited on a basement of very sodic oceanic rocks. A granitic terrane was probably the source for the quartzofeldspathic rocks because of the relative abundance of potassium-rich rocks. Some volcanic or volcaniclastic material may have been added during sedimentation, but evidence is equivocal.

Most authors believe that the Byram Intrusive Suite was derived by anatectic melting. It does not appear to have been derived from the quartzofeldspathic terrane at the current level of erosion. Therefore, the Byram rocks probably were derived from the terrane that supplied the debris that formed the quartzofeldspathic gneiss. The quartz-poor monzonite, syenite, and related pyroxene granite are also thought to be anatectic. Perhaps they originated at a greater depth than the Byram Intrusive Suite, as Young (1978) believes, but this is equivocal, as their relation to the Byram is really not known.

The rocks of the Reading Prong are very much like those of the Adirondacks and probably like those of the Honey Brook Upland. They have certain similarities to the rocks of the Green Mountains, Berkshires, and some of the small basement massifs of western Connecticut. They are not at all like the rocks of the eastern Hudson Highlands and some of the western Connecticut massifs, the Avondale-West Chester Massif, the Baltimore Gneiss, or the rocks of the northern and southern Blue Ridge. These facts probably will have some importance in microplate reconstruction, but just what is a bit murky at this time.

REFERENCES CITED

Aaron, J. M., 1975, Geology of the Nazareth quadrangle, Northampton County, Pennsylvania: U.S. Geological Survey Open-File Report, 75–92, 353 p.

Baker, D. R., and Buddington, A. F., 1970, Geology and magnetite deposits of the Franklin and part of the Hamburg quadrangle, New Jersey: U.S. Geological Survey Professional Paper 638, 73 p.

Bartholomew, M. J., 1977, Geology of the Greenfield and Sherando quadrangles, Virginia: Virginia Division of Mineral Resources Publication 4, 43 p.

Bascom, Florence, and Stose, G. W., 1932, Description of the Coatesville and West Chester quadrangles [Pa.-Del.]: U.S. Geological Survey Geologic Atlas, Folio 223, 15 p.

Bayley, W. S., 1910, Iron mines and mining in New Jersey: New Jersey Geological Survey, Final Report Series, v. 7, 512 p.

—— 1914, The Precambrian sedimentary rocks in the Highlands of New Jersey: International Geological Congress, 12th, Toronto, 1913, Compte Rendus, p. 325–334.

—— 1941, Pre-Cambrian geology and mineral resources of the Delaware Water Gap and Easton quadrangles, New Jersey and Pennsylvania: U.S. Geological Survey Bulletin 920, 96 p.

Bayley, W. S., Salisbury, R. D., and Kimmel, H. B., 1914, Description of the Raitan quadrangle [New Jersey]: U.S. Geological Survey Geologic Atlas, Folio 191, 32 p.

Blatt, Harvey, Middleton, Gerard, and Murray, Raymond, 1972, Origin of sedimentary rocks: Englewood Cliffs, N.J., Prentice-Hall, 634 p.

Buckwalter, T. V., 1959, Geology of the Precambrian rocks and Hardyston Formation of the Boyertown quadrangle: Pennsylvania Geological Survey, 4th Series, Geological Atlas 197, 15 p.

—— 1962, The Precambrian geology of the Reading 15-minute quadrangle: Pennsylvania Geological Survey, 4th Series, Progress Report, 161, 49 p.

Buddington, A. F., 1957, Interrelated Precambrian granitic rocks, northwest Adirondacks, New York: Geological Society of America Bulletin, v. 68, p. 291–306.

Buddington, A. F., and Leonard, B. F., 1962, Regional geology of the St. Lawrence County magnetite district, northwest Adirondacks, New York: U.S. Geological Survey Professional Paper 376, 145 p.

Carvalho, A. V., III, and Sclar, C. B. 1979, Gahnite-franklinite geothermometer at

the Sterling Hill zinc deposit, Sussex County, New Jersey [abs.]: Geological Society of America Abstracts with Programs, v. 11, no. 1, p. 6.

Crowley, W. P., 1976, The geology of the crystalline rocks near Baltimore and its bearing on the evolution of the eastern Maryland Piedmont: Maryland Geological Survey Report of Investigations 27, 40 p.

Darton, N. H., Bayley, W. S., Salisbury, R. D., and Kummel, H. B., 1908, Description of the Passaic quadrangle, [N.J.-N.Y.]: U.S. Geological Survey Geologic Atlas, Folio 157, 27 p.

Dodd, R. T., Jr., 1965, Precambrian geology of the Popolopen Lake quadrangle, southeastern New York: New York State Museum and Science Service Map and Chart Service Map and Chart Series No. 6, 39 p.

Drake, A. A., Jr., 1967a, Geologic map of the Easton quadrangle, New Jersey-Pennsylvania: U.S. Geological Survey Geologic Quadrangle Map GQ-594.

—— 1967b, Geologic map of the Bloomsbury quadrangle, New Jersey: U.S. Geological Survey Geologic Quadrangle Map GQ-595.

—— 1969, Precambrian and Lower Paleozoic geology of the Delaware Valley, New Jersey-Pennsylvania: in Subitzky, S., ed., Geology of selected areas in New Jersey and eastern Pennsylvania and Guidebook of Excursions: New Brunswick, N.J., Rutgers University Press, p. 51–131.

—— 1970, Structural geology of the Reading Prong in Fisher, G. W., and others, eds., Studies in Appalachian geology—Central and southern: New York, John Wiley, p. 271–291.

—— 1978, The Lyon Station-Paulins Kill nappe—the frontal structure of the Musconetcong nappe system in eastern Pennsylvania and New Jersey: U.S. Geological Survey Professional Paper 1023, 20 p.

—— 1980, The Taconides, Acadides, and Alleghenides in the central Appalachians, in Wones, D. R., ed., proceedings, "The Caledonides in the USA." I.G.C.P. Project 27-Caledonide Orogen, 1979 Meeting, Blacksburg, Virginia: Virginia Polytechnic Institute and State University Memoir 2, p. 179–187.

Drake, A. A., Jr., Epstein, J. B., and Aaron, J. M., 1969, Geologic Map and sections of parts of the Portland and Belvidere quadrangles, New Jersey-Pennsylvania: U.S. Geological Survey Miscellaneous Geologic Investigations Map I-552.

Drake, A. A., Jr., and Lyttle, P. T., 1980, Alleghanian thrust faults in the Kittatinny Valley, New Jersey, in Manspeizer, Warren, ed., Field studies of New

Jersey Geology and Guide to field trips: Newark, N.J., Rutgers University, p. 91–114.

Drake, A. A., Jr., Lyttle, P. T., and Owens, J. P., 1978, Preliminary geologic map of the Newark quadrangle, Pennsylvania, New Jersey, New York: U.S. Geological Survey Open-File Report 78-595, scale, 1:250,000.

Drake, A. A., Jr., McLaughlin, D. B., and Davis, R. E., 1961, Geology of the Frenchtown quadrangle, New Jersey-Pennsylvania: U.S. Geological Geologic Quadrangle Map GQ-133.

——1967, Geologic map of the Riegelsville quadrangle, Pennsylvania: U.S. Geological Survey Geologic Quadrangle Map GQ-593.

Drake, A. A., Jr., and Morgan, B. A., 1980, Precambrian plate tectonics in the Brazilian shield—Evidence from the pre-Minas rocks of the Quadrilatero Ferrifiro, Minas Gerais: U.S. Geological Survey Professional Paper 1119-B, 19 p.

Faill, R. T., and MacLachlan, D. B., 1980, Cross-sections of the Appalachians in Pennsylvania [abs.]: Geological Society of America Abstracts with Programs, v. 12, no. 2, p. 33–34.

Fenner, C. N., 1914, The mode of formation of certain gneisses in the highlands of New Jersey: Jour. Geology, v. 22, p. 594–612, 694–702.

Grauert, Borwin, 1974, U-Pb systematics in heterogeneous zircon populations from the Precambrian basement of the Maryland Piedmont: Earth and Planetary Science Letters, v. 23, p. 238–248.

Grauert, Borwin, Crawford, M. L., and Wagner, M. E., 1973, U-Pb isotopic analyses of zircons from granulite and amphibolite facies rocks of the West Chester Prong and Avondale anticline, southeastern Pennsylvania: Carnegie Institute of Washington Yearbook 72, p. 290–293.

Hague, J. M., Baum, J. L. Herrman, L. A., and Pickering, R. J., 1956, Geology and structure of the Franklin-Sterling area, New Jersey: Geological Society of America Bulletin, v. 67, p. 434–474.

Hall, C. E., 1881, Report of the Geology of Philadelphia and the southern parts of Montgomery and Bucks Counties: Pennsylvania Geological Survey, 2d, Reports of Progress C6, 145 p.

Hall, L. M., Helenek, H. C., Jackson, R. A., Caldwell, K. G., Mose, Douglas, and Murray, D. P., 1975, Some basement rocks from Bear Mountain to the Housatonic Highlands, in New England Intercollegiate Geological Conference, 67th Annual Meeting, 1975, p. 1–29.

Herz, Norman, 1968, The Roseland alkalic anorthosite massif, Virginia: New York Museum and Science Service Memoir 18, p. 357–368.

Hotz, P. E., 1952, Magnetite deposits of the Sterling Lake, New York-Ringwood, New Jersey area: U.S. Geological Survey Bulletin 982-F, p. 153–244.

Kamp, P. C. van de, 1968, Geochemistry and origin of metasediments in the Haliburtons-Madoc area, southeastern Ontario: Canadian Journal Earth Sciences v. 5, p. 1337–1372.

Kastelic, R. L., Jr., 1980, Precambrian geology and magnetite deposits of the New Jersey Highlands in Warren County, New Jersey: U.S. Geological Survey Open-File Report 80-789, 140 p.

Long, L. E., Cobb, J. C., and Kulp, J. L., 1959, Isotopic ages on some igneous and metamorphic rocks in the vicinity of New York City: New York Academy of Science Annals, v. 80, p. 1140–1147.

Lowe, K. E., 1950, Storm King Granite at Bear Mountain, N.Y.: Geological Society of America Bulletin, v. 61, p. 137–190.

MacLachlan, D. B., 1979a, Major structures of the Pennsylvania Great Valley and the new geologic map of Pennsylvania [abs.]: Geological Society of America Abstracts with Programs, v. 11, no. 1, p. 43.

——1979b, Geology and mineral resources of the Temple and Fleetwood quadrangles Berks County, Pennsylvania: Pennsylvania Geological Survey, 4th ser., Atlas 187ab, 71 p.

MacLachlan, D. B., Buckwalter, T. V., and McLaughlin, D. B. 1975, Geology and mineral resources of the Sinking Spring 7½-minute quadrangle, Pennsylvania: Geological Survey, 4th ser., Atlas 177d, 228 p.

McLelland, James, 1979, The structural framework of the southern Adirondacks in New England Intercollegiate Geological Conference, 71st Annual Meeting, and New York State Geological Association 51st Annual Meeting,

Troy, New York Guidebook: Troy and Albany, N.Y., Rensselaer Polytechnic Institute and New York State Geological Survey, p. 120–146.

Miller, B. L., and others, 1939, Northampton County, Pennsylvania: Pennsylvania Geological Survey, 4th ser., Bulletin C48, 496 p.

——1941, Lehigh County, Pennsylvania: Pennsylvania Geological Survey, 4th ser., Bulletin C39, 492 p.

Nockolds, S. R., 1954, Average chemical compositions of some igneous rocks: Geological Society of America Bulletin, v. 65, p. 1007–1032.

O'Connor, J. T., 1965, A classification for quartz-rich igneous rocks based on feldspar ratios: U.S. Geological Survey Professional Paper 525-B, p. B79-B84.

Offield, T. W., 1967, Bedrock geology of the Goshen-Greenwood Lake area, N.Y.: New York State Museum and Science Service Map and Chart Series No. 9, 78 p.

Pettijohn, F. J., Potter, P. E., and Siever, Raymond, 1973, Sand and sandstone: New York, Springer-Verlag, 618 p.

Rankin, D. W., 1976, Appalachian salients and recesses: Late Precambrian continental breakup and the opening of the Iapetus Ocean: Journal of Geophysical Research, vol. 81, no. 32, p. 5605–5619.

Rankin, D. W., Stern, T. W., McLelland, James, Zartman, R. E., and Odom, A. L., 1983, Correlation chart for Precambrian rocks of the Eastern United States: U.S. Geological Survey Professional Paper 1241-E, 18 p.

Ratcliffe, N. M., 1980, Brittle faults (Ramapo fault) and phyllonitic ductile shear zones in the basement rocks of the Ramapo seismic zones, New York and New Jersey, and their relationship to current seismicity, in Manspeizer, Warren, ed., Field studies of New Jersey geology and guide to field trips: Newark, N.J., Rutgers Univ., p. 278–312.

Ratcliffe, N. M., and Zartman, R. E., 1976, Stratigraphy, isotopic ages, and deformational history of basement and cover rocks of the Berkshire massif, southwest Massachusetts, in Page, L. R., ed., Contributions to the stratigraphy of New England: Geological Society of America Memoir 148, p. 373–412.

Sims, P. K., 1958, Geology and magnetite deposits of the Dover district, Morris County, New Jersey: U.S. Geological Survey Professional Paper 287, 162 p.

Smith, B. L., 1969, The Precambrian geology of the central and northeastern parts of the New Jersey Highlands, in Subitzky, Seymour, ed., Geology of selected areas in New Jersey and eastern Pennsylvania and guidebook of excursions: New Brunswick, N.J., Rutgers University Press, p. 35–47.

Spencer, A. C., and Kummel, H. B., Wolff, J. E., Salisbury, R. D., and Palache, Charles, 1908, Description of the Franklin Furnace quadrangle, [New Jersey]: U.S. Geological Survey, Geologic Atlas, Folio 161, 27 p.

Streckeisen, Albert, 1976, To each plutonic rock its proper name: Earth Science Reviews, v. 12, p. 1–33.

Swineford, Ada, Frye, J. C., and Leonard, A. B., 1955, Petrography of the late Tertiary volcanic ash falls in the central Great Plains: Journal of Sedimentary Petrology, v. 25, no. 4, p. 243–261.

Wagner, M. E., and Crawford, M. L., 1975, Polymetamorphism of the Precambrian Baltimore Gneiss in southeastern Pennsylvania: American Journal of Science, v. 275, p. 653–682.

Wherry, E. T., 1918, Precambrian sedimentary rocks in the highlands of eastern Pennsylvania: Geological Society of America Bulletin, v. 29, p. 375–392.

Wolff, J. E., and Brooks, A. H., 1898, The age of the Franklin white limestone of Sussex County, New Jersey: U.S. Geological Survey 18th Annual Report, pt. 2, p. 425–457.

Young, D. A., 1971, Precambrian rocks of the Late Hopatcong area, New Jersey: Geological Society of America Bulletin, v. 82, p. 143–158.

——1972, A quartz syenite intrusion in the New Jersey Highlands: Journal of Petrology, v. 13, p. 511–528.

——1978, Precambrian salic intrusive rocks of the Reading Prong: Geological Society of America Bulletin, v. 89, p. 1502–1514.

MANUSCRIPT ACCEPTED BY THE SOCIETY AUGUST 2, 1983

Geological Society of America
Special Paper 194
1984

The geology of the Honey Brook Upland, southeastern Pennsylvania

William A. Crawford
Department of Geology
Bryn Mawr College
Bryn Mawr, Pennsylvania 19010

Alice L. Hoersch
Department of Geology
LaSalle College
Philadelphia, Pennsylvania 19141

ABSTRACT

The Honey Brook Upland includes a southern region of upper amphibolite facies gneisses with the assemblage of plagioclase, dark green hornblende, and minor garnet and a northern region of low-pressure granulite facies rocks including an anorthosite suite plus gneisses with the assemblage hypersthene, mesoperthite, light green augite and dark green to brown hornblende. $^{40}Ar/^{39}Ar$ dates of coexisting hornblende and biotite from granulite gneisses, about 880 and 850 m.y., respectively, may represent cooling/uplift ages following Grenville metamorphism. The chemical trends exhibited by the mafic granulite facies gneisses and all the amphibolite facies gneisses show calc-alkaline affinities and are similar to those of the Cascades volcanic rocks; those of the felsic granulite facies gneisses closely resemble charnockite trends. Two-dimensional modeling of magnetic anomaly profiles suggest the anorthosite body is stocklike with vertical to steeply southeast dipping sides. The pattern of magnetic anomalies within the anorthosite massif indicates magnetite-rich hornblende gabbro segregations occur within the magnetite-poor anorthosite. A proposed geologic history is: (1) intrusion of the anorthosite into the charnockites followed by extrusion of a calc-alkaline series of volcanic rocks and deposition of volcaniclastic sediments (pre-Grenville); (2) burial and metamorphism during the Grenville orogeny; (3) mid-late Precambrian uplift followed by deposition of the Cambro-Ordovician sedimentary sequence; (4) burial and retrograde metamorphism during the Taconic orogeny; and (5) post-Taconic uplift accompanied by scissors motion on the Brandywine Manor fault.

INTRODUCTION

The Honey Brook Upland, a subdivision of the Pennsylvania Piedmont, lies some 45 km west northwest of Philadelphia (Fig. 1). Portions of 5 quadrangles comprise the bulk of the Upland, which is bounded on the north by the Juro-Triassic Lowland and on the south by the Cambro-Ordovician sedimentary rocks of the Chester Valley. Previous studies include Smith (1922), Bascom and Stose (1938), Postel (1951), O'Neill (1952), Waraska (1952), Robelen (1968), Crawford and others (1971), Huntsman (1975), Crawford and Huntsman (1976), Demmmon (1977), Thomann (1977), Organist (1978), Crawford (1979),

Raman and Hewins (1979), and Crawford and Crawford (1980). Workers prior to 1965 postulated a plutonic igneous origin for all crystalline rocks of the Upland, followed by undefined metamorphic events. Those after 1965 recognize the dominant role played by metamorphism in determining the character of these rocks and suggest a variety of protoliths.

The Honey Brook Upland contains three major rock groups: amphibolite facies gneiss, granulite facies gneiss, and an anorthosite suite (Fig. 2). The granulite gneiss terrane includes felsic, occasionally graphite-bearing, as well as mafic gneisses. Arkoses

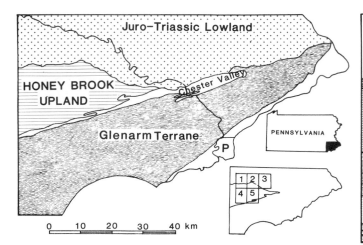

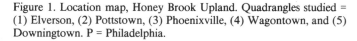

Figure 1. Location map, Honey Brook Upland. Quadrangles studied = (1) Elverson, (2) Pottstown, (3) Phoenixville, (4) Wagontown, and (5) Downingtown. P = Philadelphia.

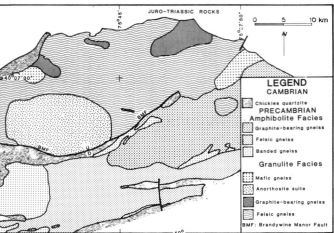

Figure 2. Geologic sketch map of the Honey Brook Upland. Compiled from maps by Demmon, (1977); Huntsman (1976); Robelen (1968); and Thomann (1977).

of the Triassic Stockton Formation or quartzites of the Cambrian Chickies Formation nonconformably overlie the granulites to the north. The ovoid anorthosite body lies within the granulite region but is bounded on the south by the Brandywine Manor fault which juxtaposes anorthosite against Chickies quartzite. This Brandywine Manor fault also separates the granulite and amphibolite facies gneiss terranes. It dies out to the east northeast, where the southern field of amphibolite facies gneisses have a transitional contact with the granulites to the north. The amphibolite facies region includes graphite-bearing and other felsic and intermediate rock types as well as banded amphibolite gneiss. An inlier of felsic granulite rocks lies within graphite-bearing amphibolite facies gneiss in the eastern portion of the Upland. Chickies Quartzite nonconformably overlies the amphibolite facies region along both its northern and southern borders.

The paucity of outcrops within the anorthosite and granulite facies terranes precludes any structural analysis. Sufficient outcrops exist within the amphibolite facies terrane for structural analysis, but such study has yet to be accomplished. The general map pattern suggests the Brandywine Manor fault (Fig. 2) raised both the anorthosite body and granulite gneiss relative to the amphibolite gneiss block. However, the lack of a distinct boundary between granulite and amphibolite-facies rocks in the eastern part of the Upland indicates that the Brandywine Manor fault probably dies out near the isolated septum of Chickies Quartzite in the center of the Downingtown quadrangle (Fig. 1 and 2). Thus, fault displacement apparently increases westward suggestive of a scissors motion.

Altered and fresh diabase dikes cut the crystalline rocks. None of these dikes have been radiometrically dated but, by tradition (Bascom and Stose, 1938), the altered ones have been termed Precambrian. The fresh ones are Triassic in age and are similar to Triassic diabase dikes elsewhere in the eastern part of the state. Minor exposures of Precambrian marble, ultramafic

rocks, and pegmatites are scattered throughout the Upland. These interesting rock types neither have been studied in detail nor are included in this report.

The Cambro-Ordovician sedimentary sequence south of the Upland, in the Chester Valley, consists of carbonates overlying the Chickies Quartzite. These sediments are overlain by the pelites of the Wissahickon Formation. The northern edge of the Wissahickon defines the northern boundary of the Glenarm Terrane of Crawford and Crawford (1980). Within this Terrane, the Cream Valley fault juxtaposes the Wissahickon schist against a southern block of Grenville age Precambrian gneisses (Fig. 3).

Two metamorphic events (M1 and M3 of Crawford and Crawford, 1980) have affected the Upland rocks and those of the northern Glenarm Terrane (Fig. 3). A Grenville event (M1) imparted high-pressure granulite facies assemblages to the Precambrian gneisses in the Glenarm Terrane and low-pressure granulite and upper amphibolite facies assemblages to the gneisses in the Upland. Following the deposition of the Cambro-Ordovician sedimentary sequence and the Wissahickon pelitic rocks, a second metamorphic event (M3) overprinted the Upland gneisses with greenschist facies assemblages. This event generated greenschist facies assemblages in the Chester Valley sequence and the northern Wissahickon pelites and also overprinted the Glenarm Terrane Precambrian granulite gneisses with amphibolite facies minerals or assemblages (Fig. 3). The M2 event of Crawford and Crawford (1980) is found only near the Wilmington Complex along the Delaware-Pennsylvania border 25 km to the south of the Upland.

The M1 event in the Glenarm Terrane occurred at about 1000 m.y. (Crawford and Crawford, 1980). Preliminary paleomagnetic studies indicate the pole for the anorthosite massif is upward and lies in a west-southwesterly direction. This is consistent with the location of other Grenville age poles in North America. The only reliable radiometric age determinations avail-

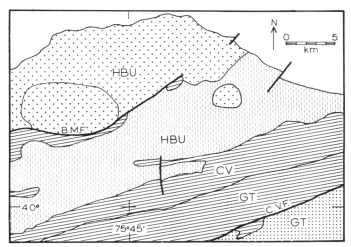

TABLE 1. MINERAL ASSEMBLAGES OF HONEY BROOK UPLAND ROCKS

	Quartz	Mesoperthite	K-feldspar	Plagioclase	Hypersthene	Augite	Hornblende	Garnet	Biotite	Muscovite	Epidote	Graphite
AMPHIBOLITE FACIES:												
Graphite-bearing Felsic gneisses	X	±	X				Tr±		X±	X±	X	X
Other Felsic gneisses	X		X±						X	X	X	
Intermediate gneisses	X		X				X	Tr±	X	X	X	
Banded mafic gneisses	X±		X	X±		X±	X	X±	X±		Tr±	Tr
GRANULITE FACIES:												
Graphite-bearing Felsic gneisses	X	X		X	X				Tr			X
Other Felsic gneisses	X	X	Tr	X	X	X±	X	Tr±	X			
Mafic gneisses	X	X±		X	X±	X±	X	Tr±	Tr±			
Anorthosite				X		X±	X	Tr±				
Intermediate types				X	Tr±	X±	X	±				
Hornblende-gabbro				X		X±	X					

Note: X = present in essential amounts; Tr = present in trace amounts; ± = may or may not be present.

Figure 3. Metamorphic facies map, Honey Brook Upland (HBU), Chester Valley (CV), and Glenarm Terrane (GT) (after Crawford and Crawford, 1980). BMF = Brandywine Manor fault. CVF = Cream Valley fault. Dots: Granulite facies, moderate to high pressure (MI), overprinted by amphibolite facies (M3). Crosses: Granulite facies; low to moderate pressure (M1), overprinted by greenschist facies (M3). Dashes: Upper amphibolite facies (M1), overprinted by greenschist facies (M3). Rules: Greenschist facies (M3).

able for the M1 event in the Upland are 880, 850 m.y. $^{40}Ar/^{39}Ar$ ages (Sutter and others, 1980) from the mafic granulite regions. These two dates of hornblende and biotite, respectively, are interpreted at cooling/uplift ages following Grenville metamorphism and they indicate no disturbance of either mineral during Paleozoic orogenesis. The M3 event occurred between 440 and 320 m.y. ago in rocks of the Glenarm Terrane (Crawford and Crawford, 1980). The sole biotite cooling age for this event is 403 m.y. (Sutter, personnal communication) from the southwestern most mafic amphibolite gneisses.

PETROGRAPHY

Introduction

The amphibolite facies rocks are distinguished from the granulite facies rocks by the presence of plagioclase and dark green hornblende and the lack of mesoperthite, hypersthene, and light green augite. Rocks of both facies contain an overprinting greenschist facies assemblage of blue-green amphibole rims around hornblende, uralitization of pyroxene, and saussuritization of plagioclase. Detailed petrographic descriptions are in the appendix to this paper.

Amphibolite Facies

Massive, slightly foliated felsic and intermediate gneisses predominate over layered, well foliated mafic gneisses (Fig. 2). Quartz, plagioclase, hornblende, biotite, and muscovite are present in varying amounts in all rocks belonging to the amphibolite facies (Table 1). Minor orthoclase is found in graphite-bearing

and other felsic rocks, but no mesoperthite occurs south of the Brandywine Manor fault. Minor hypersthene and augite occur in mafic portions of the layered gneisses immediately south of the Brandywine Manor fault. Both intermediate and mafic gneisses may contain garnet which is absent in the graphite-bearing and other felsic types. The layered amphibolite gneisses exhibit an alteration of mafic and felsic layers (Fig. 4) which are isoclinally folded and regionally trend northeast. Green hornblende and plagioclase comprise the bulk of the mafic layers, with quartz and plagioclase prevalent in the felsic layers (Table 1).

Granulite Facies

Felsic and Mafic Gneisses. All granulite facies rock types are intimately intermingled in field occurrence. Most of these rocks are medium- to coarse-grained felsic gneisses with a faint streaky foliation caused by an alignment of the sparse mafic grains and elongation of quartzofeldspathic clusters (Fig. 5). Slightly foliated intermediate and well-foliated mafic types are interlayered as subordinate varieties (Fig. 6). The mafic rocks predominate in two sub-regions (Fig. 2). The one patch of felsic granulite gneiss that occurs south of the boundary between the two major metamorphic facies may be an inlier in the amphibolite facies terrane revealing the presence of underlying granulite rocks.

Quartz, plagioclase, and anhedral medium-grained mesoperthite dominate over subordinate hornblende, augite, and hypersthene in the felsic gneisses (Table 1). Graphite-bearing felsic varieties are poor in hornblende and augite but contain up to 10 percent hypersthene. Hornblende dominates in the mafic gneisses with plagioclase, augite, and hypersthene comprising the remainder of the rock.

Anorthosite Suite. These rocks range, with increasing mafic (predominantly hornblende) content of the rock, from anorthosite, which comprises 40 percent of the anorthosite body, through leuco-hornblende gabbro (40%) to hornblende gabbro

Figure 4. Banded amphibolite facies gneiss at spillway of Marsh Creek Lake, Downingtown, 7½′ quadrangle, exhibiting isoclinal folding. A massive felsic amphibolite facies gneiss unit lies at the base of the outcrop.

Figure 5. Equigranular felsic granulite facies gneiss showing a faint foliation.

Figure 6. Medium-grained mafic granulite facies gneiss and coarse-grained, equigranular felsic granulite facies gneiss.

(20%) (Fig. 7a, b and c). All of these types belong to the anorthosite suite (Table 1 and Fig. 8).

For a detailed description of the mineralogy the reader is referred to Crawford and others (1971). In that paper, the anorthosite body was described as consisting of concentric zones becoming more gabbroic (hornblende-rich) toward the perimeter. However, further field work and magnetic studies (see the following section on magnetic profile modeling) seem to demonstrate that, although mafic content increases towards the perimeter, the mafic rocks actually occur as randomly sited segregations of irregular shape and variable composition rather than in definite zones of increasing mafic content.

The contact between the anorthosite body and its country rocks is not exposed. However, a road cut exposes an apophysis or dike of mafic anorthosite (hornblende gabbro) containing blocks of anorthosite intruding the granulite facies gneisses (Waraska, 1952; Organist, 1978) (Fig. 9 a and b).

CHEMISTRY

Introduction

Whole rock major element analyses of the Honey Brook Upland amphibolite facies gneisses, granulite facies gneisses, and anorthosite suite rocks and their Barth Catanorms are given in Tables 2, 3, and 4 respectively. Analyses were obtained through a combination of x-ray fluorescene spectroscopy, atomic absorption spectroscopy, UV-Vis spectroscopy, and wet chemical means by W. A. Crawford, assisted by Kim Ouderkirk. Other analyses used

Figure 7. A. Anorthosite; B. Leuco-hornblende gabbro; C. Hornblende gabbro. Light areas composed of plagioclase; dark areas predominantly hornblende with minor pyroxene and opaque minerals.

in the averages are from Crawford and others (1971); Huntsman (1975); Demmon (1977); and Thomann (1977).

Amphibolite Facies Gneisses

Plots of selected major element oxide values versus SiO_2

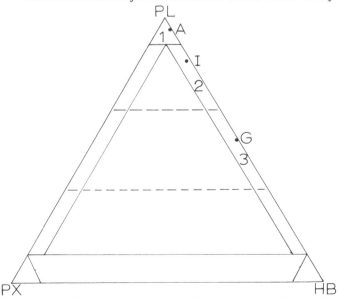

Figure 8. Classification of Anorthosite Suite rocks according to the modal Plagioclase-Pyroxene-Hornblende diagram. A = average of 9 modes of anorthosite samples; I = average of 11 modes of leuco-hornblende gabbro samples; G = average of 4 modes of gabbro samples. Field 1 = anorthosite; Field 2 = leuco-hornblende gabbro; Field 3 = hornblende gabbro, after Streckeisen (1973).

(Fig. 10) demonstrate a similarity in trends to those of the Cascades (Williams, 1942; Crawford and Huntsman, 1976). The protoliths of these amphibolite facies gneisses may, then, be volcanic rocks of the calc-alkaline series. To further test this hypothesis, these data were plotted on a series of diagrams (Fig. 11a, b, and c) proposed by Irvine and Baragar (1971) as a guide to the chemical classification of volcanic rocks.

The Honey Brook Upland amphibolite facies gneisses plot within the envelope containing the Cascades and Aleutian calc-alkaline rocks (Fig. 11a and b) and are subalkaline (Fig. 11a) and K-poor or "common" variants (Fig. 11b). The mafic amphibolite facies gneisses lie in the fields of basalt and andesite, while the felsic gneisses lie in the dacite-rhyolite fields on Fig. 11c, corresponding to similar plots of the Cascades and Aleutians. The gradational compositional change from mafic to felsic gneiss is shown by the diffuse boundary between these two fields (Fig. 12). The protoliths of some amphibolite facies rocks analyzed by us may have been locally derived volcaniclastic sedimentary rocks. Inclusion of such volcaniclastic sedimentary rocks in these plots would account for some of the observed scatter. These data suggest a calc-alkaline trend for these rocks, but alone they do not discern between a volcanic or plutonic origin. However, the intimate interlayering of mafic and felsic amphibolite gneiss layers seen in many outcrops (Fig. 4) is difficult to explain by plutonic processes. This combination of field occurrence and chemical data persuades us to propose a volcanic origin for these rocks.

The graphite-bearing felsic gneisses plot in the same field as do the other felsic amphibolite facies gneisses indicating these rocks probably do not belong to a separate unit but are merely graphite-bearing varieties of the felsic amphibolite facies gneisses.

Figure 9. A. Mafic anorthosite (hornblende gabbro) dike intrudes felsic granulite facies gneiss. Pennsylvania Turnpike (mileposts 307.6 - 307.8). B. Detailed view showing anorthosite blocks in hornblende gabbro dike with granulite country rock on lower right.

Granulite Facies Gneisses

Mafic Granulite Gneisses. Analyses of the mafic granulite plot within the fields reserved for basalt and andesites by Streckeisen (1979) (Fig. 13). These mafic gneisses lie within the same field as do the mafic amphibolite facies gneisses (Fig. 12) and overlap the calc-alkaline trends shown by the amphibolite facies gneisses (Fig. 10) as well. Thus, these granulite facies mafic rocks also are of probable volcanic origin belonging to the same series as the amphibolite facies gneisses.

Felsic Granulite Gneisses and Anorthosite Suite. The felsic granulite facies gneisses do not plot in the same field as do the felsic amphibolite facies gneisses (Fig. 14). Note, however, that the graphite-bearing granulite facies gneisses coincide with the felsic granulite gneiss field confirming the field and petrographic observations that these graphite-bearing rocks do not belong to a separate unit. All the felsic granulite facies gneisses belong to the charnockite family (Fig. 15 a and b) and not the granite-quartz monzonite suite, because they contain hypersthene and mesoperthite (Streckeisen, 1973). The anorthosite suite shows a similar trend to the one displayed by the Morin anorthosite pluton of Quebec (Papezik, 1965, p. 705) (Fig. 16). Thus the granulite facies rocks of the Honey Brook Upland exhibit the usual anorthosite-charnockite relationship

characteristic of other such suites on the east coast of North America.

Summary

The chemical data combined with petrographic and field information leads us to suggest a quite different origin for Honey Brook Upland rocks from that proposed by Bascom and Stose (1938), and further promulgated by two editions of the Geologic Map of Pennsylvania (Gray, 1960; Berg and others, 1980). Bascom believed all the graphite-bearing rocks to belong to a single meta-sedimentary unit, the Pickering gneiss, which was intruded by a co-magmatic family of plutonic rocks including gabbro, granodiorite, quartz monzonite and anorthosite.

The alternative petrogenesis we offer follows:

1. The graphite-bearing gneisses in each facies are merely graphite-bearing varieties of the felsic and intermediate gneisses and do not constitute a separate meta-sedimentary unit.

2. All the amphibolite facies gneisses and the mafic granulite facies gneisses have volcanic or volcaniclastic protoliths and are not meta-gabbros, meta-granodiorites, or mafic meta-quartz monzonites.

3. The felsic granulite facies gneisses belong to the charnockite family and are not quartz monzonites.

4. The anorthosite massif contains a suite of rocks ranging in

TABLE 2. REPRESENTATIVE INDIVIDUAL ANALYSES AND AVERAGE ANALYSES FOR EACH ROCK TYPE--AMPHIBOLITE FACIES ROCKS

	Graphite-bearing felsic gneisses Average of 4	Other Felsic gneisses			Banded mafic gneisses		
		SW-44	SW-49	Average of 23	SW-48A	SW-50	Average of 12
SiO_2	69.87	74.76	65.26	70.21	48.01	49.43	51.42
TiO_2	0.58	0.12	0.32	0.27	1.39	1.30	1.12
Al_2O_3	16.69	12.13	14.19	14.52	13.36	16.05	15.64
Fe_2O_3	0.38	0.98	2.06	0.92	4.77	3.21	1.90
FeO	1.74	0.57	4.04	2.00	8.76	8.74	7.84
MnO	0.04	0.02	0.11	0.04	0.26	0.20	0.15
MgO	1.42	0.08	2.38	1.22	5.04	5.56	5.88
CaO	2.11	0.29	5.79	1.82	12.52	9.33	8.66
Na_2O	2.33	3.28	3.11	4.21	2.42	3.10	3.33
K_2O	3.22	5.65	0.34	2.67	0.38	0.24	0.86
P_2O_5	nd	0.00	0.04	0.16	0.17	0.13	0.16
LOI	nd	1.50	1.29	1.11	1.85	1.37	1.61
Total	98.38	99.38	98.93	99.15	98.93	98.66	98.57
Barth Katanorms							
Q	34.79	31.88	27.62	27.59	1.76	0.75	0.35
C	6.24	0.11	0.00	2.00	0.00	0.00	0.00
Or	19.61	34.53	2.10	16.16	2.38	1.48	5.25
Ab	21.57	30.47	29.14	38.73	22.97	28.92	30.89
An	10.80	1.49	24.80	8.18	25.87	30.33	26.03
Hy	5.78	0.30	9.65	5.65	8.49	19.39	19.68
Di	0.00	0.00	3.93	0.00	30.88	13.51	13.82
Mt	0.41	1.06	2.25	0.99	5.27	3.49	2.06
Il	0.84	0.18	0.47	0.39	2.05	1.89	1.62
Ap	0.00	0.00	0.09	0.35	0.38	0.29	0.35
SAL	92.98	98.48	83.64	92.64	52.96	61.46	62.51
FEM	7.03	1.53	16.37	7.37	47.05	38.55	37.50

TABLE 3. REPRESENTATIVE INDIVIDUAL ANALYSES AND AVERAGE ANALYSES FOR EACH ROCK TYPE--GRANULITE FACIES ROCKS

	Graphite-bearing Felsic gneisses Average of 3	Other Felsic gneisses			Mafic gneisses		
		SW-5A	SW-52	Average of 25	SD-29B	SD-32B	Average of 12
SiO_2	66.53	65.46	75.12	67.46	60.13	53.75	53.54
TiO_2	0.56	0.56	0.20	0.49	0.89	1.31	1.77
Al_2O_3	15.90	14.61	12.25	14.94	17.28	16.13	14.88
Fe_2O_3	2.98	3.30	1.28	2.01	3.67	1.79	2.81
FeO	2.29	3.32	0.75	2.65	5.48	8.02	8.86
MnO	0.06	0.10	0.01	0.06	0.08	0.15	0.18
MgO	0.83	0.05	0.02	0.55	1.99	4.05	4.49
CaO	2.33	1.78	0.30	1.50	4.38	6.84	8.16
Na_2O	4.22	4.14	3.73	3.79	3.52	4.39	3.26
K_2O	2.70	5.54	4.89	5.33	1.08	0.91	0.77
P_2O_5	nd	0.01	0.01	0.09	0.10	0.38	0.25
LOI	nd	0.97	0.55	0.98	0.92	1.11	1.08
Total	98.40	99.84	99.11	99.85	99.52	98.83	100.05
Barth Katanorms							
Q	23.74	15.11	32.39	18.98	19.94	1.95	5.95
C	2.02	0.00	0.34	0.48	2.94	0.00	0.00
Or	16.37	33.49	29.67	32.12	6.60	5.52	4.68
Ab	38.88	38.04	34.40	34.71	32.66	40.43	30.10
An	11.86	5.05	1.47	7.00	21.78	22.19	24.38
Hy	3.16	0.85	0.08	3.71	10.66	17.28	16.38
Di	0.00	3.14	0.00	0.00	0.00	8.06	12.46
Mt	3.20	3.53	1.37	2.15	3.97	1.92	3.02
Il	0.81	0.80	0.29	0.70	1.29	1.88	2.54
Ap	0.00	0.03	0.04	0.20	0.22	0.82	0.54
SAL	92.85	91.67	98.25	93.27	83.89	70.07	65.09
FEM	7.16	8.34	1.76	6.74	16.12	29.94	34.92

TABLE 4. REPRESENTATIVE INDIVIDUAL ANALYSES AND AVERAGE ANALYSES FOR EACH ROCK TYPE--ANORTHOSITE SUITE ROCKS

	Anorthosites			Intermediate types			Hornblende Gabbros		
	SW-9B	SW-16	Average of 9	SW-14	SW-10H	Average of 11	SW-10D	SW-10E	Average of 4
SiO_2	53.08	54.53	53.86	52.29	48.88	51.56	44.93	42.60	46.90
TiO_2	0.18	0.39	0.32	0.67	1.42	0.95	1.97	2.71	1.98
Al_2O_3	26.42	27.61	26.16	23.59	18.24	23.02	15.78	13.75	17.94
Fe_2O_3	0.59	0.58	0.57	1.40	6.17	1.83	7.43	9.28	5.13
FeO	0.49	0.60	0.83	3.32	5.53	3.20	8.48	10.16	7.61
MnO	0.00	0.04	0.03	0.07	0.11	0.05	0.14	0.16	0.13
MgO	0.49	0.30	0.58	1.03	2.96	1.64	4.37	5.28	3.31
CaO	10.53	9.50	9.30	7.98	7.98	9.22	8.83	8.66	9.22
Na_2O	5.07	4.74	4.77	4.86	4.46	4.43	3.40	2.87	3.75
K_2O	0.55	0.58	0.85	1.09	1.17	0.95	1.37	1.31	1.12
P_2O_5	0.09	0.11	0.13	0.34	0.13	0.21	0.38	0.31	0.46
LOI	1.18	1.05	1.84	1.29	1.33	1.59	1.44	2.02	1.79
Total	98.67	100.03	99.24	98.94	98.38	99.05	98.52	99.11	99.34
Barth Katanorms									
Q	0.00	3.86	2.02	0.00	0.00	0.13	0.00	0.00	0.00
C	0.00	2.37	0.89	0.00	0.00	0.00	0.00	0.00	0.00
Or	3.26	3.40	5.06	6.52	7.18	5.73	8.53	8.27	6.89
Ab	43.81	42.17	43.08	44.14	41.58	40.60	30.29	27.26	35.04
An	47.80	45.99	45.56	39.81	27.31	40.97	25.05	22.19	30.0
Ne	1.08	0.00	0.00	0.00	0.00	0.00	1.14	0.16	0.00
Wo	0.12	0.00	0.00	0.00	0.00	0.00	0.00	0.00	0.00
Hy	0.00	0.89	2.11	0.73	0.74	5.39	0.00	0.00	0.44
Di	2.91	0.00	0.00	2.45	10.33	3.46	14.81	17.23	11.58
Ol	0.00	0.00	0.00	3.24	3.87	0.00	8.29	9.83	6.10
Mt	0.62	0.60	0.60	1.48	6.70	1.96	8.19	10.36	5.58
Il	0.26	0.54	0.45	0.95	2.06	1.36	2.90	4.04	2.88
Ap	0.19	0.23	0.28	0.72	0.29	0.45	0.84	0.70	1.01
SAL	95.92	97.76	96.58	90.46	76.06	87.42	65.00	57.87	71.92
FEM	4.09	2.25	3.43	9.55	23.95	12.59	35.01	42.14	28.09

composition from pure anorthosite, through leuco-hornblende gabbro, to hornblende gabbro (i.e. mafic anorthosite).

METAMORPHISM

Metamorphic minerals used as geothermometers and geobarometers are not common in the rocks of the Honey Brook Upland. Those rocks north of the Brandywine Manor Fault (Fig. 3) all contain or are associated with rocks which contain hypersthene and mesoperthite, light green augite, and dark green to brown hornblende (Table 1). This assemblage coupled with the absence of abundant garnet indicates low-pressure granulite facies (Crawford and Crawford, 1980). Valley and O'Neil (1981) found that Upland samples of graphite-bearing marbles, collected from felsic granulite gneiss terrane, show small $^{13}C/^{12}C$ fractionations (3.0 and 3.1) which are similar to those of the highest grade samples of the Adirondacks (750°C). Those gneisses south of the Brandywine Manor fault (Fig. 3) contain plagioclase in the range An_{40-65}, dark green hornblende, and pink garnet locally (Table 1), all indicative of the upper amphibolite facies of metamorphism (Crawford and Crawford, 1980). As explained in the Introduction, the high-grade metamorphism of the Upland occurred during the Grenville orogeny. A greenschist metamorphic event affected the entire Upland during the Taconic Orogeny. This event caused overprinting of blue-green amphibole rims around hornblende, decomposition of pyroxenes, and saussuritization of plagioclase.

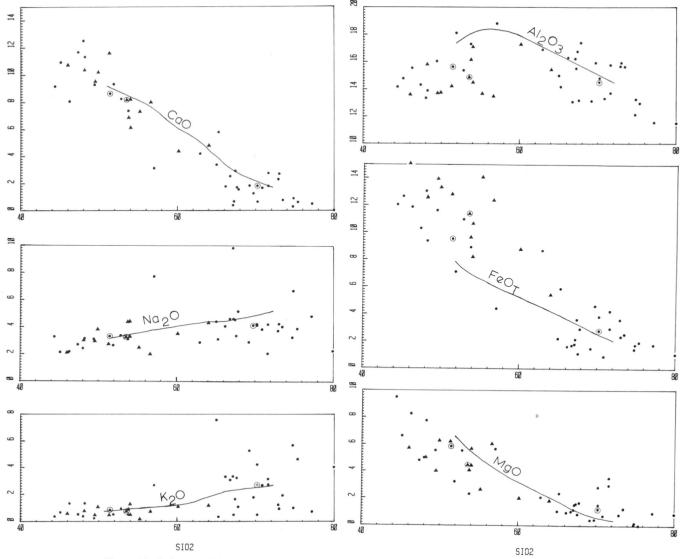

Figure 10. Oxide vs. SiO$_2$ (in weight percent) trends of the Honey Brook Upland amphibolite facies gneisses and mafic granulite facies gneisses are similar to those of the Cascade volcanic suite (solid lines after Williams, 1942). Solid circles = amphibolite facies gneisses; solid triangles = mafic granulite facies gneisses; averages enclosed in open circles.

AEROMAGNETIC MODELS

Portions of the aeromagnetic maps of the Wagontown and Downingtown quadrangles (Bromery and others, 1960a and 1960b) are superimposed on the geologic map of the Upland (Fig. 17). Socolow (1959 and 1960) points out two major patterns of the magnetic anomaly contours. First, the contact between the anorthosite body and the granulite gneiss country rock is marked by a belt of anomalies ranging in amplitude from −215 to +610 gammas. (In addition to this highlighting of the border of the anorthosite body, we notice this pattern of anomalies extends beyond the body on the northeast). Second, the ran-

dom distribution of anomalies (ranging from −70 to +155 gammas) is characteristic of the country rock gneisses.

To provide an estimate of the subsurface configuration of the anorthosite body and the attitudes of its contacts with country rock, computer-calculated two-dimensional profiles of theoretical magnetic anomalies (Fig. 18) were generated for comparison with observed magnetic profiles taken from the aeromagnetic maps. The computer program (Talwani and Heirtzler, 1964) calculates theoretical magnetic profiles for a two-dimensional cross section containing rock units of dissimilar magnetic susceptibility and treats the anomalies as though the sole factor generating them is induced magnetization. No large scale regional trends are evi-

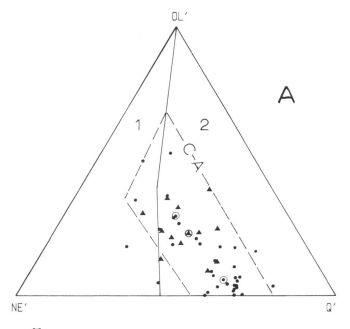

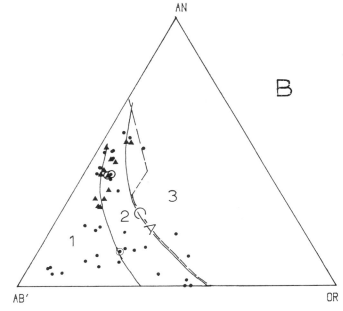

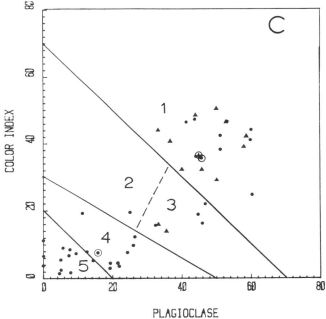

Figure 11. A. Comparison of amphibolite facies gneisses with Cascade and Aleutians suites demonstrating the subalkaline character of all 3 suites. Solid line separates the alkaline (1) and subalkaline (2) fields and the dashed line encloses the field of the Cascades and Aleutian samples (C-A) given by Irvine and Baragar (1971, Fig. 4). See explanation to Fig. 10 for meaning of symbols. $01' = 01 + 0.75Hy$; $Ne' = Ne + 0.6Ab$; $Q' = Q + 0.4Ab + 0.25Hy$. B. Comparison of amphibolite facies gneisses with Cascades and Aleutians suites demonstrating the "normal" to K-poor affinities of these rocks. Solid lines mark fields of K-poor (1), "normal" (2), and K-rich, (3) volcanic rocks; dashed line marks minimum Ab′ content of Cascades and Aleutians rocks, after Irvine and Baragar (1971, Fig. 8). See explanation to Fig. 10 for meaning of symbols. $Ab' = Ab + 1.6667Ne$. C. Volcanic nomenclature applied to amphibolite facies gneisses using the Color Index vs. Plagioclase diagram after Irvine and Baragar (1971, Fig. 7). Field: 1 = basalt; 2 = tholeiitic andesite; 3 = andesite; 4 = dacite; 5 = rhyolite. See explanation to Fig. 10 for meaning of symbols. Color Index = Di + Hy + 01 + Mt + Il + Hm; Plagioclase = An + Ab + 1.667Ne.

Figure 12. Normative Quartz-Mafic-Feldspar diagram displaying the fields of banded amphibolite gneisses and felsic amphibolite facies gneisses. Field 1: closed circles = felsic gneisses; closed squares = graphite-bearing gneisses. Field 2: closed circles = banded gneisses; closed triangles = mafic granulite gneisses. Averages enclosed in open circles. Q = Q; M = Di + Hy + O1 + Mt + Il + Hm; F = Or + An + Ab + 1.6667Ne.

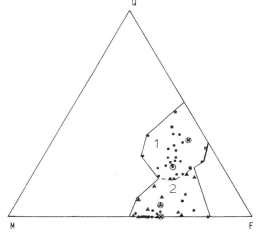

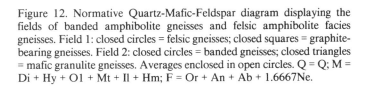

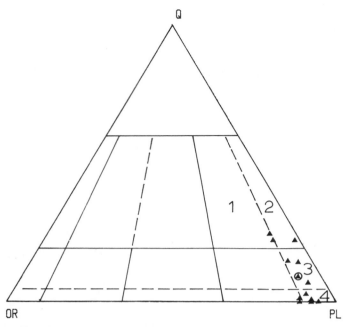

Figure 13. Mafic granulite gneiss classified as basalts and dacites on a normative quartz-orthoclase-plagioclase diagram (after Streckeisen, 1979). Fields 1 & 2 = dacite; fields 3 & 4 = basalt. Open circle = average.

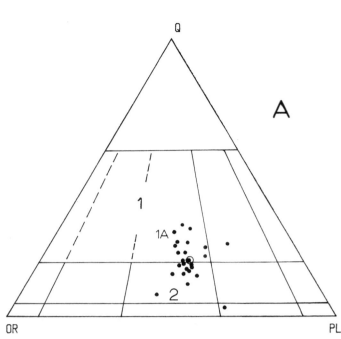

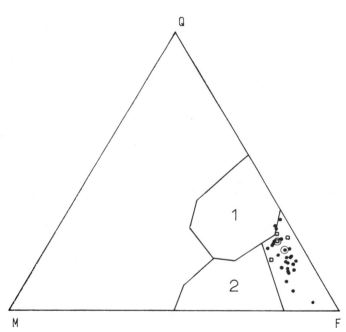

Figure 14. Felsic granulite facies gneisses lie in a field separate from those of the amphibolite facies gneisses and the mafic granulite facies gneisses. Closed circles = felsic granulite facies gneisses; open squares = graphite bearing granulite facies gneisses. From Fig. 12: Field 1 = felsic amphibolite facies gneisses; Field 2 = banded amphibolite gneisses + mafic granulite facies gneisses. Open circles = averages. See explanation to Fig. 12 for meaning of Q, M, and F.

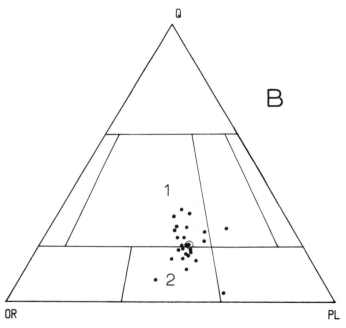

Figure 15. Felsic granulite facies gneisses show affinities toward charnockite suite rocks. Open circle = average. A. Igneous classification of charnockites (after Streckeisen, 1974). Normative quartz-orthoclase-plagioclase. Field 1 = charnockite. Most felsic granulite facies gneisses plot within the subfield of farsundite (1A) or the field of mangerite (2). B. Metamorphic classification of charnockites (after Winkler, 1979). Most felsic gneisses plot in the fields of charnockitic granolite (Field 1) or hypersthene perthiclase granolite (Field 2).

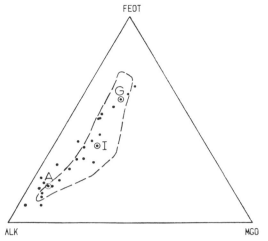

Figure 16. Total Iron (as FeO)-Alkalies-MgO diagram. Differentiation trend of the Honey Brook Upland anorthosite suite (closed circles) compared with that of the Morin anorthosite pluton field indicated by dashed line), after deWaard (1968). Open circles = average: anorthosite (A); leuco-hornblende gabbro (I); and hornblende gabbro (G).

TABLE 5. MAGNETIC SUSCEPTIBILITIES OF
HONEY BROOK UPLAND ROCKS

	Number of samples	Susceptibility (emu/cc x 10^{-6})			
		Mean	Std. Dev.	Range High	Low
Greenschist Facies:					
Chickies quartzite	2	12	4	15	10
Amphibolite Facies Gneisses:					
Low	13	57	30	99	19
Medium	13	271	173	554	102
High	7	3953	2541	8300	1023
Granulite Facies Gneisses:					
Low	18	39	19	82	18
Medium	15	199	109	473	109
High	35	3002	2765	15464	911
Anorthosite Suite:					
Low	19	37	12	66	19
Medium	11	373	276	835	127
High	12	4487	3327	10745	1253

Note: Susceptibility measurements obtained with a Bison Model 3101A Magnetic Susceptibility Meter.

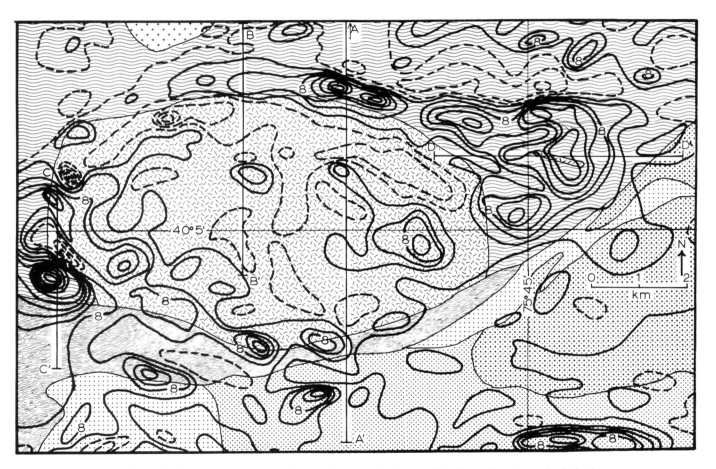

Figure 17. Aeromagnetic map superimposed on the geologic map of the Honey Brook anorthosite body and surrounding country rocks. Aeromagnetic data from Bromery and others (1960a and 1960b). Positive anomalies = solid lines; negative anomalies = dashed lines. Contour interval = 50 gammas. Solid line marked by 8 is the 8000 gamma contour. Section lines for magnetic profiles presented in Fig. 18 labeled A-A', etc. Geologic symbols the same as for Fig. 2.

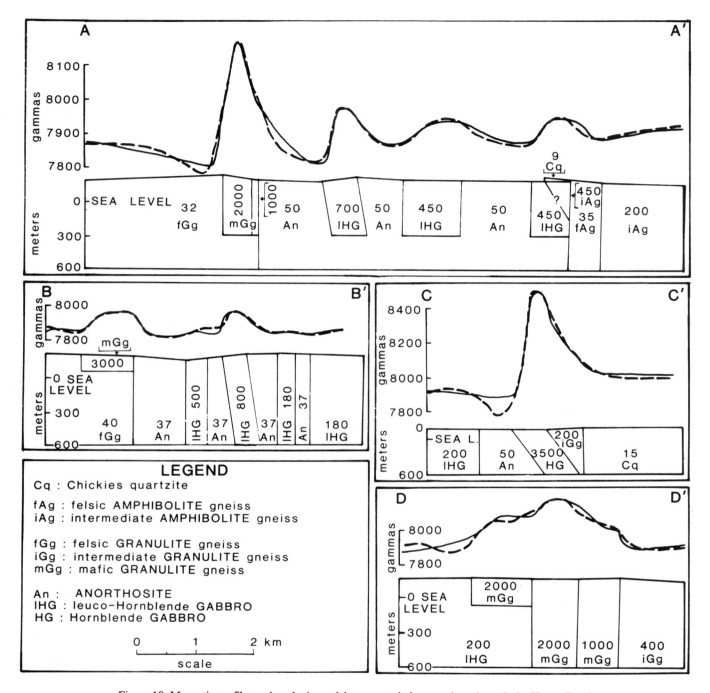

Figure 18. Magnetic profiles and geologic models computed along sections through the Honey Brook anorthosite body. Observed magnetic profiles taken from Fig. 17 are shown as solid lines; computer-generated model magnetic profiles as dashed lines. Susceptibility values (emu/cc × 10^{-6}) used in the calculations are shown within each model. Irregular upper surfaces of the geologic models reflect real topography along the cross sections.

dent on the observed profiles so that computer-calculated model profiles can be directly compared with observed ones.

The magnetic susceptibilities used in the modeling (Table 5) are grouped according to their magnitudes into three classes:

Low: susceptibilities less than 100×10^{-6} emu/cc

Medium: susceptibilities between 100 and 1000×10^{-6} emu/cc

High: susceptibilities greater than 1000×10^{-6} emu/cc

The means and standard deviations demonstrate that the above groupings differ from each other by an order of magnitude.

Fig. 18 compares the computer-calculated and observed magnetic profiles. Since all our models are first approximations, actual dimensions of the bodies producing the anomalies could deviate slightly from those presented here. For example, the dimensions of the mafic granulite gneiss producing the positive anomaly on the northern end of profile B-B′ could be modelled as being narrower and deeper than we show. However, the geologic map and field magnetic susceptibility measurements constrain both the location of geologic contacts, and the location and size of bodies of different magnetic character. Several geologic implications are derived from the profiles and the aeromagnetic map:

1. The anorthosite body is stock-like with vertical or steep boundaries (Profiles A-A′, B-B′ and C-C′). Profile D-D′ (on the east) indicates that a slab of granulite gneiss, possibly a roof pendant, overlies the anorthosite body. This suggestion is further supported by the outcrop exhibiting a dike of hornblende gabbro (mafic anorthosite) penetrating the felsic granulite facies gneisses (Fig. 9).

2. The magnetic anomalies outlining the anorthosite body probably reflect a discontinuous iron-rich thermal aureole developed in the country rock when the anorthosite suite was emplaced. Bascom and Stose (1938) postulated the existence of such an aureole.

3. The pattern of magnetic anomalies (Profiles A-A′ and B-B′) within the anorthosite body suggests that the body is not concentrically zoned as proposed by Crawford and others (1971); but, rather, that segregations of magnetite-rich gabbro are dispersed throughout a matrix of magnetite-poor anorthosite. Mapping confirms the presence of magnetite-rich gabbro in areas which are magnetically high.

4. Profile A-A′ also indicates the presence of material of high magnetic susceptibility beneath the Chickies Quartzite along the southern border of the anorthosite body. Several rock types have the necessary susceptibilities to cause this pattern. They are: leuco-hornblende gabbro, intermediate granulite facies gneiss and intermediate amphibolite facies gneiss. We chose a combination of leuco-hornblende gabbro and intermediate amphibolite facies gneiss.

CONCLUSIONS

A proposed geologic history of the Honey Brook Upland includes:

1. Stock-like intrusion by the anorthosite suite into the char-

nockitic rocks. The stock contains subordinate segregations of hornblende gabbro and leuco-hornblende gabbro in a dominant matrix of anorthosite.

2. Extrusion of a calc-alkaline volcanic suite accompanied by deposition of locally derived volcaniclastic sediments.

3. Burial and metamorphism during the Grenville orogeny (M1, Crawford and Crawford, 1980). The occurrences of graphite are most probably due to CO_2-H_2O fluid equilibrium relationships during this metamorphism. The nature of these relationships remains to be discovered.

4. Mid-late Precambrian uplift and erosion of the Precambrian rocks, and deposition of the Cambro-Ordovician clastic-carbonate sequence.

5. Burial and retrograde metamorphism during the Taconic orogeny (M3, Crawford and Crawford, 1980).

6. Post-Taconic uplift accompanied by scissors motion on the Brandywine Manor fault.

ACKNOWLEDGMENTS

We thank M. L. Crawford for her critical review of the manuscript and moral support for the project; Krishna Sinha and Mark Loisell for their instruction on XRF techniques; and Juliet C. Reed for her corrections to the manuscript. The comments of the symposium reviewers were constructive and were invaluable in improving the presentation of the information. We are indebted to the contributions to the geology of the Upland made by Floyd E. Demmon, John R. Huntsman, Gail Organist, Peter G. Robelin, and William F. Thomann. We acknowledge the support of the Ida H. Ogilvie bequest to the Geology Department, Bryn Mawr College, and the Faculty Development Fund of LaSalle College.

APPENDIX

The detailed petrographic information below is provided for the interested reader.

Felsic Amphibolite Facies Gneisses

Quartz and plagioclase ($An_{40 \pm 5}$) dominate with biotite, muscovite, and epidote comprising the remaining minerals (Table 1). Hornblende and orthoclase occur rarely. Granular quartz-rich layers alternate with and grade into quartz-plagioclase layers. Scattered foliated biotite layers may reach 2 mm in width. Clots of muscovite flakes are parallel to surrounding biotite, while epidote grains occur near plagioclase laths. Scattered grains of opaque minerals and rare garnet porphyroblasts lie mainly in biotite-rich layers. Chlorite partially replaces garnet along cracks and edges of grains and completely replaces biotite, muscovite, and epidote in these felsic gneisses. Graphite-bearing varieties are distinguished from other felsic amphibolite facies gneisses solely by the presence of graphite.

Banded Amphibolite Facies Gneisses

Within the mafic-rich layers individual hornblende blades or aggregates of blades reach 2 mm in length and surround subhedral grains of plagioclase. The predominant pleochroic scheme of these hornblendes is green-yellow but brown-green and yellow-brown varieties occur. Plagioclase ($An_{60 \pm 5}$) occurs as fresh granular anhedral crystals or as laths up to 5 mm long. Alteration to fine-grained saussurite is common, with decomposition greatest in cores of plagioclase grains. Opaque minerals occur along plagioclase-hornblende contacts or as minute inclusions within many hornblendes. Biotite flakes lie randomly in the matrix, while epidote is concentrated near altered, broken plagioclase laths. In order of decreasing abundance, apatite, sphene, and zircon are ubiquitous. Locally disseminated garnet porphyroblasts are common and chlorite replaces some mafic minerals.

The felsic layers contain considerable quartz, which, along with anhedral plagioclase ($An_{50 \pm 5}$), predominates over bladed hornblende. Muscovite, biotite, and epidote occur in close proximity to plagioclase grains. Chlorite replaces many of the mafic minerals, including pink garnet porphyroblasts. Accessory apatite is common, while sphene and zircon are rare.

Felsic and Intermediate Granulite Facies Gneisses

Anhedral, medium-grained mesoperthite consists of microcline feldspar hosting lightly saussuritized plagioclase lamellae. Microcline with no evidence of exsolution occurs in minor amounts. Fresh plagioclase ($An_{45 \pm 5}$) occurs, but cloudy, saussuritized grains are also common. The anhedral granulated quartz has undulatory extinction. Hornblende is the most abundant mafic mineral and often has borders of a blue-green amphibole or green-brown biotite. Blue-green uralite combined with chlorite is common surrounding the anhedral, pink and green hypersthene. Green to colorless augite alters along fractures, cleavage cracks, and grain margins to patches of dark yellow-orange biotite often associated with a dark cloudy material. Idioblastic garnet occurs sporadically in these rocks. Minor amounts of opaque minerals are ubiquitous. Trace accessories include zircon, sphene, and apatite. The graphite in graphite-bearing felsic granulite facies gneisses occurs as thin, irregularly-shaped, medium-grained rods or, in those gneisses once mined for graphite, comprises distinct layers up to 5 cm thick as well as irregular patches and stringers.

Mafic Granulite Grade Gneisses

Most medium-grained, equigranular, hornblende-rich granulite gneisses contain hypersthene and/or augite, but the ratio of pyroxene to hornblende varies considerably (Table 1). Many mafic granulites are strongly layered into felsic and mafic units which range in width from a few mm to tens of cm. Hornblende and subordinate plagioclase comprise the mafic layers. Albite-twinned, anhedral, medium-grained plagioclase crystals are interstitial to mafic grains or comprise distinct felsic bands along with anhedral, granular, undulatory quartz, and medium-grained anhedral mesoperthite. The plagioclases ($An_{55 \pm 5}$) exhibit light saussuritization, usually in the core of grains. The subhedral hornblende crystals define the foliation and frequently show light blue-green amphibole growing on the ends of columnar grains. The hornblende may be fresh or almost totally altered to red-brown or green biotite. Anhedral hypersthene grains form isolated clumps or strings of clumps with augite. Both pyroxene varieties often exhibit complete alteration to a cloudy, diffuse, very fine-grained aggregate containing chlorite and other products of undetermined composition. Alteration first occurs around grain boundaries and along internal fractures and cleavage. Anhedral pink garnets contain abundant quartz, some apatite, and sparse opaque mineral inclusions. Accessory opaque minerals, zircon, and apatite round out the primary minerals.

Anorthosite Suite

Plagioclase and hornblende comprise the bulk of these rocks, but augite, hypersthene, and garnet are occasionally present in amounts less than 5 percent. Accessories include apatite, zircon, and the opaque minerals magnetite and ilmenite, which contains ovoid patches of exsolved hematite. Alteration products consist of chlorite, saussurite, blue-green uralite, and biotite.

REFERENCES CITED

Bascom, F., and Stose, G. W., 1938, Geology and mineral resources of the Honeybrook and Phoenixville quadrangles, Pennsylvania: U.S. Geological Survey Bulletin 891, 145 p.

Berg, T. M., Edmunds, W. E., Geyer, A. R., and others, 1980, Geologic Map of Pennsylvania: Pennsylvania Geologic Survey, 4th series, Map 1, Scale 1:250,000, 2 sheets.

Bromery, R. W., Henderson, J. R., Jr., Zandle, G. L., and others, 1960a, Aeromagnetic map of the Wagontown quadrangle, Chester County, Pennsylvania: U.S. Geological Survey Geophysical Investigations Map GP-223, Scale 1:24,000.

Bromery, R. W., Zandle, G. L., and others, 1960b, Aeromagnetic map of the Downingtown quadrangle, Chester County, Pennsylvania: U.S. Geological Survey Geophysical Investigations Map GP-224, Scale 1:24,000.

Crawford, M. L., and Crawford, W. A., 1980, Metamorphic history of the Pennsylvania Piedmont: Journal of the Geological Society, v. 137, p. 311–320.

Crawford, W. A., 1979, Geologic remapping and metamorphic history of the Honey Brook Upland, S. E. Pennsylvania: Geological Society of America Abstracts with Programs, v. 11, p. 8–9.

Crawford, W. A., and Huntsman, J. R., 1976, Metamorphic rock types, Wagontown 7½′ quadrangle, Chester County, Pennsylvania: Geological Society of America Abstracts with Programs, v. 8, p. 156.

Crawford, W. A., Robelen, P. G., and Kalmbach, J. H., 1971, The Honey Brook anorthosite: American Journal of Science, v. 271, p. 333–349.

Demmon, F. E., 1977, Investigations of the origins and metamorphic history of Precambrian Gneisses, Downingtown 7½′ quadrangle, southeastern Pennsylvania [M.A. Thesis]: Bryn Mawr College, 71 p.

de Waard, D., 1968, The anorthosite problem: The problem of anorthosite-charnockite suite of rocks, in Isachsen, I. Y., ed., Anorthosite and related rocks: New York State Museum and Science Service, Memoir 18, p. 71–91.

Gray, Carlyle, and others, 1960, Geologic Map of Pennsylvania (3rd printing 1979): Pennsylvania Geologic Survey, 4th series, Map 1, Scale 1:250,000, 2 sheets.

Huntsman, J. R., 1975, Crystalline rocks of the Wagontown 7½ minute quadrangle [M.A. Thesis]: Bryn Mawr College, 69 p.

Irvine, T. N., and Baragar, W.R.A., 1971, A guide to the chemical classification of the common volcanic rocks: Canadian Journal of Earth Sciences, v. 8, p. 523–548.

O'Neill, B. J., Jr., 1952, Geology of the anorthosite massif in Chester County, Pennsylvania [M.S. Thesis]: California Institute of Technology, 35 p.

Organist, G., 1978, A study of the Honey Brook anorthosite country rock contact [A.B. Thesis]: Bryn Mawr College, 30 p.

Papezik, V. S., 1965, Geochemistry of some Canadian anorthosites: Geochemica et Cosmochimica Acta, v. 29, p. 673–709.

Postel, A. W., 1951, Problems of the Pre-Cambrian in the Phoenixville and Honeybrook quadrangles, Chester County, Pennsylvania: Pennsylvania Academy of Science, v. 113–119.

Raman, S. V., and Hewins, R. H., 1979, Petrologic and geochemical evidence on the origin of the Honey Brook anorthosite, Pa.: Geological Association of Canada Abstracts, v. 4, p. 73.

Robelen, P. G., 1968, The petrology of the Honeybrook anorthosite, Chester County, Pennsylvania [M.A. Thesis]: Bryn Mawr College, 30 p.

Smith, I. F., 1922, Genesis of anorthosites of Piedmont Pennsylvania: The Pan-American Geologist, v. 38, p. 29–50.

Socolow, A. A., 1959, Geologic Interpretation of aeromagnetic map Elverson quadrangle: Pennsylvania Geological Survey, 4th Series, Information Circular 35, 5 p.

Socolow, A. A., 1960, Geologic interpretation of aeromagnetic maps Pottstown, Wagontown, Downingtown, Coatesville, Unionville, Honeybrook, and Parkesburg quadrangles: Pennsylvania Geological Survey, 4th Series, Information Circular 37, 14 p.

Streckeisen, A., 1973, Plutonic rocks, classification and nomenclature: Geotimes, v. 18, p. 26–29.

Streckeisen, A., 1974, How should charnockites be named (?): Centenaire de la Société Géologique de Belgique, Géologie des Domains Crystallins, Liege, p. 349–360.

Streckeisen, A., 1979, Classification and nomenclature of volcanic rocks, lampropyres, carbonatites, and melilitic rocks: Geology, v. 7, p. 331–335.

Sutter, J. F., Crawford, M. L., and Crawford, W. A., 1980, $^{40}Ar/^{39}Ar$ age spectra of coexisting hornblende and biotite from the Piedmont of Pennsylvania: their bearing on the metamorphic and tectonic history: Geological Society of America Abstracts with Programs, v. 12, p. 85.

Talwani, M. and Heirtzler, J. R., 1964, Computation of magne-tic anomalies caused by two dimensional structures of arbitrary shape *in* Computers in the mineral industries, part 1: Stanford University publications, Geological Sciences, v. 9, p. 464–480.

Thomann, W. F., 1977, Igneous and metamorphic petrology of the Honey Brook Uplands in the Elverson, Pottstown, and Phoenixville quadrangles, southeastern Pennsylvania [M.A. Thesis]: Bryn Mawr College, 65 p.

Valley, J. W., and O'Neil, J. R., 1981, 13C/12C exchange between calcite and graphite: a possible thermometer in Grenville marbles: Geochemica et Cosmochemica Acta, v. 45, p. 411–420.

Waraska, I. R. 1952, The petrology and structural geology of the Pre-Cambrian igneous rocks near Wallace, Chester County, Pennsylvania [M.A. Thesis]: Bryn Mawr College, 17 p.

Williams, H., 1942, The Geology of Crater Lake National Park, Oregon: Carnegie Institute of Washington Publication 540, 162 p.

Winkler, H.G.F., 1979, Petrogenesis of metamorphic rocks (fifth edition): Springer-Verlag, New York, 348 p.

MANUSCRIPT ACCEPTED BY THE SOCIETY AUGUST 2, 1983

Geological Society of America
Special Paper 194
1984

Tectonic evolution of the Baltimore Gneiss anticlines, Maryland

Peter D. Muller*
Maryland Geological Survey
Johns Hopkins University
Baltimore, Maryland 21218

David A. Chapin*
Department of Geological Sciences
Lehigh University
Bethlehem, Pennsylvania 18015

ABSTRACT

Geologic mapping, structural analysis, gravity modeling, and magnetics indicate that the Phoenix, Texas, Chattolanee, and Towson gneiss anticlines near Baltimore, Maryland are part of a large refolded crystalline nappe system rooted beneath the Towson anticline. Map patterns and limited structural and geophysical data from the Woodstock, Mayfield, and Clarksville anticlines suggest they form a similar nappe system that has no clearly evident root zone. Grenville-age (1,000–1,200 m.y.) Baltimore Gneiss comprises the cores of the nappes; Lower Cambrian–Precambrian to Ordovician metasedimentary rocks (Glenarm Supergroup) compose the cover.

Three periods of deformation affected the Baltimore Gneiss (D_1, D_2, D_3). D_1 (Grenvillian) is evidenced by radiometric age data; however, except for a transposed compositional layering in certain felsic gneiss, all D_1 structures were obliterated or completely obscured by later tectonism. D_2 (Taconic-Acadian) involved three phases of folding (F_{2a}, F_{2b}, F_{2c}), amphibolite facies metamorphism, local migmatization, and the development of pervasive structural elements. D_3 (Alleghanian-Palisades) resulted in predominantly brittle faulting and open folding (F_3).

The causes of D_1 are enigmatic. D_2 was the result of early Paleozoic collision and suturing of an ocean floor/island arc terrane (Baltimore Complex) with a continental margin or fringing microcontinent (Baltimore Gneiss–Glenarm terrane). D_3 was the manifestation of early Mesozoic continental rifting, and possibly latest Paleozoic trans-current plate motions.

The tectonic evolution of the Baltimore Gneiss–Glenarm terrane near Baltimore involved a complex sequence of compressional and extensional ductile strain followed by brittle-ductile to brittle displacements. These deformations were manifested structurally in the emplacement, multiple refolding, and subsequent faulting of a large crystalline nappe system.

---

*Present address: Muller—Department of Earth Science, SUNY, College at Oneonta, Oneonta, New York 13820. Chapin—Gulf Oil Corporation, Gulf Exploration Technology Center, Houston Technical Center, P.O. Box 36506, Houston, Texas 77236.

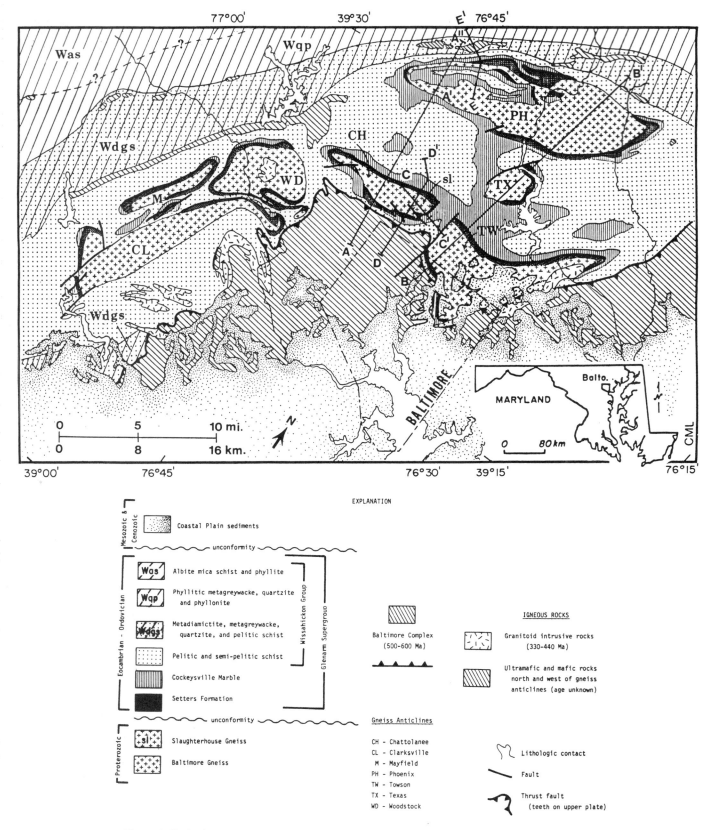

Figure 1. Geologic map of the Baltimore area, Maryland. Modified from Cleaves and others (1968).

INTRODUCTION

Geologic Setting

Layered quartzo-feldspathic gneiss and migmatite interlayered with minor hornblende gneiss and amphibolite are exposed in the cores of seven gneiss anticlines near Baltimore, Maryland (Fig. 1). These Grenville-age (1,000–1,200 m.y.) lithologies compose the Baltimore Gneiss. They are overlain by a thick sequence of quartzite, marble, schist, metagraywacke, metadiamictite, and minor intercalated amphibolite and ultramafic rock designated the Glenarm Supergroup (Crowley, 1976). Both the Baltimore Gneiss and the Glenarm Supergroup were multiply deformed (Freedman and others, 1964; Fisher, 1970; Higgins, 1973) and metamorphosed (Lapham and Basset, 1964; Tilton and others, 1970; Grauert, 1973a). Despite the present structural conformity along the gneiss-Glenarm contact, the basal metasedimentary rocks are assumed to have been deposited nonconformably on the Baltimore Gneiss during latest Precambrian–early Paleozoic time (Broedel, 1937; Hopson, 1964).

Immediately south and east of the gneiss anticlines, metamorphosed ultramafic, mafic, felsic, and pelitic rocks of the Baltimore Complex (Crowley, 1976; Morgan, 1977) lie in fault contact with Glenarm schists (Fig. 1). This imbricated belt of serpentinized peridotite, metagabbro, amphibolite, and closely associated mafic and felsic gneisses (James Run Formation of Higgins, 1972) is interpreted as an allochthonous ocean floor/island arc terrane tectonically emplaced onto the North American continental margin from the southeast (Rodgers, 1972; Crowley, 1976; Morgan, 1977). Isolated lenses of serpentinite and ultramafic schist occur in the Wissahickon Group north and west of the gneiss anticlines as well.

The crystalline rocks of the Baltimore area were intruded by a variety of granitoid plutons and pegmatites. They range in composition from granodiorite to alkali granite and in texture from gneissic to massive (Hopson, 1964). Some are concordant synkinematic intrusions, while others are late to postkinematic dikes and sills. Radiometric ages cluster at 400 to 450 m.y., but range from 300 to 600 m.y. old.

The Baltimore Gneiss anticlines lie at the culmination of the doubly plunging Baltimore-Washington anticlinorium (Fisher, 1963) along the axis of maximum curvature of a major bend in the Appalachian mountain system (Susquehanna Orocline of Wise, 1970; Pennsylvania Reentrant of Williams, 1978). They occur within the zone of highest grade metamorphism (kyanite-staurolite zone) in the Baltimore area and are associated with pegmatite swarms in the adjacent lower Glenarm metasedimentary rocks.

Although referred to as mantled-gneiss domes, the seven outcrop areas of gneiss are actually complex fold structures. Because their macroscopic form is anticlinal, they are referred to as gneiss anticlines throughout this paper. Internally, they show abundant evidence of multiple deformation, and their map patterns reflect interference due to refolding. Local migmatization

and pervasive flowage and recrystallization during post-Grenville orogenesis has made recognition of Grenville-age structures problematic.

Mylonitic zones meters to several tens of meters in width occur in several of the gneiss anticlines. They parallel the local structural grain but do not extend beyond the margins of the gneiss. High-angle faults characterized by silicified breccias locally cut all of the gneiss anticlines and truncate the Towson and Texas anticlines (Fig. 1). Small reverse faults were postulated along the overturned flanks of some of the gneiss anticlines (Mathews, 1907; Broedel, 1937; Cloos, 1964).

Several unmetamorphosed diabase dikes of probable Jurassic-Triassic age cut obliquely across all regional metamorphic structural elements. Unconsolidated clastic sediments of the Atlantic Coastal Plain overlap the crystalline rocks on the southeast.

Previous Work

After establishing the general stratigraphic sequence in the crystalline rocks of the Baltimore area, Mathews (1904, 1907) interpreted the gross structure of the area as a series of sharply crested anticlinoria cored by Baltimore Gneiss and separated from each other by tight synclinoria of the overlying metasedimentary rocks. Knopf and Jonas (1925, 1929) concurred with this interpretation and mapped the inlier of quartzite, marble, and schist in the Phoenix anticline (Fig. 1) as a simple synclinal infold on the crest of this anticline. Assuming a Precambrian age for the overlying metasedimentary rocks, they postulated that the gneiss anticlines formed during a late Precambrian deformation prior to the deposition of the Lower Paleozoic sedimentary rocks in the Frederick Valley, approximately 40 km to the west.

On the basis of the first detailed structural mapping of the Baltimore Gneiss, Broedel (1937) concluded that the gneiss formed the cores of gently plunging anticlines locally overturned to either the northwest or southeast. The complexity of folding in the gneiss led Broedel to think that pre-Glenarm (Grenville?) structures controlled the overall form of the anticlines. Eskola (1949) compared the Baltimore Gneiss anticlines with the classic mantled gneiss domes of eastern Finland. On the basis of Eskola's model, Hopson (1964) interpreted the Baltimore anticlines as remobilized basement diapirs that had punched upward into the Glenarm metasedimentary rocks along the crest of the Baltimore-Washington anticlinorium.

Geophysical studies of the Baltimore area were initiated by Bromery (1968). He studied the areal distribution of aeromagnetic and gravity anomalies in an attempt to distinguish major geologic features. Fisher and others (1979) used Bromery's (1967a) magnetic data to study large-scale Piedmont structures by delineating major magnetostratigraphic units. Chapin (1981) carried out a detailed study of the gravity anomalies over the Chattolanee anticline and used gravity modeling techniques to determine the subsurface structure of the Chattolanee and western parts of the Phoenix and Towson anticlines.

Rodgers (1970) noted that the map pattern of the Phoenix anticline resembled that of the southwesternmost gneiss anticline of the Philadelphia area, which Bailey and Mackin (1937) and Mackin (1962) convincingly argued was an arched recumbent fold. Crowley (1976) extended this interpretation to the Baltimore Gneiss anticlines and interpreted the inlier of quartzite, marble, and schist in the Phoenix anticline as an antiformal syncline. On the basis of minor structures observed in the Phoenix 7½-minute quadrangle, Moller (1979) mapped the eastern part of the Phoenix anticline as a recumbent fold refolded into a northeast-trending antiform. Fisher and others (1979) interpreted the Phoenix anticline as the tongue of a large refolded nappe rooted to the southeast in the area of the Towson anticline.

ROCK UNITS

Baltimore Gneiss

Hopson (1964) recognized five major varieties of Baltimore Gneiss: banded gneiss, including amphibolite; augen gneiss; veined gneiss; migmatite; and granitic gneiss. He considered all varieties to be gradational with each other and noted their overall mineralogic homogeneity. On the basis of primarily field- and hand-specimen criteria, Crowley (1976) distinguished four basic lithotypes of Baltimore Gneiss: layered gneiss, augen gneiss; streaked augen gneiss; hornblende gneiss and amphibolite; and a possible late Precambrian granitic intrusive which he named Slaughterhouse Gneiss. Crowley also recognized minor feldspathic quartzite and biotite schist in the Woodstock anticline.

Biotite schist is a minor, but structurally important, lithology in the Towson and Phoenix anticlines as well.

7½-minute quadrangle mapping in the Phoenix anticline (Muller, in prep.) has revealed the existence of minor calc-silicate gneiss and feldspathic quartzite. Lenses of tremolite schist and soapstone found in the central part of the Phoenix anticline are interpreted as tectonic slivers emplaced along a ductile fault zone, rather than an ultramafic lithology within the Baltimore Gneiss.

Hopson (1964) showed that interlayered felsic gneiss and amphibolite (paragneiss) cropping out in the southern part of the city of Baltimore were texturally and chemically distinct from Baltimore Gneiss to the north and west. Subsequent radiometric dating (Tilton and others, 1970) revealed that these paragneisses were much younger than the Baltimore Gneiss (Wetherill and others, 1968). They have been since reassigned to the James Run Formation (Higgins, 1972), part of the allochthonous Baltimore Complex.

Medium-grained, layered, light to dark feldspar-quartz-biotite gneiss is the dominant lithology within the Baltimore Gneiss (Table 1). It is estimated to constitute 70% to 80% of the exposed gneiss. The fabric of the gneiss grades from uniform and weakly foliated to strongly layered and migmatitic. The scale of layering ranges from millimeters to meters. The gneiss is locally mylonitic with feldspar or quartz-feldspar augen streaked into a flaser structure. Layered biotite and hornblende-biotite gneisses commonly display type 2 and 3 interference patterns (Ramsay, 1967).

Fine- to coarse-grained granitic veins and sills are common features in the gneiss. Massive to foliated, pink to buff, medium-

TABLE 1. LITHOLOGIC, TEXTURAL, AND MINERALOGIC CHARACTERISTICS OF THE BALTIMORE GNEISS

Major Lithologies	Mesoscopic Fabric	Common Mineral Assemblages	Probable Protolith
Layered gneiss-migmatitic gneiss	Medium-coarse grained, planar to highly contorted compositional layering	Plagioclase-quartz-microcline-biotite-(sphene-epidote-allanite-apatite-magnetite-zircon)*	Felsic volcanic and epiclastic rocks
Augen gneiss-mylonitic gneiss	Medium-grained, porphyroblastic (feldspar), locally mylonitic and streaked		
Hornblende gneiss and amphibolite	Medium-grained, massive to well foliated and/or lineated	1) Plagioclase-quartz-biotite-hornblende-(garnet-epidote-microcline-sphene-magnetite) 2) Plagioclase-hornblende-biotite-quartz-(epidote/clinozoisite-sphene-garnet-magnetite-pyrite) 3) Hornblende-plagioclase-epidote/clinozoisite-quartz-(diopside-garnet-sphene-magnetite)	Intermediate to mafic volcanic and epiclastic rocks
Granitic gneiss-Slaughterhouse gneiss	Fine-medium grained, massive to weakly foliated	Plagioclase-microcline-quartz biotite-(muscovite-epidote-sphene-apatite-garnet-zircon)	Felsic volcanic and intrusive rocks

* Minerals in parentheses are accessory and trace phases which may or may not be present in every sample.

grained muscovite-biotite leucogneiss occurs as sheetlike bodies in several of the gneiss anticlines. Thicknesses range from several tens of meters to more than 100 m. Crowley (1976) designated this lithology Slaughterhouse Gneiss in the Chattolanee anticline and considered it to be a pre-Glenarm intrusive unit.

The dominant layered gneiss member is composed of sub-equal amounts of oligoclase, quartz, and microcline, biotite being the primary dark mineral (Hopson, 1964; Crowley, 1976). Sphene, epidote, allanite (commonly metamict and rimmed by epidote), apatite, magnetite, and zircon are common accessory and trace minerals (Table 1). In migmatitic gneisses the neosomes are enriched in microcline and depleted in biotite relative to the paleosomes (Hopson, 1964; Olsen, 1972, 1977). Accessory muscovite generally is restricted to areas of mylonitic augen gneisses and bodies of Slaughterhouse Gneiss. Garnet and hornblende are rare except in strongly segregated gneisses or in zones of intercalated mafic gneiss and amphibolite.

The mineralogy of the intercalated mafic gneisses and amphibolites is more variable than that of the quartzo-feldspathic gneisses. Mineral assemblages vary from plagioclase-quartz-biotite with minor garnet and/or hornblende to hornblende-plagioclase-epidote/clinozoisite with minor quartz and, locally, diopside (Table 1). Apatite and opaques are the most common accessories. Epidote "fels" occurs as small lenses or pods associated with amphibolite.

All Baltimore Gneiss lithologies exhibit crystalloblastic textures. No relict igneous or sedimentary structures are recognized (Hopson, 1964; Olsen, 1972). In general, leucocratic varieties are granoblastic, while biotite-rich gneisses commonly contain porphyroblastic microcline or plagioclase within a strongly schistose matrix. Some of the augen gneisses display mylonitic textures characterized by thin zones of fine quartz-feldspar mortar and/or ribbon quartz and highly strained feldspar or quartz-feldspar porphyroclasts. Apparently, recrystallization predominated over (or continued after) cataclasis; thus, only mylonitic gneisses rather than mylonites are found. Narrow zones of cataclasite in which biotite is completely chloritized and feldspars are thoroughly sericitized compare in texture and style of deformation with the highly silicified breccias found along postregional metamorphic high-angle faults cutting the margins of the gneiss anticlines.

Radiometric age determinations (U/Pb zircon ages) indicate that the Baltimore Gneiss experienced at least two major thermal events, one ranging from 1,000 to 1,200 m.y. ago and the other from 420 to 450 m.y. ago (Tilton and others, 1970; Grauert, 1973a). Rb/Sr biotite ages reveal that temperatures remained high enough for diffusion of strontium until approximately 290 m.y. ago (Wetherill and others, 1968). Because some zircons in the Baltimore Gneiss appear detrital (Tilton and others, 1970), the primary depositional age of the gneiss is assumed to be pre-1,200 m.y. old.

Upper amphibolite facies mineral assemblages found in the Baltimore Gneiss are compatible with those of the adjacent lower Glenarm schists (Winkler, 1976; Olsen, 1977). Therefore, it is not known if both the Grenville and early Paleozoic metamor-phic events or only the Paleozoic event(s) occurred at upper amphibolite facies *P-T* conditions. Abundant textural, chemical, and structural evidence of anatexis and remobilization of the gneiss during early Paleozoic metamorphism (Hopson, 1964; Olsen, 1977) suggests that all indication of Precambrian metamorphism was obliterated by subsequent tectonism. Thus, it is possible that the grade of Grenville metamorphism was higher (granulite facies), as recognized in the Baltimore Gneiss of the Philadelphia area (Wagner and Crawford, 1975).

Glenarm Supergroup

The Baltimore Gneiss is mantled by a thick sequence of predominantly clastic metasedimentary rocks originally designated the Glenarm Series (Knopf and Jonas, 1923). The stratigraphic nomenclature of the Glenarm Series was revised several times (Southwick and Fisher, 1967; Higgins and Fisher, 1971; Higgins, 1972), and Crowley (1976) reclassified it as the Glenarm Supergroup after elevating the Wissahickon Formation to group status. The supergroup is composed from basal units upward of the Setters Formation, Cockeysville Marble, and the Wissahickon Group and is considered by all previous workers to unconformably overlie the Baltimore Gneiss. The upper limit of the Glenarm was tentatively placed at the Ijamsville Phyllite (Southwick and Fisher, 1967); however, Fisher (1978) later interpreted the Ijamsville as correlative with the uppermost Glenarm.

The age, origin, and stratigraphic relationships within the Glenarm Supergroup have been debated for nearly 70 yr (see a review in Higgins, 1972, 1976; Seiders and others, 1975; Seiders, 1976). This paper follows the informal stratigraphic nomenclature presented by Fisher and others (1979) and the age range suggested by Higgins (1972; Lower Cambrian–Precambrian to Ordovician). However, the interpretation of the stratigraphic relationships shown in Figure 2 are those of the authors.

The Glenarm Supergroup was metamorphosed regionally to staurolite and kyanite grade near the margins of the gneiss anticlines. Sillimanite, mainly fibrolite, occurs locally in this highest temperature zone (Hopson, 1964), and ilmenite-bearing quartz-feldspar veins and segregations are common. Rb-Sr mineral ages from relatively undeformed pegmatites cutting the regional Glenarm schistosity range from 425 to 470 m.y. old (Wetherill and others, 1966; Muth and others, 1979), suggesting the peak of amphibolite facies metamorphism occurred from late Middle to early Late Ordovician time. Localized retrograde metamorphism affected the Glenarm to varying degrees after the Ordovician thermal peak.

Setters Formation

Potassic, quartz-rich metasedimentary rocks of the Setters Formation make up the basal unit of the Glenarm Supergroup (Mathews, 1905; Knopf and Jonas, 1929). The Setters is composed primarily of fine- to medium-grained feldspathic muscovite quartzite and feldspathic quartz-mica schist, mica gneiss, and

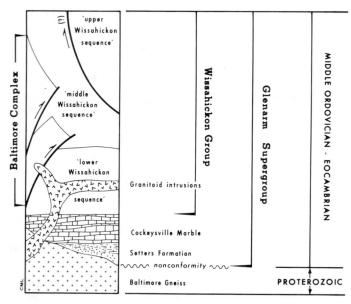

Figure 2. Generalized stratigraphic column of the crystalline rocks near Baltimore, Maryland.

medium- to coarse-grained pelitic schist. Stretched quartz-pebble conglomerate is a minor constituent. The estimated thickness of the Setters Formation ranges from 0 to 500 m (Crowley, 1976).

Schistosity/compositional layering (denotes mesoscopic parallelism of structural elements listed) in the Setters Formation parallels that in the Baltimore Gneiss at the contact between the two units. A characteristic feature of the gneiss-Setters contact is its recurrent variation from steeply dipping to overturned due to open, subhorizontally inclined folding.

Cockeysville Marble

A texturally and mineralogically variable carbonate unit, the Cockeysville Marble (Mathews and Miller, 1905), overlies the Setters Formation. It consists of six major lithologies: calc-schist, phlogopite-calcite metadolomite, quartz metadolomite, calc-gneiss, calcite marble, and calc-silicate marble (Choquette, 1960). The internal stratigraphy of the Cockeysville is still poorly understood, and the overall distribution of rock types does not conform to the outcrop patterns of the surrounding formations (Crowley, 1976). It ranges in thickness from 0 to 1.4 km (Crowley, 1976).

A recently excavated exposure of the Setters-Cockeysville Marble contact behind the new Hunt Valley shopping mall near Cockeysville, Maryland, reveals that the schistosity/compositional layering in both units is parallel at the contact.

Wissahickon Group

A thick, areally extensive sequence of pelitic and semipelitic schist, metagraywacke, metadiamictite, quartz schist, and phyllite intercalated with lenses of amphibolite and ultramafic rock

overlies the Cockeysville Marble. Originally named the Wissahickon Formation (Bascom, 1902; Mathews, 1904, 1905), this sequence was elevated to group status by Crowley (1976). Although it has been studied extensively and its stratigraphic nomenclature has frequently been revised (Southwick and Fisher, 1967; Higgins and Fisher, 1971; Higgins, 1972), the large-scale structure of the Wissahickon is still poorly understood. The aggregate thickness of the group is difficult to estimate owing to structural complexity and original facies variation in the individual formations or lithotypes (Hopson, 1964; Southwick and Fisher, 1967); it probably ranges from 5 to 10 km.

The Wissahickon Group can be divided into three basic lithostratigraphic intervals in the Baltimore region: a lower sequence of interlayered medium- to coarse-grained pelitic and semipelitic schist containing minor quartzite, marble, and amphibolite; a middle sequence of interlayered metagraywacke, quartzite, and pelitic schist grading laterally into metadiamictite, quartzite, and psammitic gneiss, both facies containing lenses of amphibolite and a discontinuous belt of schistose ultramafic rocks; and an upper sequence primarily composed of interlayered fine-grained phyllitic schist, phyllonite, phyllitic metagraywacke, micaceous quartzite, and albite schist. The designation lower, middle, and upper refers to the inferred structural positions of these intervals (Fig. 2) and does not necessarily imply the original premetamorphic stratigraphy as well.

In the one known exposure of the Wissahickon Group--Cockeysville Marble contact along the south flank of the Chattolanee anticline (Arundel Corporation Greenspring Avenue quarry), the contact appears gradational and conformable or only slightly disconformable (Choquette, 1960; Chapin, 1981). The contact between the lower and middle intervals of the Wissahickon Group appears gradational in most areas; however, the transition is locally marked by a zone of partly chloritized and strongly tectonized schist, and thus it may be tectonic. The middle interval–upper interval contact is defined by a prominent belt of intensely sheared, locally phyllonitic schist and fine-grained quartzite. This 1- to 3-km-wide deformed belt represents a major, but as yet poorly understood, tectonic feature of the Maryland Piedmont. The upper contact of the Wissahickon Group is both poorly exposed and poorly defined in Maryland and southeastern Pennsylvania. Along strike in northern Virginia, the contact recently has been mapped as an unconformity with overlying phyllite and metasiltstone (Drake and Lyttle, 1981).

Baltimore Complex

Immediately south and east of the Baltimore Gneiss anticlines is an allochthonous terrane of metamorphosed ultramafic, mafic, felsic, and pelitic rocks extending from just north of the Maryland-Pennsylvania state line to approximately 20 km southwest of Baltimore. Crowley (1976) named this terrane the Baltimore Mafic Complex, divided it into three major lithotectonic belts, and considered it to be a deformed ophiolite in thrust contact with the Wissahickon Group. Morgan (1977) used the name Baltimore Complex to describe the same rocks and considered them to be a dismembered ophiolite as well. The complex includes all rocks originally designated Baltimore Gabbro or Baltimore–State Line Gabbro–Peridotite Complex (Williams, 1886; Herz, 1950, 1951; Hopson, 1964; Southwick, 1970) and thought to intrude the Wissahickon Group. It also includes a group of closely associated metavolcanic, metaepiclastic, and plutonic rocks of the James Run Gneiss or Formation (Southwick and Fisher, 1967; Higgins, 1972), Port Deposit and Franklinville Gneisses (Hershey, 1937; Southwick, 1969; Crowley, 1976), Relay Quartz Diorite or Gneiss (Hopson, 1964; Crowley, 1976), and the Baltimore paragneiss or Carrol Gneiss (Hopson, 1964; Crowley, 1976). Morgan estimated a structural thickness of approximately 10 km for the complex; however, this can only be considered tentative, taking into account the evidence of internal imbrication and folding (Crowley, 1976; Fisher and others, 1979).

The Baltimore Complex consists of a wide variety of rock types that are tectonically interleaved, intruded by granitic to tonalitic plutons, and variably metamorphosed. The plutonic mafic-ultramafic part of the complex is composed primarily of medium- to coarse-grained metagabbro and metaperidotite cut by plagiogranite dikes and veins (Southwick, 1970; Hanan, 1980). The predominantly supracrustal part of the complex consists of fine- to medium-grained amphibolite (locally pillowed metabasalt), massive leucogneiss (locally porphyritic metadacite), metagraywacke, metadiamictite, and minor pelitic schist intruded by medium- to coarse-grained granodiorite to tonalite gneiss (Higgins, 1972; Southwick, 1979). U/Pb zircon ages from the felsic metavolcanic rocks and granodioritic to tonalitic intrusive rocks range from 500 to 650 m.y. old (Tilton and others, 1970; Higgins and others, 1977).

The few exposures of the contact between the Baltimore Complex and the Wissahickon Group are strongly sheared and concordant with the schistosity in the Wissahickon (Cohen, 1937; Southwick, 1970; Muller, personal reconn.). Muller has observed narrow, steeply dipping shear zones paralleling the contact in amphibolite cropping out along the Gunpowder River in southeastern Baltimore County. Granitic intrusive rocks in this same area are commonly sheared as well. The straight trace of the contact in southeastern Baltimore and southwestern Harford Counties argues against a moderate- to low-angle thrust contact such as occurs farther southwest (Fig. 1) and suggests that the thrust was modified by later high-angle faulting.

Crowley and others (1976) mapped finely layered epidote amphibolite intercalated with schist along much of the Wissahickon Group–Baltimore Complex contact near Baltimore as part of the Wissahickon. The local presence of pelitic material in this highly sheared mafic unit suggests it may be a tectonic melange plated to the base of the Baltimore Complex and, thus, distinct from the underlying Wissahickon schists. As such, it may have a closer relationship to the area of schist and metagraywacke found interleaved with the Baltimore Complex in Harford and Cecil Counties than the adjacent Wissahickon schists.

Granitoid Intrusive Rocks

A variety of granitoid intrusive rocks is present in the Baltimore area (Fig. 1; only the largest bodies are shown). Apart from the low-alkali granodioritic to tonalitic intrusive units characteristic of the Baltimore Complex, these intrusions are generally granodioritic to quartz monzonitic in composition (Hopson, 1964; Southwick, 1969; Sinha and others, 1980). They range from strongly deformed, conformable, synkinematic sheets (Ellicott City granodiorite; 440 m.y. old) to massive, discordant, late kinematic to postkinematic stocks or dikes (Woodstock and Guilford quart monzonite; 400–450 m.y. old). Cloos (1933) concluded that the foliation and lineation within the Ellicott City pluton was a primary igneous structure rather than a later metamorphic structure. However, the pluton was affected by folding (Fig. 1; double hooked intrusion south of northern end of the Clarksville anticline) concomitant with amphibolite facies metamorphism and the development of local schistosity parallel to the foliation in the pluton. Thus, the present foliation may be secondary, although possibly mimetic to a primary flow foliation.

Medium- to very coarse-grained pegmatite dikes and sills of alkali granite composition are closely associated with the margins of the gneiss anticlines (Hopson, 1964). They commonly occur as swarms parallel to the trend of the adjacent anticline (Broedel, 1937). These pegmatites are distinct from the fine- to coarse-grained dikes and apophyses related to the granitoid intrusions. The latter may be strongly sheared and folded, while the former are generally massive and considered to represent palingenetic fluid expelled from the Baltimore Gneiss during the waning stages of the formation of the gneiss anticlines (Hopson, 1964).

The age and origin of the granitoid intrusions in the Baltimore Gneiss–Glenarm terrane remain controversial owing to varying interpretations of discordant U/Pb age data (Davis and others, 1965; Grauert, 1973b; Higgins and others, 1977). Pb^{207}/Pb^{206} ages fall in the 400- to 600-m.y. old range; however, Grauert (1973b) reported a discordant lead age of 330 m.y. old for the Gunpowder Gneiss (Fig. 1; northeast end of the Towson anticline). Because the Gunpowder Gneiss is considered an anatectic derivative of the Baltimore Gneiss (Hopson, 1964), resolution of the 100- to 200-m.y.-age discrepancy is critical to any interpretation of the timing of peak metamorphism and gneiss mobilization in the Baltimore area.

GRAVITY AND MAGNETICS

The crystalline rocks of the Baltimore area display a wide variation in the physical properties that influence their potential field signatures. Most Baltimore Gneiss and Setters Formation lithologies have characteristically low densities and low magnetic susceptibilities, while most Wissahickon Group rocks and mafic rocks of the Baltimore Complex have high densities and relatively large magnetic susceptibilities (Bromery, 1968; Fisher and others, 1979; Chapin, 1981). The Cockeysville Marble possesses intermediate densities and low susceptibilities. Ultramafic rocks (ser-

Figure 3. Simple Bouguer anomaly map of the Baltimore area, Maryland. Modified from Bromery (1967b) and Daniels (U.S. Geological Survey unpubl. data).

pentinites in particular) have low densities and high suscepti- bilities (Bromery, 1968; Fisher and others, 1979; Chapin, 1981).

In the initial study of Bouguer and magnetic anomalies in the Baltimore area, Bromery (1968) noted that the gneiss anti- clines in Baltimore County are characterized by relatively flat magnetic lows, that the Phoenix and Texas anticlines lack a dis- tinctive gravity signature, and that the Towson and Chattolanee anticlines are associated with pronounced gravity lows (Fig. 3). He concluded that the Phoenix and Texas anticlines were part of a thinfolded sheet of low-density gneiss underlain by higher den- sity Glenarm rocks, while the Towson and Chattolanee anticlines were a protuberance on the basement surface.

Subsequent investigation of the magnetic properties of the Baltimore Gneiss and surrounding Glenarm units by Fisher and others (1979) revealed that the magnetic susceptibility contrast between the Glenarm lithologies and the gneiss is insufficient to cause the observed magnetic lows. In addition, measurements of remanent magnetic moments produced an essentially random

orientation within individual anticlines and the anticlines as a whole. They concluded that the broad featureless character of the lows indicates the presence of a rock unit in the gneiss, Setters Formation, or Cockeysville Marble, possessing reverse remanent magnetization, which was not detected in their sampling pro- gram. Using the magnetic data in conjunction with limited data on minor structures and stratigraphic arguments, they interpreted the Phoenix anticline as a large refolded nappe.

Integration of the results of two-dimensional Talwani-type (Talwani and others, 1959) gravity modeling (Chapin, 1981) with detailed surface mapping and mesoscopic structural analysis provides a further method of documenting the macroscopic struc- ture of the gneiss anticlines. Despite the non-unique interpretation of structure from gravity anomalies and the limitations of two- dimensional modeling in structurally complex areas, the gross subsurface form of the anticlines can be reasonably inferred. For a full discussion of the modeling technique used, the reader is re- ferred to Chapin (1981).

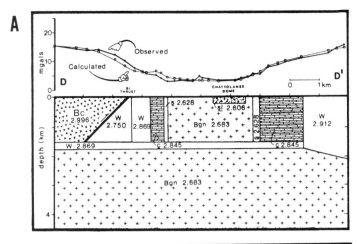

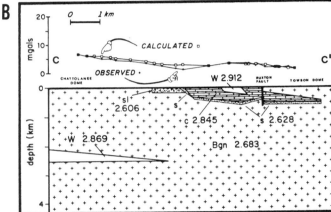

Figure 4. A. Gravity model from D to D' across the Chattolanee anticline and Baltimore Complex/Glenarm thrust contact (Chapin, 1981, Fig. 3 and 4). Bc = Baltimore Complex; W = Wissahickon Group; c = Cockeysville Marble; s = Setters Formation; Bgn = Baltimore Gneiss; sl = Slaughterhouse Gneiss. B. Gravity model from C to C' from the eastern end of the Chattolanee anticline to the western end of the Towson anticline (Chapin, 1981, Fig. 3-6 and 3-7); symbols as in A.

Chattolanee Anticline–Baltimore Complex/Glenarm Contact

Gravity modeling of the Chattolanee anticline (Chapin, 1981) indicates that the gneiss is rooted at the eastern end of the structure, but its western half is underlain by a westward-thickening wedge of higher density material (Fig. 4A, 4B). Lacking subsurface control of lithology, we are not clear as to what specific rock units are present in the wedge. It is assumed to be composed mainly of Wissahickon schist and a subordinate amount of lower density Cockeysville marble and/or Setters quartzite, schist, or gneiss. In Figure 4A the wedge was modeled with an intermediate density, labeled Wissahickon, and shown to truncate the base of the anticline. However, the actual subsurface structure is assumed to be more like that shown in Figure 6B where the lower Glenarm units are folded between rooted and unrooted gneiss. Despite the excellent fit of observed and calcu-

lated gravity in Figure 4A, the structural interpretation in Figure 6B is favored because of geologic relationships out of the plane of Figure 4A.

The Baltimore Complex–Glenarm contact immediately south of the Chattolanee anticline is best modeled as a moderately south-dipping thrust (Fig. 4A) in agreement with the interpretation of Crowley and Reinhardt (1979). The layer of Wissahickon underlying the western half of the Chattolanee anticline is modeled as continuing under the frontal portion of the allochthon. It probably represents a zone of intense shearing and attenuation related to the emplacement of the Baltimore Complex.

It is possible if not probable that the rocks immediately below the south-dipping thrust are actually a tectonic melange composed of slivers of amphibolite and Wissahickon lithologies (W = 2.750 g/cm^3 in Fig. 4A). As modeled, the contact between allochthonous rocks of the Baltimore Complex (including melange plated to its base) and autochthonous and para-autochthonous rocks of the Chattolanee anticline flattens to subhorizontal between 1- and 2-km depths at the south end of the profile (Fig. 4A).

Towson Anticline

The pronounced gravity low centered over the east-central part of the Towson anticline strongly suggests that there the gneiss is rooted basement (Fig. 3). The prominence of this low is thought to reflect the abundance of Gunpowder Gneiss (compositionally similar to the low-density Slaughterhouse Gneiss, but Paleozoic in age) intruding the Baltimore Gneiss in this area. The gravity gradient over the western third of the anticline suggests that a thin wedge of higher density rocks underlies its faulted western edge. Gravity modeling of the valley between the Chattolanee and Towson anticlines indicates a syncline of lower Glenarm units that appears to project eastward under the Towson anticline, thus accounting for this wedge of higher density rocks (Fig. 4B). The syncline of Glenarm metasedimentary rocks is overturned to the west and underlain by gneiss that extends downward indefinitely. It is cut by the subvertical Ruxton fault on the east and possibly by another high-angle fault on the west (Fig. 4B). The absence of Setters Formation on the overturned limb of the syncline under the Towson anticline (Fig. 4B) may be real. However, a thin layer of Setters may be present, which cannot be distinguished by the gravity method because of the very small density contrast between Setters quartzite and gneiss and Baltimore Gneiss. The synclinal interpretation of the valley between the gneiss anticlines is consistent with the map pattern of Crowley and Reinhardt (1979).

Phoenix Anticline

In order to further test the interpretation of the Phoenix anticline as a refolded nappe, Chapin (1981) completed a closely spaced gravity traverse across the western end of the structure (line E-E'; Fig. 1). Four different subsurface configurations con-

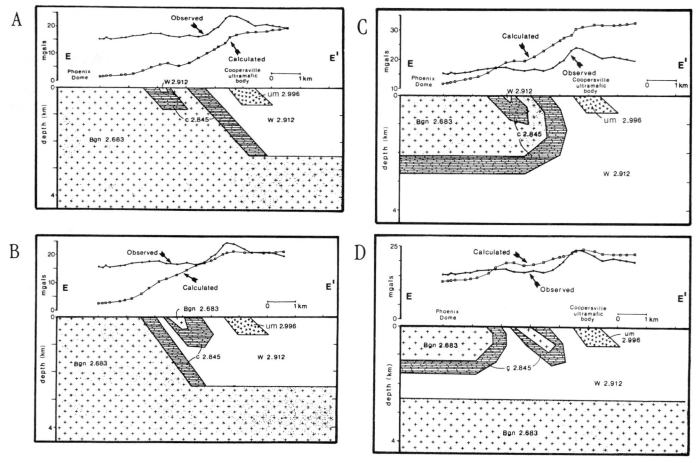

Figure 5. A–D. Gravity models from E to E′ across the western end of the Phoenix anticline (Chapin, 1981, Figs. 3-9 to 3-12); um = serpentinite and ultramafic schist; other symbols as in 4A.

sistent with surface geology (Muller, Maryland Geological Survey, unpubl. data) were modeled (Fig. 5). Model A is a rooted basement anticline with a small syncline near the apex, essentially the original interpretation of the structure given by Knopf and Jonas (1929). Model B represents a rooted anticline with a strongly attenuated crestal region overturned to the north. Model C is a refolded nappe rooted south of the profile containing a syncline of upper limb Glenarm units near the nappe closure. Model D is also a refolded nappe rooted south of the profile, but with Glenarm units exposed in an antiformal window through the lower limb of the nappe. This is essentially the interpretation of Crowley and others (1976) and Fisher and others (1979), and clearly displays the best fit of observed and calculated gravity along the profile.

The southern continuation of the Phoenix nappe is shown in Figure 6A. This interpretation differs from that of Crowley (1976) and Fisher and others (1979) in that the pre-erosion connection with the Towson anticline is direct rather than through the Texas anticline. The Texas anticline is interpreted as a culmination on a structurally lower nappe in the Towson-Chattolanee-Texas-Phoenix nappe system. Thus, lower Glenarm units in

contact with gneiss of the Texas anticline are right-side-up, in agreement with surface structural data (Crowley and others, 1975). Inliers of marble north of the western part of the Chattolanee anticline (Caves Valley) and the northeastern end of the Towson anticline (Hydes Valley) are assumed to be lesser culminations on the Texas nappe in which Baltimore Gneiss has yet to be exposed (Fig. 1).

Texas Anticline

The Texas anticline has an enigmatic geophysical signature. Like the Phoenix anticline, it causes no significant gravity anomaly (Fig. 3), but in strong contrast to all the other gneiss anticlines, it is associated with a prominent magnetic high (Bromery, 1967a; Fisher and others, 1979). Bromery (1968) interpreted this magnetic high as resulting from either a magnetic facies of Baltimore Gneiss, magnetic Wissachickon schist, or ultramafic rocks underlying the gneiss at a shallow depth (<150 m). Both Crowley (1976) and Fisher and others (1979) attributed the high to a magnetic facies of the Baltimore Gneiss. However, to account for the anomaly, this magnetite-rich lithology must constitute a

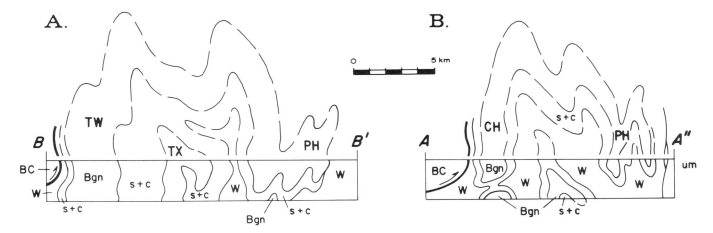

Figure 6. A. Interpretive cross section from B to B' across the Chattolanee (CH) and Phoenix (PH) anticlines; symbols as in 4A. B. Interpretive cross section from A to A' across the Towson (TW), Texas (TX), and Phoenix anticlines; symbols as in 4A.

major portion of the Texas anticline. Crowley's original field notes (Maryland Geological Survey, unpubl. data) indicate only a local cluster of six outcrops in the northeast part of the anticline. Field checking of these occurrences revealed that they were similar to magnetite-bearing gneiss in the eastern part of the Chattolanee anticline, which exhibits no associated magnetic anomaly. Furthermore, new exposures of gneiss directly beneath the peak amplitude of the high are magnetite poor. In short, magnetite-rich Baltimore Gneiss is rare and, therefore, not considered the cause of the Texas anticline magnetic high.

The cause of the magnetic high over the Texas anticline remains to be determined; however, the thinness of the gneiss suggested by its feeble gravity expression is consistent with the nappe interpretation of Figure 6A. Thus, the magnetic high can probably be attributed to Wissahickon lithologies on the lower limb of the Texas nappe.

Clarksville, Mayfield, and Woodstock Anticlines

At present, integrated geophysical interpretations of the Clarksville and Mayfield and much of the Woodstock anticlines are preliminary. The same relatively flat magnetic lows found over the Phoenix, Chattolanee, and Towson anticlines also occur over these gneiss anticlines (Fisher and others, 1979, Pl. 2).

Detailed gravity studies of this western group of gneiss anticlines remain to be completed; however, as seen in Figure 3, only the northern end of the Clarksville anticline is associated with any apparent anomaly. This weak gravity low superimposed on the westward-increasing regional gradient is similar to that centered over a swarm of Paleozoic granitic bodies in the Wissahickon Group approximately 5 km to the south, suggesting it may in fact be related to younger granitoids (Ellicott City pluton) rather than the gneiss. If the Clarksville, Mayfield, and Woodstock anticlines represent another refolded nappe structure, then their root zone is neither geologically nor geophysically evident.

STRUCTURE

Three periods of deformation affected the Baltimore Gneiss in Maryland (D_1, D_2, D_3; Table 2). The first is inferred from radiometric age data which indicate metamorphism of the gneiss at 1,000 to 1,200 m.y. ago (Hopson, 1964; Tilton and others, 1970). Folding and secondary s-surface formation are assumed to have accompanied this medium- to high-grade Grenville metamorphism. However, except for compositional layering in nonmigmatitic gneisses, the resulting structural elements were either obliterated or completely obscured by subsequent tectonism. Therefore, they were not distinguishable from pervasive D_2 elements. Broedel (1937) argued that the rounded outline of several of the gneiss anticlines could not have been produced during the deformation that generated tight northeast-trending folds in the surrounding Glenarm units and, thus, was a relict pre-Glenarm (Grenville?) feature. However, considering the abundant evidence of multiple refolding and fold interference, this argument seems invalid.

The most clearly discernable period of deformation, D_2, resulted in the majority of pervasive structural elements listed in Table 2. D_2 was divided into three phases that may represent a continuum of deformation and metamorphism spanning the entire Taconic-Acadian orogenic interval (500–360 m.y. rather than three temporally distinct thermo-structural episodes. The gross structural form of the gneiss anticlines was established during this prolonged interval.

The last structurally recognizable period of deformation, D_3, occurred well after the peak of metamorphism and resulted in the superimposition of dominantly brittle structures on the pervasive ductile structures produced during D_2. The structural form of the gneiss anticlines was modified only slightly during this poorly defined Alleghanian-Palisades deformation (260–190 m.y.).

The following discussion of structural elements is based

TABLE 2. MESOSCOPIC STRUCTURAL ELEMENTS OF THE BALTIMORE GNEISS ANTICLINES

	D_1 (Grenville 1000-1200 m.y.)	D_2 (Taconic-Acadian 360-500 m.y.)			D_3 (Alleghanian-Palisades 260-190 m.y.)
	F_1 fold generation (?)	F_{2a} fold generation	F_{2b} fold generation	F_{2c} fold generation	F_3 fold generation
Fold Style	Isoclinal (?)	Isoclinal vertical to reclined flow folds	Tight to isoclinal, plunging inclined to reclined similar and parallel folds; Z- and S-types common; locally convolute	Inclined to recumbent, open to tight parallel folds; local crenulations	Open upright to vertical parallel folds
Surfaces Folded	S_0 (?)	S_1	S_1/S_{2a}	$S_1/S_{2a}/S_{2b}$	$S_1/S_{2a}/S_{2b}$; S_{2c} (?)
Planar Structures	S_1, compositional layering	S_{2a}, axial plane foliation parallel to S_0/S_1*; fluxion structure in felsic augen gneisses	S_{2b}, secondary foliation present locally in hinge zones of F_{2b} folds; $S_{2b'}$; migmatitic segregation layering parallel to S_1/S_{2a}	S_{2c}, gently dipping spaced cleavage in hinge zones of competent quartzofeldspathic layers	Brittle faults and joint sets
Linear Structures	(?)	L_{2a}, mineral elongation and mineral streaking	L_{2b} mineral elongation and axes of boudins; A_{2b} fold axes	$L_{2c}x_{1/2a}$ intersection of S_{2c} on S_1/S_{2a}; A_{2c}, fold and crenulation axes	----
Axial Orientation	(?)	Highly variable trend and plunge	Dominantly WNW to ENE trends, variable plunge	Variable trend, gentle plunge	Variable trend and plunge
Remarks	Problematic F_1, not distinguishable from F_{2a}	Strongly appressed F_{2a}; best preserved in felsic gneisses	Associated with ductile shearing and mylonitization	F_{2b} locally transitional with F_{2c}	Silicified breccias characterize faults
Type Locality	----	Herring Run 75-100 m north of Arlington Ave., Baltimore 7-1/2' quadrangle	Davis Branch Ellicott City 7-1/2' quadrangle	Gunpowder River 50-75 m east of Upper Glencoe Rd., Hereford 7-1/2' quadrangle	unnamed stream 75-100 m upstream from junction with Gunpowder River, Hereford 7-1/2' quadrangle

*   denotes approximate parallelism except at fold hinges.

primarily on field data collected from the three best exposed gneiss anticlines: Phoenix, Towson, and Woodstock. Outcrops were examined in the other gneiss anticlines; however, because of the paucity of mesoscopic folds, little orientation data were recordable. The small outlier of gneiss along Bens Run approximately 1 to 2 km east of the eastern margin of the Woodstock anticline is considered part of that structure.

F_1/F_{2a} Folds and Associated Structures

Structural evidence of F_1 folding is at best equivocal. Tightly appressed vertical to reclined isoclinal folds in felsic gneisses along Herring Run in the city of Baltimore and Towson Run in Baltimore County may represent F_1 (Grenville) structures (Fig. 7A). However, because similar isoclines also occur in lower Glenarm lithologies, they are assumed to be flattened F_{2a} (Taconic) isoclines.

The nature of finite strain displayed in strongly attenuated,

locally disrupted F_1/F_{2a} isoclines (Fig. 7B) suggests a sequence of progressive deformation from buckle folding to homogeneous flattening and ductile boudinage. The associated pervasive s surface is a mylonitic flattening foliation (fluxion structure, Table 2), which was subsequently folded into tight F_{2b} folds. Although this progressive strain may have occurred during Grenville orogenesis, the style and sequence of deformation is one that has been widely recognized in the southern and central Appalachian Piedmont as Paleozoic in age (Tobisch and Glover, 1971; Butler, 1973; Higgins, 1973). It is consistent with an initial flowage phase of D_2 involving the emplacement and attenuation of large fold nappes followed by tight refolding during later phases of D_2 (Fig. 6).

Mesoscopic examples of F_{2a} folds exhibit long attenuated limbs and greatly thickened, angular hinges where not sheared out (Fig. 7A). Their limbs delineate compositional layering, S_1. Because F_{2a} axial trends reflect the strike of compositional layering, they are highly variable (Fig. 10A). Plunges range from moderate to subvertical. Folding of F_{2a} folds by F_{2b} folds is

Figure 7. A. Isoclinal F_{2a} folds in layered felsic gneiss along Herring Run, Towson anticline. Lens cap is 45 mm in diameter. B. Strongly flattened and necked isoclinal F_{2a} folds in veined gneiss along Herring Run, Towson anticline. Pencil is 15 cm in length.

observable in outcrop and in the map pattern of the gneiss anticlines. For example, the northeastern end of the Clarksville anticline is an isoclinal F_{2a} folded by an F_{2b} fold (Fig. 1).

The main gneissic layering of the Baltimore Gneiss, S_1, is a transposed compositional layering of presumed Grenville age (Table 2). Locally, it is coincident with a mylonitic flattening foliation in zones of streaked augen gneiss. The present structurally conformable contact between the Baltimore Gneiss and overlying lower Glenarm units is considered to be an S_{2a} structure formed as a result of intense ductile shearing along the limbs of large F_{2a} nappes. The shearing-out of isoclinal fold hinges in the Glenarm units near the gneiss contact produced transposed layering as well (Jonas, 1937). Although commonly subparallel with

migmatitic segregation layering produced during the peak of D_2 (Paleozoic) metamorphism, S_1 is a relict pre-D_2 structure. S_{2a}, an axial surface foliation, is parallel to S_1 except at uncommon F_{2a} fold closures. It is defined by aligned biotite and, locally, flattened feldspar augen. In areas of well-developed F_{2b} folding, S_{2a} is subparallel to or intersects the dominant S_{2b} foliation at a very low angle on F_{2b} limbs and wraps around F_{2b} closures. The synoptic equal-area diagram of poles to S_1/S_2 reveals the large variation in these planar elements (Fig. 10A).

The presence of refolded isoclinal folds in rocks of the Glenarm Supergroup is documented throughout the Piedmont of Maryland (Freedman and others, 1964; Fisher, 1970; Higgins, 1973). Although development of isoclinal F_{2a} folds in the

TABLE 3. FOLD-GENERATION CORRELATION CHART FOR THE EASTERN PIEDMONT OF MARYLAND

This Report	Previous Work
F_{2a}	F_1 - Freedman and others (1964) and Wise (1970); Susquehanna River region PA-MD
	First generation - Southwick (1969); Harford County, MD
	F_2 - Fisher (1970); Potomac River region, MD-VA
	Big Elk generation - Higgins (1973); Cecil County, MD
	Texas - Phoenix nappe - Crowley and others (1976); Baltimore County, MD
	F_1 - Moller (1979) - Baltimore County, MD
F_{2b}	Laminar flow folds - Choquette (1960); Baltimore County, MD
	Second generation - Southwick (1969)
	F_3 - Fisher (1970)
	Appleton and/or Rock Church generation - Higgins (1973)
	F_2 - Moller (1979)
F_{2c}	F_3 and/or F_2 - Freedman and others (1964) and Wise (1970)
	Wildcat Point generation - Higgins (1973)
	F_3 - Moller (1979)
F_3	F_4 - Fisher (1970)

Baltimore Gneiss is probably correlative with development of isoclines in the Glenarm units of the Baltimore vicinity, correlation over long distances or across major structures (for example, the fault zone separating the Baltimore Complex and Glenarm Supergroup and the phyllonitic zone between the middle and upper intervals of the Wissahickon Group) is, at best, tentative (Table 3).

F_{2b} Folds and Associated Structures

The second phase of D_2 folding resulted in a variety of fold geometries and styles (Fig. 8A). F_{2b} folds are classed as plunging inclined to reclined, tight to isoclinal, similar and flattened parallel folds. Z- and S-type parasitic folds occur on opposite limbs of many mesoscopic F_{2b} folds. All F_{2b} hinges are thickened to some extent and range from angular to rounded. Convolute F_{2b} folds are characteristic of zones of highly segregated migmatitic gneiss (Fig. 8B). Isolated bending folds related to necked and boudinaged amphibolite and hornblende gneiss layers also are examples of F_{2b} folds.

Locally, the axial orientations (axial surface and fold axes) of F_{2b} folds are transitional to those of F_{2c} folds. Thus, the two fold generations are difficult to separate and are plotted as F_{2b}/F_{2c} axes in Figure 10B. Correct identification is made even more difficult where the style of F_{2b} and F_{2c} folds overlap, such as along Herring Run in the city of Baltimore between Arlington Avenue and Argonne Drive. In other areas, however, F_{2c} folds clearly fold F_{2b} folds (Fig. 8C, 8D).

F_{2b} axes are commonly expressed as a strong rock lineation, A_{2b} (Table 2). A_{2b} also parallels the long axes of boudins formed by competent quartzo-feldspathic layers enclosed in biotite and hornblende-rich lithologies. Dimensional alignment of hornblende parallel to F_{2b} folds is common, suggesting F_{2b} folds were synchronous with amphibolite facies metamorphism.

Migmatitic segregation subparallel to S_1 compositional layering and recrystallization of biotite and hornblende parallel to the axial surfaces of F_{2b} folds resulted in a second pervasive foliation, S_{2b}. S_{2b} is subparallel to S_1/S_{2a} except at F_{2b} closures (Fig. 9A); therefore, the variation in strike and dip of S_{2b} is similar to that of S_1/S_{2a}. This variation is both a function of subsequent folding by F_{2c} and F_3 folds and the orientation of S_1/S_{2a}.

If migmatization and development of the associated metamorphic segregation layering reflected the peak of D_2 metamorphism, then the fact that F_{2b} folds deform this layering indicates that F_{2b} folding persisted beyond the thermal peak of metamorphism. Thus, the S_{2b} segregation layering and S_{2b} axial foliation are not strictly contemporaneous. S_{2b} segregation layering is essentially parallel to S_1 compositional layering, but it formed as a result of mid-D_2 migmatization (injection and metamorphic segregation), not early D_2 isoclinal folding and transposition of preexisting (Grenville) compositional layering.

F_{2b} folding was followed closely by ductile shearing along narrow (1–3 m) biotite-schist layers. F_{2b} folds in these undulating layers are sheared out to varying degrees (Fig. 9B). Locally, they are totally disrupted into zones of convolute folds and/or mylonitic gneiss. The mineralogy and texture of these biotite-rich ductile shear zones indicate syntectonic recrystallization at epidote-amphibolite facies metamorphic conditions.

The basic structural form of the Baltimore Gneiss anticlines resulted from F_{2b} refolding of large F_{2a} nappes (Fig. 6). The large inlier of Glenarm metasedimentary rocks in the Phoenix anticline is the clearest expression of macroscopic F_{2b} folding in the Baltimore area. F_{2b} folds are also the dominant mesoscopic fold structures in the lower Glenarm units surrounding the gneiss anticlines.

F_{2b} digitations in the gneiss-Setters contact in the Phoenix and Clarksville anticlines indicate that both rock units deformed as a coherent structural unit during F_{2b} folding. This is in contrast to the apparent decoupling of the gneiss and enveloping Glenarm units that occurred during the late stages of F_{2a} nappe formation and resulted in local shearing out of the Setters and/or Cockeysville Formations. The local omission of one or both of these formations along the margins of the gneiss anticlines has been attributed to nondeposition (Knopf and Jonas, 1929; Hopson, 1964), ductile shearing (McKinstry, 1961), and reverse faulting (Broedel, 1937; Cloos, 1964). Although nondeposition on intrabasinal highs (Crowley, 1976) may have occurred over various time intervals, it seems unlikely that numerous sediment-free basement highs persisted throughout deposition of the entire Setters and Cockeysville Formations, both platform-type units. Likewise, omission resulting from pre-Wissahickon erosion is considered equally untenable, as there is no structural or lithologic evidence of this event. Post-D_2 faults have cut out one or more of the Glenarm units locally (for example, the Ruxton fault along the western margin of the Towson and Texas anticlines). However, these faults are characterized by silicified breccia zones and therefore do not appear responsible for the missing units where breccias are absent. Given the widespread evidence for late

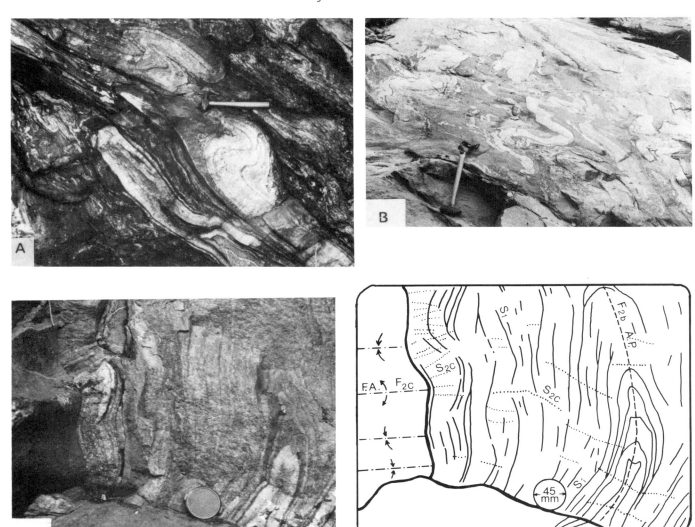

Figure 8. A. Various styles of F_{2b} folds in layered gneiss along Bens Run, Woodstock anticline. S- and Z-type folds on either side of hammer juxtaposed along a S_{2b} shear zone. Isoclinal F_{2b} folds occur in the upper right and lower left of photograph. Hammer is 40 cm in length. B. Zone of convolute F_{2b} folds in layered gneiss along Herring Run, Towson anticline. Hammer is 40 cm in length. C. and D. Photo and annotated sketch showing F_{2b} folds refolded by F_{2c} folds in layered gneiss along Charles Street, Towson anticline. Lens cap is 45 mm in diameter.

F_{2a} ductile shearing in the gneiss, a tectonic mechanism is considered to be the main cause of the local absence of these units.

As with F_{2a} folds, correlation of F_{2b} folds in the Baltimore Gneiss with folds in the Glenarm Supergroup and Baltimore Complex not within the immediate Baltimore area is considered only tentative. However, if the proposed correlations shown in Table 3 are correct, then it can be inferred that the allochthonous Baltimore Complex was emplaced onto the Glenarm terrane early in D_2, prior to the peak of metamorphism accompanying F_{2b} folding.

F_{2c} Folds and Associated Structures

The final phase of D_2 deformation resulted in folds that locally are difficult to distinguish from F_{2b} folds (Fig. 9C). However, in many exposures, F_{2c} folds clearly refold F_{2b} folds (Fig. 8C, 8D). F_{2c} folds are classed as gently plunging inclined to horizontal, open to tight parallel folds with rounded hinges (Table 2). F_{2c} crenulations were observed in some schistose lithologies.

A rough, disjunctive, spaced cleavage (Powell, 1979) is characteristic of F_{2c} folds in quartzofeldspathic layers (Fig. 9D). This gently dipping to subhorizontal cleavage results in an intersection lineation parallel to F_{2c} fold and crenulation axes. No mineral lineation is associated with F_{2c} folds; however, F_{2c} folds earlier L_{2a} and L_{2b} lineations. The lack of a pervasive F_{2c} axial foliation, together with the parallel geometry and spaced cleavage seen in many F_{2c}, folds suggest that the late stages of D_3

A

B

Figure 9. (this and facing page) A. F_{2b} folds in layered gneiss displaying incipient axial planar schistose (S_{2b}) across compositional layering (S_1/S_{2a}). Sheepfold Lane, Phoenix anticline. Pen is 12 cm in length. B. F_{2c} folds in felsic gneiss layers adjacent to a biotite schist zone containing sheared out F_{2a}/F_{2b} folds (lower left); Herring Run, Towson anticline. Pencil is 15 cm in length and is oriented parallel to F_{2c} axes, while F_{2a} and F_{2b} axes are normal to the photograph. C. Cascadelike zone of F_{2b}/F_{2c} folds in layered gneiss along Herring Run, Towson anticline. Man is 1.9 m in height. D. F_{2c} hinge zone is layered gneiss displaying well-developed disjunctive cleavage (S_{2c}); Charles Street near Greater Baltimore Medical Center, Towson anticline. Lens cap is 45 mm in diameter. E. Silicified breccia from brittle fault cutting southeastern end of the Chattolanee anticline near Greenspring Avenue. Angular fragments (dark gray) are fine-grained, vuggy silica of early hydrothermal event recemented by lighter colored, fine-grained, vuggy silica of later hydrothermal event.

were characterized by a transitional brittle-ductile style of deformation. Kinked and fractured tourmaline crystals from F_{2c} crenulations in the overlying Setters Formation also suggest a less ductile style of deformation than that which characterized F_{2b} folding.

The axial trends of F_{2c} folds are dominantly northwest to northeast with gentle westward plunges. They are basically the same as those of F_{2b} folds, although F_{2b} axes appear to have more varied plunge (Fig. 10B). The variation in trend of both F_{2b} and F_{2c} folds may be the result of subsequent F_3 folding, but it is probably largely due to the original variation in the orientation of $S_1/S_{2a}/S_{2b}$ surfaces. Outcrop-scale F_{2c} folds are primarily responsible for the repeated variation in dip of the gneiss-Glenarm contact from upright to overturned.

Pi-poles determined from S_1/S_{2a} great circle girdles from major field stations in the Phoenix, Towson, and Woodstock anticlines (Fig. 10C-10E) give orientations indicative of either F_{2b} or F_{2c} axes. As noted previously, F_{2b} and F_{2c} folds are locally difficult to distinguish where mesoscopic refolding is not visible. These relationships suggest that F_{2b} and F_{2c} folds are essentially coaxial.

The fact that F_{2c} folds are commonly open with subhorizontal axial surfaces suggests that they may be the result of vertical compression. Thus, F_{2c} folding may reflect uplift of the core of the F_{2a} nappe system during the final phase of D_2. The timing of this uplift cannot be determined precisely; however, it may coincide with terminal Acadian (360 m.y.) K/Ar and Rb/Sr mica

ages recorded throughout the Piedmont of Maryland and Pennsylvania (Lapham and Bassett, 1964; Wetherill and others, 1966). Southeasterly derived detrital staurolite and kyanite in the Pennsylvanian Pottsville Formation of the Valley and Ridge of northeastern Pennsylvania (Meckel, 1967) may have originated from the Baltimore region. If true, this suggests that D_2 ceased by Late Mississippian time.

F_3 Folds and Associated Structures

Pervasive D_2 ductile deformation was succeeded by a poorly defined, predominantly brittle deformation (D_3) which resulted in open to gentle, upright to vertical parallel folds (F_3) and narrow brecciated and hydrothermally altered fault zones. The brittle to brittle-ductile character of this period of deformation suggest D_3 occurred within entirely different thermal and stress regimes than D_2.

Mesoscopic F_3 folds are not well developed in the Baltimore Gneiss. They occur as gentle warps of the main foliation and compositional layering, and commonly exhibit jointing subparallel to their axial surfaces. F_3 folds display parallel geometry with broad, rounded hinge zones (Table 2). Cross-strike axial trends and steep to vertical plunges are common; however, orientations vary considerably. Macroscopic F_3 folds are relatively conspicuous in the map pattern of the gneiss anticlines. The prominent change in trend of the northern contact of the Towson anticline is a good example (Fig. 1). Open F_3 folds of the gneiss-Glenarm contact are found in the Phoenix, Clarksville, and Woodstock anticlines as well.

The open Z-shaped map pattern of the five southernmost gneiss anticlines (Fig. 1) may be a large F_3 structure related to a northeast-striking, right-lateral shear couple. Narrow, high-angle shear zones in the Baltimore Complex along the straight southeastern margin of the Towson anticline could represent the trace of the eastern right-lateral shear, while the discontinuous belt of schistose ultramafic rocks in the Wissahickon Group north and west of the gneiss anticlines could represent the western right-lateral shear of the couple. Right-lateral motion along this large shear couple would impose a clockwise rotation on the block between the shear zones, possibly giving rise to the observed open F_3 folds with variable orientations.

Steeply dipping to vertical faults of unknown displacements cut across the margins of most of the gneiss anticlines (Fig. 1). The largest of these, the Ruxton fault, truncates the eastern end of the Towson and Texas anticlines. These brittle faults are characterized by healed breccias that display evidence of several brecciation events (Fig. 9E). They strike from west-northwest to northeast, and slickensides suggest movements were primarily dip-slip. Distinctive silica ± chlorite ± hematite rock occurs as lenses or pods at gneiss-marble fault contacts. Elsewhere, silicified breccias composed of Baltimore Complex amphibolite and gneiss are present locally. Within the gneiss, zones of closely fractured, hydrothermally altered gneiss are inferred to represent similar D_3 brittle faults. Biotite is completely chloritized, feldspar strongly

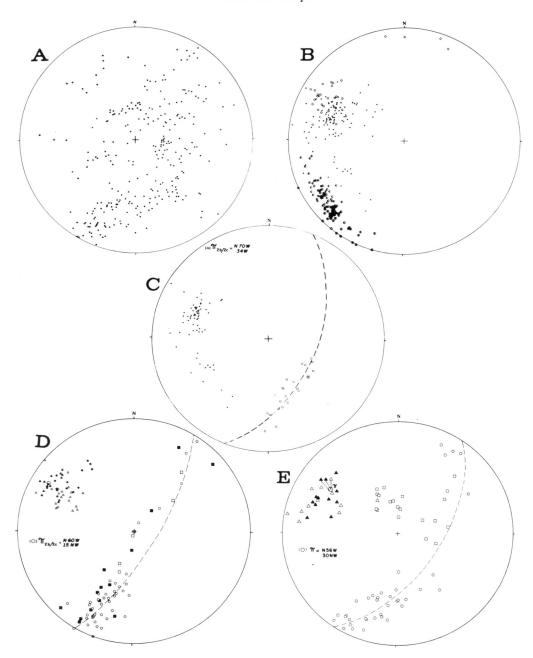

Figure 10. A. Equal-area lower hemisphere plot of poles to $S_1/S_{2a}/S_{2b}$ (compositional layering/folia-tion) from Baltimore Gneiss of the Phoenix, Woodstock, Chattolanee, and Towson anticlines. B. Equal-area lower hemisphere plot of fold axes from Baltimore Gneiss of the Phoenix, Woodstock, and Towson anticlines. Dots = F_{2b} axes; open circles = F_{2c} axes; half-filled circles = transitional F_{2b}/F_{2c} axes. C. Equal-area lower hemisphere plot of structural elements from Baltimore gneiss in Davis Branch and an unnamed stream ~0.5 km north of Woodstock anticline. Open circles = poles to $S_1/S_{2a}/S_{2b}$ (com-positional layering/foliation); dots = F_{2b} axes. D. Equal-area lower hemisphere plot of structural elements from Baltimore Gneiss in Towson Run, Towson anticline. Open circles = poles to $S_1/S_{2a}/S_{2b}$ (compositional layering/foliation); open triangles = mineral lineations, open triangles = F_{2b} axes; filled triangles = F_{2c} axes; filled squares = poles to F_{2b} axial surfaces; open squares = poles to F_{2c} axial surfaces. E: Equal-area lower hemisphere plot of structural elements from Baltimore Gneiss along Gunpowder Falls near Glencoe, Phoenix anticline. Symbols as in D.

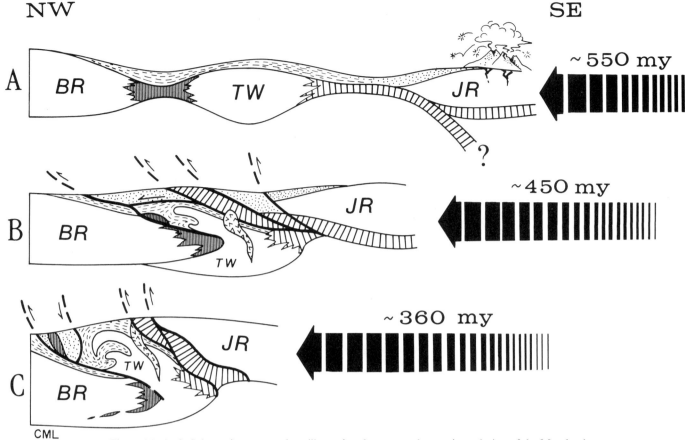

Figure 11. A–C. Schematic cross sections illustrating the proposed tectonic evolution of the Maryland Piedmont. BR = Blue Ridge Block, includes Grenville-age gneisses, latest Precambrian-Cambrian Catoctin greenstone and Chilhowee Group clastic rocks; TW = Towson Block, includes Baltimore Gneiss, Setters Formation, and Cockeysville Marble; JR = James Run island arc, includes James Run Formation and related volcanic rocks; open vertical-ruled pattern = ocean crust; dots = predominantly volcaniclastic rocks of the Wissahickon Group; dashes = predominantly pelitic and psammitic rocks of the Wissahickon Group. See text for further discussion.

seriticized, and microshears of comminuted quartz-feldspar lace the gneiss in these narrow zones.

The relationship of brittle faulting to F_3 folding is unclear. However, because the brittle faults cut the margins of the gneiss anticlines, they probably postdate the open folding, which is most clearly expressed in gently folded gneiss-Glenarm contacts. Presumably, the brittle faulting and concurrent hydrothermal alteration were the relatively near-surface expression of extensional strain during the Palisades disturbance. Poorly exposed (Triassic-early Jurassic?) diabase dikes appear to cut F_3 folds in the Phoenix and Clarksville anticlines without being folded. Hence, intrusion is considered post-F_3 folding, probably contemporaneous with brittle faulting.

PALEOZOIC TECTONIC SYNTHESIS: A SPECULATIVE OVERVIEW

Recent tectonic syntheses of the central and southern crystalline Appalachians (Rankin, 1975; Hatcher, 1978; Fisher and others, 1979; Glover, 1979; Cook and Oliver, 1981) all have advocated late Precambrian rifting of the North American continental margin which resulted in several microcontinental or peninsular blocks separating marginal seas from the main proto-Atlantic ocean (Iapetus). The Baltimore Gneiss and its mantle of Glenarm Supergroup metasedimentary rocks are considered one of these blocks. The subsequent Paleozoic evolution of these marginal Grenville-basement blocks has been interpreted differently in the various models. The following speculative model for the Maryland Piedmont differs in certain aspects from previous models but incorporates several basic concepts common to all of them. All directions refer to present geographic coordinates.

At the onset of the Cambrian, the Baltimore Gneiss-Glenarm terrane (Towson Block, Fig. 11A) formed a microcontinental fragment east of the true North American continental margin (Blue Ridge Block, Fig. 11A). It was separated from the Blue Ridge basement by a marginal sea underlain by a transitional-type crust. This crust is assumed to have been composed of attenuated continental rocks intruded by ultramafic and mafic material and overlain by fine-grained clastic sediments. East of the Towson Block, oceanic crust was being subducted beneath an island arc (James Run arc, Fig. 11A).

Convergence during Cambrian–Middle Ordovician time resulted in obduction of the Baltimore Complex (oceanic crust plus James Run arc) onto the Towson Block (Fig. 11B). The Towson Block overrode the eastern edge of the Blue Ridge Block, including diamictite shed westward in front of the advancing allochthon, as envisioned by Fisher (1979). At depth, Grenville basement (Baltimore Gneiss) and its supracrustal cover (Glenarm Supergroup) were emplaced as a major west-vergent to recumbent nappe system during the early Taconic interval. Depression of the eastern margin of the Blue Ridge Block not only initiated a westward migrating flysch basin but also led to partial melting and intrusion of granitic plutons into the nappe system about 440 to 500 m.y. ago (Fig. 11B).

The Taconic thermal peak associated with intrusion of granitoids resulted in regional amphibolite facies metamorphism and local migmatization of the Baltimore Gneiss. It was accompanied and outlasted by tight refolding of the early Taconic nappes. Final Acadian suturing of the various crustal blocks involved ductile shearing and open to tight folding as the core of the nappe system was uplifted (Fig. 11C). Ultramafic parts of the transitional basement were serpentinized and emplaced into the overlying Glenarm Group at this time.

Open folding and brittle faulting followed uplift and cooling of the sutured crustal blocks. The folding may have been related to transcurrent movements along right-lateral strike-slip faults such as occur in the Eastern Piedmont fault system in the southern Appalachians (Bobyarchick, 1981). If so, this weak deformation probably was Alleghanian and represented a basement response to thin-skinned thrusting farther west. Hydrothermal alteration and brecciation associated with high-angle faults signaled the onset of continental rifting during the Palisades disturbance.

CONCLUSIONS

1. The Baltimore Gneiss experienced three major periods of deformation (D_1, D_2, D_3). U/Pb zircon ages indicate high-temperature recrystallization of Grenville age (1,000–1,200 m.y.) designated D_1. However, except for a transposed compositional layering in some felsic gneisses, no unequivocal D_1 structural elements were recognized. The most clearly expressed period of deformation, D_2, manifests Taconic-Acadian tectonism (500–360 m.y.). D_2 was polyphase and involved amphibolite facies metamorphism and local migmatization of the gneiss. The phases may have been distinct thermal-structural events or a continuum of deformation during a gradually varying thermal regime. Three fold generations (F_{2a}, F_{2b}, F_{2c}) and associated planar structures formed as a result of pervasive ductile deformation. The final period of deformation, D_3, was nonpervasive and resulted in open folds and brittle faults within the gneiss and possibly a major high-angle right-lateral shear couple bounding a rigid

crustal block containing the gneiss anticlines. This poorly defined postregional metamorphic deformation fell within the Alleghanian-Palisades interval (260–190 m.y.).

2. The structural form of the Baltimore Gneiss anticlines developed during D_2 and was only slightly modified during D_3. The map pattern of the anticlines reflects the interference of macroscopic F_2 folds. The Baltimore Gneiss–Glenarm Supergroup contact is structurally conformable as a result of extreme attenuation and ductile shearing during F_2 isoclinal folding.

3. The Phoenix, Texas, Towson, and Chattolanee anticlines are part of a major F_{2a} nappe system rooted in the Towson anticline. The Woodstock, Mayfield, and Clarksville anticlines form a similar refolded nappe structure, but with no clearly evident root zone. The Setters Formation, Cockeysville Marble, and lower Wissahickon Group were transported northward and westward and locally inverted as a result of D_2 nappe formation. The structural framework envisioned is one with many of the characteristics of the Pennine nappes.

4. The Baltimore Gneiss represents a complexly deformed remnant of a Grenville basement block originally separated from the North American craton by a marginal sea. During early Paleozoic consumption of either a back-arc basin or the proto-Atlantic Ocean, an island arc (James Run arc) collided with and partially overrode the Baltimore Gneiss and its cover of continental margin sediments. The collision initiated remobilization of the gneiss and the formation of several crystalline nappe systems. Suturing of the island arc, microcontinental Baltimore Gneiss-- Glenarm terrane, and cratonic Blue Ridge terrane resulted in refolding and uplift of the nappe systems, now expressed as the Baltimore Gneiss anticlines.

ACKNOWLEDGMENTS

We are grateful to Mervin Bartholomew formerly of the Virginia Division of Mineral Resources for the invitation to participate in the "Grenville Terranes of the Appalachians Symposium." Jonathan Edwards of the Maryland Geological Survey helped in several aspects of the field work and reviewed early drafts of the manuscript. The first author (Muller) is indebted to George Fisher of the Johns Hopkins University, both for helpful discussions concerning the structure and tectonics of the central Appalachian Piedmont and for continued encouragement in pursuing this work.

Gravity surveying and modeling was supported in part by a Lehigh University fellowship while the second author (Chapin) was completing his master's thesis, and by a Maryland Geological Survey student technical assistantship.

REFERENCES CITED

Bailey, E. B., and Mackin, J. H., 1937, Recumbent folding in the Pennsylvania Piedmont; preliminary statement: American Journal of Science, 5th ser., v. 33, p. 187–190.

Bascom, F., 1902, The geology of the crystalline rocks of Cecil County: Cecil County Report, Maryland Geological Survey, p. 83–148.

Bobyarchick, A. R., 1981, The eastern Piedmont fault system and its relationship to Alleghanian tectonics in the southern Appalachians: Journal of Geology, v. 89, p. 335–348.

Broedel, C. H., 1937, The structure of the gneiss domes near Baltimore, Maryland: Maryland Geological Survey, v. 13, p. 149–187.

Bromery, R., 1968, Geological interpretation of aeromagnetic and gravity surveys of the northeastern end of the Baltimore-Washington anticlinorium, Harford, Baltimore, and part of Carroll County, Maryland [Ph.D. dissertation]: Baltimore, Johns Hopkins University, 124 p.

Bromery, R. W., 1967a, Aeromagnetic map of Baltimore County and Baltimore City, Maryland: U.S. Geological Survey, Geophysical Investigations Map GP-613, scale 1:62,500.

—— 1967b, Simple Bouguer gravity map of Baltimore County and Baltimore City, Maryland: U.S. Geological Survey, Geophysical Investigations Map GP-614, scale 1:62,500.

Butler, J. R., 1973, Paleozoic deformation and metamorphism in part of the Blue Ridge thrust sheet, North Carolina: American Journal of Science, v. 273A, p. 72–88.

Chapin, D. A., 1981, Geological interpretations of a detailed Bouguer gravity survey of the Chattolanee Dome, near Baltimore, Maryland [MS thesis]: Bethelem, Lehigh University, 137 p.

Choquette, P. W., 1960, Petrology and structure of the Cockeysville Formation (pre-Silurian) near Baltimore, Maryland: Geological Society of America Bulletin, v. 71, p. 1027–1052.

Cleaves, E. T., Edwards, J. Jr., Glaser, J. D., 1968, compilers, Geological map of Maryland: Maryland Geological Survey, scale 1:250,000.

Cloos, E., 1933, Structure of the Ellicott City granite, Maryland: National Academy of Science Proceedings, v. 19, p. 130–138.

—— 1964, Structural geology of Howard and Montgomery Counties: Maryland Geological Survey, The Geology of Howard and Montgomery Counties, p. 216–261.

Cohen, C. J., 1937, Structure of the metamorphosed gabbro complex at Baltimore: Maryland Geological Survey, v. 13, p. 205–236.

Cook, F. A., and Oliver, J. E., 1981, The late Precambrian–early Paleozoic continental edge in the Appalachian orogen: American Journal of Science, v. 281, p. 993–1008.

Crowley, W. P., 1976, The geology of the crystalline rocks near Baltimore, and its bearing on the evolution of the eastern Maryland Piedmont: Maryland Geological Survey Report of Investigations 27, 40 p.

Crowley, W. P., and Reinhardt, J., 1979, Geologic map of the Baltimore West quadrangle, Maryland: Maryland Geological Survey, scale 1:24,000.

Crowley, W. P., Reinhardt, J., and Cleaves, E. T., 1975, Geologic map of the Cockeysville quadrangle, Maryland: Maryland Geological Survey, scale 1:24,000.

—— 1976, Geologic map of Baltimore County and Baltimore City: Maryland Geological Survey, scale 1:62,500.

Davis, G. L., Tilton, G. R., Aldrich, L. T., Hart, S. R., and Steiger, R. H., 1965, The minimum age of the Glenarm Series, Baltimore, Maryland: Carnegie Institution on Washington Year Book 64, p. 174–177.

Drake, A. A., and Lyttle, P. T., 1981, The Accotink schist, Lake Barcroft, metasandstone, and Popes Head Formation—keys to an understanding of the tectonic evolution of the northern Virginia Piedmont: U.S. Geological Survey Professional Paper 1205, 16 p.

Eskola, P. E., 1949, The problem of mantled gneiss domes: Quarterly Journal of the Geological Society of London, v. 104, p. 461–476.

Fisher, G. W., 1963, The petrology and structure of the crystalline rocks along the Potomac River, near Washington, D.C. [Ph.D. dissertation]: Baltimore,

Johns Hopkins University, 241 p.

—— 1970, The metamorphosed sedimentary rocks along the Potomac River near Washington, D.C., *in* Fisher, G. W., and others, eds., Studies of Appalachian geology: Central and southern: New York, Wiley-Interscience, p. 299–315.

—— 1978, Geologic map of the New Windsor quadrangle, Maryland: U.S. Geological Survey Miscellaneous Investigations Series Map I-1037, scale 1:24,000.

—— 1979, Stratigraphy and structure of the central Appalachian Piedmont, Maryland and Pennsylvania: A review of recent evidence [abs.], *in* Wones, D. R., ed., Proceedings of the Caledonides in the USA: Virginia Polytechnic Institute, Department of Geologic Sciences Memoir 2, p. A6.

Fisher, G. W., Higgins, M. W., and Zeitz, I., 1979, Geological interpretations of aeromagnetic maps of the crystalline rocks in the Appalachians, northern Virginia to New Jersey: Maryland Geological Survey Report of Investigations no. 32, 40 p.

Freedman, J., Wise, D. U., and Bentley, R. D., 1964, Pattern of folded folds in the Appalachian Piedmont along the Susquehanna River: Geological Society of America Bulletin, v. 75, p. 621–638.

Glover, L., 1979, Tectonic evolution of the Virginia Piedmont and Blue Ridge [abs], *in* Wones, D. R., ed., Proceedings of the Caledonides in the USA: Virginia Polytechnic Institute, Department of Geological Sciences Memoir 2, p. 82–90.

Grauert, B. W., 1973a, U-Pb isotopic studies of zircons from the Baltimore Gneiss of the Towson Dome, Maryland: Carnegie Institution of Washington Year Book 72, p. 285–288.

—— 1973b, U-Pb isotopic studies of zircons from the Gunpowder Granite, Baltimore County, Maryland: Carnegie Institution of Washington Year Book 72, p. 288–290.

Hanan, B. B., 1980, The petrology and geochemistry of the Baltimore Mafic Complex, Maryland: [Ph.D. dissertation]: Blacksburg, Virginia Polytechnic Institute and State University, 216 p.

Hatcher, R. D., 1978, Tectonics of the western Piedmont and Blue Ridge, southern Appalachians: Review and speculation: American Journal of Science, v. 278, p. 276–304.

Hershey, H. G., 1937, Structure and age of the Port Deposit granodiorite complex: Maryland Geological Survey, v. 13, p. 109–148.

Herz, N., 1950, The petrology of the Baltimore Gabbro and the Baltimore-Patapsco Aqueduct [Ph.D. dissertation]: Baltimore, Johns Hopkins University.

—— 1951, Petrology of the Baltimore Gabbro: Geological Society of America Bulletin, v. 62, p. 979–1016.

Higgins, M. W., 1972, Age, origin, regional relations, and nomenclature of the Glenarm Series, central Appalachian Piedmont: A reinterpretation: Geological Society of America Bulletin, v. 83, p. 989–1026.

—— 1973, Superimposition of folding in the northeastern Maryland Piedmont and its bearing on the history and tectonics of the central Appalachians: American Journal of Science, v. 273a, p. 150–195.

—— 1976, Age, origin, regional relations, and nomenclature of the Glenarm Series, central Appalachian Piedmont: A reinterpretation: Reply: Geological Society of America Bulletin, v. 87, p. 1523–1528.

Higgins, M. W., and Fisher, G. W., 1971, A further revision of the stratigraphic nomenclature of the Wissahickon Formation in Maryland: Geological Society of America Bulletin, v. 82, p. 769–774.

Higgins, M. W., Sinha, A. K., Zartman, R. E., and Kirk, W. S., 1977, U-Pb zircon dates from the central Appalachian Piedmont: A possible case of inherited radiogenic lead: Geological Society of America Bulletin, v. 88, p. 125–132.

Hopson, C. A., 1964, The crystalline rocks of Howard and Montgomery Counties: Maryland Geological Survey, The Geology of Howard and Montgomery Counties, p. 27–215.

Jonas, A. I., 1937, Tectonic studies in the crystalline schists of southeastern Pennsylvania and Maryland: American Journal of Science, ser. 5, v. 34, p. 364–388.

Knopf, E. B., and Jonas, A. I., 1923, Stratigraphy of the crystalline schists of Pennsylvania and Maryland: American Journal of Science, ser. 5, v. 5, p. 40–62.

—— 1925, Map of Baltimore County and Baltimore City showing the geological formations: Maryland Geological Survey, scale 1:62,500.

—— 1929, The geology of the crystalline rocks of Baltimore County, *in* Baltimore County: Maryland Geological Survey, p. 97–199.

Lapham, D. M., and Basset, W. A., 1964, K-Ar dating of rocks and tectonic events in the Piedmont of southeastern Pennsylvania: Geological Society of America Bulletin, v. 75, p. 661–668.

Mackin, J. H., 1962, Structure of the Glenarm Series in Chester County, Pennsylvania: Geological Society of America Bulletin, v. 73, p. 403–410.

Mathews, E. B., 1904, The structure of the Piedmont plateau as shown in Maryland: American Journal of Science, ser. 4, v. 17, p. 141–159.

—— 1905, Correlation of Maryland and Pennsylvania Piedmont formations: Geological Society of America Bulletin, v. 16, p. 329–346.

—— 1907, Anticlinal domes in the Piedmont of Maryland: Johns Hopkins University Circular, v. 26, p. 615–622.

Mathews, E. B., and Miller, W. J., 1905, Cockeysville Marble: Geological Society of America Bulletin, v. 65, p. 347–366.

McKinstry, H. E., 1961, Structure of the Glenarm Series in Chester County, Pennsylvania: Geological Society of America Bulletin, v. 72, p. 557–558.

Meckel, L. D., 1967, Origin of the Pottsville Conglomerates (Pennsylvanian) in the central Appalachians: Geological Society of America Bulletin, v. 78, p. 223–258.

Moller, S. A., 1979, Geologic map of the Phoenix quadrangle, Maryland: Maryland Geological Survey, scale 1:24,000.

Morgan, B. A., 1977, The Baltimore Complex, Maryland, Pennsylvania, and Virginia, *in* Coleman, R. G., and Irwin, W. P., eds., North American Ophiolites: Oregon Department of Geology and Mineral Industries Bulletin 95, p. 41–49.

Muth, K. G., Arth, J. G., and Reed, J. C., Jr., 1979, A minimum age for high-grade metamorphic granite intrusion in the Piedmont of the Potomac River gorge near Washington, D.C.: Geology, v. 7, p. 349–350.

Olsen, S., 1977, Origin of the Baltimore Gneiss migmatites at Piney Creek, Maryland: Geological Society of America Bulletin, v. 88, p. 1089–1101.

Olsen, S. N., 1972, Petrology of the Baltimore Gneiss [Ph.D. dissertation]: Baltimore, Johns Hopkins University, 295 p.

Powell, C., McA., 1979, A morphological classification of rock cleavage: Tectonophysics, v. 58, p. 21–34.

Ramsay, J. G., 1967, Folding and fracturing of rocks: New York, McGraw-Hill Book Company, 568 p.

Rankin, D. W., 1975, The continental margin of eastern North America in the southern Appalachians: The opening and closing of the proto-Atlantic Ocean: American Journal of Science, v. 275A, p. 298–336.

Rodgers, J., 1970, The tectonics of the Appalachians: New York, Wiley-Interscience, 271 p.

—— 1972, Latest Precambrian (post-Grenville) rocks of the Appalachian region: American Journal of Science, v. 272, p. 507–520.

Seiders, V. M., 1976, Age, origin, regional relations, and nomenclature of the Glenarm Series, central Appalachian Piedmont: A reinterpretation: Discussion: Geological Society of America Bulletin, v. 87, p. 1519–1522.

Seiders, V. M., Mixon, R. B., Stern, T. W., Newell, M. F., and Thomas, C. B., Jr.,

—— 1975, Age of plutonism and tectonism and a new age limit on the Glenarm Series in the northeast Virginia Piedmont near Occoquan: American Journal of Science, v. 275, p. 481–511.

Sinha, A. K., Hanan, B. B., Sans, J. R., and Hall, S. T., 1980, Igneous rocks of the Maryland Piedmont: Indicators of crustal evolution, *in* Wones, D. R., ed., The Caledonides in the USA: Virginia Polytechnic Institute, Department of Geological Sciences Memoir 2, p. 131–136.

Southwick, D. L., 1969, Crystalline rocks of Harford County, The Geology of Harford County, Maryland: Maryland Geological Survey, p. 1–76.

—— 1970, Structure and petrology of the Harford County part of the Baltimore–State Line gabbro-peridotite complex, *in* Fisher and others, eds., Studies of Appalachian geology: Central and southern: New York, Wiley Interscience, p. 397–415.

—— 1979, The Port Deposit Gneiss revisited: Southeastern Geology, v. 20, p. 101–118.

Southwick, D. L., and Fisher, G. W., 1967, Revision of stratigraphic nomenclature of the Glenarm Series in Maryland: Maryland Geological Survey Report of Investigation 6, 19 p.

Talwani, M., Worzel, J. L., and Landisman, M., 1959, Rapid gravity computations for two-dimensional bodies with application to the Mendocino Submarine Fracture Zone: Journal of Geophysical Research, v. 64, p. 49–51.

Tilton, G. R., Doe, B. R., and Hopson, C. A., 1970, Zircon age measurements in the Maryland Piedmont, with special reference to Baltimore Gneiss problems, *in* Fisher and others, eds., Studies of Appalachian geology: Central and southern: New York, Wiley-Interscience, p. 429–434.

Tobisch, O. T., and Glover, L., 1971, Nappe formation in part of the southern Appalachian Piedmont: Geological Society of America Bulletin, v. 82, p. 2209–2230.

Wagner, M. E., and Crawford, M. L., 1975, Polymetamorphism of the Precambrian Baltimore Gneiss in southeastern Pennsylvania: American Journal of Science, v. 275, p. 653–682.

Wetherill, G. W., Tilton, G. R., Davis, G. L., Hart, S. R., and Hopson, C. A., 1966, Age measurements in the Maryland Piedmont: Journal of Geophysical Research, v. 71, p. 2139–2155.

Wetherill, G. W., Davis, G. L., and Lee-Hu, C., 1968, Rb-Sr measurements on whole rocks and separated minerals from the Baltimore Gneiss, Maryland: Geological Society of America Bulltin, v. 79, p. 757–762.

Williams, G. H., 1886, The gabbros and associated hornblende rocks occurring in the neighborhood of Baltimore, Maryland: U.S. Geological Survey Bulletin 28, 78 p.

Williams, H., compiler, 1978, Tectonic lithofacies map of the Appalachian orogen: St. Johns Memorial University of Newfoundland, scale 1:1,000,000.

Winkler, H.G.F., 1976, Petrogenesis of metamorphic rocks (4th edition): New York, Springer-Verlag, 334 p.

Wise, D. U., 1970, Multiple deformation, geosynclinal transitions and the Martic line problem in Pennsylvania, *in* Fisher and others, eds., Studies in Appalachian geology: Central and southern: New York, Wiley-Interscience, p. 317–334.

MANUSCRIPT ACCEPTED BY THE SOCIETY AUGUST 2, 1983

Geological Society of America
Special Paper 194
1984

Southern and central Appalachian basement massifs

Robert D. Hatcher, Jr.
Department of Geology
University of South Carolina
Columbia, South Carolina 29208

ABSTRACT

Basement constitutes rocks which belong to a previous orogenic cycle which have been reactivated and incorporated into a younger cycle. Basement massifs may be classified according to their relative position in an orogen as external or internal massifs. They may also be categorized according to their role in deformation, as thrust-related, fold-related and composite massifs. All Appalachian external massifs were transported following removal from the overridden edge of the ancient North American continental margin. Most of the internal massifs are also probably transported, but several (Pine Mountain and Sauratown Mountains) may exist as windows exposing parauthochthonous basement beneath the main thrust sheet. The latter reside immediately west of the low (west) to high (east) gravity gradient which probably outlines the old edge of Grenvillian crust. Reactivated crustal material generated during early Paleozoic orogeny plays the same mechanical role in reactivation as basement from the previous Grenville cycle. Basement (Grenville) massifs are distributed throughout the western Blue Ridge from Georgia to Maryland. Additionally, internal massifs are also present (Pine Mountain belt, Tallulah Falls and Toxaway domes, Sauratown Mountains anticlinorium, State Farm Gneiss dome, Baltimore Gneiss domes, and Mine Ridge anticlinorium). Basement internal massifs probably served to localize thrusts by causing them to detach and ramp over and around the massifs. Their antiformal shape may in part be as much related to thrust mechanics as to folding.

INTRODUCTION

The basement rocks of the southern and central Appalachians (Fig. 1) are well known and have been better defined in recent years owing to the completion of a number of field and geochronological studies of the respective bodies that occur within the Blue Ridge and Piedmont. For purposes of this paper, most of the discussion will be confined to relationships of Paleozoic deformation and reactivation of Precambrian basement which resides within the southern and central portions of the Appalachian orogen. The purpose of this paper is to summarize some ideas regarding the nature and structural setting as well as structural controls exerted by basement massifs in their respective positions in the orogen.

Practically all basement inliers which reside within the southern and central Appalachians may in some way be traced to the ancient Grenville crust which is precursor to sediments that were deposited prior to formation of the Appalachian orogen.

Most, if not all, of the massifs are allochthonous with respect to the rocks which underlie them. Many have been moved relative to the rocks which surround them as well.

This manuscript was improved considerably by review comments of M. J. Bartholomew. Support research on the Tallulah Falls dome, Pine Mountain belt, Bryson City and Ela domes and Sauratown Mountain anticlinorium by the National Science Foundation (Grants GA-20321, EAR 76-15564, EAR 78-26313 A01, EAR 79-11802, and EAR 81-08402) is gratefully acknowledged. Clemson University provided support for work on the Toxaway dome.

BASEMENT MASSIFS

The basement massifs of the southern and central Appalachians may be divided into two major categories: external

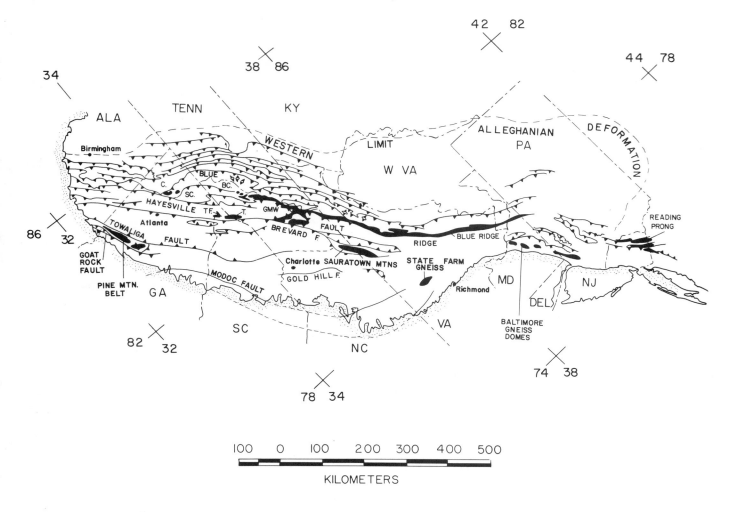

Figure 1. Map of the southern and central Appalachians showing the main subdivisions and the distribution of Grenville basement rocks (black).

massifs, which reside along the outer fringe of the metamorphic core near the western margin of the Blue Ridge, and internal massifs, which occur considerably east of the western limit of occurrence of basement rocks within the high grade core. This is a convenient way to view basement massifs, because it places them into the context of their relative position in the orogen and points up one of the important attributes of many orogenic terranes.

Basement massifs may also be classified according to their present relationships to Paleozoic structures (Fig. 2). Thrust-related massifs are primarily restricted to those external massifs which have simply been transported far enough from the internal zones to have not been appreciably deformed by other processes. Fold-related massifs are mostly internal. They should not have been transported on thrusts but are penetratively deformed and generally complexly folded. Most basement massifs are composite, since they have been folded and likely penetratively deformed during the Paleozoic, and transported on thrusts either before or after the other deformation (Fig. 3).

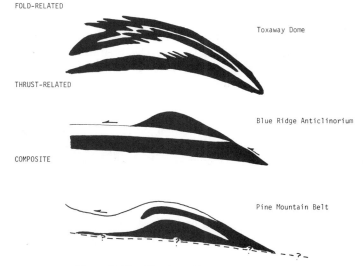

Figure 2. Classification of basement massifs.

External Massifs

The external massifs of the southern and central Appalachians consist of small masses of Precambrian basement rocks transported in thrust sheets, occurring near the western edge of the Blue Ridge, the much larger continuous mass that merges northward into Virginia within the Blue Ridge anticlinorium, and the Reading Prong in New Jersey and New York. The small group includes the masses west of the Mountain City window in the Holston-Iron Mountain thrust sheet, the Corbin and Salem Church Granites of the Blue Ridge of northeast Georgia, and several other small isolated masses of basement rocks occurring in eastern Tennessee and westernmost North Carolina. Small isolated masses also occur in domes in the western Blue Ridge, such as the Bryson City and Ela domes south of the Great Smoky Mountains National Park. The Ravensford anticline near Bryson City also is cored by a larger mass of basement rocks. All these smaller bodies occur along strike from the more continuous larger mass which lies to the northeast.

Large continuous basement massifs occur in the Blue Ridge of western North Carolina and east Tennessee, northeast of the Great Smoky Mountains National Park. These cover a wide area, are intimately associated with large thrust sheets, and are traceable northward into the rapidly narrowing Blue Ridge anticlinorium of Virginia. The latter plunges gently northeastward and terminates in Maryland and Pennsylvania as the South Mountain fold. The Reading Prong consists of a complex of basement-cover nappes emplaced by a shear folding mechanism. According to Drake (1970), they were emplaced during the Ordovician, then cut by faults again later in the Paleozoic.

All the external massifs of the western Blue Ridge and Reading Prong are closely associated with thrust faults, either directly bounded by thrusts on their contacts or are integral parts of thrust sheets.

Internal Massifs

The internal massifs may be divided into two subgroups based upon complexity of Paleozoic deformation. Included are relatively simple structures present in parts of the southern Appalachians, along with the more composite structures which are present in other areas.

Simple structures include bodies of basement rocks including the fold-related Toxaway Gneiss of the eastern Blue Ridge of North and South Carolina (Hatcher, 1977), the Tallulah Falls dome of northeast Georgia (Hatcher, 1974, 1976), and the State Farm Gneiss of Virginia (Glover and Tucker, 1979). These structures are by no means absolutely simple, but their history is apparently not composite. They are intimately involved in the Paleozoic folding sequence like the rocks which surround them. Their outcrop pattern is largely determined by large-scale isoclinal folding and refolding processes which have affected both the basement and enclosing rocks. The Baltimore Gneiss domes of Maryland and Pennsylvania are fold-related structures on the

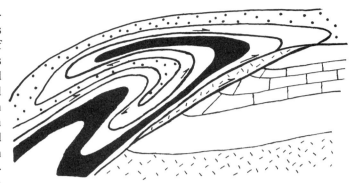

Figure 3. Hypothetical composite massif illustrating a probable maximum degree of complexity of folding, mobilization of basement, thrusting and thrust involvement with a simple thrust-related massif.

surface but geophysical data strongly suggest that these massifs have also been transported on one or more thrusts (Fisher and others, 1979). That the rocks contained within these structures are basement rocks has been documented by radiometric age determinations (Tilton and others, 1970; Odom and others, 1976; Fullagar and Odom, 1973; Glover and others, 1978).

The composite basement internal massifs include those which have had both a history of involvement with folding and a history of involvement with faulting (Fig. 2 and 3). The Sauratown Mountains anticlinorium (Fig. 4) is such a structure which has a complex folding history associated with it, and then has been overridden by, and perhaps contained within, several thrust sheets of the Piedmont and Blue Ridge respectively.

The Pine Mountain window has been documented to be a complexly folded structure (Schamel and Bauer, 1980). It has been known for a long time to be a structure involving large faults but only recently has it been confirmed as a window (Hatcher and others, 1981). The Pine Mountain belt has been known many years to be flanked by the Goat Rock and Towaliga faults (Crickmay, 1933). Large nappe structures have been recognized within the Pine Mountain belt and have been mapped in some detail in the western parts by Schamel and Bauer (1980) and Schamel and others (1980). The structure at the east end of the Pine Mountain belt has been recently shown to be a premetamorphic fault complex which involves emplacement of a thrust sheet containing Piedmont rocks over the basement rocks of the Pine Mountain belt prior to the metamorphic thermal peak affecting the rocks of this area. Portions of this early fault were reactivated later along the northwest and southeast flanks of the Pine Mountain belt by high angle faulting (Hatcher and others, 1981).

DISCUSSION

The basement massifs of the southern Appalachians appear to be part of several kinds of structures. Those in the western Blue Ridge occur as part of the thrust faulted Blue Ridge

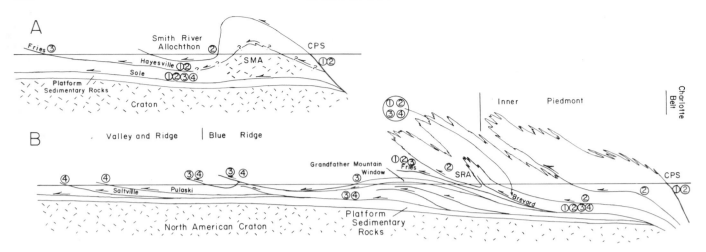

Figure 4. Cross-sections through the Sauratown Mountains anticlinorium (A) and the Grandfather Mountain window (B), showing the relationships of major thrusts to basement rocks. Numbers on faults indicate major episodes of motion. The number (1) indicates a probable Taconic event, (2) is probably late Taconic or Acadian(?), (3) is Acadian or Alleghanian, and (4) indicates Alleghanian movement. CPS - Central Piedmont suture (speculative).

anticlinorium, while those in the central to eastern Blue Ridge and western to central Piedmont occur in large anticlinal structures which are not as continuous as the Blue Ridge anticlinorium. Those in the more internal parts of the orogen have an obvious relationship to Paleozoic fold structures which have been super-imposed on both the basement rocks and the enclosing cover rocks. Interestingly, those in the internal parts of the orogen appear to be related and may possibly in some instances control large faults that occur in association with these structures. It has been pointed out by Price and Hatcher (1980, 1983) that a relationship exists between the distance from the large gravity gradient to the external massifs in the orogen and possible amounts of horizontal transport, and that this distance increases southward in the southern and central Appalachians. Another basic attribute of both external and internal massifs is that no Grenville massifs occur east of the principal gravity gradient in the Appalachians (see maps by Williams, 1978, and Haworth and others, 1981).

Milnes and Pfiffner (1977) have pointed out that there is a relationship between the localization of the Glarus overthrust in the western Alps and the underlying Aar massif of Hercynian basement rocks. Some of the large massifs in the internal parts of the Appalachians might likewise assert similar controls on positions of thrust sheets which ride over and possibly under them. The composite Grandfather Mountain window, Sauratown Mountains anticlinorium and Pine Mountain belt massifs may play a role similar to the Aar massif in the Alps.

Both the massifs in the Pine Mountain window and the Sauratown Mountains anticlinorium occur almost immediately west of the gravity gradient in the southern Appalachians. These may be the easternmost parautochthonous massifs having been transported only a minimal distance from the eastern edge of the subsurface Grenville crust. The Sauratown Mountains basement

massifs appear to have localized, at least at the present level of erosion, the position of the northeast end of the Inner Piedmont.

The Charlotte belt terminates to the northeast of the east end of the Pine Mountain belt. Likewise, Carolina Slate belt rocks turn southward and pass under the Coastal Plain about one hundred kilometers to the northeast of the east end of the Pine Mountain belt. This may or may not be significant regarding the relationship of the regional gravity gradient and hence overthrust basement to localization of these structures, but an increase in metamorphic grade occurs southwestward from the Georgia-South Carolina line, where slate belt rocks are in the greenschist facies, to the Pine Mountain belt, where the rocks are in the sillimanite zone of the east end and in the adjacent Piedmont.

CONCLUSIONS

1. Southern and central Appalachian basement massifs may be grouped into external and internal massifs based upon their positions relative to the metamorphic core. The external group resides in the westernmost part of the Blue Ridge. The internal group occurs in the eastern Blue Ridge and Piedmont.

2. Basement massifs may be classified according to the degree of involvement in deformation with their cover as thrust-related, fold-related and composite massifs. Simple thrust and fold-related massifs may involve only thrusting, folding or cleavage development. Composite massifs involve both folding and thrusting.

3. Controls of positions of thrusts may have been exerted by basement massifs in the Sauratown Mountains and Grandfather Mountain window.

REFERENCES CITED

Crickmay, G. W., 1933, The occurrence of mylonites in the crystalline rocks of Georgia: Am. Jour. Sci., v. 226, p. 161–177.

Drake, A. A., Jr., 1970, Structural geology of the Reading Prong, *in*, Fisher, G. W., Pettijohn, F. J., Reed, J. C., Jr., and Weaver, K. N., eds., Studies of Appalachian geology: Central and southern: New York, Interscience, p. 271–291.

Fisher, G. W., Higgins, M. W., and Zietz, Isidore, 1979, Geological interpretations of aeromagnetic maps of the crystalline rocks in the Appalachians, Northern Virginia to New Jersey: Maryland Geol. Survey, Rept. Inv. 32, 43 p.

Fullagar, P. D. and Odom, A. L., 1973, Geochronology of Precambrian gneisses in the Blue Ridge province of northwestern North Carolina and adjacent parts of Virginia and Tennessee: Geol. Soc. America Bull., v. 84, p. 3065–3080.

Glover, Lynn III, Mose, D. G., Poland, F. B., Bobyarchick, A. R., and Bourland, W. C., 1978, Grenville basement in the eastern Piedmont of Virginia: Implications for orogenic models: Geol. Soc. America Abs. with Programs, v. 10, p. 169.

Glover, Lynn III, Bobyarchick, A. R., Bourland, W. C., Brown, W. R., Goodwin, B. K., Mose, D. G., and Poland, F. B., 1979, Virginia Piedmont geology along the James River from Richmond to the Blue Ridge, *in* Glover, Lynn III and Tucker, R. D., eds., Guides to field trips 1–3, southeastern section meeting, Geol. Soc. America, Blacksburg, VA, p. 1–41.

Hatcher, R. D., Jr., 1974, Introduction to the tectonic history of northeast Georgia: Georgia Geol. Soc. Guidebook 13-A, 59 p.

—— 1976, Introduction to the geology of the eastern Blue Ridge of the Carolinas and nearby Georgia: Carolina Geol. Soc. Guidebook, 53 p.

—— 1977, Macroscopic polyphase folding illustrated by the Toxaway dome, eastern Blue Ridge, South Carolina-North Carolina: Geol. Soc. America Bull. 89, p. 1678–1688.

Hatcher, R. D., Jr., Hooper, R. J., and Odom, A. L., 1981, Transition from the east end of the Pine Mountain belt into the Piedmont, central Georgia: Preliminary results: Geol. Soc. America Abs. with Programs, v. 13.

Haworth, R. T., Daniels, D. L., Williams, Harold, and Zietz, Isidore, 1981, Gravity anomaly map of the Appalachian orogen: Memorial Univ. Newfoundland, map no. 3, scale 1/1,000,000.

Milnes, A. G. and Pfiffner, O. A., 1977, Structural development of the infrahelvetic complex, eastern Switzerland: Eclogae Geol. Helv., v. 70, p. 83–95.

Odom, A. L., Russell, G. S., Russell, C. W., 1976, Distribution and age of Precambrian basement in the southern Appalachians: Geol. Soc. America Abs. with Programs, v. 8, p. 238.

Price, R. A. and Hatcher, R. D., Jr., 1980, Geotectonic implications of similarities in the orogenic evolution of the Alabama-Pennsylvania Appalachians and the Alberta-British Columbia Canadian Cordillera: Geol. Soc. America Abs. with Programs, v. 12, p. 504.

——, 1983 Geotectonic implications of similarities in the orogenic evolution of the Alabama-Pennsylvania Appalachians and the Alberta-British Columbia Canadian Cordillera: Geol. Soc. America Memoir 158, p. 149–160.

Schamel, Steven and Bauer, David, 1980, Remobilized Grenville basement in the Pine Mountain window, *in*, Wones, D. R., ed., The Caledonides in the USA: International Geol. Correl. Prog. Project 27, Virginia Polytech. Inst. Geol. Sci., Mem. 2, p. 313–316.

Schamel, Steven, Hanley, T. B., and Sears, J. W., 1980, Geology of the Pine Mountain window and adjacent terranes in the Piedmont Province of Alabama and Georgia: Geol. Soc. America Southeastern Sec. Guidebook, 69 p.

Tilton, G. R., Doe, B. R., and Hopson, C. A., 1970, Zircon age measurements in the Maryland Piedmont, with special reference to Baltimore Gneiss problems, *in*, Fisher, G. W. and others, eds., Studies in Appalachian geology, central and southern: New York, Interscience, p. 429–437.

Williams, Harold, 1978, Tectonic lithofacies map of the Appalachians orogen: Memorial Univ. Newfoundland, map no. 1, scale 1/1,000,000.

MANUSCRIPT ACCEPTED BY THE SOCIETY AUGUST 2, 1983

Printed in U.S.A.

Geological Society of America
Special Paper 194
1984

The core of the Blue Ridge anticlinorium
in northern Virginia

James W. Clarke
U.S. Geological Survey
M.S. 928, National Center
Reston, Virginia 22092

ABSTRACT

Proterozoic Y and Z granitic rocks compose the core of the Blue Ridge anticlinorium in northern Virginia. The oldest units are the Flint Hill Gneiss; an augen-bearing gneiss; the "Pedlar Formation"; and a porphyroblastic granitic gneiss. These appear to have ages in the 1,000-m.y. to 1,100-m.y. range. The porphyroblastic granitic gneiss is in fault contact with the Flint Hill; the Flint Hill grades into the augen-bearing gneiss; and the "Pedlar" appears to be in gradational contact with the Flint Hill. The porphyroblastic granitic gneiss carries xenoliths of an older layered biotite gneiss that lie transverse to the foliation of the host gneiss. The 1,010-m.y.-old Marshall Metagranite intrudes the porphyroblastic granitic gneiss and the Flint Hill Gneiss. Granitic rocks of the 700-m.y.-old Robertson River Formation intrude the Flint Hill Gneiss, the augen-bearing gneiss, and the porphyroblastic granitic gneiss. All units of the core of the anticlinorium are cut by metabasaltic dikes, which are presumed to be feeder dikes for the younger metavolcanic rocks of the Catoctin Formation of the region. The presence of charnockitic rocks in the western part of the core of the anticlinorium indicates that a granulite grade of metamorphism was reached in this terrane. The absence of hypersthene in other rocks of the area may reflect only their chemical and mineral composition. The possibility exists that these disparate-facies rocks were juxtaposed tectonically.

INTRODUCTION

The Blue Ridge anticlinorium is a long, narrow uplift that extends about 880 km northeast from Georgia to Pennsylvania. In general, it separates the Piedmont geologic province on the east from the Valley and Ridge province on the west. From southern Virginia for about 250 km northeast into Maryland, the anticlinorium has a central core of largely 1-b.y.-old granitoid rocks. This core is flanked on both sides by younger Proterozoic Z metasedimentary and metavolcanic rocks.

The geology of the Blue Ridge anticlinorium in northern Virginia has been reviewed by Espenshade (1970), and this review is still the most authoritative for the region as a whole. Since it was written, however, Lukert and Nuckols (1976) have published geologic maps of the Linden and Flint Hill quadrangles; Lukert and Halladay (1980) have published a geologic map of the Massies Corner quadrangle; Espenshade (1981, unpub. data) has mapped the Marshall and part of the Rectortown quadrangles; and Clarke has mapped most of the Orlean quadrangle (Fig. 1). The detailed mapping of these 7.5-minute quadrangles in the central part of northern Virginia has provided much new information on the geology.

The present study focuses on the geology of the core of the anticlinorium in these newly mapped quadrangles (Figs. 1 and 2). The new findings in the Linden, Flint Hill, and Massies Corner quadrangles are reviewed along with unpublished information on the Orlean and Marshall quadrangles. All units present in the core are described briefly, and some inferences are drawn as to the Proterozoic geologic history of the area.

FLINT HILL GNEISS

The Flint Hill Gneiss has been mapped in the Massies Corner (Lukert and Halladay, 1980), Flint Hill, Linden (Lukert

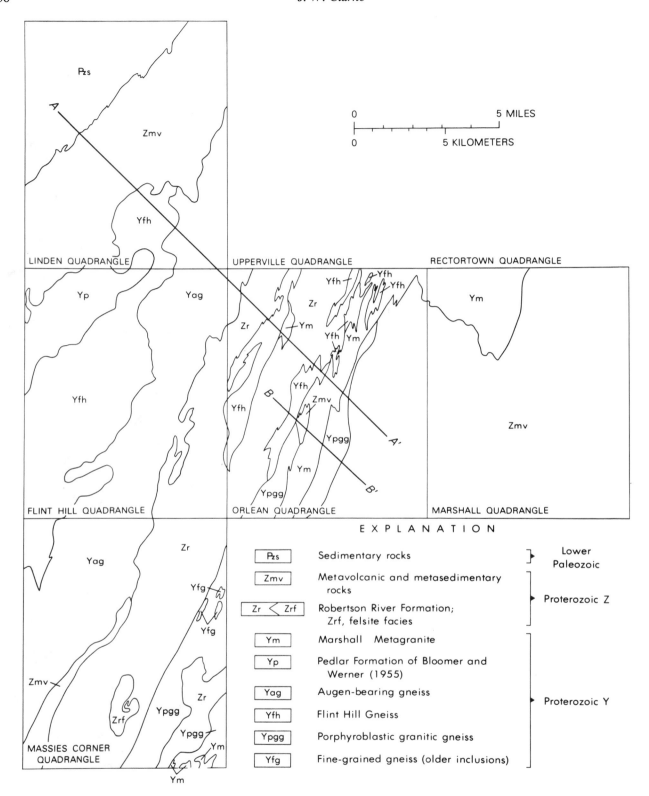

Figure 1. Generalized geologic map of the core of the Blue Ridge anticlinorium in the Massies Corner, Flint Hill, Linden, Orlean, and Marshall quadrangles of northern Virginia. Cross sections along lines A-A' and B-B' are shown in Figure 2.

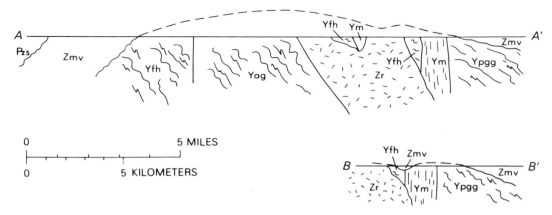

Figure 2. Cross sections of core of the Blue Ridge anticlinorium in northern Virginia. See Figure 1 for location.

and Nuckols, 1976), and Orlean (Clarke, unpub. data) quadrangles, where it underlies areas totaling about 53 km² (Fig. 1). Reconnaissance suggests that only small areas of Flint Hill Gneiss totaling less than 10 km² remain to be mapped in adjacent quadrangles.

The Flint Hill Gneiss has a distinctive appearance in outcrop (Lukert and others, 1977). It is strongly banded, and the bands are commonly contorted and kinked. Dark layers, 1 to 2 mm thick, are rich in biotite and are generally continuous. Light layers, 2 to 5 mm thick and as long as 30 mm, are discontinuous aggregates of quartz and feldspar. Grain size in both dark and light layers is 0.2 to 0.3 mm. The Flint Hill grades into the augen-bearing gneiss.

The average mode of the Flint Hill Gneiss is 30% quartz, 40% microcline, 20% oligoclase (An-25 to An-28), and 10% biotite and accessory minerals. Muscovite and chlorite are abundant in some samples and locally exceed biotite in abundance. Ilmenite grains surrounded by coronas of sphene are common. Other accessory minerals are zircon, magnetite, apatite, and epidote.

Blue-quartz–bearing leucocratic veins and dikes are widespread in the Flint Hill Gneiss; they are 1 cm to 5 m wide and a few centimeters to several meters long. The veins grade in composition from entirely blue quartz, to mixtures in which blue quartz forms the core and microcline the selvage, to pegmatitic microcline veins. The blue-quartz veins are generally lenses conformable with the banding in the gneiss. The dikes cut across the contorted and kinked bands of the gneiss without further disturbing these bands. The blue-quartz veins are ubiquitous in the Flint Hill, and the unit can be mapped in areas lacking rock exposures by the presence of blue-quartz fragments on the ground surface.

Homogeneity of the Flint Hill Gneiss and its granitic mineral composition suggest that it was originally an igneous rock. The strong banding raises a question, however, as to whether it is a granite that has been metamorphosed or remobilized metasedimentary or metavolcanic material. This question cannot be answered at present. The Flint Hill is cut by the Marshall Metagranite.

A Pb-207/Pb-206 age of 1,081 m.y. has been obtained on the Flint Hill Gneiss (Thomas W. Stern, 1976, written commun.).

AUGEN–BEARING GNEISS

An augen-bearing gneiss has been mapped in the Massies Corner (Lukert and Halladay, 1980), Flint Hill, Linden (Lukert and Nuckols, 1976), and Orlean (Clarke, unpub. data) quadrangles. The area mapped in all four quadrangles is approximately 110 km². This unit extends southwestward into the Washington quadrangle and northeastward into the Upperville quadrangle, where its distribution is not known.

The augen-bearing gneiss is a medium- to dark-gray foliated rock in which microcline megacrysts 1 to 2 cm long are present in a medium-grained matrix of quartz, microcline, plagioclase, biotite, and accessory minerals. The quartz is commonly blue and composes 10% to 40% of the rock. Microcline is present in amounts from 10% to 40% of the rock, and plagioclase from 5% to 40%. The microcline megacrysts are anhedral and equant to subhedral and elongate parallel to foliation. Biotite composes 10% to 15% of the rock. The average composition is that of adamellite.

The augen-bearing gneiss has a gradational contact with the Flint Hill Gneiss. The relationship between these two gneisses is not known. They may be facies of one another, or one may be intrusive into the other. No radiometric age data are available for the augen-bearing gneiss.

The augen-bearing gneiss differs from the porphyroblastic granitic gneiss, which occurs to the east, in that the microcline megacrysts of the former are generally single crystals, whereas the porphyroblasts of the latter are generally medium-grained aggregates of microcline and quartz. Also, the augen-bearing gneiss has the composition of adamellite, whereas the porphyroblastic granitic gneiss generally has the composition of granite.

PEDLAR FORMATION OF BLOOMER AND WERNER

The Pedlar Formation of Bloomer and Werner (1955) underlies an area of about 20 km^2 on the western margin of the core of the anticlinorium in the Flint Hill and Linden quadrangles (Lukert and Nuckols, 1976). The rock here is a greenish-gray, medium-grained, granular biotite-hypersthene adamellite gneiss. Quartz composes 10% to 20% of the rock, microcline about 40%, plagioclase 30%, hypersthene (partly or completely uralitized) 5% to 13%, and biotite 3% to 4%. Smaller amounts of hornblende, garnet, chlorite, epidote, apatite, magnetite, ilmenite, and zircon are present. The plagioclase ranges in composition from An-26 to An-32.

In the study area, the Pedlar is in gradational contact with the Flint Hill Gneiss. Near the contact, the Pedlar becomes finer grained and takes on a layering similar to that of the Flint Hill.

PORPHYROBLASTIC GRANITIC GNEISS

A porphyroblastic granitic gneiss underlies several small areas totaling about 14 km^2 in the eastern half of the Orlean quadrangle (Clarke, unpub. data). This same rock unit underlies an aggregate area of 21 km^2 in the Massies Corner quadrangle, where it is designated coarse leucogneiss by Lukert and Halladay (1980).

The rock is gray to pink and strongly foliated. The porphyroblasts are rounded aggregates of quartz and microcline that have a grain size of 1 to 5 mm. The porphyroblasts themselves are generally 2 to 3 cm in diameter, although some diameters exceed 10 cm. The matrix is composed of blue quartz, microcline, plagioclase, biotite, and accessory minerals. Composition is generally that of granite, although in places it is that of adamellite. No veins of blue quartz are present in this unit. The foliation is produced by biotite-rich septa, which form an anastomosing continuum around segregated porphyroblasts and less biotitic matrix. The foliation is commonly wavy and folded isoclinally.

The possibility was considered that these porphyroblastic aggregates of quartz and microcline are porphyroclastic relicts that remained after the rest of the rock had been "milled down" by tectonic processes. Xenoliths are present in the gneiss transverse to its foliation, however, and these xenoliths would show some deformation if the host gneiss had been deformed so pervasively as to yield a rock containing porphyroclastic relicts. However, the xenoliths are not deformed. Further, the quartz in the porpyroblastic aggregates shows no signs of cataclasis. Because quartz is very sensitive to tectonic stress, it would hardly have remained unaffected by an intense cataclastic process. Consequently, the aggregates of quartz and microcline are interpreted as porphyroblasts.

No radiometric age data are available on the porphyroblastic granitic gneiss within the area mapped. The foliation of this unit, however, is crosscut by intrusive bodies of the Marshall Metagranite, which has an age of 1,010 m.y. The porphyroblastic granite gneiss may be much older than the Marshall Metagranite because it appears to have had its present foliation at the time of emplacement of the Marshall.

FINE-GRAINED GNEISS (OLDER INCLUSIONS)

Inclusions have been found in the porphyroblastic granitic gneiss that appear to be older than this gneiss. Three of these in the Massies Corner quadrangle are large enough to be mapped and are designated fine-grained gneiss by Lukert and Halladay (1980). A xenolith of layered biotite gneiss 50 cm by 10 cm was found in the porphyroblastic granitic gneiss in the Orlean quadrangle. It lies transverse to the foliation of the host rock and carries porphyroblasts similar to those of the host. The xenolith appears to have already been foliated when it was included in the rock that is now the porphyroblastic granitic gneiss. These older inclusions may be fragments from terrane that is older than any of the other rock units recognized here.

MARSHALL METAGRANITE

The Marshall Metagranite has been mapped in the Marshall, Rectortown, and Orlean quadrangles, where it underlies areas totaling more than 20 km^2 (Clarke, unpub. data; Espenshade, 1981, written commun.). Other rocks that seem definitely to be Marshall but have not been traced by continuous mapping from the type locality are present to the north in Londoun County (Jeffrey L. Howard, 1981, personal commun.) and also to the southwest in the Massies Corner quadrangle where they are designated leucogneiss (Lukert and Halladay, 1980). More than a dozen additional references to Marshall are found in the literature, but these correlations are yet to be demonstrated.

The Marshall Metagranite is a fine- to medium-grained, light- to dark-gray gneissic biotite granite to adamellite. The quartz has a bimodal distribution in grains 0.2 to 0.3 mm long and 1 to 2 mm long. The coarser quartz, generally blue, is not everywhere present. Quartz and microcline are commonly platy and parallel to foliation. This platy habit and the occurrence of biotite in septa give the rock a weak foliation. Both microcline and plagioclase are commonly sericitized to the extent that the quartz may appear as disseminated grains embedded in a uniform continuum of altered feldspar. Anastomosing shear surfaces commonly give the rock the appearance of an augen gneiss, the augen being unreduced clasts of granite. Where the rock is more sheared, it grades into a flaser gneiss.

A typical mode for Marshall Metagranite is 25% quartz, 45% microcline, 25% plagioclase, and 5% biotite. Muscovite, sericite, and epidote are commonly present. Accessories are ilmenite, magnetite, zircon, and apatite.

Aplite dikes 20 to 30 cm wide cut the Marshall Metagranite at high angles to its foliation. The aplites are massive to slightly foliated. This foliation is in some places continuous with that of the host rock, and in other places it is parallel to the dike walls. The aplite dikes commonly have cores of blue quartz.

Bodies of the Marshall Metagranite cut across the foliation of the Flint Hill Gneiss and the porphyroblastic granitic gneiss. This crosscutting relationship and the granitic composition indicate that the Marshall was formed by consolidation of an intrusive magma. The presence of bimodal quartz in Marshall rocks that have not been sheared suggests that the rock has been metamorphosed. The larger quartz grains are interpreted as having grown during the metamorphism.

The Pb-207/Pb-206 age of zircon from the type locality of the Marshall Metagranite at Marshall is 1,010 m.y. (Thomas W. Stern, 1977, written commun.).

ROBERTSON RIVER FORMATION

The Robertson River Formation underlies an aggregate area of 65 km^2 in the Massies Corner quadrangle (Lukert and Halladay, 1980), 21 km^2 in the Flint Hill quadrangle (Lukert and Nuckols, 1976), and 42 km^2 in the Orlean quadrangle. This unit has been traced in reconnaissance southward from the Massies Corner quadrangle to its type locality in Madison County, Virginia. Its extent to the north in the Upperville quadrangle is not known.

Four different rock types are included in the Robertson River Formation: magnetite granite, biotite-hornblende granite, alkali granite, and felsite. This listing is the order of decreasing abundance in the Orlean quadrangle. To the west in the Massies Corner and Flint Hill quadrangles, the biotite-hornblende variety predominates. All varieties are gray to pinkish and have little or no foliation.

The magnetite granite is composed of highly perthitic microcline, quartz, and fine-grained aggregates of magnetite and quartz. Some of these aggregates appear to be pseudomorphic after pyroxene. The rock is interpreted as having been originally a pyroxene granite in which the pyroxene altered deuterically to magnetite and quartz under conditions of high oxygen fugacity. The felsite appears to be a fine-grained phase of the magnetite granite because samples of it are strongly magnetic, as is the magnetite granite itself. Within one pluton in the Orlean quadrangle, magnetite granite grades into biotite-hornblende granite. The alkali granite characteristically contains aegirine and riebeckite.

All four varieties of the Robertson River are closely associated in space and have general mineralogical characteristics in common. All clearly postdate the orogenic events that have affected the rocks that are at least 1 b.y. old.

A Pb-207/Pb-206 age of 730 m.y. was found for zircon from a sample of the biotite-hornblende granite (Lukert and Halladay, 1980) and a Pb-207/Pb-206 age of 701 m.y. for magnetite granite (Clarke, 1981). A Pb-207/Pb-206 age of 650 m.y. was obtained for zircon from one of the alkali granites (Rankin, 1975). The granitic rocks of the Robertson River Formation seem to be a series of related intrusions that were not necessarily contemporaneous.

AGE RELATIONSHIPS

The fine-grained gneiss inclusions in the porphyroblastic granitic gneiss are the oldest rocks in the area; they contain a foliation that is transverse to the foliation of the host gneiss. The porphyroblastic granitic gneiss and the 1,081-m.y.-old Flint Hill Gneiss are cut by the 1,010-m.y.-old Marshall Metagranite. As both these gneisses are cut by Marshall, they may be of more or less the same age. Where these two gneisses are juxtaposed in the southwest part of the Orlean quadrangle, the contact is a fault; consequently, their age relationship here is unclear. The augen-bearing gneiss appears to be the same age as the Flint Hill.

The granitic rocks of the Robertson River Formation intrude the Flint Hill Gneiss, augen-bearing gneiss, and porphyroblastic granitic gneiss. Metabasaltic dikes, presumed to be feeder dikes for the metavolcanic rocks of the Catoctin Formation, cut all the principal rock units of the core of the anticlinorium as well as the unconformably overlying meta-arkose of the Fauquier Formation of Furcron (1939).

The age relationship between the Pedlar Formation of Bloomer and Werner (1955) and the other rocks of the core of the anticlinorium is not clear; the two terranes appear to be juxtaposed tectonically, although the geometry of this juxtaposition is not clear.

DISCUSSION

The time span 1,000 to 1,100 m.y. ago appears to have been the period of crustal consolidation for most of the units of the core of the Blue Ridge anticlinorium in the region of the Linden, Flint Hill, Massies Corner, Orlean, and Marshall quadrangles. Older ages cited above are highly speculative. The presence of charnockite along strike a short distance away to the northnorthwest (Jeffrey L. Howard, personal commun.) and southsoutheast (Herz, 1968), as well as the occurrences of hypersthene-bearing rocks of Bloomer and Werner's Pedlar Formation within the study area itself, suggests that a granulite grade of metamorphism may have been attained in the area. The absence of any signs of granulite facies in the other rocks within the study area may reflect only the chemical and mineralogical composition of the rocks. If these rocks had already consolidated as granites, a superposed granulite-facies metamorphism might leave them essentially unchanged, except perhaps for the development of blue quartz veins in the Flint Hill Gneiss and larger grains of blue quartz in the Marshall Metagranite.

Tectonic juxtaposing of disparate-facies rocks is also a possibility. All the rocks of the area were subjected to greenschist-facies metamorphism during the Taconic orogeny (Stromberg, 1978). This metamorphism was not strong enough to have erased evidence of a 1,000 to 1,100-m.y.-old granulite metamorphism in these rocks. Further, the gradational rocks between the Flint Hill and Bloomer and Werner's Pedlar have a mineralogy of higher grade than the Taconic greenschist facies, and consequently must result from the 1,000 to 1,100-m.y.-old event. As the Pedlar cuts the 1,138-m.y.-old Old Rag Granite of Furcron (1934; Gathright, 1976), it must also have consolidated during the event 1,000 to 1,100 m.y. ago. Therefore, Pedlar rocks showing granulite-facies attributes and rocks not showing such attributes consolidated during the same event 1,000 to 1,100 m.y. ago. Perhaps they were metamorphosed under different conditions and then brought together along the Pedlar–Flint Hill contact 1,000 to 1,100 m.y. ago.

REFERENCES CITED

Bloomer, R. O., and Werner, H. J., 1955, Geology of the Blue Ridge region in central Virginia: Geological Society of America Bulletin, v. 66, no. 5, p. 579–606.

Clarke, J. W., 1981, Billion-year-old rocks of the Blue Ridge anticlinorium of northern Virginia: A review: Virginia Journal of Science, v. 32, no. 3, p. 127.

Espenshade, G. H., 1970, Geology of the northern part of the Blue Ridge anticlinorium, *in* Fisher, G. W., and others, eds., Studies in Appalachian geology—central and southern: New York, Interscience Publishers, p. 199–211.

Furcron, A. S., 1934, Igneous rocks of the Shenandoah National Park area: Journal of Geology, v. 42, p. 400–410.

——1939, Geology and mineral resources of the Warrenton quadrangle, Va.: Virginia Geological Survey Bulletin 54, 94 p.

Gathright, T. M., II, 1976, Geology of the Shenandoah National Park, Virginia: Virginia Division of Mineral Resources Bulletin 86, 93 p.

Herz, Norman, 1968, The Roseland alkalic anorthosite massif, Virginia, *in* Origin of anorthosite and related rocks: New York State Museum and Science Service Memoir 18, p. 357–367.

Lukert, M. T., and Halladay, C. R., 1980, Geology of the Massies Corner quadrangle, Virginia: Virginia Division of Mineral Resources Publication 17, text and 1:24,000 scale map.

Lukert, M. T., and Nuckols, E. B., III, 1976, Geology of the Linden and Flint Hill quadrangles, Virginia: Virginia Division of Mineral Resources Report of Investigations 44, 83 p.

Lukert, M. T., Nuckols, E. B., and Clarke, J. W., 1977, Flint Hill Gneiss—A definition: Southeastern Geology, v. 19, no. 1, p. 19–28.

Rankin, D. W., 1975, The continental margin of eastern North America in the southern Appalachians: The opening and closing of the proto-Atlantic Ocean: American Journal of Science, v. 275-A, p. 298–336.

Stromberg, M. J., 1978, K-Ar ages of muscovite and biotite from the Blue Ridge province, northern Virginia [B.A. thesis]: Ohio State University.

MANUSCRIPT ACCEPTED BY THE SOCIETY AUGUST 2, 1983

Geological Society of America
Special Paper 194
1984

Geology and age of the Robertson River Pluton

Michael T. Lukert
Department of Geosciences
Edinboro University of Pennsylvania
Edinboro, Pennsylvania 16444

P. O. Banks
Department of Earth Sciences
Case Western Reserve University
Cleveland, Ohio 44106

ABSTRACT

Granitic rocks of the Robertson River Pluton intrude Precambrian Y gneisses in the core of the Catoctin-Blue Ridge Anticlinorium in northern Virginia. Cobbles and boulders of lithologically similar granite are present in the overlying Mechum River and Fauquier (Lynchburg) formations. Dikes of metabasalt, texturally and lithologically similar to the flows of the Catoctin Formation cut the Robertson River Pluton. The Catoctin metavolcanic rocks themselves were extruded following deposition of the Fauquier and correlative clastic sedimentary rocks.

Zircons from four samples of the Robertson River Pluton yield a discordant U-Pb age of 730 m.y., indicating that the Catoctin Formation is considerably younger than the 820 m.y. previously assigned.

INTRODUCTION

The granitic rocks of the Robertson River Pluton were named and described by Allen (1963) for exposures along the Robinson (*sic*) River in Madison County, Virginia. The discrepancy in the formation name apparently was due to the incorrect designation of the stream as the Robertson River on the 1933 edition of the Madison quadrangle. The unique character of these rocks, however, was first noted by Thiesmeyer (1938), who referred to the unit as the "Crest Hill Granite" for exposures near the settlement of Cresthill in western Fauquier County, Virginia.

Rocks of the Robertson River Pluton are exposed in the core of the Blue Ridge anticlinorium continuously from near Charlottesville in Albemarle County, northeastward to the vicinity of Upperville in Loudoun County, a distance of approximately 130 km (Fig. 1). This elongate outcrop zone ranges up to about 8 km in width. Smaller bodies of similar granitic rocks separated from the principal outcrop belt also are common in northern Virginia.

The most characteristic lithology of the Robertson River Pluton is hornblende granite. Variations within the pluton, however, include biotite granite, magnetite granite, riebeckite granite,

alaskite, and felsite. Coarse granitoid textures are typical of most of these granitic rocks, but considerable variation exists, even within a single outcrop. The lack of distinctive foliation is the textural characteristic that distinguishes the rocks of the Robertson River Pluton from the adjacent Precambrian Y rocks.

The Robertson River Pluton is intrusive into the gneissic basement rocks as indicated by: (1) the sinuous nature of the contacts, (2) the presence of hornblende granite dikes cutting the country rocks, and (3) the presence of xenoliths and inliers of gneissic rocks (roof pendants?) within the principal outcrop area of the granite, (Lukert and Nuckols, 1976).

Fault contacts were mapped between the granite and the gneissic rocks in Albemarle and Rappahannock counties (Nelson, 1962; Lukert and Halladay, 1980). The traces of these faults are rather straight, and the fault surfaces are inferred to be nearly vertical.

In Rappahannock County, the Robertson River Pluton is, for the most part, in fault contact with metamorphosed clastic sedimentary rocks of the Mechum River Formation (Lukert and Halladay, 1980). The temporal relationship between these two

EXPLANATION

5   Metasedimentary rocks

4   Robertson River Pluton

3   Charnockitic rocks

2   Old Rag Granite

1   Gneissic rocks

∿   Contact between rock units

∿   County lines

•A   Sample localities

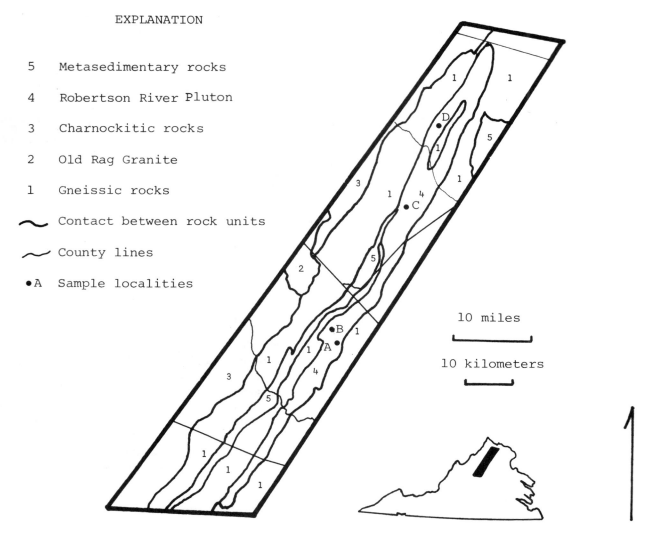

10 miles

10 kilometers

Figure 1. Generalized geologic map showing the distribution of Precambrian and Precambrian (?) rocks in part of the core of the Catoctin-Blue Ridge anticlinorium. Faults not shown. Modified from Johnson and Gathright (1978).

units is indicated by the presence of cobbles and boulders of granite in the Mechum River metasedimentary rocks that are texturally and lithologically similar to the granite of the Robertson River Pluton. Dikes of amphibolite and metabasalt cut the Robertson River Pluton. The metabasalts are texturally and lithologically similar to the metabasalts of the Catoctin Formation.

Certain textural features are common to the gneissic basement rocks and the Robertson River Pluton. These features, called ductile deformation zones, are attributed to late Paleozoic deformation (Mitra, 1979). The Robertson River rocks lack the distinctive gneissic textures that characterize the contiguous basement rocks indicating that the granite was intruded following the event(s) which altered the basement rocks approximately 1 b.y. ago.

The granite boulders present in the Mechum River Formation and the metabasalt dikes that cut the Robertson River rocks

indicate that the granite pluton is older than both the Mechum River and Catoctin Formations (Clarke, 1976; Lukert and Halladay, 1980). These field relations are in contradiction to previously published isotopic age determinations. Because of textural and mineralogical similarities, Lukert and Halladay (1980), included a riebeckite granite in the Robertson River Pluton. Rankin (1976) reported a 650 m.y. Pb^{207}/Pb^{206} age for zircons from this riebeckite granite. Rankin and others (1969), however, suggested that the Catoctin volcanics were extruded about 820 m.y. ago.

The purposes of this paper are: (1) to define as closely as possible the time of intrusion of the Robertson River Pluton and (2) to relate this intrusive event to the chronostratigraphy of the anticlinorium.

Four samples were collected from the Robertson River Pluton (Fig. 1, see appendix for specific locations). Samples A and B are from Madison County within the type area. Sample A has considerable textural variation, ranging from nearly felsitic to coarse granitoid. In addition, the outcrop is near a contact with gneissic rocks and the contact itself is somewhat migmatitic. Sample B was collected from a slightly weathered outcrop of medium granitoid hornblende granite. Sample C was collected from Rappahannock County, near Amissville, and consists of sheared biotite granite. The outcrop is approximately 0.7 km northwest of Rankin's (1976) dated riebeckite granite locality. Sample D was collected in Fauquier County, near the village of Hume. The rock at this locality is primarily a medium- to coarse-grained biotite granite.

ANALYTICAL RESULTS

Zircons were separated from the rock samples and split into subpopulations by standard crushing, heavy liquid, sieving, and magnetic techniques. Samples A, B, and D yielded abundant, clear, pale brown zircons dominated by stubby prisms and simple dipyramids. Zircons from sample C were appreciably less abundant and differed from the others in showing variable cloudiness of color and in having somewhat greater irregularity of grain shapes. Analyzed splits were 95-99 percent pure, the chief impurity being discrete grains of clinozoisite (?), as well as the usual inclusions of apatite, ilmenite, etc. in the zircons themselves. The morphological characteristics of the zircons appear to be consistent with a homogeneous magmatic origin. No unambiguous evidence for contamination with older (pre-Robertson River) zircon was seen.

Isotopic analyses followed conventional procedures for the times at which they were done. Work on samples A and B was begun in 1968, using the then-current method of borax fusion, with Pb loaded in the spectrometer in the sulfide form. Pb and U spikes were added independently for concentration measurements. Sample C and D were analyzed more recently by the HF bomb dissolution technique, with Pb loaded on a silica gel-phosphoric acid substrate. A combined Pb-U spike was used for these samples. In addition to the modifications in chemical procedures, instrumentation evolved from analog to digital methods of data acquisition in the interval between sample pairs A,B and C,D, with a corresponding improvement in accuracy of readout.

The analyses are thus not all comparable in terms of amount of sample analyzed, internal precision, potential for mass discrimination, amount and composition of blank, or overall reproducibility. A detailed discussion of these various analytical uncertainties is not warranted in this short paper. Based on our lab experience we assign an overall precision (one standard deviation) of 2 percent to the Pb/U ratios and 0.5 percent to the radiogenic Pb^{207}/Pb^{206} ratios. The latter are significantly influenced by the common Pb correction because most analyses have sufficiently high contents of common Pb that small errors become magnified by the correction process. The correction for samples A and B was based on analyses of reagent blank Pb with 204:206:207:208 = 1:18.3:15.61:38.2, and for samples C and D

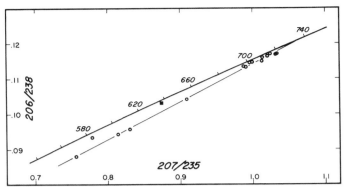

Figure 2. Portion of concordia diagram showing data in Table 1. For clarity, data points are shown as small circles; these are not intended to represent error limits. See text for discussion of uncertainties. Square is Rankin's (1976) analysis from the riebeckite granite.

was based on analyses of reagent blank Pb with 204:206:207:208 = 1:19.0:15.65:38.2. Uncertainties deriving from these different choices are within the limits stated above. In view of the wide variations in chemical procedures and amounts of reagents used for different samples at different times it seems likely that the majority of the common Pb is associated in some fashion with the samples. Analyses of feldspar Pb from samples A and B showed that use of these leads for the common Pb correction would not significantly alter the age calculations.

Analytical results are listed in Table 1, and the data are plotted on an expanded portion of the concordia diagram in Fig. 2. All points except the +100NM fraction of sample C lie close to a straight line. Neglecting for the moment this anomalous point, a best fit straight line through the remainder gives an upper intercept age of 732 ± 5 m.y. and a lower intercept of 172 ± 50 m.y. These figures are derived from the regression method of York (1966), using 206/238 and 207/206 as independent variables in the calculation. The majority of points lie within 0.5 percent of the fitted chord for any of the ratios Pb^{206}/U^{238}, Pb^{207}/U^{235}, or Pb^{207}/Pb^{206}. The poorest fit (0.7 percent) occurs for two of the early PbS runs where variable mass fractionation may have been a more significant problem. Because the lower intercept of the chord is not far from the origin, the closeness of fit of an individual data point is determined mainly by its Pb^{207}/Pb^{206} ratio. Thus, within the assigned standard deviation for this ratio the data are internally consistent with the assumption of a linear distribution on the concordia diagram.

Departure of the one anomalous point is problematical. This analysis was conducted in routine fashion since there was no reason to anticipate difficulty. However, immediately thereafter work on an unrelated sample showed anomalous results which were traced to weak mass beams in the vicinity of mass 204 causing an erroneous lowering of the Pb^{206}/Pb^{204} ratio in that sample. If this effect was present during analysis of the +100NM fraction of sample C, it would explain its low 207/206 age and justify its exclusion from the fitted chord. It is not possible to establish that the other fractions of sample C are completely free of this effect. However, there is no reason to expect that samples

TABLE 1. ANALYTICAL DATA FOR ZIRCON FRACTIONS FROM FOUR SAMPLES OF ROBERTSON RIVER GRANITE

Zircon Fraction[1]	Pb Isotope Ratios			Concentrations (ppm)		Atom Ratios			Apparent Ages[2]		
	$\frac{206}{204}$	$\frac{206}{207}$	$\frac{206}{208}$	Pb*	U	$\frac{206}{238}$	$\frac{207}{235}$	$\frac{207}{206}$	$\frac{206}{238}$	$\frac{207}{235}$	$\frac{207}{206}$
Sample A											
100/200	1159	13.22	6.762	38.57	331.3	0.1147	0.9993	0.06317	700	703	714
100/200M	703.6	11.95	5.167	41.72	352.2	0.1134	0.9874	0.06314	692	697	713
-200/NM	954.0	12.67	6.886	32.87	283.4	0.1151	1.013	0.06380	702	710	735
-200M	1909	14.15	6.737	45.11	384.0	0.1145	0.9964	0.06310	699	702	712
Sample B											
100/200	243.1	8.099	3.872	20.87	176.9	0.1169	1.032	0.06407	713	720	744
-200	547.2	11.06	5.698	21.69	183.5	0.1171	1.034	0.06403	714	721	743
-200NM	632.6	11.57	6.071	17.64	150.4	0.1165	1.021	0.06360	710	714	729
-200M	591.5	11.38	5.973	25.58	219.3	0.1159	1.014	0.06345	707	711	724
Sample C											
+100NM	215.3	7.824	3.081	89.96	916.8	0.09336	0.7784	0.06047	575	585	640
100/200	145.9	6.186	2.449	102.2	1023	0.09436	0.8142	0.06256	581	605	694
-200M	233.1	8.039	3.118	124.3	1331	0.08800	0.7562	0.06232	544	572	685
Sample D											
+100	388.5	9.953	5.216	24.75	37.8	0.1042	0.9092	0.06328	639	657	718
100/200NM	1270	13.36	7.704	21.37	182.2	0.1170	1.025	0.06350	713	716	725
-200	1173	13.24	7.565	31.82	280.1	0.1133	0.9910	0.06341	692	699	722
-200M	423.5	10.47	5.601	41.28	432.3	0.09576	0.8304	0.06289	589	614	705

Note: [1]Figures refer to mesh size range; M and NM refer to magnetic and non-magnetic splits.
[2]Based on tables of Stacey et al., 1973.
*Radiogenic lead.

A, B, and D are affected because they were analyzed at completely different times using different batches of reagents.

DISCUSSION

The distribution of the data on the concordia diagram (Fig. 2) would conventionally be interpreted to mean that the Robertson River Pluton crystallized at approximately the time given by the upper intercept age and that the zircons subsequently were subjected to a process or processes that induced variable amounts of discordance, presumably by some mechanism involving loss of Pb. We see no evidence in the data to contradict this interpretation; thus we conclude that the crystallization age of the Robertson River Pluton is approximately 730 m.y. (rounded off to avoid the appearance of unwarranted precision). The discordance pattern could be explained by any of several mechanisms which, unfortunately, cannot be distinguished from one another in the absence of data points at extreme values of discordance. We would argue, however, that the lower intercept age (172 ± 50 m.y.) is probably a geometric artifact and is not likely to have direct geologic significance.

The crystallization age of 730 m.y. inferred here conflicts seriously with Rb/Sr results. Mose and Nagel (1984) obtained an isochron age of 563 ± 14 m.y. for several samples from interior portions of the Robertson River Pluton. They suggest that the zircon age is open to question because of possible admixtures of older inherited zircons in the analyzed populations, combined with an overprint of secondary discordance. While this combina-

tion can easily lead to spuriously "rotated" discordance arrays, we find no independent confirmation of its validity for the samples analyzed here. Morphological characteristics of the zircons display a limited range of variability with no unusual grain shapes or colorations that could be attributed to contamination by older material. The data, again excepting the one anomalous point, show no resolvable departures from linearity within the assigned uncertainties, implying that if a heterogeneous history is to be postulated its results must closely mimic a normal discordance pattern. The ultimate test—compatibility with other geologic information—is unfortunately not applicable because the age of the Robertson River Pluton is only broadly constrained by field relationships to be between about 1 b.y. and early Cambrian (ca. 530-560 m.y.), and both results are consistent with these limits.

Irrespective of the above considerations, the multicomponent model cannot bring the two methods into agreement without calling on unusual primary ages or unusual trends of discordance in the supposed components of the zircon suites. Essentially this is because the data for samples A and B are clustered rather closely to the concordia curve, and therefore they must lie near one of the vertices of any reasonably chosen discordance polygon. For example, suppose a three component polygon is chosen, with the oldest component being a wall rock age of about 1000 m.y. and the youngest being an apparent discordance "event" near zero age. Then the third component (i.e. the Robertson River crystallization age) cannot be less than about 660 m.y. without causing some of the data to fall outside the polygon.

Prior to work on samples C and D, Lukert (1973) used the

zircon data from samples A and B to fit a chord with a lower intercept of about 670 m.y. and an upper intercept of about 1050 m.y. These two samples taken alone thus appeared to behave according to a two component model in which a small admixture of wall rock material is present, even though morphological evidence for contamination in the zircon populations was not seen. Lukert therefore interpreted the Robertson River crystallization age to be about 670 m.y.

It must be recognized, however, that the data points for samples A and B are tightly clustered. The total range of Pb/U ratios spans about 2 percent and of Pb^{207}/Pb^{206} ratios about 1.5 percent. If analytical uncertainties are viewed conservatively these ranges are too small to construct any chord with confidence. Unless analytical errors happened to have cancelled themselves out to a greater extent than our estimate of precision suggests is likely, it is probably fortuitous that the chord fitted by Lukert (1973) produced an upper intercept age that looks geologically reasonable. Samples C and D provide a much greater range of discordance values from which to construct a chord. The chord so constructed is consistent with samples A and B in that the departures of individual data points behave as expected for the stated uncertainties. Thus we believe that our present interpretation should supersede the earlier one, and that the best estimate of the age of the Robertson River Pluton from zircon data is 730 m.y.

In our view the difference between Rb/Sr and U/Pb results should continue to be regarded as an unresolved problem requiring further investigation. Both methods rely on goodness of fit to a preconceived linear model as justification for their conclusions. Perhaps, in the final analysis, neither will be entirely correct.

Rankin (1976) reported a 650 m.y. Pb^{207}/Pb^{206} age for zircon from a riebeckite granite near our sample D. His analysis plots distinctly off the chord shown in Fig. 2, even though general similarities in mineralogy and texture suggest a possible genetic relation between the riebeckite granite and the Robertson River Pluton. Considering the meagerness of the data, we can make little of this except to note, as one possibility, that the riebeckite granite appears to be resolvably younger than the Robertson River Pluton. Alternatively, if a multicomponent model can be shown to be valid for the zircon data, the two units could conceivably be of the same age.

The most problematic aspect of our result is its implication for the age of the Catoctin volcanic rocks exposed along the Blue Ridge. The Robertson River Pluton is cut by dikes of greenstone, which are reasonably interpreted to be feeder dikes for the Catoctin Formation. Rankin and others (1969) reported a zircon intercept age of 820 m.y. for felsic volcanic rocks from North Carolina, southern Virginia, and Pennsylvania which were presumed to be equivalent to part of the Catoctin Formation in northern Virginia. Correcting for the differences in half-lives used by Rankin and others as compared to the more recently accepted values used herein, the intercept age would be reduced to about 790 m.y. This is definitely older than the 730 m.y. age reported here for the Robertson River Pluton and hence conflicts with the

most reasonable interpretation of field evidence for relative ages. The 790 m.y. age is not truly representative of the Catoctin Formation, because it is largely controlled by samples from the Grandfather Mountain and Mount Rogers formations, whose equivalence to the Catoctin cannot be precisely established.

If this correlation is in error, the only data pertinent to this discussion is the 700 m.y. Pb^{207}/Pb^{206} age they reported for a Catoctin rhyolite from Pennsylvania. Adjusting this analysis to more recently accepted half-life values reduces their Pb^{207}/Pb^{206} age to 686 m.y. This age is remarkably consistent both with our isotopic data from the Robertson River pluton and age relationships we interpret from field evidence.

The clastic rocks which lie stratigraphically beneath the Catoctin volcanic rocks provide further support for our interpretation of the relative ages of the Robertson River and Catoctin units. Cobbles and boulders lithologically and texturally similar to the Robertson River granite are present near the base of both the Fauquier (Lynchburg) Formation (Lukert and Clarke, 1981) and its probable correlative, the Mechum River Formation (Lukert and Halladay, 1980). This indicates not only that the Catoctin Formation is younger than the Robertson River Pluton, but that the granitic rocks were uplifted, eroded, and subsequently buried beneath the clastic sediments prior to extrusion of the Catoctin volcanic rocks.

Conclusions

The Precambrian Y gneisses of the northern Virginia Piedmont were intruded by the granitic rocks of the Robertson River Pluton approximately 730 m.y. B.P. These granitic rocks subsequently were uplifted and eroded, contributing sediment to the Mechum River and Fauquier (Lynchburg) Formations. Following this depositional event, the volcanic rocks of the Catoctin Formation were extruded. We suggest that the 650 m.y. old riebeckite granite may be a late derivative of the Robertson River magma because of textural and lithologic similarities to other rocks of the pluton.

ACKNOWLEDGMENTS

Facilities for the U-Pb isotopic work were supported by National Science Foundation grants GP 3638 and GA 10129.

APPENDIX: LOCATIONS OF DATED SAMPLES

Sample A—Outcrop in roadcut on east side of secondary road 638 on north side of Robinson River in Madison County, Virginia, Brightwood 7½' quadrangle, about 3.6 km north of Madison, Virginia *39° 24.7' latitude, 78° 14.4' W longitude).

Sample B—Outcrop on hillside 1.4 km southwest of intersection of secondary roads 603 and 609 at the village of Haywood in Madison County, Virginia, Madison 7½' quadrangle (38° 26.7' N latitude, 78° 15.8' W longitude).

Sample C—Outcrop 30 m north of U.S. Highway 211 in Rappahannock County, Virginia, Massies Corner 7½' quadrangle, about 21 km west of Warrenton, Virginia (38° 41.9′ N latitude, 78° 2.4′ W longitude).

Sample D—Large pavement outcrop in field 430 m southwest of intersection of secondary roads 635 and 688 at the village of Hume in Fauquier County, Virginia, Flint Hill 7½′ quadrangle (38° 49.7′ N latitude, 78° 15.8′ W longitude).

REFERENCES CITED

Allen, R. M., Jr., 1963, Geology and mineral resources of Greene and Madison Counties: Va. Div. Mineral Resources Bull. 78, 98 p.

Clarke, J. W., 1976, Blue Ridge anticlinorium in northern Virginia: Virginia Journal of Science, v. 27, no. 2, p. 77.

Johnson, S. S., and Gathright, T. M., II, 1978, Geophysical characteristics of the Blue Ridge anticlinorium in central and northern Virginia, *in* Contributions to Virginia Geology—III: Va. Div. Mineral Resources Publication 7, p. 23–36.

Lukert, Michael T., 1973, The petrology and geochronology of the Madison Area, Virginia, [Ph.D. Dissertation]: Cleveland, Case Western Reserve University, 218 p.

Lukert, M. T., and Clarke, J. W., 1981, Age relationships in Proterozoic Z rocks of the Blue Ridge anticlinorium of northern Virginia (Abs.) Geol. Soc. America Abstracts with Programs, v. 13, p. 29.

Lukert, M. T. and Halladay, C. R., 1980, Geology of the Massies Corner quadrangle, Virginia: Va. Div. Mineral Resources Publication 17, map with accompanying text.

Lukert, M. T., and Nuckols, E. B., III, 1976, Geology of the Linden and Flint Hill quadrangles, Virginia: Va. Div. Mineral Resources Rept. Investigations 44, 83 p.

Mitra, Gautam, 1979, Ductile deformation zones in Blue Ridge basement rocks and estimation of finite strain: Geol. Soc. America Bull., v. 90, p. 935–951.

Mose, Douglas G., and Nagel, Susan, 1984, Rb-Sr age for the Robertson River pluton and its implication on the age of the Catoctin Formation in Virginia (this volume)

Nelson, W. A., 1962, Geology and mineral resources of Albemarle County, Va. Div. Mineral Resources Bull. 77, 92 p.

Rankin, D. W., 1976, Appalachian salients and recesses: Late Precambrian continental breakup and the opening of the Iapetus Ocean: Jour. Geophy. Res., v. 81, p. 5605–5619.

Rankin, D. W., Stern, T. W., Reed, J. C., and Newell, M. F., 1969, Zircon ages of felsic volcanic rocks in the upper Precambrian of the Blue Ridge, Appalachian Mountains: Science, v. 166, p. 741–744.

Stacey, J. S., and Stern, T. W., 1973, Revised tables for the calculation of lead isotope ages, U.S. Dept. Commerce, Nat'l Tech. Inf. Service, PB-220 919, 35 pp.

Thiesmeyer, L. R., 1938, Plutonic rocks of northwestern Fauquier County, Virginia: Gol. Soc. America Bull., v. 49, p. 1963–1964.

York, D., 1966, Least squares fitting of a straight line, Can. Jour. Physics, v. 44, p. 1079–1086.

MANUSCRIPT ACCEPTED BY THE SOCIETY AUGUST 2, 1983

Geological Society of America
Special Paper 194
1984

Rb-Sr age for the Robertson River pluton in Virginia and its implication on the age of the Catoctin Formation

Douglas G. Mose
Susan Nagel
Department of Chemistry and Geology
George Mason University
Fairfax, Virginia 22030

ABSTRACT

The Robertson River pluton is a granitic pluton that intruded Grenville rocks in the Blue Ridge province of western Virginia. Rb-Sr whole-rock analyses show that the pluton was intruded at 570 ± 15 m.y. ago with a relatively high initial $^{87}Sr/^{86}Sr$ ratio of 0.712. The Robertson River pluton and the Grenville host rocks are cut by numerous metabasalt dikes that are probably feeder dikes related to the Catoctin Formation, a widespread plateau basalt that once covered this part of the Blue Ridge. The cross-cutting relationship of the metabasalt dikes to the Robertson River pluton considered along with the early Cambrian age of the Chilhowee Group, which overlies the Catoctin Formation, indicate that the Catoctin Formation was extruded at a time close to the Precambrian-Cambrian boundary at about 570 m.y. ago.

INTRODUCTION

The Blue Ridge in western Virginia is composed mainly of Precambrian rocks, the most ancient of which yield radiometric ages of 1,000 to 1,200 m.y. (Bartholomew and others, 1981; Lukert, 1982). This ancient terrane is composed mostly of granulite and upper amphibolite facies rocks, and it commonly is related to the Grenville province in southeastern Canada. Post-Grenville rocks in the Blue Ridge are mostly latest Precambrian in age, and include both sedimentary and volcanic strata and several granitic plutons. One of these plutons, the Robertson River pluton, is the subject of this report.

Grenville rocks in central Virginia were slowly uplifted after an interval of high-grade metamorphism about 1,000 m.y. ago, and the present erosional level in the Grenville terrane must have been reached in latest Precambrian time when the high-grade metamorphic rocks were covered successively by sedimentary strata of the Mechum River and Lynchburg formations, Catoctin volcanic rocks, and the Chilhowee Group (the upper part of the Chilhowee Group is known to be early Cambrian in age based on fossils). The ages of the Mechum River and Lynchburg formations are not well known, since fossils have been found in neither

of these units. The Catoctin Formation is younger than the Mechum River and Lynchburg formations, and the Chilhowee Group overlies the Catoctin volcanic rocks along the western side of the Blue Ridge in central Virginia. Since the age of the Catoctin has been known only as younger than about 1,000 m.y. and older than early Cambrian (about 550 m.y.), the Catoctin has been studied using the U-Pb and Rb-Sr dating techniques. The results of these studies are compared to the age of the Robertson River pluton in this report.

PREVIOUS CHRONOLOGY STUDIES

The volcanic rocks of the Catoctin Formation are part of a group of post-Grenville, but latest Precambrian igneous rocks that are found in Great Britain, Newfoundland, Cape Breton, New Brunswick, and New England (Rast and others, 1976; Williams, 1979). Similar igneous rocks also are found in the Blue Ridge of the southern Appalachians (Rankin and others, 1969; Rankin, 1975, 1976). The rocks, including the Catoctin metabasalt in central Virginia, are thought to have formed in an environment of

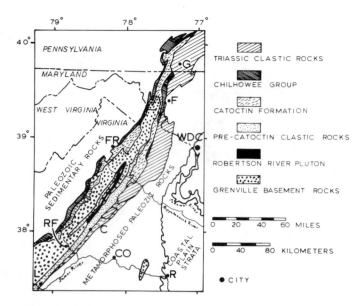

Figure 1. Geologic map of the Blue Ridge province and adjacent areas in Virginia. Key to symbols: G = Gettysburg, F = Frederick, WDC = Washington (District of Columbia), FR = Front Royal, C = Charlottesville, CO = Columbia, L = Lynchburg, R = Richmond, RF = small exposure of Rockfish River pluton sampled for Rb-Sr study.

crustal tension along rifted continental margins. The post-Grenville but pre-Catoctin strata may therefore contain the record of the vertical crustal movements, rifting, and sedimentological phenomena associated with the earliest history of the Proto-Atlantic.

Rankin (1975, 1976) associated granitic plutonic activity with this rifting environment, and commented on the uncertainty of the timing of this event in the Blue Ridge. In short, the few U-Pb data from these plutons yield ages of about 700 to 800 m.y. (Rankin and others, 1969; Lukert, 1973; Davis, 1974). However, some of these plutons in North Carolina have also been studied by the Rb-Sr whole-rock technique and yield well defined isochron ages of about 600 to 700 m.y. (Odom and Fullagar, 1971).

The lack of agreement between the results of the U-Pb and Rb-Sr techniques when applied to these plutons has been of some concern. We think that the U-Pb ages are of uncertain value because preliminary studies have shown that these granitic plutons contain multicomponent zircon populations (Odom and Fullagar, 1971; Eckelmann and Mose, 1981). Two-component populations, composed of an older host-rock component and a younger plutonic component, have been utilized to generate meaningful U-Pb ages (Bickford and others, 1981). One recently studied example of this situation is the Rockfish River pluton in the Blue Ridge of Virginia (Fig. 1). Rb-Sr data from this pluton yield an age of 646 ± 55 m.y., (Tables 2 and 3). Although U-Pb data yield an upper intercept age of about 820 m.y., some of the zircons have visible cores (Davis, 1974). All the U-Pb data are discordant, and all fall in a triangular field that is defined by the age of the host rock (about 1,000 m.y.), the Rb-Sr age, and 0 m.y. on a U-Pb concordia diagram. It is interesting to interpret in this

fashion U-Pb data from all the post-Grenville but Precambrian plutons in the Blue Ridge. However, a general paucity of U-Pb data from these plutons and the apparent complexity in their zircon populations, which may exceed simple two-component systems, have resulted in U-Pb concordia diagrams whose usefulness for age determinations is unclear.

Rb-Sr AGE OF THE ROBERTSON RIVER PLUTON

The Robertson River pluton (Fig. 1) is the largest known post-Grenville pluton in the Blue Ridge of Virginia (Allen, 1963; Lukert, 1973). The pluton is a nonfoliated to weakly foliated hornblende or biotite granitic rock that is locally rich in magnetite. The pluton contains xenoliths of Grenville rock, and cobbles of lithologically similar granite are present in the Mechum River Formation (Lukert and Halladay, 1980).

Nineteen samples from the Robertson River pluton were collected for this study (Fig. 2, Table 3). Ten samples came from along the central axis of the pluton exposure area. Samples A1 to A6 were collected from along the eastern margin of the pluton and samples R1 to R3 were collected from along the western margin. The Rb-Sr technique used in this study is based on the radioactive decay of ^{87}Rb to ^{87}Sr by beta emission. The half-life for the decay is estimated to be about 4.89×10^{10} yr, and the decay constant is 1.42×10^{-11} yr^{-1} (Steiger and Jager, 1977). The

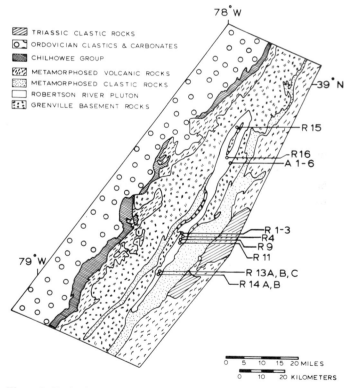

Figure 2. Geologic map of the Robertson River pluton and adjacent area in the Blue Ridge of Virginia (map modified from Johnson and Gathright, 1978). Sample locations for the Rb-Sr study of the Robertson River pluton are shown.

samples of the Robertson River pluton and other rocks that were examined in this study were each about 10 kg. Each was crushed and split to a 10-g portion which was powdered, and about 0.3 g of each sample was isotopically analyzed. Each analysis was done using ^{84}Sr and ^{87}Rb spikes and ultrapure HF, $HClO_4$ and HCl.

The strontium analyses were made using a Nier-type, 6-in. radius, single filament mass spectrometer with a programmable automatic data acquisition system at the Department of Terrestrial Magnetism of the Carnegie Institution in Washington, D.C. The Rb isotopic analyses were made using a 12-in. radius mass spectrometer at Florida State University. All the strontium isotopic compositions were calculated from analyses of sample plus spike mixtures. The ^{85}Rb/^{87}Rb ratio was taken to be 2.593 (Steiger and Jager, 1977).

Analyses of the National Bureau of Standards 70a standard feldspar were performed over the course of this study, and the data[1] are in close agreement with the data for this standard reported by others, indicating that there are no significant systematic errors in the isotope tracer calibration used in this study. Analyses of the Eimer and Amend strontium standard[2] are also in agreement with the certified value for this standard, indicating that there is no significant systematic error in the mass spectrometer calibration. The ^{87}Rb blanks have averaged 0.0013 ppm, and the ^{86}Sr blanks have averaged 0.0041 ppm. Compared to the Rb and Sr concentrations in the samples analyzed in this study, these blank values are not significant.

The Rb-Sr age and initial ^{87}Sr/^{86}Sr ratios on the isochron diagrams were calculated using the regression treatment described by York (1966). The one-standard-deviation experimental error on ^{87}Rb/^{86}Sr was calculated to be 2%, and the one-standard-deviation experimental error in ^{87}Sr/^{86}Sr was calculated to be 0.05%. These error estimates were derived from an examination of duplicate analyses done over the past seven years, and these estimates include sample-splitting errors. The estimates of one-standard-deviation experimental errors that do not include an error increment related to sample-splitting are 1% for ^{87}Rb/^{86}Sr and 0.02% for ^{87}Sr/^{86}Sr. The isotopic data are presented in Tables 1, 2, and 4; sample locations are given in Tables 3 and 5. The errors assigned to the reported age and initial ^{87}Sr/^{86}Sr ratio on the isochron diagrams are given at the 68% confidence level (1 σ). Visual examinations of the fit of the Rb-Sr isotopic data to isochron lines are provided by Figures 3 and 4; the data on these diagrams are presented as ±2 σ errors boxes (±4% for ^{87}Rb/^{86}Sr and ±0.10% for ^{87}Sr/^{86}Sr).

As shown in Figure 2 and Table 3, nine of the Robertson River samples are from the margin of the pluton (open symbols on Fig. 3), and ten samples are from along the central axis of the pluton (solid symbols on Fig. 3). The maximum slope line that could be drawn using all the data corresponds to an age of about 720 m.y. (dashed line on Fig. 3). It seems unlikely that this

720-m.y. line has any geologic significance, since the pluton margin contains many xenoliths of Grenville rocks that probably contributed radiogenic ^{87}Sr to the Robertson River magma. This problem has been discussed elsewhere (Ellwood and others, 1980), and it is known that pluton-margin samples often yield data points that plot above the isochron generated by less contaminated pluton-center samples.

Samples from along the central axis of the pluton yield a well-defined isochron corresponding to an age of 570 ± 15 m.y.; the pluton was intruded with a relatively high initial ^{87}Sr/^{86}Sr ratio of 0.7121 ± 0.0004. Incorporation of ancient country rock enriched in radiogenic ^{87}Sr is commonly thought to cause high initial ^{87}Sr/^{86}Sr ratios in plutons, and the only late Precambrian and Cambrian plutons in the Appalachians that have high initial ^{87}Sr/^{86}Sr ratios are located in areas of ancient (1-b.y.-old) country rock (Mose, 1981). In the case of the Robertson River pluton, our interpretation is that while the entire pluton probably increased its overall ^{87}Sr/^{86}Sr ratio by incorporation of melted host rock, the magma convection was not sufficient to completely mix the strontium in the pluton margin with the strontium in the pluton center. In any case, while the data might be construed to yield an intrusion age of about 720 m.y. (most of the Rb-Sr data points would fall below this 720-m.y. "isochron" because of isotopic changes due to Paleozoic metamorphism), it seems more reasonable that the plutons was intruded at 570 ± 15 m.y. ago (based on the pluton-center samples).

TABLE 1. ISOTOPIC DATA FROM SAMPLES OF THE ROBERTSON RIVER PLUTON

Sample Number	^{87}Rb/^{86}Sr Atomic ratio	^{87}Sr/^{86}Sr Atomic ratio	^{86}Sr (ppm)	^{87}Rb (ppm)
Robertson River Pluton: Central Axis				
R-4	1.205	0.7204	26.379	32.161
R-9	0.081	0.7131	9.075	0.739
R-11	8.137	0.7777	4.213	34.679
R-13A	0.536	0.7166	30.371	16.478
R-13B	0.533	0.7162	27.738	14.947
R-13C	0.761	0.7185	21.722	16.713
R-14A	2.750	0.7343	11.069	30.792
R-14B	2.466	0.7322	12.161	30.335
R-15	10.750	0.7963	1.163	34.607
R-16	2.709	0.7365	7.996	21.917
Robertson River Pluton: Central Axis				
A-1	7.950	0.7944	4.211	33.862
A-2	10.115	0.8118	3.446	35.260
A-3	4.095	0.7526	6.963	28.840
A-4	10.467	0.7982	3.010	31.871
A-5	13.064	0.8277	2.525	33.373
A-6	5.576	0.7571	5.419	30.543
Robertson River Pluton: Western Margin				
R-1	4.372	0.7528	6.541	28.930
R-2	5.109	0.7565	6.883	35.572
R-3	4.298	0.7531	8.206	35.677

TABLE 2. ISOTOPIC DATA AND ISOCHRON FROM SAMPLES OF THE ROCKFISH RIVER PLUTON

Sample Number	^{87}Rb/^{86}Sr Atomic Ratio	^{87}Sr/^{86}Sr Atomic Ratio	^{86}Sr (ppm)	^{87}Rb (ppm)
RF-1	5.012	0.7661	6.394	32.419
RF-1	5.713	0.7658	5.431	31.383
RF-3	3.499	0.7530	6.245	22.104
RF-4	3.722	0.7533	5.880	22.138
RF-5	7.350	0.7829	5.175	38.479
RF-12	2.836	0.7410	8.458	24.262
RF-13	7.058	0.7818	4.772	34.070
RF-14	4.451	0.7566	5.641	25.402

Note: See Figure 5.

---

[1]NBS 70a (4 analyses): ^{87}Rb = 149.9 ± 0.6 (1 σ) ppm, ^{87}Sr = 6.02 ± 0.003 (1 σ).

[2]E and A (34 analyses): (^{87}Sr/^{86}Sr)$_n$ = 0.70794 ± 0.00006 (1 σ).

TABLE 3. LOCATION OF SAMPLES EXAMINED FOR Rb-Sr STUDY OF THE
ROBERTSON RIVER AND ROCKFISH RIVER PLUTONS

Sample Number	Virginia 7 1/2-min quadrangle	N. lat.	W. long.
Robertson River Pluton: Central Axis			
R-4	Brightwood	38°24'28"	78°13'42"
R-9	Madison	38 23 28	78 15 13
R-11	Madison	38 22 54	78 15 1
R-13A	Rochelle	38 15 48	78 21 22
R-13B	Rochelle	38 15 48	78 21 22
R-13C	Rochelle	38 15 48	78 21 22
R-14A	Barboursville	38 14 48	78 21 48
R-14B	Barboursville	38 14 48	78 21 48
R-15	Orlean	38 51 24	78 56 15
R-16	Massies Corner	38 44 16	78 2 15
Robertson River Pluton: Eastern Margin			
A-1	Massies Corner	38 40 31	78 0 1
A-2	Massies Corner	38 40 29	78 0 1
A-3	Warrenton	38 40 29	79 59 57
A-4	Massies Corner	38 40 31	78 0 1
A-5	Massies Corner	38 40 29	78 0 1
A-6	Warrenton	38 40 25	79 59 57
Robertson River Pluton: Western Margin			
R-1	Brightwood	38 24 47	78 14 21
R-2	Brightwood	38 24 46	78 14 21
R-3	Brightwood	38 24 45	78 14 21
Rockfish River Pluton			
RF-1	Lovingston	37 48 56	78 46 48
RF-2	Lovingston	37 49 5	78 46 35
RF-3	Lovingston	37 49 7	78 46 35
RF-4	Lovingston	37 49 8	78 46 35
RF-5	Lovingston	37 48 58	78 46 48
RF-12	Lovingston	37 49 6	78 46 35
RF-13	Lovingston	37 49 6	78 46 35
RF-14	Lovingston	37 49 7	78 46 35

Note: Samples R-13A, B, and C were collected at intervals of about 3 m over a 40-m-long outcrop. Samples 14A and B were collected about 5 m apart from a 10-m-long outcrop. Samples A-1 to 6 were collected at intervals of about 10 m from a 70-m-long outcrop. Samples R-1 to 3 were collected at intervals of about 15 m from a 50-m-long outcrop. Samples RF-2, 3, 4, 12, 13, and 14 were collected at intervals of about 7 m over a 80-m-long outcrop. Samples RF-1 and 5 were collected about 10 m apart from a 100-m-long outcrop.

TABLE 4. ISOTOPIC DATA FROM METABASALT OF THE
CATOCTIN FORMATION AND FROM A METABASALT DIKE IN THE
ROBERTSON RIVER PLUTON (V-12)

Sample Number	$^{87}Rb/^{86}Sr$ atomic ratio	$^{87}Sr/^{86}Sr$ atomic ratio	^{86}SR (ppm)	^{87}Rb (ppm)
Catoctin Formation in Shenandoah National Park				
CT-1	0.040	0.7044	37.670	1.529
CT-2	0.910	0.7101	10.863	9.997
CT-3	0.296	0.7061	8.186	2.453
CT-4	0.214	0.7057	5.201	1.124
CT-5	0.239	0.7050	10.386	2.509
CT-7	0.072	0.7042	20.800	1.508
CT-8	0.012	0.7040	56.302	0.656
Metabasalt Dike in Granite				
V-12	0.361	0.7066	28.583	10.434

TABLE 5. LOCATION OF SAMPLES EXAMINED FOR AN ISOTOPIC
AND CHEMICAL STUDY OF THE CATOCTIN FORMATION AND
A METABASALT DIKE

Sample Number	Virginia 7 1/2-min quadrangle	N. lat.	W. long.
Catoctin Formation in Shenandoah National Park			
CT-1	Old Rag Mountain	38°35'12"	78°21'41"
CT-2	Old Rag Mountain	38 36 22	78 22 0
CT-3	Big Meadows	38 31 18	78 25 37
CT-4	Old Rag Mountain	38 35 53	78 21 47
CT-5	Big Meadows	38 33 8	78 23 7
CT-7	Big Meadows	38 33 30	78 22 55
CT-8	Big Meadows	38 34 12	78 22 41
Metabasalt Dike in Granite			
V-12	Massies Corner	38 43 58	78 0 10

Note: The Catoctin samples were collected using the geologic map of Gathright (1976). Metabasalt sample V-12 is from the center of a dike about 10 m wide in the Robertson River pluton.

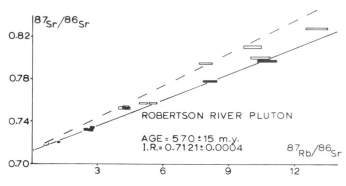

Figure 3. Rb-Sr isochron diagram showing analyses of samples from along the margin of the Robertson River pluton (open symbols) and from along the central axis of the pluton (solid symbols). The maximum slope line which could be defined by these data (dashed line) corresponds to an age of about 720 m.y. An isochron generated by only the samples from along the pluton center yield an isochron age of 570 ± 15 m.y. The data that fall above the pluton-center isochron are thought to be from samples that crystallized with relatively high initial $^{87}Sr/^{86}Sr$ ratios due to assimilation of ^{87}Sr from the host rocks.

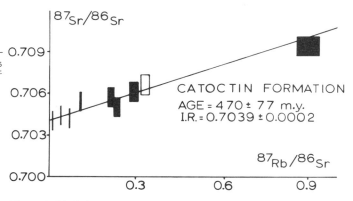

Figure 4. Rb-Sr isochron diagram derived from samples of the Catoctin Formation (solid symbols). These data yield an age of 470 ± 77 m.y. Also shown is sample V-12 (open symbol), a metabasalt dike in the Robertson River pluton. Inclusion of sample V-12 in the regression analysis yields an age of 491 ± 54 m.y. and an initial $^{87}Sr/^{86}Sr$ ratio of 0.7039 ± 0.0002.

AGE OF THE CATOCTIN FORMATION

The Catoctin Formation in Virginia is overlain along the western side of the Blue Ridge by the Chilhowee Group. Because the Chilhowee is composed of nonmarine clastic strata that grade upward into shallow marine strata (Brown, 1970), and because the upper part of the Chilhowee contains early Cambrian fossils (Walcott, 1981; Woodward, 1949), the lower Chilhowee is considered earliest Cambrian or latest Precambrian in age. Attempts to define an absolute time scale (based on radiometric techniques) for the geologic time scale (based on fossils) have yielded increasingly more accurate estimates for the early Cambrian. The most recent paper, using the presently accepted decay constants for the radiometric techniques, indicates that the early Cambrian began about 560 m.y. ago (Cowie and Cribb, 1978). The stratigraphic

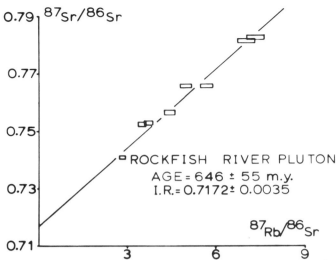

Figure 5. Rb-Sr isochron diagram derived from samples of the Rockfish River pluton.

relationship between the Chilhowee and the Catoctin thus indicates that the Catoctin Formation is no younger than about 560 m.y. old.

Many workers noted that no widespread structural discor-

dance occurs between the Chilhowee and the underlying Catoctin, and they considered the clastic strata of the lower Chilhowee as well as the underlying Catoctin to be early Cambrian or latest Precambrian in age (Bloomer and Werner, 1955; Whitaker, 1955; Nickelsen, 1956). However, U-Pb data obtained from one zircon sample from the Catoctin in Pennsylvania and other zircon samples from possibly related volcanic rocks in southwestern Virginia and western North Carolina yield a concordia U-Pb age of about 820 m.y. (Rankin and others, 1969). This inferred Proterozoic age for the Catoctin supported other studies that cited evidence for an erosional unconformity between the Catoctin and the overlying Chilhowee (King, 1950; Toewe, 1966; Gathright and Nystrom, 1974; Rader and Biggs, 1975; Lukert and Nuckols, 1976; Gathright, 1976).

The Grenville rocks, the Robertson River pluton, and the Mechum River and Lynchburg formations are all cut by numerous metabasaltic dikes. In the absence of any other widespread metabasaltic units in the Virginia Blue Ridge except the Catoctin Formation, and because of the close proximity of the Catoctin flows to these dikes, the dikes are generally interpreted as feeder dikes for Catoctin basalt flows (Rankin, 1975; Lukert and Nuckols, 1976; Lukert and Halladay, 1980). This apparent relationship between Catoctin flows and metabasalt dikes is important, be-

TABLE 6. CHEMICAL ANALYSES IN WEIGHT PERCENT

Sample Number	SiO_2	Al_2O_3	Fe_2O_3	FeO	MgO	CaO	Na_2O	K_2O	TiO_2	Data Source
Catoctin Metabasalt Flows										
CT-1	48.3	13.9	5.1	7.2	5.9	8.1	3.2	0.5	2.3	New data
CT-3	44.7	14.6	6.8	9.0	7.8	4.4	3.3	0.8	2.7	New data
CT-5	45.7	14.9	6.8	8.4	8.0	5.9	4.3	0.4	2.5	New data
CT-7	48.0	14.1	5.4	8.0	6.2	9.3	2.9	0.4	2.2	New data
631	45.3	14.4	6.6	8.3	6.0	6.0	4.6	0.6	4.1	Reed and Morgan, 1971
632	48.9	14.9	6.5	6.8	5.7	6.4	4.5	0.6	2.3	same
633	46.0	14.9	6.5	9.1	6.9	4.9	4.6	0.1	2.6	same
634	48.1	14.1	3.1	7.6	5.8	9.0	3.9	0.9	2.0	same
673B	49.8	14.6	5.4	6.0	6.3	7.7	4.1	0.6	1.9	same
674A	48.8	13.9	5.1	8.4	6.7	7.4	3.2	0.1	2.3	same
677B	49.7	13.3	6.1	8.9	5.0	6.4	2.5	0.1	2.8	same
1	45.3	14.4	6.6	8.3	6.0	6.0	4.6	0.6	4.1	Reed, 1964
2	48.9	14.9	6.5	6.8	5.7	6.4	4.5	0.6	2.3	same
3	46.0	14.9	6.5	9.1	6.9	4.9	4.6	0.1	2.6	same
4	48.1	14.1	3.1	7.6	5.8	9.0	3.9	0.9	2.0	same
Average:	47.3	14.4	5.7	8.0	6.3	6.8	3.9	0.5	2.6	
Range:	44-50	13-15	3-7	6-9	5-8	4-10	2.5	0.1-0.9	2-4	
Metabasalt Dike										
V-12	47.2	13.4	4.6	9.4	4.7	8.8	2.3	0.9	3.1	New data

Note: The new data were obtained by Deborah Kay using the techniques described in Shapiro (1975). There appears to be no significant difference between the composition of metabasalt dike (V-12) and the composition of the Catoctin metabasalt flow rocks.

cause if the relationship is correct, the Robertson River pluton, which has been dated at 570 ± 15 m.y., provides a maximum age for the Catoctin flows that are related to the apparent Catoctin feeder dikes.

A Rb-Sr study of Catoctin metabasalt was conducted during the study of the Robertson River pluton. Samples were collected from over a distance of about 10 km, using a sampling interval of about 1 km (Tables 4 and 5). The metabasalts yielded an isochron age of 470 ± 77 m.y. (Fig. 4). Although the isochron slope is strongly determined by one point and the study yielded an age with a large uncertainty due largely to the low ^{87}Rb contents of the samples, the isochron age is clearly younger than the minimum age for the Catoctin based on the early Cambrian age of the overlying Chilhowee Group. Since volcanic rocks are known to isotopically respond to metamorphic events, it is reasonable to compare the Rb-Sr age determination of the Catoctin to the time of regional metamorphism in the Blue Ridge. Despite a paucity of radiometric data in western Virginia, the major metamorphic event (Taconic orogeny) is thought to have occurred about 480 m.y. ago based on ^{40}Ar/^{39}Ar studies of similar rock in North Carolina (Dallmeyer, 1975a, 1975b). It thus appears that although the Rb-Sr age of the relatively coarse-grained Robertson River pluton was not "reset to 480 m.y." by the Taconic orogeny, the fine-grained Catoctin basalts did respond by strontium isotopic homogenization at about 480 m.y. ago.

An important observation is that metabasalt sample V-12 from a 10-m-wide dike in the Robertson River pluton is isotopically (Table 4) and chemically (Table 6) identical to the Catoctin metabasalt. This similarity is compatible with the interpretation that this metabasalt dike is a feeder dike related to the Catoctin volcanic rocks. It therefore seems reasonable to use the age of the Robertson River pluton (570 ± 15 m.y.) as a maximum age for Catoctin volcanic rocks related to this feeder dike. This age, combined with the early Cambrian age of the Chilhowee Group which overlies the Catoctin, shows that these basalt flows were

deposited near the Precambrian-Cambrian boundary at about 560 to 580 m.y. ago.

CONCLUSIONS

The Robertson River pluton in the Blue Ridge of western Virginia is a granitic body emplaced in latest Precambrian or earliest Cambrian time (Rb/Sr age = 570 ± 15 m.y.). Metabasalt dikes, which are thought to be feeder dikes for the metabasalt flows of the Catoctin Formation, cut the Robertson River pluton, and the Catoctin volcanic rocks are overlain by the Chilhowee Group, which contains early Cambrian fossils. The ages of the Robertson River pluton and the Chilhowee Group therefore indicate that the Catoctin volcanic flows were extruded at some time between about 560 and 580 m.y. ago.

The emplacement of large masses of granitic magma such as the Robertson River pluton was a major event in the latest Precambrian history of the Virginia Blue Ridge. Rankin (1976) has shown that felsic flows and tuffs in the Catoctin Formation are probably related to granitic plutons such as the Robertson River pluton. The high initial ^{87}Sr/^{86}Sr ratios of the Robertson River pluton and other plutons of similar age and composition in the Grenville of North Carolina (Odom and Fullagar, 1971), Virginia (Rockfish River pluton: Table 2), and southeastern New York (Mose, 1981), and the lack of evidence for a latest Precambrian metamorphic event indicate that these granitic magmas were formed by the melting of Grenville continental crust under anorogenic conditions. The heat content of the upward-moving Catoctin basaltic liquids perhaps generated these granitic melts, or alternately the granitic magmas and the basaltic liquids are independent products of a major thermal event in the crust during latest Precambrian time.

ACKNOWLEDGMENTS
We thank Leo Hall, Nick Ratcliff, Roy Odom, and Jim Couley for their helpful comments.

REFERENCES CITED

Allen, R. M., 1963, Geology and mineral resources of Greene and Madison counties: Virginia Division of Mineral Resources Bulletin, 78, 102 p.

Bartholomew, M. J., Gathright, T. M., and Henika, W. S., 1981, A tectonic model for the Blue Ridge in central Virginia: American Journal of Science, v. 281, p. 1164–1183.

Bickford, M. E., Chase, R. B., Nelson, B. K., Shuster, R. D., and Arruda, E. C., 1981, U-Pb studies of zircon cores and overgrowths, and monazite: Implications for age and petrogenesis of the northeastern Idaho batholith: Journal of Geology, v. 89, p. 433–457.

Bloomer, R. O., and Werner, H. J., 1955, Geology of the Blue Ridge region in central Virginia: Geological Society of America Bulletin, v. 66, p. 579–606.

Brown, W. R., 1970, Investigations of the sedimentary record in the Piedmont and Blue Ridge in Virginia, in Fisher and others, eds., Studies of Appalachian geology, central and southern: New York, Interscience Publishers, p. 335–349.

Cowie, J. W., and Cribb, S. J., 1978, The Cambrian system, in Cohee and others, eds., Contributions to the geologic time scale: American Association of Petroleum Geologists, Studies in Geology no. 6, p. 355–362.

Dallmeyer, R. D., 1975a, Incremental ^{40}Ar/^{39}Ar ages from biotite and hornblende from retrograded basement gneisses of the southern Blue Ridge: Their bearing on the age of Paleozoic metamorphism: American Journal of Science, v. 275, p. 444–460.

Dallmeyer, R. D., 1975b, ^{40}Ar/^{39}Ar ages of biotite and hornblende from a progressively remetamorphosed basement terrane: Their bearing on interpretation of release spectra: Geochimica et Cosmochimica Acta, v. 39, p. 1655–1669.

Davis, R. G., 1974, Pre-Grenville ages of basement rocks in central Virginia: A model for the interpretation of zircon ages [M.S. thesis]: Blacksburg, Virginia Polytechnic Institute and State University, 46 p.

Eckelmann, F. D., and Mose, D. G., 1981, Search for xenocrystic zircons in latest Precambrian plutonic rocks in the Blue Ridge (abs.): Geological Society of America Abstracts with Programs, v. 13, p. 7.

Ellwood, B. B., Whitney, J. A., Wenner, D. B., Mose, D. G., and Amerigian, G., 1980, Age, paleomagnetism, and tectonic setting of the Elberton granite, northeast Georgia Piedmont: Journal of Geophysical Research, v. 85, p. 6521–6533.

Gathright, T. M., 1976, Geology of the Shenandoah National Park, Virginia: Virginia Division of Mineral Research Bulletin, 86, 93 p.

Gathright, T. M., and Nystrom, P. G., 1974, Geology of the Ashby Gap quadrangle, Virginia: Virginia Division of Mineral Research Report of Investigations 36, 55 p.

Johnson, S. S., and Gathright, T. M., 1978, Geophysical characteristics of the Blue Ridge anticlinorium in central and northern Virginia, *in* Contributions to Virginia geology III: Virginia Division Mineral Research Publication 7, p. 23–36.

King, P. B., 1950, Geology of the Elkton area, Virginia: U.S. Geological Survey Professional Paper 230, 82 p.

Lukert, M. T., 1982, Uranium-lead isotope age of the Old Rag granite, Northern Virginia: American Journal of Science, v. 282, p. 391–398.

——1973, The petrology and geochronology of the Madison area, Virginia [Ph.D. thesis]: Cleveland, Case Western Reserve University, 218 p.

Lukert, M. T., and Halladay, C. R., 1980, Geology of the Massies Corner quadrangle, Virginia: Virginia Division of Mineral Research Publication 17 (geologic map).

Lukert, M. T., and Nuckols, E. B., 1976, Geology of the Linden and Flint Hill quadrangles, Virginia: Virginia Division of Mineral Research Report of Investigations 44, 83 p.

Mose, D. G., 1981, Avalonian igneous rocks with high initial $^{87}Sr/^{86}Sr$ ratios: Northeastern Geology, v. 3, p. 129–131.

Nickelsen, R. P., 1956, Geology of the Blue Ridge near Harpers Ferry, West Virginia: Geological Society of America Bulletin, v. 67, p. 239–269.

Odom, A. L., and Fullagar, P. D., 1971, A major discordancy between U-Pb zircon ages and Rb-Sr whole-rock ages of late Precambrian granites in the Blue Ridge Province (abs.): Geological Society of America Abstracts with Programs, v. 3, p. 663.

Rader, E. K., and Biggs, T. H., 1975, Geology of the Front Royal quadrangle, Virginia: Virginia Division of Mineral Research Report of Investigations 40, 90 p.

Rankin, D. W., 1975, The continental margin of eastern North America in the southern Appalachians: The opening and closing of the Proto-Atlantic ocean: American Journal of Science, v. 275-A, p. 298–336.

——1976, Appalachian salients and recesses: Late Precambrian continental break-up and the opening of the Iapetus ocean: Journal of Geophysical Research, v. 81, p. 6505–5619.

Rankin, D. W., Stern, T. W., Reed, J. C., and Newell, M. F., 1969, Zircon ages of felsic volcanic rocks in the upper Precambrian of the Blue Ridge, Appalachian mountains: Science, v. 166, p. 741–744.

Rast, N., O'Brien, E. H., and Wardle, R. J., 1976, Relationships between Precambrian and lower Paleozoic rocks of the 'Avalon Platform' in New Brunswick, the northeast Appalachians and the British Isles: Tectonophysics, v. 30, p. 315–338.

Reed, J., 1964, Chemistry of greenstone of the Catoctin Formation in the Blue Ridge of central Virginia: U.S. Geological Survey Professional Paper 501-C, p. C69–C73.

Reed, J., and Morgan, B., 1971, Chemical alteration and spilitization of the Catoctin greenstones, Shenandoah National Park, Virginia: Journal of Geology, v. 79, p. 526–548.

Shapiro, L., 1975, Rapid analysis of silicate, carbonate and phosphate rocks (revised edition): U.S. Geological Survey Bulletin, 1401, 76 p.

Steiger, R. H., and Jager, E., 1977, Subcommission on geochronology: Convention on the use of decay constants in geo- and cosmochronology: Earth and Planetary Science Letters, v. 36, p. 359–363.

Toewe, E. C., 1966, Geology of the Leesburg quadrangle, Virginia: Virginia Division of Mineral Research Report of Investigations 11, 52 p.

Walcott, C. D., 1891, Correlations papers—Cambrian: U.S. Geological Survey Bulletin, v. 81, p. 133–144.

Whitaker, J. C., 1955, Geology of Catoctin Mountain, Maryland and Virginia: Geological Society of America Bulletin, v. 66, p. 435–462.

Williams, H., 1979, Appalachian orogen in Canada: Canadian Journal of Earth Sciences, v. 16, p. 792–807.

Woodward, H. P., 1949, Cambrian system of West Virginia: West Virginia Geological Survey, v. 20, 317 p.

York, D., 1966, Least-squares fitting of a straight line: Canadian Journal of Physics, v. 44, p. 1079–1086.

MANUSCRIPT ACCEPTED BY THE SOCIETY AUGUST 2, 1983

Geological Society of America
Special Paper 194
1984

Evolution of the Grenville terrane in the central Virginia Appalachians

A. K. Sinha
Department of Geological Sciences
Virginia Polytechnic Institute and State University
Blacksburg, Virginia 24061

M. J. Bartholomew
Virginia Division of Mineral Resources Office
Virginia Polytechnic Institute and State University
Blacksburg, Virginia 24061

ABSTRACT

Along a transect across the central Virginia Blue Ridge complex, the terrane is divisible into two discrete massifs—the Pedlar and the Lovingston, which are separated by the mid-Paleozoic Rockfish Valley fault. The Pedlar Massif is characterized by two major lithologies—an older volcanic sequence (1130 m.y.) intruded by the younger Pedlar River Charnockite Suite (1070 m.y.). The Lovingston Massif has a more complex association of rock types that include an older paragneissic sequence (Stage Road Layered Gneiss) with approximately 1870 m.y. old detrital zircons and intruded by quartz monzonite, pegmatite and charnockite of the Archer Mountain Suite, as well as anorthosites (1100 m.y.). The entire Blue Ridge complex terrane was subjected to granulite grade metamorphism approximately 920 m.y. ago. Later igneous activity is recorded at ~700 m.y. as the Rockfish River granodiorite and Catoctin greenstones.

INTRODUCTION

The Precambrian basement rocks of the Blue Ridge anticlinorium in central Virginia underwent a very complex history of sedimentation, plutonism and metamorphism. Although a significant volume of literature exists on basement rocks, enough data have been accumulated only recently to propose models relating field data to geochemistry and geochronology. This paper describes the geology of a transect across the Blue Ridge (Fig. 1) and presents chemical and isotopic data used to model the history of the area.

This east-west transect covers portions of 12, 7.5′ quadrangles, across the Pedlar and Lovingston massifs in their classic areas originally described by Bloomer and Werner (1955) (who had divided basement rocks only into three major formations—Lovingston, Marshall and Pedlar), and from which the massif names were taken by Bartholomew and others (1981). Two of the quadrangles (Sherando and Greenfield) were published ear-

lier (Bartholomew, 1977). The others have been mapped mostly by Sinha and his students.

Recently, Bartholomew and others (1981) proposed a model for the evolution of the basement rocks in central Virginia, in which two basement massifs, consisting largely of crystalline Grenville-age rocks, were juxtaposed during middle Paleozoic thrusting. The mineral assemblages of the Lovingston Massif (eastern block) and the Pedlar Massif (western block) are dominated by Grenville-age metamorphism of the shallower (lower T and P) and deeper (high T and P) granulite facies, respectively. During Paleozoic orogenesis the Lovingston Massif was thrust westward over the Pedlar Massif along the Rockfish Valley Fault. The intensity of deformation related to thrusting is best developed in the Lovingston block where intersection of Grenville and Paleozoic s-surfaces produced typical abundant augen. The same deformation in the western part of the Pedlar Massif is reflected in

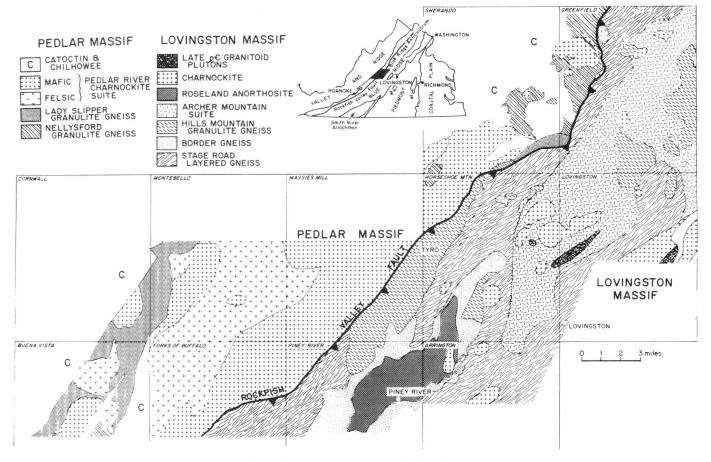

Figure 1. Generalized geologic map along the Blue Ridge transect.

development of conjugate fractures whose bisectrix defines the Paleozoic schistosity.

The following discussion on areal geology will cover each of these massifs and the fault zone separately. Detailed locations and descriptions of the type outcrops of herein-named units are included in appendix A.

PEDLAR MASSIF

The rocks in this massif were mapped as hypersthene granodiorite (Jonas, 1935) and Pedlar Formation (Bloomer and Werner, 1955). The two most important rock types in this area are herein named the Lady Slipper Granulite Gneiss and the Pedlar River Charnockite Suite. Farther north in the Sherando quadrangle, a third important rock type, the Nellysford Granulite Gneiss, is found (Bartholomew and others, 1981). The Lady Slipper Granulite Gneiss (Fig. 2A) is restricted in occurrence to the western portions of the Buena Vista and Cornwall quadrangles. Lithologically, it is a light gray, medium-grained, layered to massive gneiss with quartz, plagioclase, hypersthene, garnet, biotite, minor K-spar, chlorite, apatite and magnetite-ilmenite. A red staining of the rock results from hematite which occurs along

microfractures. The Nellysford Granulite Gneiss is a dark green, medium-grained gneiss with well developed mm or cm thick segregation layers and a granoblastic texture. The gneiss characteristically contains orthopyroxene, perthitic feldspar, plagioclase and quartz with or without garnet. This gneiss typically represents a more intermediate composition than the associated Lady Slipper Gneiss.

The Pedlar River Charnockite Suite (Fig. 2B) includes a number of mappable rock bodies containing both coarse-grained massive and foliated charnockites. Typically they show granoblastic and relict hypidiomorphic-granular textures and contain quartz, perthite, plagioclase, hypersthene, ilmenite with minor hornblende, apatite, biotite and chlorite. Along the eastern margin of the Pedlar River Charnockite Suite, near the mid-Paleozoic Rockfish Valley fault, rocks show evidence of ductile deformation as indicated by reduction in grain-size and development of a weak to strong mylonitic fabric and lineation. Locally, the rocks have undergone sufficient deformation to have developed fabrics ranging from protomylonite to ultramylonite. Typically the mineral assemblage of clinozoisite, quartz, biotite, minor muscovite and sphene show the prograde tectonic fabric, while in some areas abundant chlorite replacing biotite reflects retrogression

Figure 2. Rocks of the Pedlar Massif. A. Lady Slipper Granulite Gneiss at type locality along U.S. Highway 60. Grenville foliation is approximately vertical. B. Typical coarse-grained mesocharnockite exposed along Virginia state road 633 about 30 m west of the junction with Virginia state road 634 near Alto in the Montebello 7.5′ quadrangle.

after mylonitization. This mineralogy of the inhomogeneously tectonized zone suggests greenschist facies P, T conditions during mylonitization.

ROCKFISH VALLEY FAULT ZONE

This zone of Paleozoic ductile deformation ranges in width from one to 10 km and separates the Pedlar and Lovingston Massifs. Although a detailed analysis of this "within Blue Ridge" fault system is beyond the scope of this paper, some of its characteristics are pertinent. The most important feature of this deformation zone is the nonuniformity in the development of mylonitic fabric. Both field and petrographic observations show inhomogeneous distribution of strain resulting in thin blasto-mylonite and ultramylonite zones within less deformed proto-mylonitic charnockitic gneiss. The intensely sheared zones are characterized by a mineral assemblage of clinozoisite, biotite, quartz, muscovite, while the less deformed gneiss contains highly fractured feldspars indicative of brittle cataclasis. Late replacement of biotite by chlorite suggests syn- to post-tectonic retrogression.

LOVINGSTON MASSIF

As this terrane is the more complex part of the traverse, the numerous lithologic units are described individually: (1) the herein-named Stage Road Layered Gneiss (Fig. 3C and D), which includes most augen gneisses and schists of the old Lovingston Formation of Jonas (1935) and Bloomer and Werner (1955); (2) the Border Gneiss (Fig. 3A and B) of Hillhouse

(1960); (3) the Roseland Anorthosite complex; (4) Hills Mountain Granulite Gneiss of Bartholomew and others (1981); (5) the Archer Mountain Suite (Fig. 4A and B), formerly Archer Mountain Pluton of Bartholomew and others (1981), Lovingston Formation of Bartholomew (1977) and Lovingston granite gneiss of Davis (1974); (6) pegmatite gneiss (Fig. 4C and D) of Davis (1974); and (7) the Rockfish River Pluton (Fig. 5) of Davis (1974).

1. Stage Road Layered Gneiss

This thickly layered unit includes coarse-grained augen gneisses, fine-grained biotite gneisses, medium-grained biotite gneisses with garnets and layered schists and gneisses. The augen gneiss, the most common lithology, consists of porphyroblasts and/or porphyroclasts of potassium feldspar, blue quartz, albitic plagioclase and biotite. Minor minerals are zircon, apatite, ilmenite, muscovite, chlorite, epidote, zoisite, garnet, and calcite. The distribution of size and abundance of augen is controlled by the degree of mylonitization; schistose bands locally lack augen. The fine- and medium-grained biotite gneisses are generally interlayered with the coarser augen gneisses and mineralogically are very similar to the augen gneisses. The layered gneiss typically consists of perthite, oligoclase, hypersthene, quartz, ilmenite and magnetite. Other metamorphic minerals include biotite, clinozoisite and hornblende.

2. Border Gneiss of Hillhouse (1960)

This gneiss comprises a group of rocks that separates the

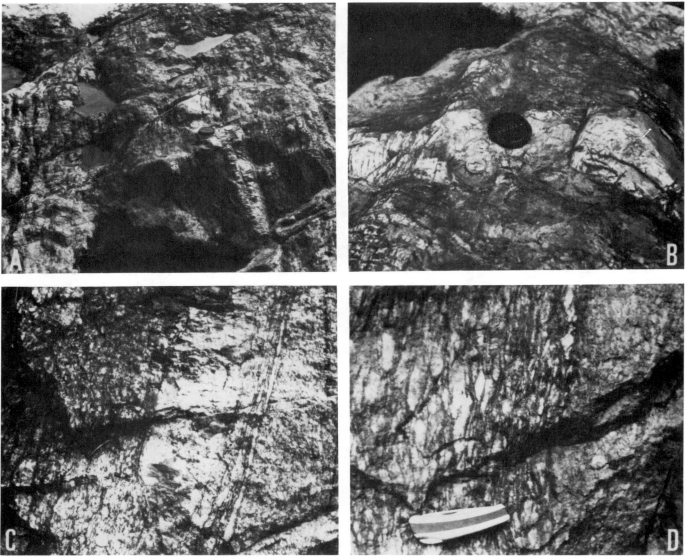

Figure 3. Layered gneisses of the Lovingston Massif. A. Border Gneiss exposed at type locality in the bed of the Tye River. B. Closeup of Border Gneiss which consists of lenses of leucogranulite gneiss within graphitic garnet-pyroxene-bearing mesogranulite gneiss. C. Stage Road Layered Gneiss at type locality along U.S. Highway 29. Grenville layering defined by lenses of augen gneiss and leucogneiss within biotite gneiss. Quarter in lower left corner for scale. D. Closeup of Stage Road Layered Gneiss. Detrital pre-Grenville zircons obtained from augen gneisses (left) here in contact with biotite gneiss (right). Knife is approximately 6 cm long.

Roseland Anorthosite from the other units (Fig. 1). Individual lithologies are difficult to map, hence the rocks are grouped in the manner proposed by Hillhouse (1960) as the Border Gneiss. The Border Gneiss is herein raised to formational rank. Lithologically the most important rocks are feldspathic and garnet graphite gneisses. Minor lenses of mafic rocks (amphibolites) and meta-anorthosite also are present. The garnet-graphite gneiss is best exposed along the Tye River and consists of andesine, hypersthene, quartz, garnet, graphite, and K-feldspar. The feldspathic gneiss generally consists of microperthite and blue quartz with minor biotite, epidote, apatite and chlorite.

3. Roseland Anorthosite

This rock unit has been mapped extensively in association with efforts to mine the nearby nelsonite (for titanium) and the anorthosite itself (for feldspar aggregate). The rock is typically coarse grained (crystals 30 cm), although locally in sills it becomes medium grained (2-5 cm). The primary feldspar is antiperthite with intergrown microcline (An_{30} with near 25 mole percent microcline). Other minerals include quartz and orthopyroxene, with the more common metamorphic minerals being tremolite and muscovite. Associated with the anorthosite and

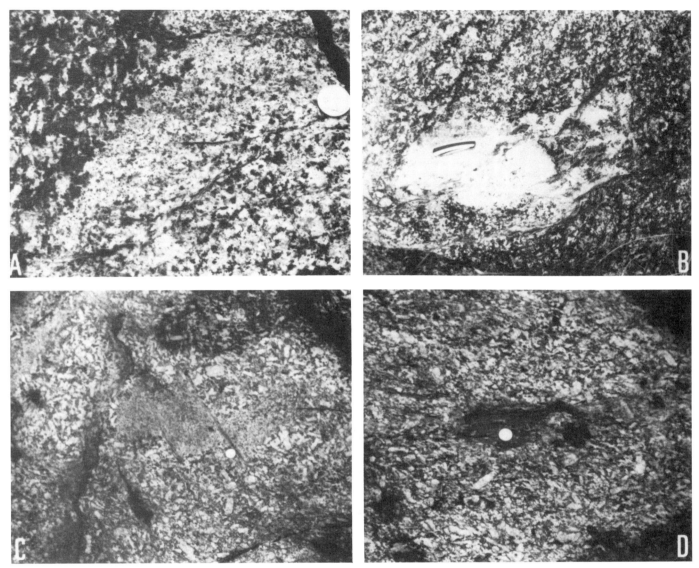

Figure 4. Intrusive rocks of the Archer Mountain Suite in the Lovingston Massif. A. Typical coarse-grained biotite quartz monzonite phase in contact with typical coarse-grained, biotite-bearing granite phase at type locality of Archer Mountain Suite along U.S. Highway 29. The contact between the phases is marked by a 2-4 cm wide zone of fine-grained, biotite-bearing granitoid. B. Typical felsic-gneiss xenolith within coarse-grained, biotite quartz monzonite at Woods Mill. Knife is approximately 6 cm long. C. Typical felsic-gneiss xenolith within pegmatitic granitoid phase of Archer Mountain Suite at representative locality along Virginia state road 617 about 30 m north of the junction with Virginia state road 623 (Stage Road), thence about 30 m up small drainage north of road within the Lovingston 7.5′ quadrangle. Nickel near center denotes scale. D. Typical mafic-gneiss xenolith within pegmatitic granitoid phase at same locality as Fig. 4C. Euhedral zircons obtained from outcrops along road at this locality. Nickel in xenolith denotes scale.

Border Gneiss, especially in margins of the anorthosite, are dikes/sills of norite and nelsonite (Hillhouse, 1960; Herz, 1969).

4. Hills Mountain Granulite Gneiss

According to Bartholomew and others (1981) and Bartholomew (1977), the Hills Mountain Granulite Gneiss is a medium gray, medium-grained rock with a poorly developed segregation layering. These streaky mafic lenses containing orthopyroxene are typically less than 1 cm in width. Feldspar porphyroblasts (1-2 cm in length) are commonly found in the more leucocratic material that forms the bulk of the rock. In addition to orthopyroxene, perthitic feldspar, plagioclase and quartz, with or without amphibole, constitute the major minerals.

Figure 5. Mafic-gneiss xenolith in granite of Rockfish River Pluton at type locality along Virginia state road 617. Euhedral zircons with cores obtained from outcrops at this locality. Penny in lower left part of xenolith denotes scale.

5. Archer Mountain Suite

This plutonic suite covers large portions of the Lovingston, Horseshoe Mountain and Greenfield 7.5′ quadrangles and is comprised primarily of biotite-quartz-monzonite intruded by small massive charnockite plutons (Fig. 1). The western margin of the Archer Mountain Suite shows a marked mylonitic fabric and was interpreted by Bartholomew (1977) to represent Paleozoic deformation. Mineralogically, the biotite-quartz-monzonite facies of the Archer Mountain Suite contains perthite, plagioclase, microcline and biotite. Paleozoic metamorphic mineral assemblages are characterized by biotite, sphene and clinozoisite. The massive charnockite stocks within the suite are characterized by an igneous assemblage of perthitic feldspar porphyroblasts, microcline, plagioclase and orthpyropyroxene. Paleozoic retrogression produced actinolite and chlorite. Also present in this suite are very coarse-grained pegmatites (andesine anorthosite of Watson and Cline, 1916; Lovingston pegmatite of Davis, 1974) made up primarily of andesine antiperthite, perthitic microcline, blue quartz and secondary muscovite. This rock is very similar to some phases of the Roseland Anorthosite described earlier.

6. Rockfish River Pluton

This granodiorite pluton intrudes pegmatite and quartz monzonite gneiss of the Archer Mountain Suite and is typical exposure is along Virginia state road 617. The rock is coarse grained with a weak foliation and mineralogically is characterized by andesine, perthitic microcline, biotite and quartz with minor sphene, allanite and fluorite. The Paleozoic metamorphic overprint is characterized by biotite, chlorite, epidote, sphene and albite.

GEOCHEMISTRY

Sixty-eight new chemical analyses were done using the technique of Sinha and Merz (1977) on selected samples across the traverse and averages are given in Table 1 and shown as Harker variation diagrams in Figs. 6 and 7.

Pedlar Massif

The Lady Slipper Granulite Gneiss shows an apparent bimodal SiO_2 content (66% and 76%) with a gap between 67 and 75 percent SiO_2. This may be either a result of sampling bias or represent a valid observation. However, the layered nature of the gneiss as well as calc-alkaline nature coupled with the bimodality would suggest a volcanic or volcaniclastic origin for this unit. Al_2O_3/SiO_2 ratios usually are below 0.3, suggesting an origin from a low Al magma. K_2O/Na_2O ratios usually range from 1.4 to 2.8, with the higher ratios common in the more silicic samples,

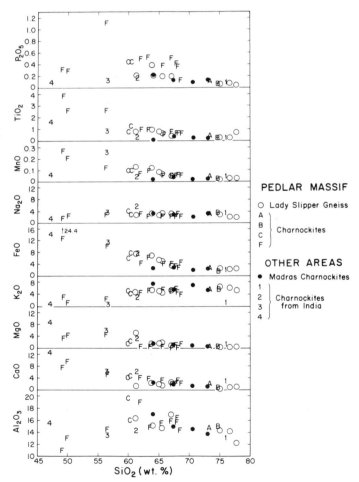

Figure 6. Harker variation diagram for all samples analyzed from the Pedlar Massif. The Madras Charnockites from Table 2 of Weaver and others (1970) and types 1, 2, 3 and 4 are from Table 1 of Iyer and Kutty (1978). The data are shown for reference only.

TABLE 1. AVERAGE MAJOR ELEMENT COMPOSITION OF THE MOST COMMON LITHOLOGIES
DISCUSSED IN THE TEXT

SiO_2	A	B	C	D	E	F	G	H	I	J
SiO_2	65.44	72.05	53.81	72.08	58.87	63.08	71.06	65.36	75.42	62.44
Al_2O_3	15.53	15.30	13.54	14.01	15.59	15.12	15.09	14.92	13.27	15.81
CaO	2.93	0.61	10.52	1.00	4.64	3.35	1.68	2.29	0.76	4.27
MgO	1.16	0.50	8.11	2.01	2.34	1.66	0.31	1.37	0.41	1.40
K_2O	4.72	4.88	1.30	3.43	3.51	4.50	5.59	5.07	5.80	4.40
FeO	5.12	2.34	9.54	6.09	8.45	6.50	3.11	5.37	2.48	6.64
Na_2O	3.24	4.78	2.55	1.60	3.24	2.86	3.29	3.10	2.37	2.86
MnO	0.08	0.05	0.11	0.08	0.13	0.09	0.05	0.08	0.03	0.10
TiO_2	0.86	0.13	1.16	0.57	1.18	1.31	0.45	0.70	0.21	1.28
P_2O_5	0.41	0.04	0.25	0.08	1.14	0.61	0.15	0.29	0.60	0.60
Total	99.49	100.68	100.89	100.88	99.09	99.08	100.78	98.54	100.81	99.80
Samples averaged=	5	9	3	1	5	13	2	4	3	23

Note: A = Archer Mountain Suite; B = Rockfish River Pluton; C = Roseland Anorthosite; D = graphite granulite of the Border Gneiss; E = Broder Gneiss; F = augen gneiss phase of Stage Road Layered Gneiss; G = Hills Mountain Granulite Gneiss; H = mafic phase of the Lady Slipper Granulite Gneiss; I = felsic phase of the Lady Slipper Granulite Gneiss; J = Pedlar River Charnockite Suite.
See Appendix B for analytical methods.

where potassium feldspar predominates over plagioclase in the mode. Sericitization of the plagioclase suggests addition of K_2O.

The Pedlar River Charnockite Suite shows a large range in SiO_2 (49 to 70%). Although our preliminary mapping indicates that the suite is made up of discrete bodies (shown in Fig. 1 but not labelled as such), chemically they appear to be very similar. Data as shown in the Harker variation diagram show smooth curves, indicative of fractional crystallization. The rocks of the Pedlar River Charnockite Suite have low Al_2O_3/SiO_2 ratio (<0.3), although they are consistently higher in CaO and P_2O_5 than samples from the Lady Slipper Granulite Gneiss. The K_2O/Na_2O + CaO ratios are less than one (average of 0.6) and are significantly lower than the averages of 0.9 for the mafic, and 1.9 for the felsic, Lady Slipper Granulite Gneiss. The rather limited variation in FeO/MnO ratios of 60 to 65 for a given plutonic suite argues strongly for the magmatic trend. This observation is similar to one documented by Singhinolfi (1971) for granulites from Brazil. The relatively high P_2O_5 content (0.5%) reflects the large modal volume percent of apatite (1%) in the intermediate charnockites (60-66% SiO_2) and probably reflects the cumulate nature of apatite. While all the other major elements show predictable variations with SiO_2, potassium shows no significant variation (64-69% SiO_2) and may have been mobilized during granulite facies metamorphism.

Lovingston Massif

The more intense Paleozoic deformational overprint in this block makes it more difficult to map lithologies over large distances. As suggested by Bartholomew (1977) and in this report, some gneisses are clearly related to igneous bodies, while others probably have a sedimentary protolith. As such, the geochemical data are of necessity generalized and discussions are limited to major lithologic units. The Stage Road Layered Gneiss and the Border Gneiss show a large range in SiO_2 (55% to 68%), with a maxima in Al_2O_3 (16.5% at 61% SiO_2). A high SiO_2 (72.8%) sample is the interlayered garnet-graphite gneiss. The geochemical trends are similar to those observed for the charnockites of the Pedlar Massif, but gross layering of lithologies and the presence of rounded detrital zircons suggest sedimentary or sedimentary/volcanic precursors for these lithologies (Sinha and Bartholomew, 1982). The apparent igneous trends of smoothly decreasing CaO, MgO, FeO and TiO_2 with increasing SiO_2 perhaps reflect the volcanic nature of assemblages of original volcanic compositions. The obvious plutonic rocks include the Archer Mountain Suite, Roseland Anorthosite and the younger Rockfish River Pluton. The Archer Mountain Suite, which includes quartz monzonites, granodiorites and granites, shows a relatively narrow range in Al_2O_3 (14 to 16%) and curved fractionation trends for CaO,

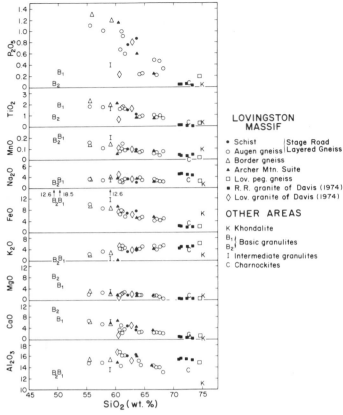

Figure 7. Harker variation diagram for all samples analyzed from the Lovingston Massif. Points labelled K, B₁, B₂, I, and C are from Table 1 of Weaver and others (1978).

MgO, FeO and TiO_2. The scatter in K_2O is probably due to variations in the ratio of volumes of retrograde to relict igneous mineralogies. The smaller igneous body (the Lovingston quartz monzonite gneiss mapped by Davis, 1974) is very similar to the bulk of the Archer Mountain Suite and is interpreted as a stock of the main mass. Adjacent to the quartz monzonite (along strike) is the coarse-grained porphyritic pegmatite gneiss which is characterized by both a high SiO_2 (74%) and high K_2O (6%) content. Except for Al_2O_3, K_2O and FeO, it is very similar to khondalites from Madras, India (Weaver and others, 1978). The pegmatite gneiss is unlikely to be a metasediment, inasmuch as intrusive contacts (it locally intrudes the quartz monzonite), xenoliths (Fig. 4C and D) and alignment of feldspar crystals are common. Although a limited amount of chemical data is available for the Roseland Anorthosite, its close association with charnockites (in the margins of the anorthosites and in the adjacent Border Gneiss as well as in the Pedlar Massif) suggests a possible genetic relationship. The charnockite data shown earlier (Figs. 6 and 7) define fairly smooth fractionation trends on which can be superimposed the noritic anorthosite and norite analyses from Herz (1969). The andesine anorthosites with only plagioclase do not plot on the charnockite curves for Al_2O_3 and P_2O_5, indicating the cumulate monomineralic compositions. The exact genetic

relationship for the charnockites and the anorthosite, and indeed, the relationship of charnockites of the Pedlar and Lovingston massifs, are as yet unclear, but Nd-Sm isotopic studies underway should help in the resolution.

The Rockfish River Pluton is a high SiO_2 (71-73%), high total alkali ($Na_2O + K_2O \simeq 10\%$) granitoid body that probably is derived from the granulites in a manner similar to the origin of the Pike's Peak Batholith (Barker and others, 1975). Details of this are to be published elsewhere.

GEOCHRONOLOGY

Although a number of scattered radiometric analyses are available for these rocks of Grenville age (Tilton and others, 1960), we present additional U/Pb ages of zircons across our traverse. With varying lithologies and metamorphic grades even zircon ages do not clearly identify igneous events.

All analyses plotted on Fig. 8, with the exception of one (Tilton and others, 1960), were done at Virginia Polytechnic Institute and State University. Analytical techniques are described in appendix B.

Pedlar Massif

Samples from the Lady Slipper Gneiss and the Pedlar River Suite give concordia intercept ages (assuming recent lead loss) of 1130 m.y. and 1075 m.y., respectively (Fig. 8). The 55 m.y. time difference is greater than analytical uncertainty and is considered to reflect a real difference in age. This difference agrees with the

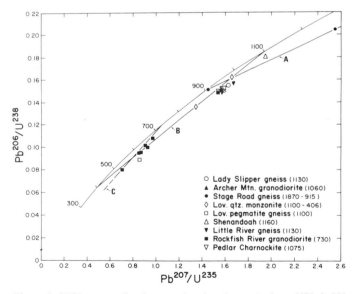

Figure 8. U-Pb concordia diagram showing data only from VPI & SU laboratory with the exception of sample labelled Shenandoah, which is from Tilton and others (1960). The concordia intercept ages are given in brackets. Line A is discordia from 915 m.y. to 1860 m.y.; line B is discordia from 406 m.y. to 1100 m.y. for the Lovingston quartz monzonite and line C is discordia from 0 m.y. to 730 m.y. for Rockfish River granodiorite.

TABLE 2. U-Pb DATA ON ZIRCONS FOR SELECTED ROCK TYPES

	Pb	U	$\frac{Pb^{206}}{Pb^{204}}$	$\frac{Pb^{207}}{Pb^{206}}$	$\frac{Pb^{206}}{Pb^{208}}$	$\frac{Pb^{207*}}{Pb^{206*}}$	$\frac{Pb^{206*}}{U^{238}}$	$\frac{Pb^{207*}}{U^{235}}$
Lady Slipper Granulite Gneiss								
100-200	75.0	488	5977	0.07670	13.42	0.07467 (1060)	0.1505 (904)	1.592 (967)
200-325	84.2	539	9495	0.07614	14.80	0.07464 (1059)	0.1546 (927)	1.629 (928)
Pedlar River Charnockite								
100-200	43.6	275	8677	0.07527	10.02	0.07362 (1031)	0.1582 (916)	1.588 (965)
200-325	45.4	287	9522	0.07498	9.87	0.07348 (1027)	0.1521 (913)	1.579 (962)
Archer Mountain Granodiorite								
100-200	23.2	142	4117	0.07700	7.59	0.07339 (1024)	0.1521 (913)	1.621 (962)
200-325	23.1	142	1650	0.08169	7.03	0.07305 (1015)	0.1492 (896)	1.539 (949)

Note: Additional data given in Davis (1974).
*Lead and uranium concentrations are in ppm; non-magnetic fraction samples split by mesh size; ages given in parentheses below calculated ratios.

field relationships which indicate that the charnockites are intrusive into the older volcanic/volcaniclastic sequence. Additional analyses from other phases of the Lady Slipper Gneiss and the Charnockite Suite may provide ranges in ages which will allow us to understand the full thermal history.

Lovingston Massif

In accordance with our field and petrographic observations, the augen gneiss lithology of the Stage Road Layered Gneiss is a metasedimentary unit containing two zircon populations, an older, rounded detrital fraction and a metamorphically produced euhedral fraction. The rounded grains give a Pb^{207}/Pb^{206} age of 1420 m.y., and the more euhedral grains give an essentially concordant age of 915 m.y. Although details of alternative models are given in Davis (1974), our preferred model is similar to Davis's in interpreting the augen-bearing gneisses as a clastic sequence that was metamorphosed to granulite facies at 915 m.y. The discordia intercept at 1870 ± 200 m.y. (Fig. 8) probably reflects the age of the protolith that must have provided clastic components to the Stage Road Layered Gneiss. The ages from the rocks of the Archer Mountain Suite may not reflect the effects of the 915 m.y. event because insufficient time elapsed between crystallization and metamorphism to generate significant amounts of radiogenic lead. This suggestion also is supported by the low uranium concentration (Table 2). However, the effects of later (Paleozoic) retrogression is evident for the quartz monzonite data—a discordia gives ages of 1100 and 406 m.y. The younger age of 406 m.y. (with a large error of ±35 m.y.) is similar to Rb/Sr ages of 350 m.y. (Dietrich and others, 1969), and these

ages are similar to those measured on the Brevard fault zone (Sinha and Glover, 1978; Odom and Fullagar, 1973; Bond and Fullagar, 1973). It is tempting to suggest that these ages mark movement along the Brevard fault zone in the southern Appalachian correlative with the Rockfish Valley fault zone in the central Appalachians.

The Rockfish River Pluton, which intrudes the older Grenville terrane, was interpreted to be 820 ± 20 m.y. old (Davis, 1974). Reevaluation of the morphology of the zircons (as stated by Davis, 10% have cores) suggests that the 820 m.y. obtained by Davis is the maximum age; by using the lower Pb^{207}/Pb^{206} ratios, an age of 730 m.y. is indicated. This age is similar to ages reported for the Robertson River Pluton by both Lukert and Clarke (1981), and Lukert and Banks (1984), and is similar to the Pb^{207}/Pb^{206} age of 700 m.y. for the Catoctin rhyolite (Rankin and others, 1969).

SUMMARY

Reconnaissance mapping, minimal numbers of chemical analyses and limited geochronology restrict us from making very elaborate petrologic evolutionary models. However, the gross distribution of lithologies and ages (Fig. 9, Table 3) suggest that this corridor essentially can be characterized as a cross section of the lower crust. Granulite grade para- and ortho-gneisses are spatially and temporally associated with charnockites, suggesting a very dry environment most likely to occur in the lower crust. Work underway on oxygen, Nd, Pb and Sr isotopes coupled with geochemical studies will provide more information about the genesis of these rocks.

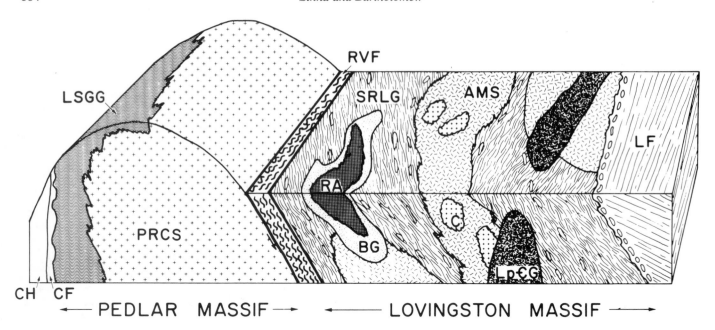

Figure 9. Schematic cross section of the terrane. CH = Chilhowee Group; CF = Catoctin Formation; LSGG = Lady Slipper Granulite Gneiss; PRCS = Pedlar River Charnockite Suite: RVF = Rockfish Valley Fault; SRLG = Stage Road Layered Gneiss; RA = Roseland Anorthosite; BG = Border Gneiss; AMS = Archer Mountain Suite; C = charnockites; LPCG = Late Precambrian granites; LF = Lynchburg Formation.

TABLE 3. SCHEMATIC SUMMARY OF EVENTS DETERMINED FROM ZIRCON U-Pb AGES AND RECONNAISSANCE STRUCTURAL DATA

Age (m.y.)	Pedlar Massif	Rockfish Valley Fault	Lovingston Massif
200-300	Brittle deformation associated with thrusting of Blue Ridge over Valley and Ridge		
360-430	Conjugate fractures, bisectrix of which becomes s-surface in cover rocks	Metamorphism and thrusting, s_2 dominant	Augens produced by intersection of Grenville (s_1) and Paleozoic (s_2) schistosities. Local mylonite zones reflect dominant s_2
700-730	Catoctin volcanic rocks extruded, 700 m.y.*		Intrusion of Rockfish River pluton, only s_2 found, 730 m.y.
	Erosion		Erosion
915-940	Peak of granulite grade metamorphism as recorded by metamorphic zircons. Intense deformation is recorded as s_1 schistosity.		
1060-1100	Emplacement of Pedlar River Charnockite Suite		Emplacement of Archer Mountain Suite
1130	Deposition of volcaniclastic protolith for the Lady Slipper Granulite Gneiss		Deposition of protolith or Stage Road Gneiss, source terrane contains 1870 m.y. old zircons

*Pb^{207}/Pb^{206} age obtained on the Catoctin volcanic rocks by Rankin and others (1969).

ACKNOWLEDGEMENTS

The senior author acknowledges the support of NSF grant EAR 323040-1 for support of this project and the operation of the geochemistry laboratory. Field effort of John Reynolds and Robert Davis is appreciated. Hal Pendrak, Don Bodell, Nhury Schurig, Sharon Chiang and Jean Perdue provided technical support. The authors wish to acknowledge the help of Dr. Anita Andrew for stimulating discussions.

APPENDIX A. LOCATION OF TYPE SECTIONS FOR VARIOUS LITHOLOGIES DISCUSSED IN THE TEXT.

1. Lady Slipper Granulite Gneiss: Buena Vista 7.5′ quadrangle, 37°44′30″N, 79°18′W; approximately 400 m east on U.S. highway 60 past junction with Blue Ridge Parkway. This outcrop was sampled for zircon U-Pb age.

2. Pedlar River Charnockite Suite (used informally as Pedlar River Pluton in Bartholomew and others, 1981): Montebello 7.5′ quadrangle, 37°45′30″N, 79°11′30″W; 400 m directly S10°W of Hog Camp Gap where Appalachian Trail reaches crest of Cole Mountain (elevation = 3940′). This outcrop sampled for zircon U-Pb age.

3. Nellysford Granulite Gneiss: see Bartholomew and others (1981).

4. Stage Road Layered Gneiss: Lovingston 7.5′ quadrangle, 37°51′N, 78°48′W; on U.S. highway 29, 50 m west of intersection with Virginia state road 617.

5. Border Gneiss: Horseshoe Mountain 7.5′ quadrangle, 37°45′30″N, 78°59′30″W; on Tye River, 650 m north of junction with Virginia highways 151 and 56.

6. Roseland Anorthosite: Piney River 7.5′ quadrangle, 37°43′N, 79°1′30″W, approximately 1300 m west of Virginia highway 56, near Piney River but north of Virginia Blue Ridge Railroad.

7. Hills Mountain Gneiss: see Bartholomew and others (1981).

8. Archer Mountain Plutonic Suite: (a) Lovingston 7.5′ quadrangle, 37°51′N, 78°49′30″W, 100 m south, along U.S. highway 29, from junction with Virginia highway 6 north where it crosses Rockfish River at Woods Mill. (b) Lovingston 7.5′ quadrangle, 37°50′15″N, 78°47′15″W, intersection of Virginia highways 623 and 617.

9. Rockfish River Pluton: Lovingston 7.5′ quadrangle, 37°49′N, 78°46′15″W, on Virginia state road 617, 1 km northwest from junction with Virginia state road 620. This location was used for zircon U-Pb age by Davis (1974).

APPENDIX B

1. Analytical procedures for U-Pb analyses of zircons

Zircons were separated from 5-10 kg size samples using standard techniques of the Wilfley table, heavy liquids and magnetic separator. All analyzed zircons were nonmagnetic (1 to 1.6 amp) using horizontal feed and 5° tilt for the Frantz separator. The samples were sized, washed in warm 8N HNO_3 acid and after rinsing in water, the final rinse was made using distilled acetone. Samples (2-8 mg) were dissolved in Krogh bombs, although separate splits were used for isotopic composition and U, Pb concentrations. Repeat analyses (three) of a zircon population gave variations in uranium and lead concentrations of 0.58 percent and 0.5 percent, respectively. Blank corrections of 600 pg (range from 300-900 pg) for lead were applied to all measured data. This correction does not remove all common lead, suggesting that some original component is present. Mass spectrometry was done on a 35 cm, 60° sector mass spectrometer at VPI & SU. Replicate analyses of NBS SRM 983, CIT and NBS SRM U500 gave the following results:

	Pb^{206}/Pb^{204}	Pb^{207}/Pb^{206}	Pb^{206}/Pb^{208}	U^{238}/U^{235}
SRM 983	2751	.071213	.013622	
CIT	16.582	.92954	.45932	
SRM U500				1.001

All analyses were normalized to absolute values using appropriate mass fractionation correction factors.

2. Analytical procedures for Rapid Rock chemical analyses.

In an effort to increase regional coverage by analyzing large numbers of samples, two different methods are used at the VPI & SU Geochemistry Laboratory. For high accuracy and precision the fused pellet technique of Norrish and Chappel (1977) is used; for reconnaissance geochemistry pressed powder pellets described by Sinha and Merz (1977) is used. Usually, the analytical error is within two percent of the fused pellet technique. All analyses reported in this paper were done by the pressed powder technique.

REFERENCES CITED

Barker, F., Wones, D. R., Sharp, W. N., and Desborough, G. A., 1975, The Pikes Peak batholith, Colorado Front Range, and a model for the origin of the gabbro-anorthosite-syenite-potassic granite suite: Precambrian Research, v. 2, p. 97–160.

Bartholomew, M. J., 1977, Geology of the Greenfield and Sherando quadrangles, Virginia: Virginia Division of Mineral Resources Publication 4, 43 p.

Bartholomew, M. J., Gathright, T. M., Henika, W. S., 1981, A tectonic model for the Blue Ridge in central Virginia: American Journal of Science, v. 281, no. 11, p. 1164–1183.

Bloomer, R. O., and Werner, H. J., 1955, Geology of the Blue Ridge in central Virginia: Geological Society of America Bulletin, v. 66, p. 579–606.

Bond, P. A. and Fullagar, P. D., 1973, Origin and age of the Henderson augen gneiss and associated cataclastic rocks in southwestern North Carolina: Geological Society of America, Abstracts with Programs, v. 6, p. 336.

Davis, R. G., 1974, Pre-Grenville ages of basement rocks in central Virginia: a model for the interpretation of zircon age, [M.S. thesis]: Blacksburg, Virginia Polytechnic Institute and State University, 47 p.

Dietrich, R. V., Fullagar, P. D., and Bottino, M. L., 1969, K/Ar and Rb/Sr dating of tectonic events in the Appalachians of southwestern Virginia: Geological Society of America Bulletin, v. 80, p. 307–314.

Herz, N., 1969, The Roseland alkalic anorthosite massif, Virginia: New York State Museum and Science Memoir 18, p. 357–367.

Hillhouse, D. N., 1960, Geology of the Piney-River-Roseland Titanium area, Nelson and Amherst counties, Virginia, [Ph.D. dissertation]: Blacksburg, Virginia Polytechnic Institute and State University, 129 p.

Iyer, A.G.V. and Kutty, T.R.N., 1978, Geochemical comparison of Archaean granulites in India with Proterozoic granulites in Canada, *in* Windley, B. F. and Naqvi, S. M., *Archaean Geochemistry:* Elsevier Scientific Publishing Company, p. 269–288.

Jonas, A. I., 1935, Hypersthene granodiorite in Virginia: Geological Society of America Bulletin, v. 46, p. 47–60.

Lukert, M. T., and Banks, P. O., 1984, Geology and age of the Robertson River Pluton, *in* Bartholomew, M. J., and others, editors, *The Grenville Event in the Appalachians and Related Topics:* Geological Society of America Special Paper 194 (this volume).

Lukert, M. T. and Clarke, J. W., 1981, Age relationships in Proterozoic Z rocks of the Blue Ridge anticlinorium of northern Virginia: Geological Society of America Abstracts with Programs, v. 13, no. 1, p. 29.

Norrish, K. and Chappel, B. W., 1977, X-ray fluorescence spectrometry, *in* Zussman, J., *Physical methods in Determinative Mineralogy:* Academic Press, p. 201–273.

Odom, A. L. and Fullagar, P. D., 1973, Geochronologic and tectonic relationships between the Inner Piedmont, Brevard zone, and Blue Ridge belts, North Carolina: American Journal of Science, v. 273-A, p. 133–149.

Rankin, D. W., Stern, T. W., Reed, J. C., Jr., and Newell, M. F., 1969, Zircon ages of felsic volcanic rocks in the upper Precambrian of the Blue Ridge, Appalachian Mountains: Science, v. 166, p. 741–744.

Sighinolfi, G. P., 1971, Investigations into deep crustal levels: Fractionating effects and geochemical trends related to high-grade metamorphism: Geochimica et Cosmochimica Acta, v. 35, p. 1005–1021.

Sinha, A. K. and Bartholomew, M. J., 1982, Evolution of the Grenville terrane in central Virginia: Geological Society of America Abstracts with Programs, v. 14, nos. 1 & 2, p. 82.

Sinha, A. K. and Merz, B. A., 1977, Analytical procedures for XRF analyses: *in* Evaluation and Targeting of Geothermal Energy Resources in the Southeastern United States, Progress Report VPI&SU-5103-5, p. B-11-B-16.

Sinha, A. K. and Glover, L., III, 1978, U/Pb systematics of zircons during dynamic metamorphism: Contributions to Mineralogy and Petrology, v. 66, p. 305–310.

Tilton, G. R., Wetherill, G. W., Davis, G. L., and Bass, M. N., 1960, 1000-million-year-old minerals from the eastern United States and Canada: Journal Geophysical Research, v. 65, p. 4173–4179.

Watson, T. L. and Cline, J. H., 1916, Hypersthene syenite and related rocks of the Blue Ridge region: Geological Society of America Bulletin, v. 27, p. 193–234.

Weaver, B. L., Tarney, J., Windley, B. F., Sugavanam, E. B., and Rao, V. V., 1978, Madras granulites: Geochemistry and P-T conditions of crystallization, *in* Windley, B. F. and Naqvi, S. M., *Archaean Geochemistry:* Elsevier Scientific Publishing Company, p. 177–203.

MANUSCRIPT ACCEPTED BY THE SOCIETY AUGUST 2, 1983

Geological Society of America
Special Paper 194
1984

Rock suites in Grenvillian terrane of the Roseland district, Virginia

Part 1. Lithologic relations

Norman Herz
U.S. Geological Survey
University of Georgia
Athens, Georgia 30602

Eric R. Force
U.S. Geological Survey
Reston, Virginia 22092

ABSTRACT

The Roseland district of Nelson and Amherst Counties, Virginia, is a typical Grenvillian terrane, analogous to similar terranes of eastern Canada. The oldest rocks in the Roseland district are layered granulites, quartz mangerites, and quartzo-feldspathic gneisses. Ages on zircon from the Shaeffer Hollow Granite, a leucocratic granite related to these oldest rock types, are discordant but apparently pre-Grenvillian. The Roseland Anorthosite intrudes these older rocks and consists of andesine antiperthite megacrysts and blue quartz in a finer grained oligoclase-K feldspar matrix. The Roseland Anorthosite is more alkalic and silicic than massif anorthosites elsewhere. Pyroxene megacrysts and rutile are found in its border areas.

After the emplacement of the anorthosite, the Roses Mill and Turkey Mountain ferrodiorite-charnockite plutons (of the Roses Mill Plutonic Suite) were intruded. Layered diorite and nelsonite, an ilmenite-apatite rock, are found near the bases of these plutons. These rocks may have formed, in part, by liquid immiscibility. Ages determined on the Roses Mill Pluton are about 970 m.y. The largest part of the plutons is altered to biotitic granitic augen gneiss. The Rockfish Valley deformation zone crosses the district from northeast to southwest. Northwest of the deformation zone are ferrodioritic charnockitic rocks of the Pedlar massif, which are slightly older than the Roses Mill Pluton. These ages may have been modified by metamorphism.

The Mobley Mountain Granite is a fine- to medium-grained, subsolvus two mica granite and has been dated at 650 m.y. Various mafic and ultramafic dikes, of different ages, are present throughout the district.

The major structure of the district is the Roseland dome, cored by the Roseland Anorthosite, and trending northeast for at least 22 km. Three periods of deformation are evident; one is possibly pre-Grenvillian and is seen only in the oldest rocks. The Roses Mill Plutonic Suite was deformed and underwent retrograde metamorphism to lower amphibolite assemblages in Proterozoic Z time. Paleozoic deformation was responsible for a reactivation of the Rockfish Valley deformation zone, which originated as a Precambrian feature and a selective overprinting of retrograde greenschist facies metamorphism on the granulite- to upper amphibolite-facies assemblages of the country rock.

The principal resources of the district are rutile and ilmenite. Both are present as hard rock and saprolite deposits. The rutile formed at the anorthosite border, and ilmenite is contained largely in nelsonite and mafic ferrodiorite bodies.

INTRODUCTION

Grenvillian terranes of eastern Canada are characterized by the association anorthosite-ferrodiorite-charnockite-granulite. The Roseland district of Nelson and Amherst Counties in central Virginia, best known for its titanium deposits and alkalic anorthosite, is the only terrane of the southern Appalachians known to have this association. The titanium deposits include rutile in the anorthosite border zone and ilmenite in nelsonite and ferrodioritic rocks. The anorthosite is unusually alkalic and has a combined $Na_2O + K_2O$ content of 8.4% (Table 1), about 1½ times the world average for massif anorthosites (Nockolds, 1954).

The Roseland district is in the border area of the Blue Ridge physiographic province to the northwest and the Inner Piedmont to the southeast (Fig. 1). The Virginia Blue Ridge Complex of Brown (1958) is basement for the region and includes igneous and high grade metasedimentary rocks of Proterozoic Y age. Brown's complex extends some 500 km from Pennsylvania to North Carolina and is considered to be an anticlinorium having different rock sequences on each limb (Conley, 1978, p. 123). The complex is bounded by rocks of Proterozoic Z and early Paleozoic age; to the southeast is the James River Synclinorium, and to the northwest, the complex is imbricately thrust over Paleozoic rocks.

TABLE 1. CHEMICAL COMPOSITIONS OF IGNEOUS ROCKS FROM THE ROSELAND DISTRICT, VIRGINIA

	1	2	3	4	5	6	7	8	9
SiO_2	70.59	74.4	60.4	54.5	60.71	54.9	51.9	3.6	69.21
TiO_2	0.47	0.04	0.15	0.52	2.00	2.4	1.5	30.0	0.37
Al_2O_3	14.13	14.7	23.6	25.7	14.05	15.6	16.4	3.0	14.63
Fe_2O_3	*	0.15	0.29	0.83	*	3.4	2.7	13.4	*
FeO	3.55	0.12	0.21	1.46	8.71	7.6	7.0	22.2	4.93
MgO	0.59	0.07	0.13	0.83	1.43	2.7	6.1	2.0	0.26
CaO	1.26	1.2	5.0	9.62	4.68	5.9	8.4	11.6	1.26
Na_2O	3.01	3.9	5.6	4.66	2.61	3.0	3.4	0.0	3.00
K_2O	5.39	4.2	2.8	1.06	3.32	2.4	1.3	0.68	5.38
MnO	0.04	0.00	0.02	0.02	0.14	0.14	0.18	0.33	0.08
P_2O_5	0.14	0.00	0.08	0.11	0.89	0.9	0.4	9.8	0.11
Sum	99.17	98.78	98.28	99.31	98.54	98.94	99.28	96.61	99.23

Note: Analytical methods: 1, X-ray fluorescence; 2,3,5,6,8,9, U.S. Geological Survey, Reston, Va., rapid chemical analysis methods.
Column
1. Leucocharnockite of layered granulite unit, average of 12 analyses, Ames (1981).
2. Shaeffer Hollow Granite.
3. Roseland Anorthosite, average of 4 analyses.
4. World average, massif anorthosite, Nockolds (1954).
5. Charnockite of Pedlar massif, average 4 analyses.
6. Charnockitic ferrodiorite of Roses Mill Plutonic Suite, average of 10 analyses.
7. World average, diorite, Nockolds (1954).
8. Nelsonite.
9. Mobley Mountain Granite, average of 2 analyses.
* All Fe included together calculated as FeO.

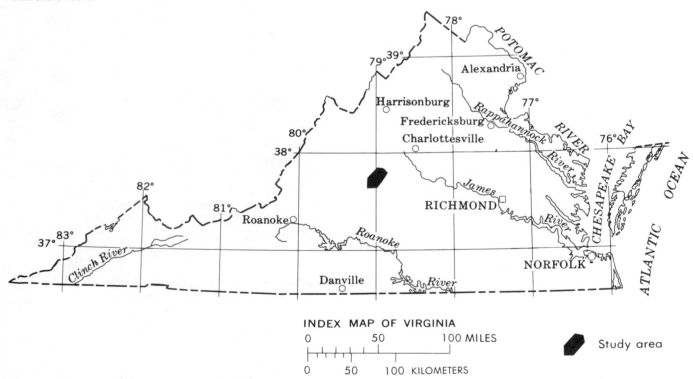

Figure 1. Index map of Virginia showing location of the study area, the Roseland district.

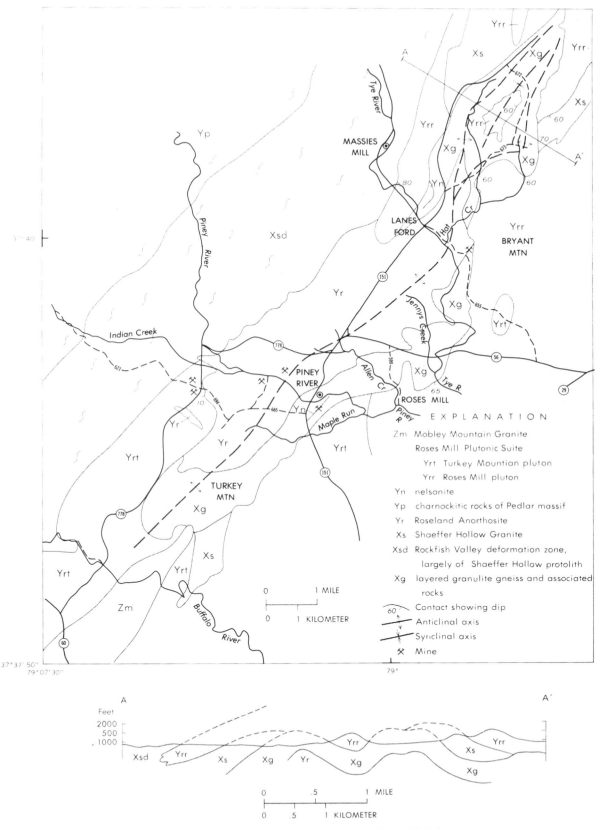

Figure 2. Geologic map of the Roseland district, Virginia.

The basic structure of the Roseland district appears to be an elongate, northeast-trending dome, cored by the anorthosite pluton and adjacent to a major shear zone to the northwest (Fig. 2). The rocks are largely Grenvillian and pre-Grenvillian in age, and most have been metamorphosed to the granulite facies. A pervasive later shear foliation correlative with amphibolite-facies minerals dips steeply southeast. Paleozoic metamorphism producing greenschist facies, presumably of Taconic age, is not pervasive and is related to regional thrusting and small-scale folding.

Previous Work

Watson and Taber (1913) described the principal rock types in the area, which included basement monzonite gneiss and schist, syenite (approximately contemporaneous and gradational with the basement, called andesine anorthosite in a footnote), gabbro, nelsonite, and Triassic and/or Jurassic diabase. Bloomer and Werner (1955) proposed a stratigraphy that became the standard for more than two decades for the Blue Ridge. It included, from oldest to youngest: (1) basement complex gneiss—elongate inclusions in younger rocks only; (2) "Lovingston formation" gneiss containing large K-feldspar porphyroblasts and granite; (3) "Marshall formation"—gradational upper and lower contacts, strongly layered quartzo-feldspathic bands without prominent augen; (4) "Pedlar formation"—largely hypersthene granodiorite; and (5) Roseland Anorthosite.

The most recent mapping by Bartholomew (1977) and our-

selves (Herz and Force, unpub. data) has resulted in major changes in concepts of both the basic stratigraphy and the mode of formation of the Grenvillian (Table 2). The area in the vicinity of Roseland previously shown as Marshall Formation on the geologic map of Virginia (1963) represents a major deformation zone, the Rockfish Valley zone, about 2 to 5 km wide (Fig. 2). "Marshall" lithologies are consistent with mylonitic rocks, and the gradational contacts previously described for the "Marshall" are due to varying intensity of deformation within the zone. The zone separates two distinctive terranes, which in the area immediately north of the Roseland district, have been called the Lovingston massif (east of the Rockfish Valley zone) and the Pedlar massif (west of the zone) (Bartholomew and others, 1981).

Proposed Lithologic Units

As shown in Table 2, lithologies observed in this study area are similar to those found by Bartholomew (1977). We prefer to use the terms "Lovingston" and "Pedlar" informally, as "massifs," following Bartholomew and others (1981).

The oldest rocks are layered granulite gneisses, which include quartzo-feldspathic granulite-gneiss and other varieties of gneiss. Garnet, graphite, blue quartz, and pyroxene are characteristic minerals. Leucocratic coarse-grained granite and charnockite, mangerite, and quartz mangerite intrude the granulites. Granulites are found largely on the southeast side of the anorthosite. In the northwestern part, granulite was caught up in the Rockfish Valley deformation zone and is not distinguishable as a

TABLE 2. LITHOLOGIC UNITS AND EVENTS, ROSELAND DISTRICT AND NEARBY AREAS, VIRGINIA

M.y. ago	Erathem	Rock units	Metamorphic and tectonic events; lithologies	Bartholomew (1977)
	Mesozoic	Diabase dikes		diabase
	Paleozoic		formation of mylonite; greenschist-facies metamorphism	
	Proterozoic Z		formation of mylonite	pyroxenite and mafic dikes
650		Mobley Mountain Granite; mafic and ultramafic dikes	amphibolite facies, augen gneiss formation	
	upper	Roses Mill Plutonic Suite	charnockite-ferrodiorite suite, minor cumulates, nelsonite	massive charnockite
1000	Proterozoic Y	Charnockitic rocks of Pedlar massif	granulite facies; partial anatexis	Lovingston Formation
		Roseland Anorthosite	alkalic andesine anorthosite	
	lower			
1800	Proterozoic X	Shaeffer Hollow Granite and some leucocharnockite	blue quartz-perthite granite	
	?	quartz mangerite and mangerite		
+1800		layered granulite gneiss		layered granulite gneiss

TABLE 3. ZIRCON ANALYTICAL DATA AND AGES FOR SAMPLES
FROM THE ROSELAND DISTRICT, VIRGINIA

	1	2	3	4
Pb, ppm	68.3	16.2	42.6	92.5
U, ppm	437.9	93.3	252.5	268.0
Th, ppm	174.0	38.1	120.0	74.0
$^{208}Pb/^{206}Pb$	0.14211	0.14192	0.10385	0.57623
$^{207}Pb/^{206}Pb$	0.07481	0.07801	0.07538	0.27120
$^{207}Pb/^{206}Pb$	0.00021	0.00046	0.00010	0.01192
$^{206}Pb/^{238}U$	0.148473	0.163660	0.166173	0.167665
age, m.y.	892	977	991	999
$^{207}Pb/^{235}U$	1.469967	1.611882	1.694359	2.525860
age, m.y.	918	975	1006	1279
$^{208}Pb/^{232}Th$	0.048633	0.048589	0.033883	0.087479
age, m.y.	960	959	674	1695
$^{207}Pb/^{206}Pb$	0.071810	0.071435	0.073955	0.109268
age, m.y.	980	970	1040	1787

Note: Analysts: T. W. Stern and M. F. Newell, U.S. Geological Survey, Reston, Virginia. Decay constants and isotopic abundances from Steiger and Jeager (1978).
1. Leucosome in layered granulite in the Tye River, just above Lanes Ford.
2. Ferrodiorite from the type locality of the Roses Mill Pluton in the Arrington quadrangle.
3. Massive charnockitic ferrodiorite of the Pedlar massif near Jack's Hill, Massies Mill quadrangle.
4. Shaeffer Hollow Granite exposed in a rock slide on the west slope of Byrant Mountain ridge, Horseshoe Mountain quadrangle.

mappable unit. U/Pb determinations on zircon from the Shaeffer Hollow Granite, a leucocratic granite associated with granulite, yield a discordant $^{207}Pb/^{206}Pb$ age of 1787 m.y. (Table 3, column 4). An anatectic leucosome of the granulite has been dated at 980 m.y. (slightly discordant) by U/Pb on zircon (Table 3, column 1), suggesting that some of its features are Grenvillian in age.

The Roseland Anorthosite intrudes granulite and mangerite. The anorthosite is a megacrystic andesine antiperthite rock containing abundant blue quartz, pyroxene, and, locally, rutile in its border areas. It composes the core of the dome that is the major structural feature of the district.

Some time after the emplacement of the anorthosite and the consolidation of this basement terrane, the Roses Mill and Turkey Mountain Plutons were intruded. Both plutons are largely ferrodioritic in composition and grade into charnockite. Near the presumed base of the plutons, layered diorite and impure nelsonite, an ilmenite-apatite rock, are found. The largest part of the Roses Mill and much of the Turkey Mountain are altered to biotitic granitoid augen gneiss. Zircon ages determined on the Roses Mill Pluton are concordant at about 970 m.y. (Table 3, column 2).

The Pedlar massif underlies the area northwest of the Rockfish Valley deformation zone. It consists largely of ferrodioritic charnockite in the study area and also includes some poorly layered granulite. The former contains abundant perthitic and antiperthitic feldspars, quartz, and orthopyroxene and varying amounts of clinopyroxene, hornblende, biotite, ilmenite, and apatite. The zircon age determined for ferrodioritic Pedlar is nearly concordant at 1,040 m.y. (Table 3, column 3).

The youngest Proterozoic rocks in the Roseland district are the Mobley Mountain Granite and mafic dikes, possible feeder dikes for the Catoctin metabasalt, found both east and west of the district. The Mobley Mountain Granite is subsolvus, fine to medium grained, and contains accessory zircon, sphene, apatite, and fluorite. It has been dated at 652 ± 22 m.y. by the Rb/Sr wholerock method (Herz and others, 1981). The Irish Creek tin district is about 23 km north-northwest of the Roseland district. Its greisen cuts the Pedlar massif and has been dated at about 635 ± 11 m.y. by concordant $^{40}Ar/^{39}Ar$ age spectra (Hudson and Dallmeyer, 1981). The tin greisens and the Mobley Mountain Granite appear to be genetically related.

The latest igneous events in the area are represented by a variety of mafic dikes, including ultramafic rocks and lamprophyres of unknown age, and Triassic and/or Jurassic diabase.

GRANULITES AND RELATED ROCKS

The lithologies that compose the unit mapped as "layered granulite gneiss and associated rocks" (Fig. 2) include: (1) layered granulite gneiss that is generally leucocratic and that contains blue quartz and perthitic feldspar (Table 1, column 1), is mafic in places, and contains accessory to abundant orthopyroxene, clinopyroxene, garnet, graphite, and sulfides; (2) mesocratic mangerite and quartz mangerite; (3) blue quartz-perthite leucocharnockite that is generally concordant and that forms layers of an apparent injection or anatectic origin, and is thus somewhat younger than (1 and 2). The Shaeffer Hollow Granite is related but is mapped separately.

The granulite unit is in contact with the anorthosite on its structurally lower side and largely with augen gneiss of the Roses Mill and Turkey Mountain Plutons on its upper side. Anorthosite dikes intrude the granulite as do dikes of a Roses Mill lithology, although the upper contact is generally mylonitic in detail. Granulite underwent Grenvillian age granulite-facies metamorphism and shows an additional deformation stage that is missing from the younger rocks. Paleozoic deformation resulted in local retrograde mineral formation and mylonitization.

Layered Granulite Gneisses

Although impressive exposures of layered granulites are found in streambeds, the units cannot be traced for any great distance, and individual lithologies cannot be mapped separately. Most of the units appear to be metasedimentary or volcaniclastic in origin, on the basis of geochemistry, mineralogy (including garnet-graphite), and probable relict bedding features. The layering, however, appears to be caused by a combination of sedimentary layering, lit-par-lit injection, metamorphic segregation or anatexis, and deformation that obliterated most of the original features of the unit.

The metamorphic mineral assemblage in the layered granulite gneiss formed under granulite-facies *P-T* conditions and includes blue quartz, perthitic feldspars, orthopyroxene, clinopyroxene, garnet, rutile, ilmenite, pyrrhotite, graphite, and apatite. Locally, red biotite and amphibole are prograde phases; retrograde minerals include greenish biotite, actinolitic amphibole, white micas, epidote minerals, sphene, and chlorite.

Equant to platy blue quartz composes up to 65% of the layered granulite gneiss and in many places shows little strain.

Feldspars compose as much as 45% of the layered granulite gneiss and range from perthite through mesoperthites to andesine antiperthite. Microprobe data (Ames, 1981) indicate that K-feldspar lamellae have a composition of Or_{96-98}, and optical and microprobe data indicate that plagioclase lamellae are An_{33-40}. Garnet, in grains as long as 5 cm, ranges from 0% to 50% of the layered granulite gneiss. Orthopyroxene compositions are Fs_{33-36} En_{64-66} Wo_{1-2} and clinopyroxenes Fs_{13-28} En_{29-41} Wo_{43-46}, according to microprobe data of Ames (1981).

Pyrrhotite and subsidiary pyrite form as much as 8% of the layered granulite. Ilmenite is the most common oxide, although rutile is also present near anorthosite or leucocharnockite dikes. Together, they form as much as 12% of some granulites. Coarse flake graphite constitutes as much as 5%.

Field relationships and geochemical evidence suggest that the granulites represent a packet of thinly bedded marine sediments and possibly volcaniclastic rocks that have undergone dehydration reactions at *P-T* conditions of the granulite metamorphic facies.

Mangerite and Leucocharnockite

Mangerite and leucocharnockite are metamorphic rocks that we believe to have formed from an igneous precursor. Igneous names have been used for them. Mangerites (pyroxene-bearing monzonites) and related rocks are especially common near the anorthosite margin as large metamorphosed dikes, sills, and possibly flows, now interlayered with the layered granulites and having sharp or gradational contacts. The mangerites are typically hypidiomorphic granular, locally having layering of pyroxene and feldspar that may be of metamorphic and/or cumulate igneous origin. The variation in amount of mafic and felsic minerals is great, and the rocks grade into quartz mangerites, gabbronorites, and charnockites. The most common mangeritic rock types, however, are poor in quartz and K-feldspar and rich in plagioclase and orthopyroxene. Plagioclase (antiperthitic Na-andesine) is the most abundant mineral in the mangerite, making up 15 to about 70 modal percent. Plagioclase is largely equant, 0.5 to 1 mm across, and shows kink-banding and zoning. Orthopyroxene and subsidiary clinopyroxene are commonly tabular, are aligned and uralitized to some degree, and make up 10% to 30% of the mangerites. Brown hornblende, also alined and uralitized to some degree, constitutes 10% to about 30%. Garnet is only locally present but forms as much as 30% of some mangerites as coarse pale anhedral grains. Accessory minerals include deep-brown biotite containing rutile needles, magnetite, ilmenite, rutile, apatite, graphite, and pyrrhotite. Retrograde alteration products, including white micas, epidote, amphiboles, and sphene, are common.

Leucocharnockites are abundant in the unit mapped as "layered granulite gneiss and associated rocks" (Fig. 2) and consist largely of platy blue quartz and feldspar. They generally form concordant layers of an apparent injection or anatectic origin in layered granulite and mangerite but in places appear to grade into dikes of anorthositic composition. Principal minerals in the leucocharnockite are blue quartz, commonly unstrained in plates as long as 3 cm forming 10 to 65 modal percent;[1] plagioclase, antiperthitic andesine (An_{30}) or oligoclase, 2% to 45%; and mesoperthite and perthitic K-feldspars, 0% to 70% (Ames, 1981). Accessory minerals include rutile, ilmenite, biotite, pyroxenes, amphibole, and zircon. Retrograde alteration of plagioclase to clinozoisite and white micas is common.

Both leucocharnockites and mangerite are probably igneous in origin (possibly the graphitic mangerite was pyroclastic) but were modified by granulite-facies metamorphism. The platy texture locally cuts banding or is parallel to fold axes; however, it must be a high-temperature feature, as both the quartz and feldspars involved in the texture formed at high temperatures. Mesoperthite implies formation temperatures of more than 600 °C. Strain in platy blue quartz is completely annealed, and included rutile needles form a true hexagonal array. Some leucocharnockite may have been produced as an anatectic leucosome of granulite, approximately in Grenvillian time.

Shaeffer Hollow Granite (here named)

Leucocratic rocks having tabular feldspars 2 to 5 cm in length, and locally having recognizable porphyritic igneous textures, are found as roof pendants and screens between ferrodiorite intrusive bodies. The tabular phenocrysts grade into augen, and in most exposures, bands of flaser gneiss or mylonite are also seen. Xenoliths of platy-textured leucocharnockite occur in the Shaeffer Hollow Granite, which is cut by dikes of, and forms xenoliths in, the Roses Mill Pluton. Discordant U/Pb age determinations on zircon from the Schaeffer Hollow give a $^{207}Pb/^{206}Pb$ age of 1,787 m.y.[2] (Table 3, column 4).

The granite is best exposed in the northeastern part of the mapped area (Fig. 2), in and north of Shaeffer Hollow, in Horseshoe Mountain quadrangle. The granite is generally coarse grained and contains blue quartz (10%–20%), perthitic microcline

---

[1]Gordon L. Nord, Jr., U.S. Geological Survey (1981, unpub. data), studied blue quartz from a leucocharnockite-anorthosite transition by use of an analytical transmission electron microscope. He found a high density of fine particles less than 250 nm (0.25 μm) in diameter in a thin section containing the most intensely blue quartz. X-ray energy dispersive analysis of these particles showed that some contain titanium and iron, suggesting that they are ilmenite or a spinel (more likely the former judging from their tabular hexagonal habit). Nord also determined that the quartz was blue only in reflectance but brown in transmission, which suggests that the blue color is due to a scattering phenomenon. The submicron particles are the obvious candidates for the scattering centers. The authors and Nord speculate that the presence of the particles is a high-temperature exsolution phenomenon and that blue quartz may be a potential geothermometer in metamorphic terranes. Abundant dauphine twin boundaries indicate that the quartz cooled through the alpha-beta transition at 573 °C.

[2]A more nearly concordant age of 980 m.y. was later obtained by Tom Stern from another outcrop of this unit. Details will be presented elsewhere.

(30%–50%), antiperthite, about An$_{30}$ (15%–30%), and aggregates of biotite and uralitic hornblende, which generally make up less than 10%. Accessory minerals include apatite, sphene (after ilmenite), and zircon. The retrograde minerals clinozoisite, chlorite, and white micas are also common. Chemical composition of a representative sample of the Shaeffer Hollow Granite is given in Table 1 (column 2).

ROSELAND ANORTHOSITE

The Roseland Anorthosite forms the core of an elongate dome, the basic structural element of the district. The anorthosite body is 14.4 km long, trends northeast, and covers about 35 km^2. Texturally, the rock is similar to massif anorthosites elsewhere; megacrysts of andesine antiperthite, 10 to 20 cm in diameter, are separated by a granulated matrix of finer grained oligoclase and minor microcline and blue quartz. The megacrysts are light blue, whereas the granulated areas are cream to white. Some megacrysts were originally 1 m or more in length, as indicated by exposures where broken fragments appear to have nearly identical orientations and suggest a larger original crystal. The granulated zones were produced by protoclasis of both megacrysts and original groundmass and are most obvious near the margins of the anorthosite body. Granulation probably accompanied anorthosite emplacement.

Border facies of the anorthosite are also megacrystic, but they are quartz mangerite in composition, containing abundant blue quartz, orthopyroxene, and feldspar megacrysts and common rutile and ilmenite. Possibly this coarse impure anorthosite is the result of high-temperature re-equilibration (and partial melting?) of ilmenite-bearing mangerite with iron-poor anorthosite to form rutile plus coarse feldspar, pyroxene, and quartz, as grain size and rutile content decrease into mangerite. Alteration of the plagioclase to clinozoisite is especially evident in anorthosite border zones, in the granulated groundmass, and along planar surfaces, and is presumably a Paleozoic retrogressive effect.

Anorthosite dikes are seen in older rocks of the Roseland district, and xenoliths of anorthosite are seen in younger rocks. Dikes of anorthosite intrude layered granulites and mangerites; where swarms of anorthosite sills and dikes occur, as in Allen Creek, important rutile resources are also present. Plagioclase in the dikes is oligoclase (An$_{20-25}$), fine to medium grained, and locally antiperthitic. Thus, plagioclase in the dikes has a composition similar to that of granulated groundmass of the anorthosite, both presumably being the last phases of the pluton to crystallize.

Anorthosites are commonly found in Grenville-type terranes worldwide, but little agreement exists on either their genesis or classification (see Middlemost, 1970). The Roseland pluton is nonstratiform and domical in outline, typical of Adirondack-type massifs (Buddington, 1961, p. 422). The Roseland Anorthosite, having megacrysts of andesine antiperthite Or$_{14-21}$Ab$_{51-59}$An$_{27-28}$, hypersthene, and hemo-ilmenite, is similar to other andesine-type anorthosites (Anderson and Morin, 1969). However, important characteristics set it apart; the Roseland has more

SiO$_2$ and Na$_2$O and more than double the K$_2$O of other massif anorthosites (Table 1, columns 3 and 4), and it contains abundant modal and normative K-feldspar and quartz. CaO is correspondingly low at Roseland. Alkalic anorthosites at Pluma Hidalgo, Mexico (Paulson, 1964), and St. Urbain, Canada (Mawdsley, 1927), are similar chemically and mineralogically to the Roseland and also contain hemo-ilmenite and rutile as oxide phases.

ROSES MILL PLUTONIC SUITE (HERE NAMED)

The name "Roses Mill Plutonic Suite" is introduced for rocks mostly of ferrodioritic composition and applies to two plutons in the study area, the Roses Mill and Turkey Mountain Plutons. A roadcut at Roses Mill on the east side of the district is designated its type section. The Roses Mill Pluton forms the southeast and northwest flanks of the granulite and anorthosite, toward the northern end of the district (Fig. 2). On the western side of the district, it is truncated by the Rockfish Valley deformation zone. The Turkey Mountain Pluton is related to the Roses Mill Pluton and is in the southern part of the district.

Both plutons contain ilmenite and apatite, are ferrodioritic in composition, and are altered in varying degrees from a dark charnockitic rock grading into biotite augen gneiss. The type locality at Roses Mill shows many stages in this transition. The Turkey Mountain rocks differ from those of the Roses Mill only in being more magnetic, less deuterically altered, locally more granitic, and in places showing compositional layering. The following discussion pertains to both plutons.

Four dominant facies are present: (1) primary charnockitic ferrodiorite (Table 1, column 6), (2) secondary biotite augen gneiss, (3) dike rocks in older formations, and (4) nelsonite (Table 1, column 8). Field relationships clearly show that the plutons are igneous and younger than the anorthosite-granulite terrane, even though their contacts are generally tectonized. Chemical composition of the plutons is distinctive and diagnostically igneous. Xenoliths of the older rocks are found in the Roses Mill Pluton, and Roses Mill dikes occur in the anorthosite and granulite. A concordant U/Pb date of 970 m.y. was obtained from zircon in charnockite at the Roses Mill type locality (Table 3, column 2).

Charnockitic Ferrodiorite

The charnockitic rocks, that is, those of a granitic affinity but containing primary pyroxenes, are considered the "parent" of the more widespread biotite augen gneiss. Rock compositions range from charnockite and quartz mangerite through ferrodiorite and more mafic rocks; ferrodiorite is predominant. Charnockitic ferrodiorite forms Mars Knob, the ridge north of Bryant Mountain, and much of the high area around Turkey Mountain (Fig. 2). In most samples, the pyroxenes have been partially to completely uralitized.

The average chemical composition of the charnockitic ferrodiorite (Table 1, column 6) is similar to that of ferrodiorites

TABLE 4. SOME TRACE-ELEMENT AND ISOTOPIC DATA FOR ROSELAND DISTRICT ROCK SUITES

	Well-layered granulite		Leucocharnockite		Roseland Anorthosite	Roses Mill Plutonic Suite				Charnockite of Pedlar massif		Mobley Mountain Granite
						Roses Mill Pluton		Turkey Mountain Pluton				
	mean	σ	mean	σ	mean	mean	σ	mean	σ	mean	σ	mean
Sr (ppm)	562	365	164	54	1140	535	81	546	184	364	110	360
Rb (ppm)	48	37	198	40	15.5	73	14	49	38	186	47	102
Zr (ppm)	535	607	291	159	uk	617	190	303	265	599	407	uk
La (ppm)	357	384	654	849	uk	88	104	132	119	568	283	93
K/Rb	1034	1095	233	67	2400	441	54	555	244	234	57	437
Rb/Sr	0.19	0.20	1.31	0.46	0.008	0.14	0.02	0.15	0.19	0.59	0.33	0.28
$+\delta^{18}O$ (ppm)	8.8	1.0	8.8	0.4	7.7	7.6	0.4	8.2	0.9	9.3	0.5	
$^{87}Sr/^{86}Sr$	uk		uk		0.7052*	uk		uk		0.7061**		0.7045**
n=	6		12		3	7		8		7		2

Note: Data from Herz and Force (1982) and Ames (1981), except as noted.
uk, unknown; n, number of samples.
*Heath and Fairbairn, 1969.
** Herz and others, 1981.

(including some termed "mangerites" and "jotunites") found near anorthosites and stratiform complexes elsewhere; it differs from the chemical composition of diorite as shown in Table 1, column 7. The Roseland ferrodiorites show strong enrichment in iron and titanium, unlike typical calc-alkalic rock series. The ratio of iron oxides to magnesia is large, more than 4; the SiO_2/K_2O ratio is low, about 23; and average content is high for TiO_2 (2.4%), P_2O_5 (0.9%), Zr (620 ppm), and rare earth elements (Table 1, column 6, and Table 4). Igneous textures are apparent in fresh rocks, that contain feldspars about 3 to 5 mm across, ranging up to 10 to 20 mm. Primary foliation is generally absent, but a crude mafic-feldspar compositional layering, including layers of impure nelsonite, is commonly seen near the base of the plutons, and a fluxion foliation is shown elsewhere by trains of xenoliths and xenocrysts.

The charnockitic ferrodiorite is hypidiomorphic granular in texture and is characteristically coarse grained, having 3 to 5 mm feldspars and 1 to 3 mm pyroxenes. The unit typically has conspicuous ilmenite. Primary minerals of the unaltered rocks are colorless quartz, perthitic K-feldspar, and antiperthitic andesine (An_{31-35}) containing equant inclusions of blue quartz, and pyroxene. Pyroxene in the Roses Mill Pluton is generally corroded and uralitized, but in the Turkey Mountain Pluton, it is fresher, and orthopyroxene is 2 to 5 times more abundant than clinopyroxene. Accessory minerals are common in the more pyroxene-rich zones and include ilmenite (with magnetite intergrowths in the Turkey Mountain), apatite, and zircon; garnet has formed, apparently as a result of reactions among oxides, pyroxenes, and feldspars. Blue quartz and perthites with blue quartz inclusions, especially where coarser than 1 cm, we believe to be xenocrysts.

Biotite Augen Gneiss

Biotite augen gneiss is by far the most common rock type of both plutons. It contains light-colored, quartz-feldspar augen, 10 to 80 mm in length, surrounded by darker foliae of finer grained white quartz, feldspar, biotite; accessory minerals in the biotite-rich layers are sphene-rimmed ilmenite, apatite, garnet, and zircon. Actinolitic amphibole occurs in the dark layers either as pseudomorphs after pyroxene or as part of the dark pasty matrix. The feldspars are plagioclase (An_{0-15}), containing sericite and clinozoisite as alteration products, and perthitic microcline; both feldspars contain abundant inclusions of equant blue quartz and biotite. Widely scattered large angular grains of antiperthitic andesine or blue quartz are present in biotite augen gneiss and are similar to those seen in the granulites or anorthosite; the grains are presumably xenocrysts. Chlorite, muscovite, and epidote minerals are abundant later alteration minerals in many samples, especially near zones of cataclasis.

Charnockitic ferrodiorite grades into massive augen gneiss by the formation of foliated clots of biotite after pyroxene; further deformation destroys the interlocking texture but does not completely alter the shapes of xenoliths or the orientation of xenolith trains. This textural transition corresponds to gradual mineralogic transitions that can be traced in suites of thin sections from areas of gradational field relations; perthite becomes rimmed by albite, ilmenite becomes rimmed by sphene, and pyroxene is replaced by clots of foliated biotite and amphibole.

Mylonitic rocks commonly occur within the biotite gneiss and form zones as much as tens of metres wide. In these, feldspars are rolled, have granulated tails, and float in a biotite-epidote matrix.

Dike Rocks

Roses Mill dike rocks are common in all older rocks. In general, the dike rocks cannot be traced back into their parent bodies but can be recognized in the field by the assemblage ilmenite-apatite–colorless quartz–green amphibole and by the presence of abundant feldspar xenocrysts. The dikes range from tiny apophyses to bodies tens of metres in width, as at Jennys Creek. Contacts between dikes and wall rock are sharp, planar, or irregular but not chilled. Mineralogically, most dike rocks are similar to the augen gneiss facies but contain more amphibole; chemically, they are ferrodioritic.

Nelsonite

Nelsonite is an equigranular rock consisting of about two-thirds ilmenite and one-third apatite, in crystals averaging 1 to 2 mm across. Watson and Taber (1913) documented several varieties of nelsonite that contain rutile, magnetite, or silicate impurities. In outcrop, they form dikelike bodies as wide as 20 m and, as at Lanes Ford, as long as 1 km. We conclude that the nelsonite is a facies of the Roses Mill Plutonic Suite as indicated by the following evidence: (1) nelsonite mineralogy, ilmenite-apatite-zircon, is ubiquitous in lithologies of the plutons; (2) some pyroxene-bearing nelsonite bodies are not dikes but cumulate-like rocks that formed at the base of the plutons; (3) nelsonite dikes are always found below ferrodiorite but do not persist at depth within the underlying host rocks; (4) all gradations between ferrodiorite dikes and impure nelsonite dikes are present, and some contain apparently cumulate nelsonite at their bases.

The cumulatelike impure nelsonites are mostly found at the base of the Roses Mill Pluton intrusive sheet. They have a texture that we believe represents immiscible liquids. Elliptical quartzo-feldspathic domains are separated by a mafic matrix (Fig. 3a). This matrix is itself net-textured and has euhedral to subhedral pyroxene engulfed by ilmenite and apatite (Fig. 3b). Three original phases must have been present: (1) Quartzofeldspathic domains represent depleted silicate magma. (2) Pyroxene represents cumulate crystals from that magma. (3) Ilmenite and apatite represent a liquid immiscible in the silicate melt, which accumulated with cumulate crystals at the base of the melt. The process is further discussed in part 2 and by Force and Herz (1982).

PEDLAR MASSIF

The Pedlar Formation of Bloomer and Werner (1955) underlies the northwestern part of the district and most of the area westward at least to the crest of the Blue Ridge. In view of the great heterogeneity of the unit, we prefer to use the name informally as the Pedlar massif for this terrane.

In this area, the Pedlar is made up largely of igneous-appearing, massive or faintly foliated, dark-green charnockitic ferrodiorites that grade into granitic rocks; minor well-layered mafic and granulitic rocks also occur in this area. Toward the

southeast boundary of the massif, the Pedlar rocks become deformed in the Rockfish Valley deformation zone.

Bartholomew and others (1981) used the term "Pedlar massif" as we do; all our units east of the Rockfish Valley deformation zone occur within their "Lovingston massif." Good correlations across this zone are difficult. A slightly discordant zircon U/Pb radiometric age of the Pedlar, 1040 m.y. for $^{207}Pb/^{206}Pb$ (Table 3, column 3), is slightly older than that determined for the Roses Mill Pluton, but chemical and mineralogic compositions suggest that ferrodioritic Pedlar lithologies and the Roses Mill Plutonic Suite are closely related.

The Pedlar is generally hypidiomorphic granular and has grain sizes of 0.4 to 1 mm; feldspar tends to be coarser grained than mafic minerals. The most common rock types are ferrodioritic to charnockitic (Table 1, column 5), and they have either an equant or porphyritic texture. Associated rock types include leucocratic granites, lithologically similar to the Shaeffer Hollow Granite, and layered granulites. Feldspars are perthitic, and plagioclase is An_{25-38}, forming phenocrysts 10 to 20 mm in length, but as long as 10 cm in the porphyritic rocks. Feldspars occur largely in felsic aggregates, about 30 to 50 mm across. Two generations of quartz are evident: strained blue quartz containing rutile needles and, later, colorless quartz. Mafic minerals include orthopyroxene and clinopyroxene uratilized to varying degrees, deep-reddish-brown biotite containing rutile needles, and brown oxyhornblende. Common accessory minerals are zircon, apatite, ilmenite, magnetite, white micas, and sulfides. Retrograde minerals, including albite, epidote, clinozoisite, chlorite, and white micas, are especially common in mylonitic zones. Unakite, a retrograded Pedlar charnockite, is widespread; commonly, it is found away from any obvious zone of deformation and has its original texture preserved.

MOBLEY MOUNTAIN GRANITE (HERE NAMED)

The Mobley Mountain Granite (here named) underlies Mobley Mountain in the southwestern corner of the Roseland district in Amherst County, is about 6 × 1.5 km, elongate northeast-southwest, and forms abundant irregular smaller bodies, crosscutting veins, and migmatites for 2 km or more in each direction away from the mountain. Gravity modeling suggests that it is plug shaped and reaches a maximum depth of 7.2 km (Eppihimer, 1978). The granite clearly has an intrusive relationship in the Turkey Mountain Pluton; rare xenoliths and a migmatite are present at the contact.

The granite is typically fine to medium grained and massive or weakly foliated, except near the contact with country rock where thin biotite-rich layers impart a good foliation. Typically, the granite has a pepper and salt appearance resulting from biotite patches disseminated through granular quartz and feldspar (Table 1, column 9).

The principal minerals in the granite are microcline (about 30%); quartz (25%); plagioclase, sodic oligoclase to almost pure albite (25%); biotite, pleochroic from olive brown to yellowish

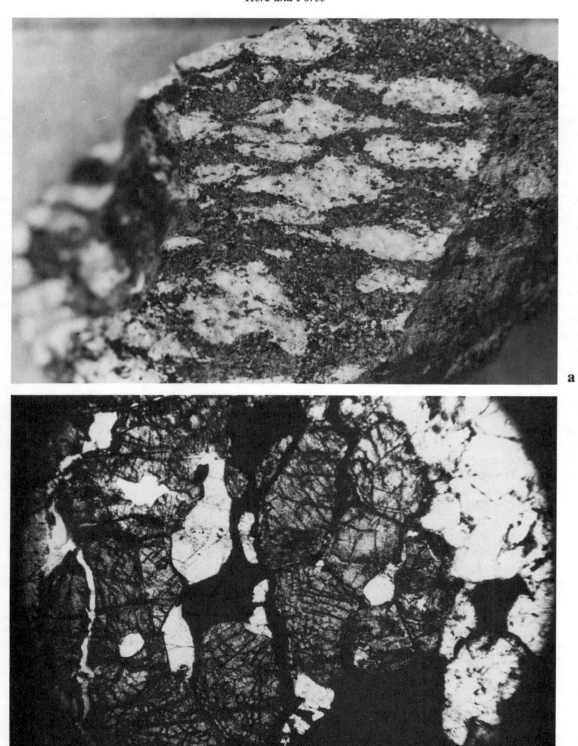

Figure 3. Textures of rock at the base of the Roses Mill Pluton. (Also see Force and Herz, 1982, Fig. 2.) (a) Hand specimen showing elliptical quartzo-feldspathic domains in mafic matrix. Field of view 5 × 8 cm. (b) Photomicrograph of pyroxene-ilmenite-apatite relations showing net texture of opaque matrix. Field of view 3 × 5 mm.

green or reddish brown (11%); clinozoisite-epidote, both primary and as plagioclase alteration products (2% at the borders of the pluton to 10% at the core); and white micas that are either a fine-grained alteration product or coarse primary-appearing flakes (2% at the borders to 0.5% at the core). Accessory minerals include zircon, apatite, sphene, fluorite, florencite, and possibly perrierite in associated pegmatites.

The Mobley Mountain Granite appears to be part of a belt of mesozonal granites emplaced about 650 m.y. ago. The Mobley Mountain Granite has been dated at 652 ± 22 m.y. by Rb/Sr whole-rock analysis (Herz and others, 1981); the initial $^{87}Sr/^{86}Sr$ ratio is 0.7045 ± 0.0005. The Irish Creek tin district is 23 km north-northwest of Mobley Mountain. Its geisen cuts the Pedlar charnockite and has been dated at about 635 ± 11 m.y. by concordant $^{40}Ar/^{39}Ar$ age spectra (Hudson and Dallmeyer, 1981). The greisen is near a granite that is mineralogically similar to and may be genetically related to the Mobley Mountain (Hudson, 1982). The greisen might have been a late-stage, volatile-element–rich product formed in the cupola zone of the granite.

MAFIC DIKES

Metamorphosed and unmetamorphosed mafic to ultramafic dikes of several ages intrude older Precambrian units and include:

1. Rare pyroxenites containing orthopyroxene (commonly altered to talc-serpentine), amphiboles, epidote, ilmenite with sphene rims, skeletal magnetite, and, in places, garnet and apatite. All but one of the dikes are adjacent to anorthosite. Blue quartz veins in some pyroxenites suggest that the dikes are pre-Roses Mill.

2. Ferrodiorites and nelsonites, which are Roses Mill equivalents.

3. Fine-grained foliated and altered mafic dikes, probably Proterozoic Z to early Paleozoic in age, which intrude all other Precambrian rocks, including mylonites, but locally are cut by the Mobley Mountain Granite. Dike thicknesses range from a few centimetres to 10 m, and some are longer than 1 km; they contain mineral assemblages produced by greenschist-facies metamorphism such as albite, biotite, white micas, epidote, chlorite, carbonate, and actinolite. Some of the dikes may be feeder dikes for the Catoctin metabasalt.

4. Diabasic dikes, ranging from 0.3 to 10 m in thickness, trend from 15°E to 15°W of north and are vertical. The diabase is fine to medium grained and has well-developed ophitic textures. Some dikes are altered; all have chilled margins or are aphanitic. Mineral assemblages include lathlike phenocrysts and groundmass plagioclase (An_{40}), fine-grained hornblende and chlorite after relict pyroxenes, ilmenite, and sphene. These rocks are presumably Triassic and/or Jurassic in age and related to the widespread diabase dike emplacement at that time in the Appalachians.

5. Gabbro. A north-trending post metamorphic gabbroic rock was found in Crawford Creek, at the south end of the Piney River quadrangle. The trend of the gabbro can be followed as a high on both the aeromagnetic and gravity maps (Eppihimer, 1978); gravity modeling suggests that it is as wide as 500 m and extends to a depth of at least 4.5 km. The geophysical signature terminates in the Pedlar formation, just below the crest of the Blue Ridge. The gabbro is medium grained (2–3 mm) and ophitic and contains stubby and lathlike plagioclase (An_{65}), orthopyroxene, clinopyroxene, relict olivine, and scattered flakes of deep-reddish-brown biotite. The trend of the dike, its primary mineralogy, and undeformed textures suggest that it belongs to the period of Mesozoic mafic-dike emplacement.

STRUCTURE

Only the most fundamental aspects of the structure are touched on here. The structural evolution can be divided into four periods: (1) pre-Grenvillian structures limited to the layered granulite gneisses; (2) Grenvillian structures that penetrate the granulites and anorthosite and cut their mutual contacts—correlative mineral assemblages are those of the granulite facies; (3) post-Grenvillian Precambrian structures, correlative with lower amphibolite-facies mineral assemblages; and (4) Paleozoic structures, typically local and correlative with greenschist-facies mineral assemblages.

The two most important structural features of the area, the dome and mylonitic rocks, probably both were initiated in the third and reactivated in the fourth period.

Pre-Grenvillian and Grenvillian Structures

Many granulite outcrops show fold sets that are not mappable. These are mostly tight or isoclinal; as many as three sets can be seen in some outcrops of compositionally layered granulites. Blue quartz veins and platy-textured leucocharnockite segregations are parallel to axial planes and cut compositional layering in older folds but are also folded themselves. Platy blue quartz has been annealed of strain features at high temperature. Foliation marked by platy textures locally cuts granulite-anorthosite contacts and must therefore be post-anorthosite.

Dome

The structure most apparent from the map pattern of units (Fig. 2) is an elongate dome oriented northeast-southwest, measuring at least 22×5 km. At Mars Knob is a doubly plunging medial syncline. The lowest unit of the dome is the anorthosite, which intrudes the structurally intermediate but oldest unit, the layered granulites; both are intruded by the highest unit, the ferrodiorites. The lower anorthosite-granulite contact is irregular in detail and has abundant anorthosite dikes and sills, but the anorthosite dips away along the flanks of the dome. Compositional layering and foliation in the granulite are generally parallel to the regional dome, despite older tight folding, to the extent that where the axis is in granulite, its positions and plunge can be determined.

The structurally highest unit, the Roses Mill Plutonic Suite, is also the youngest and intrudes both lower units. In places the plutons are in direct contact with the anorthosite or granulite, but in most places this lower contact is sheared, and mylonites containing chlorite, epidote, and calcite separate the units. However, the presence of some unsheared contacts, dikes of the Roses Mill Pluton in older rocks, and xenoliths of granulite in the Roses Mill all suggest that any relative movement on this contact must have been on a small scale.

Mylonitic Rocks and Related Deformation

A belt of mylonitic rocks, 2 to 4 km in width and part of an extensive zone that can be traced both north and south of the district, separates charnockitic rocks of the Pedlar massif to the northwest from the anorthosite-granulite–Roses Mill terrane to the southeast. The predominant protoliths of the mylonitic rocks are apparently leucocharnockite, quartzo-feldspathic granulite, and Shaeffer Hollow Granite. Bartholomew (1977) mapped an extension of the same zone having a width of about 10 km in the Sherando and Greenfield quadrangles to the northeast, and Bartholomew and others (1981) called it the Rockfish Valley deformation zone.

The age of formation of the various features of the zone is not clear. A Proterozoic Z deformation probably formed much of the steeply dipping shear foliation, as the metamorphic conditions are lower amphibolite facies and thus too high grade for the later Paleozoic deformation. A slightly foliated and recrystallized greenstone dike cuts this foliation at a low angle at Lowesville. The dike appears to be part of the Catoctin; if so, then the steep foliation is older than about 650 to 700 m.y. (Conley, 1978, p. 131). Within thin zones, however, the mineral assemblage is chlorite, albite, white micas, and epidote, suggesting a Paleozoic age. Bartholomew (1977) presented evidence of predominantly Paleozoic movement in his section of the zone. Thus, it apparently formed in Precambrian time but was reactivated during the Paleozoic.

Formation of a distributed shear foliation accompanied the metamorphism of the ferrodiorite charnockite assemblages to biotite augen gneiss in the Roses Mill Plutonic Suite and is probably related to mylonitization in the Rockfish Valley deformation zone. The foliation ranges from weak, where biotite is oriented but the igneous texture is still largely preserved, to extremely strong in thick zones of mylonite. In all but the most mylonitic augen gneisses, xenoliths are recognizable. This shear foliation is not folded by the major domal fold and is roughly parallel to its axial plane. The mineral assemblage accompanying the shear foliation is lower amphibolite facies biotite-sphene-garnet-epidote-albite, presumably of a Proterozoic Z age; later mylonites containing greenschist-facies mineral assemblages are also locally present.

Shear foliation generally dips steeply southeasterly and contains a strong downdip lineation commonly shown by biotite crinkles and granulated feldspar. At Roses Mill and elsewhere, it appears to be folded or refracted around relict massive charnockite so that its dips are inward.

Possible Diapirism

Many features of the dome are consistent with a diapiric emplacement of the anorthosite into a lower crustal granulite terrane and the subsequent emplacement of this terrane into the higher crustal level of the Roses Mill–Turkey Mountain units. The concordance of older structures may be a result of emplacement similar to nearly concordant diapiric emplacements of anorthosites elsewhere (Martignole and Schrijver, 1970). Older components of shear foliation along the domal flanks may have accompanied this emplacement. Later mylonites at the base of the Roses Mill probably formed in Paleozoic time.

GEOCHRONOLOGY

A preliminary program of U/Pb age dating on zircons has been carried out with the cooperation of T. W. Stern, U.S. Geological Survey, to help establish a chronology for the Roseland district. To aid in interpretation, we have compared the Roseland district to some other Grenvillian terranes where extensive data are available and chronological models are better established.

Ages of Other Grenvillian Terranes

Emslie (1980) has compiled chronological information for the Labrador-Quebec area straddling the Grenville front, where overprinting of events is particularly well documented (Table 5).

In the Adirondacks, a variety of dates on granulites and other rocks suggests a metamorphic peak at 1,020 to 1,100 m.y. ago (Silver, 1969). Dates older than 1,200 m.y. are considered to represent incompletely remobilized Archean crust or relict zircons derived from Archean terrane to the west (Baer and others, 1974). In Newfoundland, peak metamorphism, as seen in the anorthosite-granulite suite of the Long Range, appears to have taken place 1,100 to 1,150 m.y. ago, and hornblende and biotite $^{40}Ar/^{39}Ar$ cooling ages are 825 to 880 m.y. ago (Dallmeyer, 1978).

Anorthosite suites elsewhere in the northern hemisphere appear to be alined on a belt, in a predrift reconstruction (Herz, 1969a). Throughout this belt they have remarkable similarities, such as having associated quartz mangerites and relatively narrow time limitations on their emplacement; dates range from 1,100 to 1,700 m.y. ago and cluster around 1,300 ± 200 m.y. ago. The anorthosite-quartz mangerite suites in Canada were emplaced before the onset of the Grenville orogen and appear to be about 1,450 m.y. old north of the front but are about 1,200 m.y. old or younger south of the front.

Age Determinations In and Near the Roseland District

Past Work. The oldest dates thus far obtained in the southern Appalachians are in the range of 1,600 to 1,800 m.y. ago.

TABLE 5. RADIOMETRIC DATES OF PRECAMBRIAN EVENTS OF LABRADOR AND QUEBEC

Terrane, event, or unit	Age (m.y.)	Methods	Correlative features
Archean gneiss	2665-2390	K/Ar (hb,bi)	Kenoran orogeny; amphibolite- to granulite-facies metamorphism
Aphebian gneiss; Hudsonian orogeny	1750-1590	K/Ar (bi)	Amphibolite- to granulite-facies metamorphism
Elsonian orogeny	1480-1300 1460-1290 1420-1320	K/Ar (bi,hb) U/Pb (zir) Rb/Sr (isochron)	Anorthosite, adamellite, and massive granite
Grenville terrane	1185-950	K/Ar (bi, musc)	

Note: Data from Emslie (1980). hb, hornblende; bi, biotite; zir, zircon; musc, muscovite.

U/Pb determinations on zircon from the Baltimore Gneiss give a minimum age of 1,200 m.y. and a probable primary age of 1,600 to 1,700 m.y. (Grauert, 1972, p. 305). Zircon separated from a coarse fraction (>150 μm) of the "Lovingston augen gneiss" collected near the study area yielded a date of 1,422 to 1,870 m.y. by U/Pb (Davis, 1974). From the fine fraction (<75 μm) of the same rock, a date of 913 m.y. was obtained. Both dates define a chord from about 915 to about 1,870 m.y. ago; the later date is concordant, but the earlier is highly discordant. The dated sample is aluminous gneiss that shows an outcrop pattern of a narrow septum parallel to the regional trend, from near the town of Lovingston to the Rockfish River, a distance of about 15 km. The augen gneiss contains porphyroblasts of K-feldspar and, less commonly, of blue quartz, as much as 5 cm in diameter and surrounded by biotite-rich folia.

Davis (1974) obtained U/Pb dates of 1,080 m.y. on zircon from the "Lovingston quartz monzonite" as well as from the "Lovingston pegmatite." Both dates are considered minima. The quartz monzonite has a mineralogy similar to that of the augen gneiss but lacks the metamorphic fabric. The pegmatite is a coarse-grained rock consisting of andesine antiperthite, perthite, and blue quartz.

Lithologies in the Pedlar massif have yielded zircon ages of 1,070 to 1,150 m.y. by U/Pb (Tilton and others, 1960). These dates were apparently obtained from granitoid rocks that are related to the Old Rag Granite of Furcron (1934) dated by U/Pb on zircon at 1,140 m.y. (Lukert, 1977) and the Crozet Granite of Nelson (1962), both to the north of this district. Pedlar charnockitic ferrodiorite in the district has been dated at 1,042± 59 m.y. by whole-rock Rb/Sr (Herz and others, 1981).

The youngest dates obtained in the district are on the Mobley Mountain Granite, 652 ± 22 m.y. by the Rb/Sr whole-rock method (Herz and others, 1981). The Irish Creek tin greisen, in nearby Rockbridge County, has yielded 635 ± 11 m.y. by concordant ⁴⁰Ar/³⁹Ar spectra (Hudson and Dallmeyer, 1982) and is probably part of the same igneous event.

Our Results. U/Pb zircon ages have been determined on four samples from the Roseland district (Table 3). The samples are a leucosome in layered granulite (sample 1), ferrodiorite from the type locality of the Roses Mill Pluton (sample 2), charnockitic ferrodiorite of the Pedlar massif (sample 3), and Shaeffer Hollow Granite (sample 4) The interpretations are preliminary and should be supplemented with more work.

Sample 1, 980 m.y. old by $^{207}Pb/^{206}Pb$ (Table 3), is from a layered granulite outcrop in the Tye River, just above Lanes Ford. It is a banded rock made up of apparently anatectic segregations in granulite. The more mafic part of the granulite contains garnet, graphite, pyrrhotite, ilmenite, blue quartz, perthitic feldspars, and pyroxene. Zircons are found only in the leucocratic segregations. This occurrence is about 600 m from the anorthosite contact. The date is slightly discordant.

Sample 2, dated at 970 m.y. by $^{207}Pb/^{206}Pb$ (Table 3), is a ferrodiorite from the type locality of the Roses Mill PLuton in the Arrington quadrangle. It is a massive, medium- to coarse-grained rock consisting of quartz (12%), perthitic K-feldspar (15%), antiperthitic andesine (35%), orthopyroxene and clinopyroxene (9%), uralitic hornblende (17%), garnet (7%), ilmenite with sphene rims (3.4%), apatite (2.9%), and zircon (as much as 0.6%). This rock type grades into biotitic augen gneiss. The date is concordant.

Sample 3, dated at 1,040 m.y. by $^{207}Pb/^{206}Pb$ (Table 3), is a massive charnockitic ferrodiorite of the Pedlar massif near Jack's Hill, Massies Mill quadrangle. It is coarse grained, and orthopyroxene poikilitically encloses other phases including quartz, clinopyroxene, perthitic K-feldspar, and antiperthitic Na-andesine; accessory minerals include ilmenite, apatite, and zircon. The chemical composition of this rock is ferrodioritic. The date is almost concordant.

Sample 4, 1,787 m.y. old by $^{207}Pb/^{206}Pb$ (Table 3), is Shaeffer Hollow Granite exposed in a rock slide on the west slope of Bryant Mountain ridge, in the Horseshoe Mountain quadrangle. It is medium to coarse grained, contains elongate and alined feldspar phenocrysts, and consists of quartz, antiperthitic andesine An₃₀, perthitic K-feldspar, uralitic hornblende, ilmenite having sphene rims, apatite, and zircon. The date is discordant.

Proposed Geochronology

Radiometric dating has shown that Grenvillian and possibly Hudsonian events affected this area. No relict Archean ages have been found anywhere in the Blue Ridge; however, platy-textured leucocharnockites that cut metasedimentary rocks are present as xenoliths in the Shaeffer Hollow Granite, dated at about 1,800 m.y. Protoliths of the metasedimentary rocks are of indeterminate age, but certainly older than the granite, so the existence of Archean rocks cannot be discounted. In Canada, undisturbed Archean ages are found only west of the Grenville front. The lack of any Archean dates in the Roseland district may thus be related to its position 400 km east of the Grenville front (as located by Muehlberger and others, 1967).

The oldest dates obtained near the Roseland district, including our discordant date of 1,787 m.y. and others of 1,600 to 1,700 m.y. and 1,870 m.y. described above, were obtained on coarse-grained blue-quartz granite, granitic gneiss, or aluminous-biotite augen gneiss. The event roughly 1,800 m.y. ago is equivalent to the Hudsonian orogeny of Canada, which produced metamorphic assemblages of the granulite facies including widespread blue quartz, perthitic feldspars, and pyroxene, in the Aphebian gneiss terrane of eastern Canada.

Grenville ages are reported here for samples 1, 2, and 3 (Table 3). The age for sample 1 may record either generation of a leucocratic neosome in granulite or later metamorphism. The ages for samples 2 and 3 probably date intrusion of those igneous rocks.

The last dated event in the Roseland district was the intrusion of the Mobley Mountain Granite and formation of associated greisen, at about 635 to 650 m.y. ago. This event was part of a widespread (from northern Virginia to North Carolina) episode of granitic magmatism that appears to have been anorogenic. The Catoctin metabasalt, just outside this area to the northwest and southeast, is apparently a little younger (Mose, 1981).

No dates representing the widespread Paleozoic orogeny have been determined in this area. Paleozoic metamorphism here produced low-grade greenschist assemblages in deformation zones, but was not penetrative. It may have preceded or accompanied the final thrusting that produced the Blue Ridge anticlinorium in the latest Paleozoic.

MINERAL RESOURCES

The titanium resources of the Roseland district have been famous since the classic works of Watson and Taber (1913) and Ross (1941) and have traditionally been considered one of the country's largest hard-rock reserves of rutile and ilmenite. Mining of titanium minerals in the district started in 1878 but failed, as did later attempts to produce iron and phosphate. In 1907 the General Electric Company began development of rutile, which was continuously produced from 1910 to 1949. Ilmenite was produced starting in 1930 from saprolite deposits; mining ended with the closing of a pigment plant in 1971. The rock types hosting the ore deposits are anorthosite (rutile in the contact area with older rocks) and nelsonite (an ilmenite-apatite rock).

Anorthosite has been quarried for many years as commercial "aplite," used in the manufacture of container glass and glass-wool insulation. Limited development work, but no mining, has been done for kaolin and phosphate. Graphite in granulite may be a resource.

Part 2. Igneous and metamorphic petrology

Norman Herz

ABSTRACT

The anorthosite-charnockite-ferrodiorite assemblage at Roseland is similar to that of "Grenville" terranes elsewhere. In Canada, this magmatism may be Elsonian, about 1,400 m.y. old, but only Grenvillian dates, about 1,000 m.y., have been obtained in the Roseland district. The different plutons do not appear to be comagmatic, although their close spatial relationships and identical emplacement sequence worldwide suggest that they are cogenetic. Initial $^{87}Sr/^{86}Sr$ ratios of 0.7052 in the anorthosite and 0.7061 in charnockite of the Pedlar massif indicate that no simple and direct relationship existed between the two magmas.

The oldest rocks are the granulites, metamorphosed about 1,800 m.y. ago. Data on alkali elements suggest that the associated feldspathic gneisses and granites could have been derived from the granulites by partial melting. Oxygen isotopic data show a strong homogenization of original protoliths and later igneous rocks at about $\delta^{18}O$ of $+8.8^o/_{oo}$.

The anorthosite may have formed either by fractionation from a liquid or by partial melting. Workers do not agree on which process was dominant nor on the composition of either the original liquid or the protolith. Both the anorthosite and spatially related rocks of a ferrodiorite composition are enriched in Sr, indicating that they probably are not related by fractionation from the same magma; early segregation of Sr in the anorthosite would have left a Sr-depleted ferrodioritic magma. The differing K/Rb ratios also suggest different environments of formation; the ferrodiorites formed in a relatively shallow crustal level compared to the anorthosite. Once formed in a deep environment, the anorthosite, because of its low density, would be gravitationally unstable compared to lower crustal rocks and would rise, diapirlike, into a higher crustal level containing the ferrodiorite.

The AFM diagram shows a trimodal distribution: anorthosite data plot in the A-corner, data on ferrodiorites and nelsonites of the Roses Mill pluton plot near F-M and trend toward F, and data on most Pedlar and some Roses Mill rocks plot near a calc-alkaline trend. These rocks could not have been derived from the granulites and older rocks, although the Pedlar, Roses Mill, and Turkey Mountain assemblages appear to be chemically and lithologically similar. The nelsonites and some mafic ferrodiorites might be a product of liquid immiscibility produced during cooling of ferrodiorite magma.

The district shows an intermingling of earlier pyroxene-granulite and later amphibolite metamorphism facies. Physical conditions of metamorphism were determined by analyses of (1) coexisting orthopyroxenes and clinopyroxenes, (2) coexisting alkali and plagioclase feldspars, and (3) oxygen isotopic ratios in a quartz-rutile pair. Indicated temperatures are about 800 °C and indicated total pressures are about 8 kbar. P_{H_2O} was generally considerably less than P_{total}, and local variations in water content were responsible for the differences in mineral assemblages between adjacent layers. Typical assemblages include perthitic to antiperthitic feldspars, pyroxene (largely hypersthene), quartz (commonly deep blue), rutile and ilmenite, and hornblende and biotite rich in titanium.

Retrograde amphibolite-grade metamorphism might have accompanied the emplacement of the Mobley Mountain Granite about 650 m.y. ago. A widespread greenschist metamorphism, which accompanied Paleozoic deformation of Appalachian age throughout the Appalachians, was not pervasive here, except in deformation zones where assemblages of chlorite, carbonate, epidote, quartz, and albite were formed.

INTRODUCTION

The anorthosite pluton of the Roseland district and its associated granulites and plutons of charnockite, quartz mangerite, and ferrodiorite compositions form an assemblage typical of terranes associated with the Canadian Grenville province. The province forms a belt of 150 to 300 km wide extending from the Adirondacks to the coast of Labrador; anorthosite constitutes 15% to 20% of the terrane (Kranck, 1961, p. 300). The Roseland district is farther east of the Grenville front, as extended into the United States by subsurface and geophysical data (Muehlberger and others, 1967), than any Canadian anorthosite. The Grenville front passes through western Ohio and central Kentucky, some 400 km west of the Roseland. Other important differences between Roseland and the Canadian terranes are that the Roseland Anorthosite is unusually K rich and that its titanium ores are rutile rich, but all petrologic units are analogous (Herz, 1969b).

Anorthosite-charnockite plutons north of the Grenville front are examples of Elsonian anorogenic magmatism (Emslie, 1980), that is, they were emplaced in the interval between the Hudsonian (1,700–1,800 m.y. ago) and the Grenville orogeny (1,200–1,000 m.y. ago). The anorthosite-charnockite-ferrodiorite group at Roseland may also be anorogenic, emplaced in a tectonically stable craton environment but with all radioactive systems reset in the penetrative Grenville orogeny. However, no ages yet determined here are Elsonian (Table 3); they are only definite Grenvillian ages of about 1,040 to 980 m.y. ago and discordant zircon U-Pb Hudsonian dates of about 1,800 to 1,700 m.y.

The largest single pluton in the district is the Roseland Anorthosite, which intrudes layered granulite and quartzofeldspathic gneiss and granite. Charnockite, quartz mangerite, and ferrodiorite plutons of the younger Roses Mill Plutonic Suite intrude the anorthosite and older rocks and cover the largest area east of the anorthosite. The Mobley Mountain Granite and Irish

Creek tin gneisens intruded about 650 m.y. ago (Herz and others, 1981; Hudson and Dallmeyer, 1982) when the region had been uplifted to higher crustal levels than it had held previously. In a general way, the lithologic sequence of anorthosite-quartz mangerite-charnockite-ferrodiorite-granite also corresponds to sequentially younger ages. As in similar terranes in Canada, ultramafic rocks are exceptionally rare and formed after the main intrusive events. The concentration of mafic rocks, including nelsonites, found at or near the base of the Roses Mill Pluton may be due in large part to gravity segregation of an immiscible and relatively dense liquid.

A basic problem here, as with other Grenvillian terranes, is the explanation of this lithologic association. If all the pre-Grenvillian rocks had a common origin, (1) Were they the products or the cause of a granulite metamorphic event? (2) Did a single early parent gabbroic magma produce plagioclase cumulates that formed the anorthosite massif and result in Na- and Al-depleted ferrodioritic liquids (Table 1, columns 3 and 6)? (3) Was the original magma of ferrodiorite composition, and did it undergo massive crustal contamination? A hybrid magma might have produced some southern Norwegian anorthosites (Michot and Michot, 1969) as well as andesine anorthosite, similar to that of Roseland (Duchesne and Demaiffe, 1978). If the pre-Grenvillian rocks did not have a common origin, (4) Were the different rock suites formed from unrelated magmas at about the same time but by partial fusion and/or fractional crystallization of materials at different crustal or upper mantle levels?

Whether or not they are comagmatic, the rock suites must have some close relationship. Throughout all Grenvillian terranes, anorthosite through ferrodiorite are closely associated spatially and have the same temporal sequence, in which anorthosite is the oldest unit and granite is the youngest (Emslie, 1978; Philpotts, 1967). In addition, nelsonite, a rare rock outside this terrane, has been described from anorthosite terranes of the Roseland district, the Adirondacks, Canada, and Norway (Kolker, 1980).

GRANULITES AND RELATED ROCKS

In comparing the major-element determinations (Table 1) and some of the trace-element and isotopic data (Table 4), certain tentative conclusions can be made. The well-layered granulites (Table 6) probably represent the oldest lithologies in the district and include both sedimentary and igneous protoliths (Ames, 1981). The data indicate that the feldspathic gneisses, including the Shaeffer Hollow Granite types, could have been derived by partial melting of the well-layered granulites. Partial melting of mixed igneous-sedimentary protoliths will produce a liquid that is enriched in Rb and depleted in K and Sr compared to the protolith, especially in these granulite terranes (Sighinolfi, 1969). Rb has a much lower distribution coefficient in mineral/melt pairs than either K or Sr for plagioclase, amphiboles, and clinopyroxene (Cox and others, 1979, p. 334). Partial melting of the layered granulites to produce the Shaeffer Hollow Granite would also

TABLE 6. CHEMICAL COMPOSITIONS OF WELL-LAYERED GRANULITES OF THE ROSELAND AREA

	Mean	Standard deviation
SiO_2	62.42	5.94
TiO_2	1.48	0.53
Al_2O_3	14.21	1.46
Fe_2O_3	7.06	3.35
MgO	1.68	0.64
CaO	3.54	1.16
Na_2O	3.49	0.74
K_2O	3.19	0.86
MnO	0.09	0.04
P_2O_5	0.63	0.58
H_2O	1.41	0.76
Sum	99.2	

Note: Data from Ames (1981). N=6; X-ray fluorescence analysis; all Fe calculated as Fe_2O_3.

lead to the differences seen in K/Rb and Rb/Sr ratios, as well as the absolute abundances of these elements (Table 4).

The extreme variability of the K/Rb ratios of the well-layered granulites is due to (1) their variety of protoliths and (2) whether they represent restite + anatectite or did not undergo any partial melting. Field relations of the granulites and their mineralogy suggest that they were derived from a packet of sedimentary and volcaniclastic rocks. Major-element discriminant function analysis (Shaw, 1972) also suggests that the granulites were derived from sedimentary and volcanic rocks (Ames, 1981). The low K/Rb ratio of 233 for the feldspathic gneiss (Table 4) is typical of many granulites, as is the high Rb/Sr ratio of 1.31 (Hanson, 1978). The similar $\delta^{18}O$ of +8.8‰ in both granulite and gneiss (Table 4) is good evidence for an isotopic homogenization during the granulite metamorphic event (Ames, 1981). The variation in $\delta^{18}O$ in the well-layered granulites is much greater than that in the feldspathic gneiss and provides more evidence for a varied protolith that included igneous rocks having originally lower values and sediments having originally higher values. The low $\delta^{18}O/^{16}O$ ratios are similar to those in gabbros and basalts (Taylor, 1969) but are within the lower part of the total range for all the Roseland rock suites. Homogenization of the O isotopes during granulite-facies metamorphism may have raised an initially lower $\delta^{18}O$. The metamorphic conditions suggest temperatures of 750 °C and higher and total pressures of 8 ± 1 kb, or 25 km of burial; the anorthosite formed deeper than the level at which this metamorphic event took place.

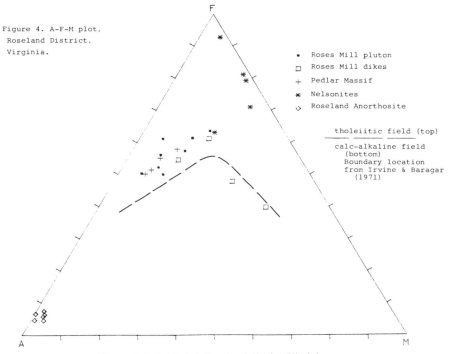

Figure 4. A-F-M plot,
Roseland District,
Virginia.

- Roses Mill pluton
□ Roses Mill dikes
+ Pedlar Massif
* Nelsonites
◇ Roseland Anorthosite

tholeiitic field (top)

calc-alkaline field
(bottom)
Boundary location
from Irvine & Baragar
(1971)

Figure 4. A-F-M plot, Roseland district, Virginia.

ANORTHOSITES, FERRODIORITES, AND RELATED ROCKS

The anorthosites could not have been derived in any way from the granulites. A partial melt of the granulites would not produce a rock as enriched in Sr and depleted in Rb as the anorthosite is. The anorthosite formed predominantly by the accumulation of plagioclase from melts formed in the uppermost mantle, or lower crust, as suggested by the anorthosite's low $^{87}Sr/^{86}Sr$ ratios ranging from 0.7047 to 0.7056 (Heath and Fairbairn, 1969) and K/Rb ratio of 2400 (Table 4), similar to the ratio obtained in an achondritic-type mantle (Gast, 1965).

Strong arguments have been advanced for and against a common parental magma for the anorthosite and charnockite suites (Isachsen, 1969). Strontium isotopic studies on the Morin anorthosite complex (Barton and Doig, 1977) indicate an initial $^{87}Sr/^{86}Sr$ ratio of 0.7043 and an age of 1,124 m.y. for the anorthosite and a ratio of 0.7050 and an age of 1,030 m.y. for mangerites of the surrounding terrane, suggesting either that the two suites are not derived from the same parent or that if they are closely related, at least one has been massively contaminated. The average initial $^{87}Sr/^{86}Sr$ ratio of the Roseland Anorthosite is 0.7052 (Heath and Fairbairn, 1969; this report, Table 4); that of the charnockite of the Pedlar massif is 0.7061 (Herz and others, 1981). These ratios indicate that a simple and direct relationship between the two magmas is unlikely.

A plot of the chemical analyses of the different rocks of the Roseland district (from Herz and Force, unpub. data) on an AFM diagram (Fig. 4) reveals at least a trimodal distribution. Data from the anorthosites are clustered in the A-corner; data on many Roses Mill and Turkey Mountain nelsonites plot near the F-M join, trending toward F, and data from the Pedlar and most other Roses Mill rocks plot in a tholeiitic field but near the calc-alkaline field. The diagram clearly shows that neither a simple crystallization-fractionation model based on a normal calc-alkaline trend nor a model based on an extreme Skaergaard Fe-enrichment trend can be used to explain the origin of these rock types. The diagram also suggests that although the rocks are closely associated in space and time, they are not comagmatic and that more than one parent magma must have existed.

The anorthosite and the Roses Mill Pluton probably are not genetically related. Field relationships clearly show that the Roses Mill intrudes the anorthosite. The anorthosite is greatly enriched in Sr and depleted in Rb. Any magma produced after the segregation of the anorthosite would have to be depleted in Sr and enriched in Rb. Values in the Roses Mill show the contrary: (1) Of all rock types in the district, the anorthosite has the lowest Rb/Sr ratios, and the Roses Mill has the second lowest. (2) The Roses Mill is also enriched in Sr and depleted in Rb. The K/Rb ratio of 441 for the Roses Mill Pluton suggests that it is a product of a relatively deep crustal environment (Sighinolfi, 1969) that was higher than that of the anorthosite.

Another problem that is not resolved by field data is the relationship between the Pedlar charnockite and the Roses Mill Pluton. They are presently separated by the Rockfish Valley deformation zone, and nowhere can contact relations be observed. Major- and trace-element analyses of both the Pedlar and the Roses Mill are strikingly similar (Table 1; Herz and Force, unpub.

TABLE 7. COMPOSITIONS OF COEXISTING FELDSPARS AND PYROXENES OF THE ROSELAND DISTRICT, VIRGINIA,
AND INDICATED TEMPERATURES

Sample number	Plagioclase Or:Ab:An	Orthoclase Or:Ab:An	Orthopyroxene En:Fs:Wo	Clinopyroxene En:Fs:Wo	Plagioclase (T°C)	Pyroxenes (T°C)
195			63.1:35.5:1.4	41.0:14.2:44.8		928
105	1.3:65.2:33.5	38.2:37.9:23.9	61.9:36.2:1.8	41.0:15.6:43.4	1043	945
18d	3.4:57.8:38.8	26.3:41.8:31.9	59.6:38.9:1.5	36.1:19.0:44.9	1220	878
11	35.7:39.0:25.3	95.7: 2.5: 1.8	57.5:40.7:1.8		<500	±870
12a			54.7:43.5:1.8	34.7:19.2:46.1		865
112			42.6:55.6:1.8	30.3:26.3:43.4		853*
88a	1.5:62.1:36.5	36.8:39.3:23.9	40.5:57.8:1.8	31.9:26.3:41.8	1110	872
20d	1.6:64.1:34.3	80.9:14.8: 5.1	31.4:66.6:2.0	23.0:33.6:43.4	680	815
**			29.7:60.5:9.8			±1010
13b			32.8:65.1:2.1	24.5:30.2:45.3		755

*Integrated composition obtained on orthopyroxene megacryst with exsolution lamellae.
Temperature estimated from diagram of Malcolm Ross and J. S. Huebner, U.S. Geological Survey (1976 written commun.).
**Temperature determined nearby in Roseland rutile pit by $^{18}O/^{16}O$ on coexisting plagioclase and rutile gave 750° (see part 1).
Note: Pyroxene temperatures by method of Wood and Banno (1973); feldspar temperatures by method of Stormer and Whitney (1977).
Sample descriptions:
195. Arrington quadrangle, drill hole 195 ft (59.4 m) depth, 1000 m northeast of Rose Union Church. Granulitic gneiss having weak compositional layers 1-3 mm thick. Sample contains antiperthite and twinned plagioclase, large rutilated quartz, granular pyroxenes, apatite, rutile, and amphibole and carbonate associated with coarse vein quartz.
105. Similar to 195, at 105 ft (32.0 m) depth.
18d. Arrington quadrangle, compositionally layered ferrodiorite of the Roses Mill Pluton texture suggests immiscibility with felsic-rich blobs within pyroxene-ilmenite-apatite-rich rock. Sample caontains antiperthitic plagioclase and orthopyroxene containing thin exsolution lamellae.
11. Arrington quadrangle, orthopyroxene megacryst facies at anorthosite border containing finer grained feldspars and clinopyroxene.
12a. Horseshoe Mountain quadrangle, massive charnockite of the Roses Mill Pluton containing quartz (20%), perthite (29%), antiperthite (29%), orthopyroxene (6%), clinopyroxene (3.3%), ilmenite (2.2%), apatite (2.1%), biotite, uralite, zircon, clinozoisite.
112. Arrington quadrangle, drill hole 112 ft (34.1 m) depth, ferrodiorite(?), of the Roses Mill pluton containing poikilitic pyroxene containing exsolution lamellae, plagioclase twinned and not antiperthitic, quartz, K-feldspar, magnetite, ilmenite, apatite, biotite.
88a. Piney River quadrangle, course-grained massive charnockite of the Turkey Mountain Pluton, containing large abundant perthite, platioclase, quartz, pyroxenes, ilmenite having biotite rims, apatite, garnet.
20d. Horseshoe Mountain quadrangle, quartz mangerite of the Roses Mill Pluton, massive, containing antiperthite (35%), perthite (28%), coarser orthopyroxene and pleochroic pink-green finer grained clinopyroxene (16%), quartz (14%), ilmenite (4.5%), apatite (2.7%), and zircon (0.3%).
13b. Massies Mill quadrangle, coarse-grained charnockite of the Pedlar massif, massive, containing somewhat uralitized pyroxene, perthite and antiperthite, quartz, apatite.

data). Alkali elemental ratios (Table 4) also show similarities that strongly suggest a comagmatic origin for the two rocks. Most of the chemical data show that the Pedlar is similar in composition to expected later differentiates of the Roses Mill. The average Sr content is lower and Rb and La contents are higher in the Pedlar than those in the Roses Mill (Table 4). The $\delta^{18}O$ in the Pedlar is +9.3‰, which is much higher than the $\delta^{18}O$ of +7.6‰ in the Roses Mill, suggesting the possibility of contamination by country rock in the Pedlar. The Pedlar pyroxenes are generally more Fe rich, and the Roses Mill pyroxenes are more Mg rich (Table 7). In associated rock series derived from the same fractionating magma stem, iron enrichment in pyroxenes generally indicates minerals that formed later and at lower temperatures. Tie lines of coexisting orthopyroxenes and clinopyroxenes in both the Roses Mill and the Pedlar are subparallel (Ames, 1981) and do not cross; thus, these tie lines are also consistent with a common

origin from the same or closely related magmas. Compared to the Roses Mill, the Pedlar appears to be a later differentiate and may have formed in the upper part of a common or closely related magma chamber.

ANORTHOSITE EMPLACEMENT

Once large-scale segregation of anorthosite had taken place in a deep katazonal environment, the relatively low density anorthosite would be subject to strong buoyant forces and tend to rise into higher crustal levels (Ramberg, 1967). Extensive autoclasis in the anorthosite suggests that at the time of emplacement the body was largely crystalline, so that high ambient temperatures were necessary to overcome its plastic-viscous drag. Diapiric rising and later lateral spreading took place in higher levels, as the dome-shaped mass came to rest where further upward movement was impossible. Gravity data (Eppihimer, 1978) do not show an associated positive anomaly, indicating that no large mafic mass is associated with the anorthosite. If any parental mafic magma source or refractory mafic residuum from partial melting existed, the anorthosite dome has since become dissociated from it. The absence of xenoliths in the anorthosite suggests that it was largely solid by the time of emplacement. Evidence that temperatures in the anorthosite must have fallen far below the liquidus by that time is seen in the sharp contacts with the leucocratic granulites and lack of melting at the contacts. However, abundant anorthosite dikes and coarse-grained rutile-bearing quartz mangerite at the anorthosite-granulite contact suggest that some residual liquid and considerable heat were still present at the time of emplacement.

LIQUID IMMISCIBILITY IN THE ROSES MILL PLUTONIC SUITE

Until fairly recently, liquid immiscibility was not considered an important process in the formation of magmas, largely because of Bowen's (1928) experimental work. He demonstrated (1) that immiscible liquids formed only in compositions that either are not found or are exceedingly rare in nature (i.e., they had total alkalis + alumina of about 5%) and (2) that immiscibility took place at unrealistically high temperatures of about 1700 °C. Since then, Roedder (1978) and others have documented the occurrence of immiscibility in natural rock systems.

Liquid immiscibility arises as magma of a given composition becomes unstable upon cooling and splits into two or more melts having, for example, granitic and ferrobasaltic compositions. Major and minor elements will partition into each melt (Ryerson and Hess, 1978). Elemental distribution is largely determined by the relative state of polymerization of each magma: the ferrobasaltic melt is relatively depolymerized because of a low Si/O ratio which allows highly charged cations, such as REE, Ti, P, Mn, and Zr, to form stable configurations with oxygen; the granitic melt is highly polymerized and preferentially allows low-charge cations, the alkalis, to enter. Phosphorus and titanium are greatly enriched in the ferrobasaltic liquid. Both elements increase the width of the solvus; P presumably complexes metal cations with PO_4^{-3} groups in the ferrobasaltic melt. Weibe (1979) reported natural occurrences of immiscibility in anorthosite-ferrodiorite in the Nain Complex, Labrador, where very late stage differentiation produced separate FeO- and SiO_2-rich liquids.

Granitic and ferrobasaltic liquids have greatly different densities and would separate; the denser basaltic liquid would concentrate toward the bottom of the magma chamber. Rock densities in the Roseland district (Eppihimer, 1978) are Roses Mill augen and granitic gneiss, 2.801 g/cm^3; nelsonites, 3.545 g/cm^3; and Roses Mill dike rocks, 2.936 g/cm^3. The dike rocks may approximate an early magma composition, which upon cooling, formed liquid immiscibility on a massive scale.

In a few places, a relict immiscible texture is evidenced by ellipsoidal felsic globs, completely surrounded by mafic areas (Fig. 3). In the best examples, flattened felsic areas, averaging 10 × 5 mm, are completely enclosed in more irregular anastomosing mafic-rich area. The mineralogy in each is identical, but the mineral abundances are not; andesine antiperthite-rich felsic areas contrast with pyroxene-rich mafic areas. A layering is seen by the flattened felsic blobs and the compositional banding of the mafic minerals. A sample from near Little Zion Church in the Arrington quadrangle, taken from near the bottom of the Roses Mill sheet, consists of felsic areas rich in andesine antiperthite, perthitic K-feldspar, and quartz and mafic areas rich in orthopyroxene, clinopyroxene, hornblende, ilmenite, apatite, and zircon. The felsic and mafic areas have the same mineral assemblages but different proportions, and both show a cumulatelike texture. Immediately overlying this rock, typical ferrodiorite appears without any evidence of liquid immiscibility.

Roedder (1978) proposed the use of a pseudoternary Greig diagram to demonstrate the possibilities of low-temperature-liquid immiscibility in natural systems (Fig. 5). The apices are SiO_2, $Na_2O+K_2O+Al_2O_3$, and $CaO+MnO+MgO+FeO+Fe_2O_2$ $+TiO_2+P_2O_5$. Shown in Figure 5 is the field of low-temperature immiscibility in the system leucite-fayalite-SiO_2. Earliest Roses Mill liquid compositions are approximated by the dike rocks, for which data points cluster near the center of the diagram. A gap in the femic element direction below these points suggests that immiscibility was possible and that, upon cooling, the single liquid split into two immiscible liquids. One trend was then toward femic-element enrichment, and the other was toward alkali and especially silica enrichment. The first trend could have produced the more mafic mangerites and ferrodiorites and eventually, the nelsonite; the second could have produced more normal charnockites.

The presence of two liquids after the emplacement of the dike rocks can be illustrated in a K_2O-TiO_2-P_2O_5 diagram (Fig. 6). Data on the dike rocks and some Roses Mill rocks plot in an area indicating a comparatively uniform and low K_2O content of about 20% to 35%, but a high TiO_2 content. Post-dike compositions appear to show two distinct trends: one trend toward strong K enrichment represents the normal Roses Mill–Pedlar ferrodiorite-charnockites, and the other trend of an extreme K depletion led to the nelsonites. A simple differentiation-fractionation model for liquid compositions represented by the early dike rocks would lead only to a single trend, that of K enrichment, and not two such disparate trends. Similar trends have been seen in the anorthosite pluton of the Nain Complex, Labrador (Weibe, 1979). Roedder (1979) has pointed out some physical conditions under which immiscible melts have formed in both natural and experimental systems that approximate these compositions: (1) two liquids formed in the leucite-fayalite-SiO_2 system upon cooling to about 1180 °C; (2) the immiscibility field increased in this system and in many others as fugacity of oxygen increased with total pressures up to about 15 kbar, after which the immiscible field diminished; (3) a density contrasts of 0.4 to 0.87 g/cm^3 existed between the immiscible liquids; and (4) two

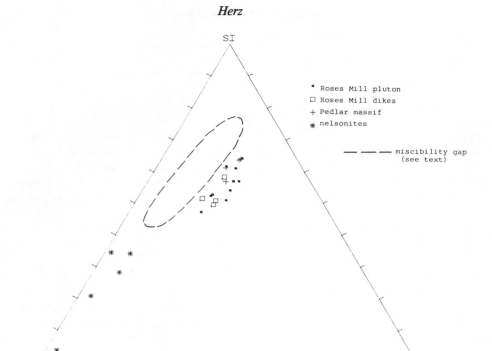

Figure 5. Greig Diagram: SiO_2–Na_2O+K_2O+Al_2O_3–MnO+MgO+FeO+Fe_2O_3+TiO_2+P_2O_5. Roses Mill Pluton and Pedlar massif.

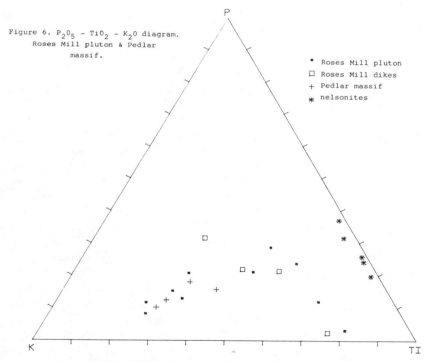

Figure 6. P_2O_5-TiO_2-K_2O diagram, Roses Mill pluton and Pedlar massif.

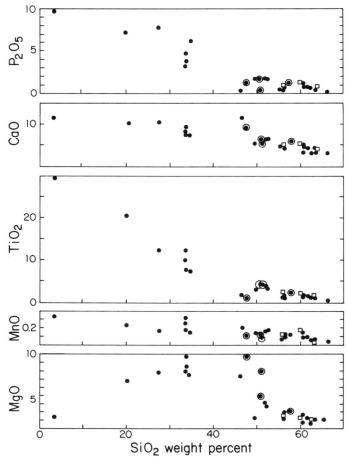

Figure 7. Harker variation diagram for selected elements. Roses Mill Pluton, black circles; Roses Mill dikes, circled circles; Pedlar massif, open squares.

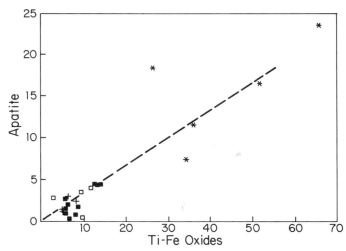

Figure 8. Normative weight percent, apatite versus Ti-Fe oxides, Roses Mill Pluton and Pedlar massif. Symbols described in Figure 4. Line is 1:3 ratio.

immiscible liquids formed in the magma remaining in a terrestrial basalt after about 56% of crystallization. All these conditions, that is, certainly temperature, pressure, contrast of density, and possibly percentage of crystallinity, could have been appropriate at the time postulated for the formation of the two immiscible liquids in the Roses Mill magma.

Evidence for fractional crystallization should be seen in Harker variation diagrams. As early formed plagioclase, pyroxenes, oxides, and apatite separate, the remaining liquid shows either impoverishment or enrichment in most elements. MgO especially is sensitive to fractionation of pyroxenes and decreases markedly as SiO_2 content increases. The diagrams in Figure 7 actually show that MgO content is high when SiO_2 content is 50%, that is, in the dike rocks, and that MgO content is strongly depleted as SiO_2 content both increases and decreases from 50%. A lack of analyzed rocks having SiO_2 contents of 38% to 47% suggests that this is the range of immiscible compositions.

Two elements whose behavior suggests liquid immiscibility are P, partitioned into basic melts by a factor of about 10 over acid melts, and Ti, partitioned into basic melts by a factor of about 3 (Watson, 1976). The Harker diagrams (Fig. 7) show that

both elements are strongly enriched in the low-SiO_2 rocks compared to the high-SiO_2 rocks.

Nelsonites should be the extreme product of liquid immiscibility. Philpotts (1967) found that a 1:2 ratio of apatite to oxides, by volume, was most common and suggested that this composition represented the eutectic point of an immiscible liquid that was associated with a liquid of dioritic composition. On a diagram of normative apatite:Ti-Fe oxide minerals (Fig. 8), most Virginia nelsonites and ferrodiorites approximate this ratio, except for those most enriched in oxide or apatite. Kolker (1980, 1982) calculated the mean ratio for oxides to apatite in magnetite and ilmenite nelsonites, worldwide, to be 1.98, with sigma 0.64.

CRYSTAL SETTLING AND CURRENT DEPOSITION

Layering and cumulate features are seen in Turkey Mountain rocks. In some leucodiorites, a vertical segregation is apparent in which mafic minerals are concentrated toward the bottoms of layers and plagioclase toward the tops. Some sorting appears to have taken place on the basis of settling velocities, controlled by differences in mineral and liquid densities. Density differences among the various rock types are striking (Eppihimer, 1978) and permit differentiation by settling. Recently, Irvine (1978, p. 725) suggested that crystal sorting by current deposition might also be an important factor in a magma chamber, especially in the segregation of large amounts of plagioclase. The end results of such a process would be structures and textures similar to those formed by crystal settling controlled by gravity.

Mineral layering in these Turkey Mountain rocks is generally not rythmic, that is, "right-side up" mineral graded, suggesting that accumulation was controlled to a greater extent by magmatic density currents than by crystal settling. However, fractionation-differentiation did take place, at least on a local scale. In exposures at the Smith Farm, Piney River quadrangle,

TABLE 8. COMPOSITIONS OF PYROXENES FROM SMITH FARM
         OUTCROP OF TURKEY MOUNTAIN FERRODIORITE

Sample No.	En	Fs	Wo
		(weight percent)	
1. lowest outcrop			
orthopyroxene	58.2	40.6	1.2
clinopyroxene	35.6	18.9	45.5
2. 50 m upslope (higher in section)			
orthopyroxene	55.2	43.6	1.2
clinopyroxene	33.7	19.5	46.8
3. 100 m upslope (highest in section)			
orthopyroxene	34.6	63.4	2.0
clinopyroxene	25.3	33.1	41.6

Note: Associated feldspars are hypersolvus in the first
two samples: (1) $An_{39.4}Ab_{59.8}Or_{0.8}$ and (2) $An_{39.8}Ab_{58.3}$
$Or_{1.9}$, and separate phases in the third: $An_{34.8}Ab_{63.8}$
$Or_{1.4}$, which agrees with the pyroxene data and with a
fractionation-differentiation origin with cooling toward
the top of the section.

cumulatelike features are seen in mineral grains. The country rock is a layered Turkey Mountain ferrodiorite, exposed on a hillside, in which mafic-rich layers are millimetres to a few centimetres in width, and feldspar-rich layers are a few to tens of centimetres in width. At the base of the section, pyroxene-rich layers are enstatite rich, and at the top, they are ferrosilite rich (Ames, 1981) (Table 8).

CHEMICAL PARAMETERS COMPARED

On the AFM projection (Fig. 4), Irvine and Baragar's (1971) divider separating calc-alkaline from tholeiitic suites shows the Roses Mill to have largely tholeiitic characteristics. A normative feldspar projection can distinguish sodic, normal, and potassic suites (Irvine and Baragar, 1971). Data on the anorthosite plot in the "normal" or K-rich andesite-basalt fields, and most Roses Mill rocks are potassic (Fig. 9). This pattern is similar to that of the Fennoscandian and Adirondack suites (Emslie, 1973, Fig. 4). As shown in an SiO_2 versus K_2O plot (Fig. 10), most of the Roseland rocks have higher K_2O contents than the reference calc-alkaline suites. This pattern is similar to the patterns seen in most anorthosite suites (Emslie, 1973, Fig. 3), though anorthosites themselves contain 52% to 56% SiO_2 and 0.5% to 1.5% K_2O. The Roseland anorthosite is slightly enriched in SiO_2 and considerably enriched in K_2 compared to anorthosites elsewhere.

MOBLEY MOUNTAIN GRANITE

More data are needed before the relationships between the Mobley Mountain Granite and these earlier Grenvillian and pre-Grenvillian rocks can be established. Geophysical evidence (Eppihimer, 1978) suggests that the Mobley Mountain Granite forms a plug-shaped body having a maximum depth of 7.2 km. Its

$^{87}Sr/^{86}Sr$ ratio of 0.7045 is lower than that determined for any rock in the district (Table 4), and the granite is corundum-normative (Herz and Force, unpub. data). These data suggest that the granite may be the product of a later crustal attenuation or rifting event that took place about 650 m.y. ago and that it may not be genetically related to the older rocks.

SUMMARY AND CONCLUSIONS—IGNEOUS ROCKS

The Roses Mill and related rocks plot as both calc-alkaline and tholeiitic trends, and there is a gap between the trends (Fig. 4). Anorthosites cluster around the A apex of the AFM diagram and are not part of any trend. The gap in the Roses Mill analyses may have been controlled by a field of liquid immiscibility which allowed the Si-alkali-rich liquid to follow an apparent calc-alkaline trend and the Ti- and P-rich liquid to show extreme alkaline depletion. Thus, the nelsonites and mafic ferrodioritic rocks are the products of a magma stem that formed by liquid immiscibility during cooling from the main Roses Mill magma. Earliest approximations to original magma compositions are assumed to be best represented by the dike rocks that intrude the older anorthosite and granulite. Thus, they were the first Roses Mill rocks emplaced.

The anorthosite appears to be a product of a deep crustal environment; any relationship to the Roses Mill magma must be tenuous, as the two do not appear to be comagmatic. The Roses Mill, Turkey Mountain, and Pedlar charnockite have many strong chemical and mineralogical similarities, which suggest either that they were comagmatic or that they were produced from magmas of similar composition. Although they plot in Irvine and Baragar's (1971) tholeiitic field, their main stem is toward enrichment in alkalies, that is, a calc-alkaline trend (Fig. 4).

The high K_2O content of the anorthosites suggests a great depth of formation. K_2O content in calc-alkaline magmas shows a positive correlation with depth to the Benioff zone (Dickinson and Hatherton, 1967). Such magmas are also emplaced at considerable horizontal distance from the trench or active part of an orogen. If the analogy is correct, then the anorthosites were emplaced in an unusual thick segment of continental crust and at some distance away from any trench, presumably within a stable craton. Evidence for a great thickness of crust at the time of formation is seen in the high K/Rb ratio and relatively low $^{87}Sr/^{86}Sr$ ratio (Sighinolfi, 1969). Grenvillian metamorphic conditions were also deep seated, as pressures were about 8 kbar (equivalent to depth of burial of about 25 km) and temperatures were about 850 °C. The anorthosite must have formed in a deeper environment than that of Grenvillian metamorphism before it rose into the granulite terrane where it was also metamorphosed. Chemical compositions similar to the Roses Mill are found today in environments of crustal attenuation and rifting. This finding suggests that the ferrodiorite magmas were formed somewhat later than the anorthosite and at a time when the tectonic environment had changed significantly. The earlier environment producing the anorthosite might have resulted from

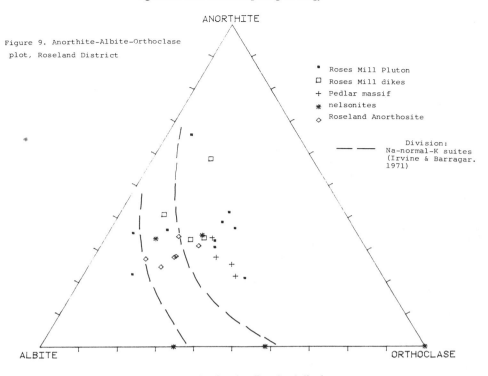

Figure 9. Anorthite-Albite-Orthoclase plot, Roseland District

■ Roses Mill Pluton
□ Roses Mill dikes
+ Pedlar massif
✳ nelsonites
◇ Roseland Anorthosite

Division:
Na-normal-K suites
(Irvine & Barragar.
1971)

Figure 9. An-Ab-Or plot, Roseland district.

FIGURE 10. WEIGHT %, SiO₂ vs K₂O. SYMBOLS AS FIGURE 3. OUTLINED IS IRVINE & BARRAGAR'S CALC-ALKALIC FIELD (1971).

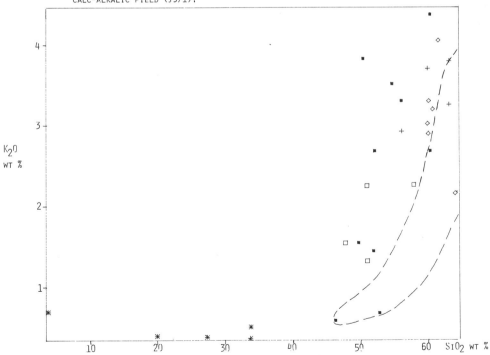

Figure 10. Weight percent, SiO₂ versus K₂O, Roseland district. Symbols described in Figure 9.

continental collision and an overthickened crust; the later ferro-diorites formed at a time of crustal attenuation on the craton.

METAMORPHISM

Description of Facies

The granulite facies is generally defined as beginning in the prograde metamorphism of pelitic rocks with the appearance of orthopyroxene (Turner, 1968, p. 327). In the Adirondacks, temperatures of 550 °C to 625 °C are inferred for the start of granulite-facies conditions and the transition from the upper amphibolite facies (Buddington and Lindsley, 1964, p. 337). Typical minerals in a granulite assemblage include hornblende, pyroxenes, perthitic feldspars, and quartz (Turner, 1968, p. 327). Closely associated with metamorphic rocks of the granulite facies are the charnockites, igneous rocks of a granitic composition having a mineral composition identical to that of granulites, which also lack hydrous mineral phases; their mineral assemblages suggest regional temperatures exceeding 700 °C at depths greater than 15 km (Carmichael and others, 1974, p. 590).

The amphibolite facies is defined by assemblages containing oligoclase-andesine and hornblende (Turner, 1968, p. 307) and indicates medium to high grades of metamorphism. The greenschist facies includes assemblages containing albite, white micas, biotite, and epidote (Turner, 1968, p. 368).

In the Roseland district, as in many other Grenvillian granulite terranes, granulite- and amphibolite-facies metamorphic rocks appear to intermingle. In the Adirondacks, for example, some granulite- and amphibolite-facies metamorphic assemblages were apparently reversed if their principle control had been falling temperature gradients away from igneous contacts. Assemblages of upper amphibolite to pyroxene-granulite metamorphic facies are also controlled by P_{H_2O}, which in typical granulite terranes is much less than P_{load}. Thus, under conditions of uniform total load pressure and temperature, granulite terrane metamorphic reactions will be dehydration reactions and will be controlled by steady state and steep P_{H_2O} gradients. Adjacent lithologic units show pyroxene-granulite- and amphibolite-facies conditions, thus demonstrating a fine-scale sensitive control by P_{H_2O}. The pyroxene granulite facies, however, is clearly predominant in the original rocks, although many rocks show the effects of a widespread retrograde amphibolite-grade metamorphic overprint. This overprint may have been caused in the waning stages of the Grenville orogeny or later when falling temperature and uplift into higher crustal levels promoted hydration reactions. A Paleozoic retrograde greenschist metamorphism here is part of a widespread Appalachian event that has been correlated with deformation throughout the Appalachians (Odom and Fullagar, 1973) and may be related to the formation of the Blue Ridge anticlinorium.

Grenvillian Metamorphism

The packet of quartzo-feldspathic rocks now represented by

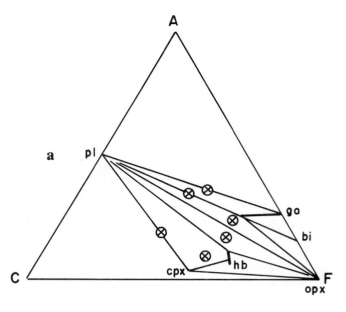

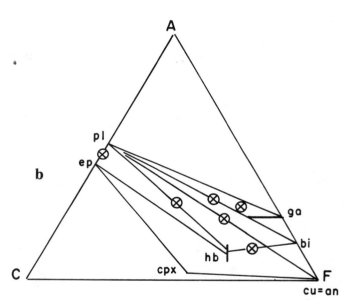

Figure 11. (a) Schematic ACF diagram for granulite-facies assemblages. Rocks have excess SiO_2 and K_2O; quartz and K-feldspar are additional possible phases (Turner, 1968). Circled X = assemblages seen in the Roseland district; pl = plagioclase; ga = garnet, bi = biotite, hb = hornblende, cpx = clinopyroxene, opx = orthopyroxene. (b) Schematic ACF diagram for amphibolite-facies assemblages. Rocks have excess SiO_2 and K_2O; quartz and K-feldspar are additional possible phases (Turner, 1968) Circle X = assemblages seen in the Roseland district; pl = plagioclase; ep = epidote; ga = garnet; bi = biotite; hb = hornblende; cpx = clinopyroxene; cu-an = cummingtonite-anthophyllite.

granulites, as well as the Roseland Anorthosite and the charnockitic rocks, underwent regional metamorphism at the highest grade, that of the hypersthene zone (Winkler, 1974, p. 245). The rocks are now characterized by mineral assemblages (Fig. 11) typical of the granulite facies and by an aspect typical of granu-

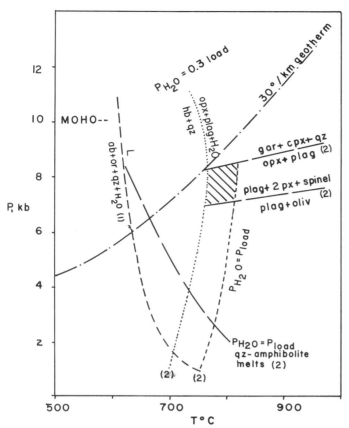

Figure 12. Metamorphic grid for granulite facies, Roseland district. (1) Bowen and Tuttle, 1950; (2) Oliver, 1977.

litic rocks (Mehnart quoted in Winkler, 1974), that is, "a fine- to medium-grained metamorphic rock composed essentially of feldspar with predominantly anhydrous ferromagnesian minerals; granoblastic texture with gneissose to massive structures. Some quartz is lenticular or in lenticular aggregates ("disc-like quartz")."

The Roseland district granulites show mineralogical characteristics that are associated with the regional hypersthene zone of metamorphism (Winkler, 1974, p. 249): (1) Alkali feldspars are typically perthitic to mesoperthitic, and plagioclase (calcic oligoclase to andesine) is antiperthitic. (2) Orthopyroxene (hypersthene or ferrohypersthene) is pleochroic greenish and pink, and clinopyroxene is a light-green diopside-hedenbergite. (3) Hornblende and biotite, when present, are relatively rich in Ti, and at highest grades, Ti goes into oxides or pyroxenes. (4) Quartz is typically deep blue and disclike and contains abundant oriented rutile needles.

Garnet is less abundant in the Roseland district than in many other granulite terranes; its sparsity can be attributed, in part, to compositions here rich in Ca-Fe components (Fig. 11) and, in part, to moderate rather than high total pressures (Fig. 12). At 800 °C, pressures of at least 8.5 kbar are needed to stabilize assemblages of clinopyroxene+garnet+quartz (Oliver,

1977); total pressures here were apparently lower than 9 kbar and closer to 8 kbar.

Orthopyroxenes and clinopyroxenes are abundant in the Turkey Mountain pluton and Pedlar massif, as well as in the granulites and border facies of the anorthosites; they are seen to a lesser extent in the Roses Mill Pluton. Pairs of coexisting pyroxenes from the anorthosite border rocks and charnockites were analyzed by electron microprobe, and their indicated temperatures were calculated (Table 7). These temperatures have possible errors of 50 °C to 150 °C (Bohlen and Essene, 1979). An integrated composition of an orthopyroxene with exsolved lamellae showed about 1010 °C, which is probably igneous. Other analyses ranged from about 930 °C to 940 °C at the anorthosite contact to 815 °C 0.75 km to the east.

The lowest temperature, 755 °C, from Pedlar charnockite, is also the farthest sample from the contact. Temperatures do not appear to fall off in any systematic way, away from the contact. However, a compositional control by original igneous pyroxenes is suggested by high En contents near the anorthosite contact, which may also correspond to the bottom of the Roses Mill magma chamber.

Temperatures of formation were also calculated by electron microprobe analysis of coexisting plagioclase and alkali feldspars as well as coexisting orthopyroxenes and clinopyroxenes (Table 7). Many feldspars, unfortunately, are perthitic and so their bulk compositions are difficult to approximate by microprobe techniques. Calculated temperatures on some feldspar pairs of about 1050 ° to 1200 °C must be indicative of original igneous conditions; most values are about 650 °C or lower and only indicate reequilibration under later metamorphic conditions. A temperature of 750 °C was determined by $^{18}O/^{16}O$ ratios in coexisting rutile and plagioclase.

Crystallization of sphene, ilmenite, and rutile were controlled by bulk chemical compositions as well as by metamorphic *P-T* conditions. The presence of ilmenite and rutile in the granulites is consistent with the general trend in which titanium silicates in low-grade rocks give way to titanium oxides in higher grade rocks (Ramberg, 1952; Force, 1976). In rocks that are rich in CaO (rocks whose compositions plot below the join pl-opx [plagioclase-hypersthene] on the ACF diagram in Fig. 11a), sphene may form (Martignole, 1975, p. 291) although granulite assemblages characteristically contain ilmenite and rutile because calcium is taken up by An-rich plagioclase. Rocks rich in Al_2O_3 (i.e., rocks whose compositions plot above the pl-opx join) will crystallize rutile instead (Martignole, 1975, p. 291). Rutile and pyroxene will also form when Ti-rich biotite or hornblende becomes unstable during progressive metamorphism. In retrograde metamorphic hydration reactions, plagioclase reacts with fluid to form mica and liberates CaO, which then combines with rutile or ilmenite to form sphene.

Oxygen isotopic ratios from coexisting rutile and plagioclase, which indicate 750 °C, and the lower 2 feldspar temperature determinations may represent reequilibration after peak metamorphism and so are minima (Table 7). Pyroxene determination may have errors of 50 °C to 150 °C, so that 800 °C ± 50 °C may be the best approximation to actual peak metamorphic temperature. The metamorphic assemblage 2 pyroxenes + garnet + quartz + plagioclase at these temperatures implies total pressures of about 8 ± 1 kbar (Fig. 12); below about 7 kbar, plagioclase + olivine would form in these rocks, and above about 10 kbar, the reaction orthopyroxene + plagioclase = garnet + clinopyroxene + quartz would take place.

At about 8 kbar total pressure and with P_{H_2O} = 0.3 P_{load}, Ti-rich biotite and hornblende are stable up to about 760 °C; if P_{H_2O} = P_{load}, they are stable up to about 820 °C. In the Roseland area, amphibolite facies in these dry rocks implies local hydration reactions brought about by a simple increase in P_{H_2O} at constant P_{total} and temperature conditions. Assemblages containing abundant amphiboles and biotite mark the transition from pyroxene-granulite to upper amphibolite facies. Metamorphism reactions seen in Roses Mill rocks illustrate some phase changes involved: pyroxene + plagioclase = hornblende + garnet; pyroxene + perthite = biotite; and ilmenite + perthite + plagioclase = albite + biotite + sphene. Although these hydration effects are seen throughout the granulitic rocks, they are most pervasive in Roses Mill augen gneisses and granitoid rocks. Typical assemblages in the Roses Mill (part I) show a great abundance of quartz, plagioclase, calcic oligoclase, or retrograde albite and brown biotite, accompanied by augen gneiss.

The Mobley Mountain Granite was emplaced about 650 m.y. ago, after the peak of Grenvillian metamorphism but before the low-grade early Paleozoic deformational event. The granite is not surrounded by any metamorphic contact aureole; Roses Mill rocks appear to have the same mineral assemblages whether found as xenoliths at the granite contact or at any distance from the contact. The granite formed at some great depth (Eppihimer, 1978) and was emplaced in a terrane that was still at amphibolite-facies conditions. Thus, assemblages of hornblende, biotite, epidote, K-feldspar, sphene, and oligoclase-andesine were stable at the time of emplacement of the Mobley Mountain Granite. Conditions were suitable for the formation of anatectic melts, seen in migmatites along the granite contact, as well as for a diffusion of alkalis carried along in pervasive aqueous intergranular films that circulated through the heated country rock. Volatile-element solutions from the Mobley Mountain magmas helped these processes along by increasing the partial pressure of water at a time when P_{H_2O} < P_{total}.

P-T conditions of the amphibolite facies at the time of the emplacement of the Mobley Mountain Granite suggest that the terrane may have been at about 550 °C to 600 °C and 6 kbar pressure and that P_{H_2O} in some places approached P_{total}. These conditions, compared to the original ones of granulite formation, perhaps 350 m.y. earlier, are consistent with a geothermal gradient of about 30 °C/km and an uplift rate of 0.05 mm/yr, or 17 km total uplift from Grenvillian to Mobley Mountain time. Wherever sufficient water was available, retrograde metamorphic reactions from granulite to amphibolite facies may have taken place.

Paleozoic Retrograde Metamorphism

A widespread greenschist-facies deformational-metamorphic event was overprinted throughout this part of the Appalachians 520 to 583 m.y. ago (Fullagar and Dietrich, 1976, p. 358–359). Throughout the district, cataclasis took place in narrow zones accompanied by greenschist mineralization. The Rockfish Valley deformation zone, which may have formed initially at the end of Grenvillian time, was reactivated, as indicated by a universal northeast-trending blastomylonitic fluxion structure.

The greenschist regional metamorphism overprinted the older higher grade assemblages in many places. Retrograde assemblages in the anorthosites, granulites, and charnockitic rocks are present in virtually every outcrop in restricted zones, such as epidote-quartz-sericite veins; patches of uralitization of pyroxene; chloritization of amphiboles, biotite, and pyroxenes; sphene partially replacing titanium oxides in many rocks; the disappearance of perthitic and antiperthitic feldspars; and the change of oligoclase-andesine to albite + sericite + calcite + clinozoisite. In the Roses Mill augen gneisses and granitic rocks, which already possessed hydrous mineral phases, especially biotite, greenschist metamorphic mineral assemblages are more widely distributed. However, the persistence of higher grade assemblages in all undeformed rocks shows that the retrograde mineral reactions did not go to completion in any large area, except within the Rockfish Valley deformation zone. Mineral assemblages suggest that temperatures were 400 °C to 500 °C and that total pressures were 4 to 6 kbar (Turner, 1968, p. 286).

REFERENCES CITED

Ames, R. M., 1981, Geochemistry of the Grenville Basement rocks from the Roseland district, Virginia [M.S. thesis]: University of Georgia, 91 p.

Anderson, A. T., and Morin, M., 1969, Two types of massif anorthosites and their implications regarding the thermal history of the crust, *in* Isachsen, Y. W., ed., Origin of anorthosite and related rocks: New York State Museum and Science Service Memoir 18, p. 57–69.

Baer, A. J., Emslie, R. F., Irving, E., and Tanner, J. G. 1974, Grenville geology and plate tectonics: Geoscience Canada, v. 1, p. 54–61.

Bartholomew, M. J., 1977, Geology of the Greenfield and Sherando quadrangles, Virginia: Virginia Division of Mineral Resources Publication 4, 43 p.

Bartholomew, M. J., Gathright, T. M., and Henika, W. S., 1981, A tectonic model for the Blue Ridge in central Virginia: American Journal of Science, v. 281, p. 1164–1183.

Barton, J. M., and Doig, Ronald, 1977, Sr isotopic studies of the origin of the Morin anorthosite complex, Quebec, Canada: Contributions to Mineralogy and Petrology, v. 65, p. 219–230.

Bloomer, R. O., and Werner, H. J., 1955, Geology of the Blue Ridge region in central Virginia: Geological Society of America Bulletin, v. 66, p. 579–606.

Bohlen, S. R., and Essene, E. J., 1979, A critical evaluation of two-pyroxene thermometry in Adirondack granulites: Lithos, v. 12, p. 335–345.

Bowen, N. L., 1928, The evolution of the igneous rocks: Princeton, New Jersey, Princeton University Press, 332 p.

Bowen, N. L., and Tuttle, O. F., 1950, The System NaAl Si_3O_8-KAlSi$_3$O$_8$-H$_2$O: Journal of Geology, v. 58, p. 489–511.

Brown, W. R., 1958, Geology and mineral resources of the Lynchburg quadrangle, Virginia: Virginia Division of Mineral Resources Bulletin 74, 99 p.

Buddington, A. F., 1961, The origin of anorthosite re-evaluated: Geological Survey of India Records, v. 86, pt. 3, p. 421–432.

Buddington, A. F., and Lindsley, D. H., 1964, Iron-titanium oxide minerals and synthetic equivalents: Journal of Petrology, v. 5, p. 310–357.

Carmichael, I.S.E., Turner, F. J., and Verhoogen, John, 1974, Igneous petrology: New York, McGraw Hill, 739 p.

Conley, J. F., 1978, Geology of the Piedmont of Virginia—interpretations and problems: Virginia Division of Mineral Resources Publication 7, p. 115–149.

Cox, K. G., Bell, J. D., and Pankhurst, R. J., 1979, The interpretation of igneous rocks: London, Allen & Unwin, 450 p.

Dallmeyer, D. D., 1978, $^{40}Ar/^{39}Ar$ incremental-release ages of hornblende and biotite from Grenville basement rocks within the Indian Head Range complex, southwest Newfoundland: Their bearing on Late Proterozoic-early Paleozoic thermal history: Canadian Journal of Earth Sciences, v. 15, p. 1374–1379.

Davis, R. G., 1974, Pre-Grenville ages of basement rocks in central Virginia: A model for the interpretation of zircon ages [M.S. thesis]: Virginia Polytechnic Institute, 46 p.

Dickinson, W. R., and Hatherton, T., 1967, Andesitic volcanism and seismicity around the Pacific: Science, v. 157, p. 801–803.

Duchesne, J. C., and Demaiffe, D., 1978, Trace elements and anorthosite genesis: Earth and Planetary Science Letters, v. 38, p. 249–272.

Emslie, R. F., 1973, Some chemical characteristics of anorthosite suites and their significance: Canadian Journal of Earth Sciences, v. 10, p. 54–71.

—— 1978, Anorthosite massifs, rapakivi granites, and late Proterozoic rifting of North America: Precambrian Research, v. 7, p. 61–98.

—— 1980, Geology and petrology of the Harp Lake Complex, central Labrador: An example of Elsonian magmatism: Canada Geological Survey Bulletin, v. 293, 136 p.

Eppihimer, R. M., 1978, A geophysical study of the Roseland anorthosite-titanium district, Nelson and Amherst Counties, Virginia [M.S. thesis]: University of Georgia, 115 p.

Force, E. R., 1976, Metamorphic source rocks of titanium placer deposits—A geochemical cycle: U.S. Geological Survey Professional Paper 959B, 16 p.

Force, E. R., and Herz, Norman, 1982, Anorthosite, ferrodiorite, and titanium deposits in Grenville terrane of the Roseland district, central Virginia, *in* Lyttle, Peter, ed., Central Appalachian geology: American Geological Institute, p. 109–119.

Fullagar, P. D., and Dietrich, R. V., 1976, Rb-Sr isotopic study of the Lynchburg and probably correlative formations of the Blue Ridge and western Piedmont of Virginia and North Carolina: American Journal of Science, v. 276, p. 347–365.

Furcron, A. S., 1934, Igneous rocks of the Shenandoah National Park area: Journal of Geology, v. 42, p. 400–410.

Gast, P. W., 1965, Terrestrial ratio of K to Rb and the composition of the Earth's mantle: Science, v. 147, p. 858–860.

Grauert, B., 1972, New U-Pb isotopic analyses of zircons from the Baltimore Gneiss and the Setters Formation: Carnegie Institution of Washington Yearbook 71, p. 301–305.

Hanson, G. N., 1978, The application of trace elements to the petrogenesis of igneous rocks of granitic composition: Earth and Planetary Science Letters, v. 38, p. 26–43.

Heath, S. A., and Fairbairn, H. W., 1969, $^{87}Sr/^{86}Sr$ ratios in anorthosites and some associated rocks, *in* Isachsen, Y. W., Origin of anorthosite and related rocks: New York State Museum and Science Service Memoir 18, p. 99–110.

Herz, Norman, 1969a, Anorthosite belts, continental drift, and the anorthosite event: Science, v. 164, no 3882, p. 944–947.

—— 1969b, The Roseland alkalic anorthosite massif, Virginia, *in* Origin of anorthosite and related rocks: Isachsen, Y. W., ed., New York State Museum and Science Service Memoir 18, p. 357–367.

Herz, Norman, Mose, D. C., and Nagel, M. S., 1981, Mobley Mountain granite and the Irish Creek tin district, Virginia: A genetic and temporal relationship: Geological Society of America Abstracts with Programs, v. 13, p. 472.

Hudson, T. A., 1982, The Irish Creek tin district *in* Lyttle, Peter, ed., Central Appalachian geology: American Geological Institute, p. 118–119.

Hudson, T. A., and Dallmeyer, R. D., 1982, Age of mineralized greisens in the Irish Creek tin district, Virginia Blue Ridge: Economic Geology, v. 77, p. 189–192.

Irvine, T. N., 1978, Density current structure and magmatic sedimentation: Carnegie Institution of Washington Yearbook 77, p. 717–725.

Irvine, T. N., and Baragar, W.R.A., 1971, A guide to the chemical classification of the common volcanic rocks: Canadian Journal of Earth Sciences, v. 8, p. 523–548.

Isachsen, Y. W., 1969, ed., Origin of anorthosite and related rocks: New York State Museum and Science Service Memoir 18, 466 p.

Kolker, Allan, 1980, Petrology, geochemistry, and occurrence of iron-titanium oxide and apatite (nelsonite) rocks [M.S. thesis]: Amherst, University of Massachusetts, 157 p.

Kolker, Allan, 1982, Mineralogy and geochemistry of Fe-Ti oxide and apatite (nelsonite) deposits and evaluation of the liquid immiscibility hypothesis: Economic Geology, v. 77, p. 1146–1158.

Kranck, E. H., 1961, The tectonic position of the anorthosites of eastern Canada: Geologic Commission of Finland Bulletin, v. 196, p. 299–320.

Lukert, M. T., 1977, Discordant zircon age of the Old Rag granite, Madison County, Virginia: Geological Society of America Abstracts with Programs, v. 9, p. 162.

Martignole, Jacques, 1975, Le Precambrian dans le sud de la province tectonique de Grenville (Bouclier Canadien): University of Montreal, 405 p.

Martignole, J., and Schrijver, K., 1970, The level of anorthosite and its tectonic pattern: Tectonophysics, v. 10, p. 402–409.

Mawdsley, J. B., 1927, St. Urbain area, Charlevoix district, Quebec: Canada Geological Survey Memoir 152, 58 p.

Michot, Jean, and Michot, Paul, 1969, The problem of anorthosites: The South Rogaland igneous complex, southwestern Norway, *in* Isachsen, Y. W., ed., Origin of anorthosite and related rocks: New York State Museum and Science Service Memoir 18, p. 399–410.

Middlemost, E.A.R., 1970, Anorthosite: A graduated series: Earth Science Reviews, v. 6, p. 257–265.

Mose, D. G., 1981, Cambrian age for the Catoctin and Chopawamsic Formations in Virginia: Geological Society of America Abstracts with Programs, v. 13, p. 31.

Muehlberger, W. R., Denison, R. E., and Lidiak, E. G., 1967, Basement rocks in continental interior of United States: American Association of Petroleum Geologists Bulletin, v. 51, p. 2351–2380.

Nelson, W. A., 1962, Geology and mineral resources of Albemarle County: Virginia Division of Mineral Resources Bulletin 77, 92 p.

Nockolds, S. R., 1954, Average chemical composition of some igneous rocks: Geological Society of America Bulletin, v. 65, p. 1007–1032.

Odom, A. L., and Fullagar, P. D., 1973, Geochronologic and tectonic relationships between the Inner Piedmont, Brevard Zone, and Blue Ridge belts, North Carolina: American Journal of Science, v. 273–A, p. 133–149.

Oliver, G.J.H., 1977, Feldspathic hornblende and garnet granulites and associated anorthosite pegmatites from Doubtful Sound, Fiordland, New Zealand: Contributions to Mineralogy and Petrology, v. 65, p. 111–121.

Paulson, E. G., 1964, Mineralogy and origin of the titaniferous deposit at Pluma Hidalgo, Oaxaca, Mexico: Economic Geology, v. 59, p. 753–767.

Philpotts, A. R., 1967, Origin of certain iron-titanium oxide and apatite rocks: Economic Geology, v. 62, p. 303–315.

Ramberg, Hans, 1952, The origin of metamorphic and metasomatic rocks: Chicago, Illinois, University of Chicago Press, 317 p.

—— 1967, Gravity, deformation, and the Earth's crust—As studied by centrifuged models: New York, Academic Press, 205 p.

Roedder, Edwin, 1978, Silicate liquid immiscibility in magmas and in the system $K_2O-FeO-Al_2O_3-SiO_2$, an example of serendipity: Geochimica et Cosmochimica Acta, v. 42, p. 1597–1617.

—— 1979, Silicate liquid immiscibility in magmas, *in* Yoder, H. S., Jr., ed., The evolution of the igneous rocks: Princeton, New Jersey, Princeton University

Press, p. 15–57.

Ross, C. S., 1941, Occurrence and origin of the titanium deposits of Nelson and Amherst Counties, Virginia: U.S. Geological Survey Professional Paper 198, 59 p.

Ryerson, F. J., and Hess, P. C., 1978, Implications of liquid-liquid distribution coefficients to mineral-liquid partitioning: Geochimica et Cosmochimica Acta, v. 42, p. 921–932.

Shaw, D. M., 1972, The origin of the Ansley Gneiss, Ontario: Canadian Journal of Earth Sciences, v. 8, p. 301–310.

Sighinolfi, G. P., 1969, K-Rb ratios in high grade metamorphism: A confirmation of the hypothesis of a continual crustal evolution: Contributions to Mineralogy and Petrology, v. 21, p. 346–356.

Silver, L. T., 1969, A geochronologic investigation of the anorthosite complex, Adirondack Mountains, New York, *in* Isachsen, Y. W., ed., Origin of anorthosite and related rocks: New York State Museum and Science Service Memoir 18, p. 233–251.

Steiger, R. H., and Jaeger, E., 1978, Contribution to the geologic time scale: American Association of Petroleum Geologists, Studies in Geology, no. 6, p. 67–71.

Stormer, J. C., Jr., and Whitney, J. A., 1977, Two-feldspar geothermometry in granulite facies metamorphic rocks, sapphirine granulites from Brazil: Contributions to Mineralogy and Petrology, v. 65, p. 123–133.

Taylor, H. P., Jr., 1969, Oxygen isotope studies of anorthosites, with particular reference to the origin of bodies in the Adirondack Mountains, New York, *in* Isachsen, Y. W., ed., Origin of anorthosite and related rocks: New York State Museum and Science Service Memoir 18, p. 111–134.

Tilton, G. W., David, G. L., Wetherill, G. W., and Bass, M. W., 1960, 1000-million-year-old minerals from the eastern United States and Canada: Journal of Geophysical Research, v. 65, p. 4173–4179.

Turner, F. J., 1968, Metamorphic petrology, mineralogical and field aspects: McGraw Hill, New York, 403 p.

Watson, E. B., 1976, Two-liquid partition coefficients: Experimental data and geochemical implications: Geochimica et Cosmochimica Acta, v. 56, p. 119–134.

Watson, T. L., and Taber, Stephen, 1913, Geology of the titanium and apatite deposits of Virginia: Virginia Geological Survey Bulletin 3A, 308 p.

Weibe, R. A., 1979, Fractionation and liquid immiscibility in an anorthositic pluton of the Nain Complex, Labrador: Journal of Petrology, v. 20, p. 239–269.

Winkler, H.G.F., 1974, Petrogenesis of metamorphic rocks (3rd edition): New York, Springer-Verlag, 320 p.

Wood, B. J., and Banno, S., 1973, Garnet-orthopyroxene and orthopyroxene-clinopyroxene relationships in simple and complex systems: Contributions to Mineralogy and Petrology, v. 42, p. 109–124.

MANUSCRIPT ACCEPTED BY THE SOCIETY AUGUST 2, 1983

Geological Society of America
Special Paper 194
1984

The Goochland granulite terrane: Remobilized
Grenville basement in the eastern Virginia Piedmont

Stewart S. Farrar

Orogenic Studies Laboratory
Department of Geological Sciences
Virginia Polytechnic Institute and State University
Blacksburg, Virginia 24061

ABSTRACT

The Goochland terrane is defined here by the areal extent of a granulite-facies metamorphic sequence in the Piedmont of eastern Virginia. The sequence of units in this terrane includes the State Farm Gneiss of Grenville age, and the overlying Sabot amphibolite and Maidens gneiss. Also included is the Montpelier metanorthosite which intrudes the other units. The granulite-facies event (M_g) produced orthopyroxene + plagioclase and clinopyroxene + garnet + plagioclase assemblages in intermediate to mafic rocks, and K-feldspar + sillimanite in interlayered pelitic gneisses. An amphibolite-facies event (M_1) produced biotite + garnet and biotite + hornblende assemblages from the intermediate to mafic granulite gneisses, and muscovite + quartz ± kyanite ± staurolite schists from the pelitic gneisses.

Because the entire Goochland terrane was subjected to the same granulite-facies event as the Grenville-age State Farm Gneiss, and because of the similarity of the Montpelier metanorthosite to the Grenville-age Roseland metanorthosite of the Blue Ridge, the entire Goochland terrane is interpreted to be Grenville in age, or older. A Paleozoic tectonic event, at least in part Alleghanian, remobilized this terrane under amphibolite-facies conditions. The Goochland terrane is the largest and easternmost internal basement massif of the southern Appalachians.

INTRODUCTION

A Rb-Sr whole-rock age of 1031 ± 94 m.y. (Glover and others, 1978, 1982) was determined for the State Farm Gneiss west of Richmond in the eastern Virginia Piedmont. This is the easternmost dated Grenville rock unit in the southern Appalachians. All previously described Grenville terranes of the southern Appalachians, including those in the Blue Ridge, Pine Mountain belt, Sauratown Mountains, Baltimore Gneiss domes, West Chester Prong, and Honey Brook Upland (Fig. 1), comprise deep crustal Grenville basement rocks, exposed by erosion and nonconformably overlain by upper Precambrian to lower Paleozoic sedimentary and/or volcanic rocks. Because the State Farm Gneiss domes (Fig. 2) are overlain by units of sedimentary and probably volcanic origin, this terrane could be interpreted in a similar fashion (Poland, 1976; Reilly, 1980). However, the entire sequence of State Farm Gneiss, overlain by Sabot amphibolite

and Maidens gneiss was intruded by the Montpelier metanorthosite and then metamorphosed to granulite facies. The areal extent of granulite assemblages and associated pelitic assemblages is used here to define the Goochland granulite terrane. This granulite metamorphic event, discovered in this study, is interpreted to be a Grenville event, thus confirming the suggestion of Glover and others (1978), based on lithology, that the entire sequence is Grenville in age.

The Goochland terrane, as defined by granulite-facies assemblages, comprises an area of the Piedmont of eastern Virginia lying between the Hylas mylonite zone on the east, and the Chopawamsic Formation on the west (Fig. 2). Reconnaissance mapping suggests that the Goochland terrane may continue into the Po River metamorphic suite of Pavlides (1980, 1981). To the south it extends for an undetermined distance toward the Raleigh

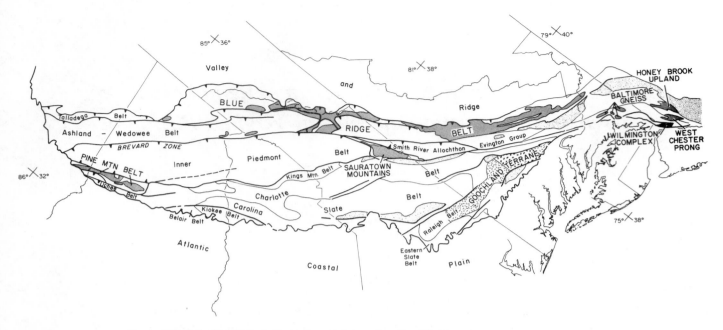

Figure 1. Distribution of Grenville terranes in the southern Appalachians, including the Blue Ridge belt, Pine Mountain belt, Sauratown Mountains, Baltimore Gneiss domes, Honey Brook Upland, West Chester Prong, and Goochland terrane. Also shown is the granulite grade Wilmington complex. The Goochland terrane is shown by double-dash pattern where mapped. Possible further extent is shown by a single-dash pattern. Modified from Glover and others (1983) and Williams (1978).

belt rocks of North Carolina, where textural evidence indicates an early (Grenville?) high-grade event (Fig. 1). This report discusses the Goochland terrane between latitudes 38° and approximately 37°15′.

Brown (1937) first described a lithologic sequence for much of this area, and Poland (1976), Bobyarchick (1976), Bourland (1976), Reilly (1980), Weems (1974), and Goodwin (1970) mapped various portions of this region. Bobyarchick and Glover (1979) discussed the role of the Hylas mylonite zone in the regional tectonics. A detailed description of the age and stratigraphy of the Goochland terrane is currently being prepared by L. Glover, III, D. Mose, S. Farrar, F. Poland, and J. Reilly. The lithologic sequence of the Goochland terrane, as best exposed in the doubly plunging State Farm antiform, consists of the State Farm Gneiss overlain by the Sabot amphibolite, which in turn is overlain by the Maidens gneiss. The Montpelier metanorthosite, first described by Clement and Bice (1982) and Bice and Clement (1982), is interpreted to have intruded the State Farm, Sabot and Maidens gneisses (Fig. 2). The State Farm and Sabot gneisses protrude through the Maidens gneiss in a subsidiary dome northeast of the State Farm antiform (Fig. 2). The felsic gneiss at Moseley, near the southeastern edge of the Goochland terrane (Fig. 2), may either interfinger with the Maidens gneiss or nonconformably overlie it. The State Farm Gneiss was intruded by two biotite granite plutons of probable Paleozoic age, that at Fine Creek Mills (Poland, 1976; Reilly, 1980) and that at Flat Rock (Reilly, 1980).

State Farm Gneiss

The State Farm Gneiss (Brown, 1937; Poland, 1976; Reilly, 1980) (Fig. 2) is medium- to coarse-grained, massive to moderately laycred, and foliated. It is a biotite-garnet-hornblende-quartz-K feldspar-plagioclase gneiss. Modal analyses given by Reilly (1980) indicate a granodiorite to tonalite composition. Relatively high titanium content, as indicated by abundant clusters of titanite grains, is characteristic of this unit. Least deformed parts of the State Farm Gneiss are massive, coarse-grained, plutonic rocks with granulite-facies mineral assemblages. Elsewhere, relict, coarse-grained, plutonic textures are common, with clusters of plagioclase and microcline replacing mesoperthite, and garnet + hornblende + biotite replacing pyroxene. Less common is a more leucocratic biotite-garnet-quartz-plagioclase gneiss, and thin pelitic schists are a minor facies (Reilly, 1980; Poland, 1976).

The State Farm Gneiss has a whole-rock Rb/Sr isochron age of 1031 ± 94 Ma (2σ) (Glover and others, 1982).

Sabot Amphibolite

The Sabot amphibolite, as described in the theses of Poland (1976) and Reilly (1980) (Fig. 2), structurally overlies the State Farm Gneiss. It is a medium- to coarse-grained amphibolite with minor interlayers of quartz-biotite-plagioclase gneiss, quartz-feldspar leucogneiss, and rare, thin, pelitic layers. The Sabot is in the form of a sheet 0.7-1.0 km thick (Poland, 1976) exposed

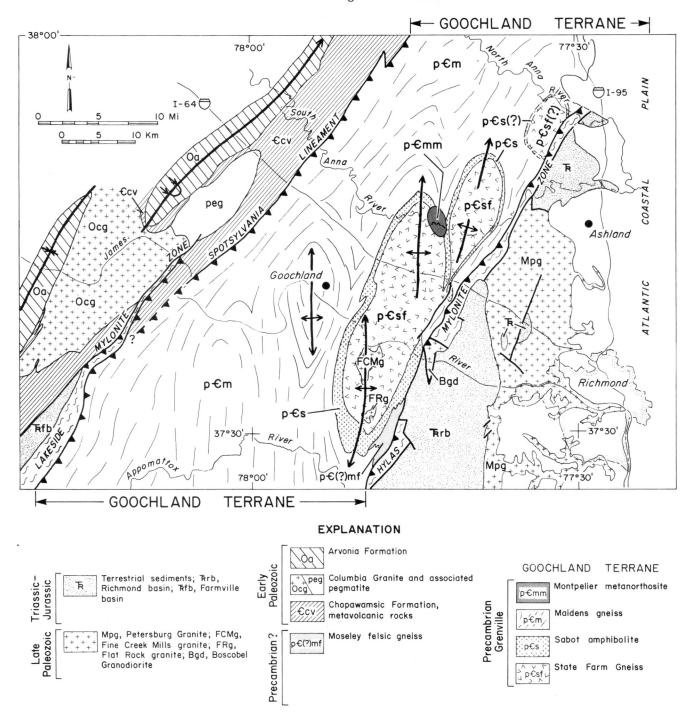

EXPLANATION

Triassic-Jurassic

[ℝ] Terrestrial sediments; ℝrb, Richmond basin; ℝfb, Farmville basin

Late Paleozoic

[+ + +] Mpg, Petersburg Granite; FCMg, Fine Creek Mills granite; FRg, Flat Rock granite; Bgd, Boscobel Granodiorite

Early Paleozoic

[Oa] Arvonia Formation

[peg / Ocg] Columbia Granite and associated pegmatite

[Єcv] Chopawamsic Formation, metavolcanic rocks

Precambrian ?

[p-Є(?)mf] Moseley felsic gneiss

GOOCHLAND TERRANE

Precambrian Grenville

[p-Єmm] Montpelier metanorthosite

[p-Єm] Maidens gneiss

[p-Єs] Sabot amphibolite

[p-Єsf] State Farm Gneiss

Figure 2. Geology of the Goochland terrane and surrounding eastern Virginia Piedmont. The Goochland terrane extends between the Spotsylvania lineament-Lakeside mylonite zone on the west, and the Hylas mylonite zone on the east. Modified from Glover and Tucker (1979), Reilly (1980), and Weems (1974).

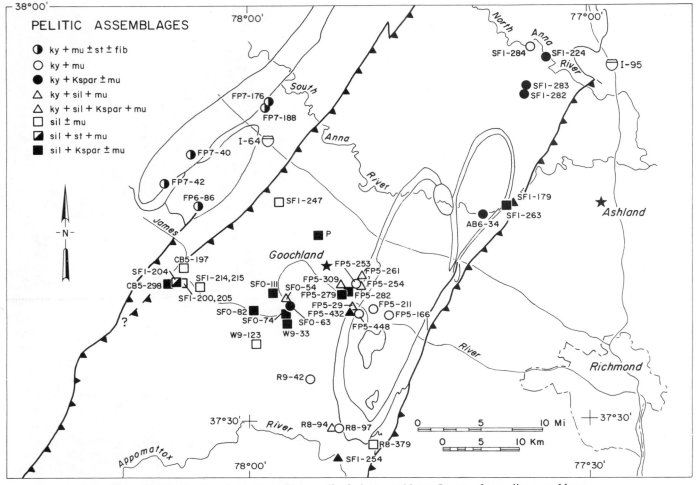

Figure 3. Distribution of outcrops with granulite-facies assemblages. Lenses of granulite assemblages occur in all map units of the Goochland terrane. Complete assemblages are given in appendix 1.

around the periphery of the State Farm antiform and the subsidiary dome to the northeast.

Maidens Gneiss

The heterogeneous Maidens gneiss, as described in the thesis of Poland (1976) (Fig. 2), structurally overlies the Sabot amphibolite. The dominant lithologies are a garnet-biotite-quartz-plagioclase gneiss containing lenses of intermediate to mafic gneiss with granulite-facies assemblages, and biotite-quartz-plagioclase-K feldspar augen gneiss. These gneisses characteristically lack the abundant titanite of the State Farm Gneiss. Other lithologies include numerous, discontinuous layers of garnet-plagioclase-quartz-K-feldspar leucogneiss, and clinopyroxene-hornblende-plagioclase amphibolite, as well as biotite-garnet-sillimanite-quartz-K feldspar pelitic gneiss. The pelitic gneisses were partially to nearly completely recrystallized to muscovite + quartz-bearing schist. Minor marble layers reacted with adjacent gneisses, producing numerous, thin, calcsilicate layers. The thickness of the Maidens gneiss, which crops out in a wide belt overlying the Sabot amphibolite, cannot be estimated readily because gently undulating folds result in repetition of the sequence.

Montpelier Metanorthosite

The metanorthosite near Montpelier (Clement and Bice, 1982; Bice and Clement, 1982) appears to be intrusive into the State Farm, Sabot, and Maidens gneisses at the northern end of the State Farm antiform (Fig. 2). The metanorthosite occurs as two major textural varieties: (1) extremely coarse grained and (2) recrystallized coarse grained. The extremely coarse-grained metanorthosite comprises antiperthitic plagioclase megacrysts up to 25-30 cm with interstitial 10-15 cm clinopyroxene, apatite, ilmenite, and rutile. Clinopyroxene is replaced by amphibole + biotite + quartz, with garnet coronas. Locally, recrystallization of the antiperthite formed individual plagioclase and microcline grains. Ilmenite commonly has coronas of rutile surrounded by titanite + garnet. This extremely coarse-grained variety is not foliated. The foliated, recrystallized metanorthosite is dominantly medium- to coarse-grained plagioclase with relatively minor interstitial microcline and quartz. Large ilmenite grains which resisted recrystallization are fractured and have coronas of rutile surrounded by titanite, and then garnet + quartz in contact with surrounding plagioclase.

The metanorthosite has large xenoliths of titanite-quartz-biotite-garnet-plagioclase gneiss with garnets as large as 8-10 cm in diameter. Small metagabbro-metanorite bodies, comprising clinopyroxene + garnet and orthopyroxene + garnet granulite gneisses, may be mafic portions of the Montpelier intrusion, or xenoliths. If they are xenoliths, the biotite-garnet-plagioclase gneiss could be from the State Farm Gneiss, and the metagabbro-metanorite could be from the Sabot amphibolite.

METAMORPHISM

The Goochland terrane was subjected to two high-grade metamorphic events: (1) a granulite-grade event (M_g), and (2) an amphibolite-grade (M_1) remobilization of this granulite terrane. The granulite event produced orthopyroxene, two-pyroxene, and garnet + clinopyroxene granulite gneisses, and K-feldspar + sillimanite-bearing pelitic gneisses. The amphibolite-grade event reintroduced H_2O in substantial amounts, recrystallizing the granulite gneisses to garnet + biotite + hornblende gneisses, and the pelitic gneisses to muscovite ± biotite ± kyanite ± staurolite schists.

Evidence of a Granulite Event

Petrography of Intermediate to Mafic Granulites. Intermediate to mafic granulite gneisses occur in all map units of the Goochland terrane (Fig. 3). These granulites commonly occur in large lenses, as much as 10 m or more in thickness, surrounded by amphibolite-grade gneisses. Some of these lenses (R8-633, FP6-1) are massive, weakly foliated, intermediate to mafic granulites of apparent plutonic origin. Other lenses are thinly layered (10s of cm) and isoclinally folded under granulite-facies conditions. These thin-layered granulite gneisses include: orthopyroxene granulite interlayered with clinopyroxene + garnet granulite (SF1-280); clinopyroxene + garnet granulite interlayered with K-feldspar + sillimanite metapelite (SF1-263); and orthopyroxene + clinopyroxene and orthopyroxene + garnet granulites interlayered with K-feldspar + sillimanite metapelite (SF0-111).

Granulite assemblages are separated into four groups by their anhydrous mafic minerals (Fig. 4 and appendix 1): orthopyroxene; orthopyroxene + clinopyroxene; orthopyroxene + garnet; and clinopyroxene + garnet. All of these coexist with plagioclase. Orthopyroxene + plagioclase granulite occurs as fine- to medium-grained, thin-layered intermediate- to mafic-granulite gneiss. It has a moderate to strong (S_g) foliation of orthopyroxene and biotite parallel to the layering. K-feldspar, where present, is concentrated in orthoclase augen surrounded by myrmekite and a concentration of biotite. Orthopyroxene + clinopyroxene + plagioclase granulite gneiss makes up the most common orthopyroxene-bearing group of assemblages (appendix 1). They occur

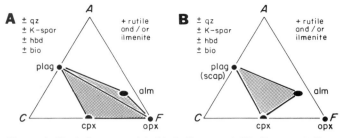

Figure 4. Shaded areas and bold tie lines on ACF diagrams indicate assemblages occurring in the Goochland terrane which are used to define granulite facies conditions. Changes in tie lines from (a) to (b) appear to be controlled by composition rather than changes in pressure and temperature, because these assemblages are intermingled throughout the terrane. Complete assemblages appear in appendix 1.

as massive, intermediate to mafic, medium- to coarse-grained granulite gneiss and as fine- to medium-grained, thinly layered granulite gneiss. K-feldspar augen in these rocks have continuous myrmekite rims and commonly are surrounded by biotite concentrations. Orthopyroxene + garnet + plagioclase granulite gneiss has been found only at locality SF0-111, occurring as thin layers within fine-grained orthopyroxene + clinopyroxene granulite. Clinopyroxene + garnet + plagioclase granulite gneisses are quite common in this terrane (appendix 1), occurring as thin to thick layers with weak to moderate foliation. Scapolite is common in clinopyroxene + garnet granulites which lack K-feldspar. The K-feldspar, which is more common than in the orthopyroxene-bearing rocks, commonly has rims of myrmekite.

Mineral Chemistry of Intermediate to Mafic Granulites. Compositions of mineral phases in polished thin sections were analyzed with a nine-spectrometer, automated ARL-SEMQ electron microprobe using silicates and oxides as standards. The data were converted to oxide weight percentages by a computer program based on the alpha factor correction scheme of Ziebold and Ogilvie (1964) as extended by Bence and Albee (1968) using the correction factors of Albee and Ray (1970). Mineral formulas and components were calculated using the computer program *SUPERRECAL* of Rucklidge (1971). The formula also calculated a stoichiometric amount of water and added it to the oxide weight percentages for the hydrous minerals.

Clinopyroxene ranges from $Wo_{46}En_{42}Fs_{12}$ to $Wo_{47}En_{30}Fs_{23}$. Orthopyroxene ranges from $Wo_2En_{62}Fs_{36}$ to $Wo_2En_{47}Fs_{51}$. Pyroxene in K-feldspar-bearing assemblages and in clinopyroxene + garnet assemblages tends to be more Fe-rich than pyroxene in the more mafic tonalitic to noritic orthopyroxene + clinopyroxene-bearing assemblages. Thin exsolution lamellae, which occur in some of the pyroxenes, were not examined in this study. The range of pyroxene compositions, and tielines between coexisting orthopyroxenes and clinopyroxenes, are shown in Fig. 5.

Plagioclase compositions range widely (appendix 1), from An_{50} in an orthopyroxene + plagioclase assemblage to An_{45-82} in orthopyroxene + clinopyroxene + plagioclase assemblages, and An_{21-43} in clinopyroxene + garnet + plagioclase assemblages.

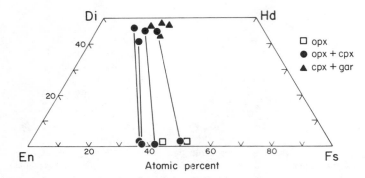

Figure 5. Pyroxene compositions in granulite rocks of the Goochland terrane. Tie lines connect coexisting pyroxenes. The most iron-rich pyroxenes are from K-feldspar-bearing assemblages.

Compositions are constant in any one thin section. From this small sampling, it appears that plagioclase in clinopyroxene + garnet assemblages or K-feldspar-bearing assemblages is somewhat more sodic than in orthopyroxene-bearing assemblages without K-feldspar.

K-feldspar occurs in some samples of each of three of the granulite assemblage groups (appendix 1). The one analyzed K-feldspar is orthoclase of composition $An_1Ab_8Or_{89}Cn_2$. Orthoclase in these rocks is only rarely perthitic. It occurs as multigrain, recrystallized augen, which generally are surrounded by rims of myrmekite.

Garnet in these granulite rocks is almandine-rich and unzoned, ranging from $Al_{54}Py_{26}Sp_2Gr_{18}$ in coronas of a metanorite to $Al_{50}Py_{11}Sp_2Gr_{37}$ in a clinopyroxene-garnet-plagioclase-scapolite gneiss. Elsewhere it is generally close to $Al_{57}Py_{17}Sp_4Gr_{22}$.

Biotite in the granulites is titanium-rich, ranging from 3.5 to 5.3 wt percent TiO_2. Biotite Fe/(Fe+Mg) generally follows the trend of the other mafic minerals, ranging from 0.27 in metanorite to 0.51 in K-feldspar-bearing, quartz-rich granulite. Tetrahedral Al ranges from 2.39 to 2.61 (Fig. 6). The metanorite contains the only fluorine-rich biotite, with 1.25 wt percent F; the others range between 0.3 and 0.5 wt percent F.

The amphibole is hornblende, ranging from ferroedenitic hornblende to edenite, according to the classification of Leake (1978). Hornblende generally forms coronas of hornblende and hornblende + quartz around orthopyroxene, and, more rarely, around clinopyroxene. These corona textures suggest that most of this hornblende was formed in the second metamorphic event. Fe/(Fe+Mg) varies with the pyroxene being replaced, ranging from 0.23 to 0.54.

Petrography of Pelitic Gneisses. Two of the best exposures of layered granulite rocks in the Goochland terrane have interlayered metapelites (SF0-111 and SF1-263; Fig. 7; appendix 2). The metapelite assemblage is K-feldspar + sillimanite + garnet + biotite + quartz + plagioclase. Minor muscovite occurs as a late alteration along sillimanite + K-feldspar contacts. Elsewhere in the central part of the Goochland terrane (Fig. 7; appendix 2), dry (Fig. 8), or nearly dry pelitic gneisses occur, with only trace amounts of biotite and late-forming muscovite in as-

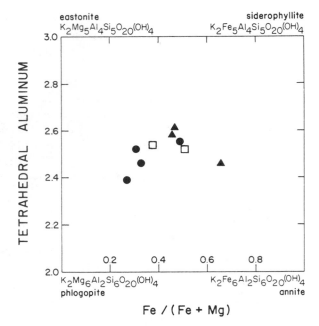

Figure 6. Compositions of biotite coexisting with orthopyroxene + plagioclase (squares); orthopyroxene + clinopyroxene + plagioclase (circles); and clinopyroxene + garnet + plagioclase (triangles). Fe/(Fe+Mg) varies with that of coexisting pyroxenes. No attempt was made to differentiate between generations of biotite, and compositions appear to be constant in a given thin section.

semblages of K-feldspar + sillimanite + quartz ± garnet ± plagioclase. The K-feldspar in these rocks is perthitic orthoclase and is commonly the most plentiful mineral. Sillimanite is fine- to medium-grained, but never fibrolitic. No kyanite occurs in any of these nearly dry assemblages. The sillimanite + K-feldspar-bearing assemblages are interpreted to have formed during the granulite event.

Evidence of an Amphibolite Facies, Second Event

The granulite facies event was followed by a distinct amphibolite facies metamorphic event. Probably more than 90 percent of the formerly granulite facies assemblages of the Goochland terrane have been recrystallized under amphibolite facies conditions. The most obvious effects in felsic to mafic assemblages are reactions involving hydration of pyroxenes to biotite and amphibole. In pelitic rocks it involves hydration of the assemblage sillimanite + K-feldspar to muscovite + quartz and formation of kyanite or staurolite.

Felsic to Mafic Gneisses. Intermediate to mafic rocks which preserve the orthopyroxene and clinopyroxene + garnet granulite assemblages commonly show progressive recrystallization to amphibolite-facies assemblages. In metanorite, orthopyroxene + clinopyroxene + plagioclase + H_2O have partially reacted to form nearly continuous rims of hornblende around the pyroxenes (Fig. 9a), or well-developed coronas of hornblende + quartz and garnet (Fig. 9b). In intermediate, K-feldspar-bearing rocks, continuous, thick myrmekite rims surrounding orthoclase, and

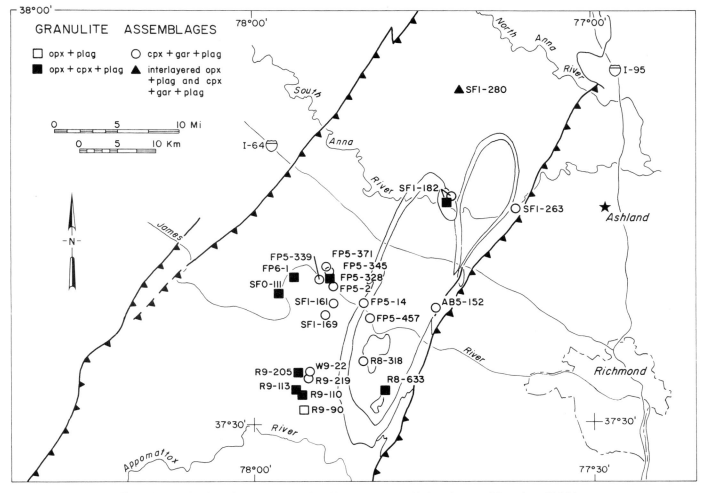

Figure 7. Distribution of outcrops with Al_2SiO_5 polymorphs. Sillimanite and sillimanite + K-feldspar are interpreted to have formed during the granulite event. Kyanite, muscovite, staurolite, and fibrolitic sillimanite formed during the second, amphibolite-grade event. Abbreviations: ky, kyanite; mu, muscovite; st, staurolite; fib, fibrolitic sillimanite; Kspar, K-feldspar; sil, sillimanite.

concentration of biotite as a replacement of pyroxene (Fig. 9c) in adjacent layers, suggest a reaction in which pyroxene + K-feldspar + H_2O go to biotite + myrmekite. Rare corona textures in more felsic rocks suggest similar reactions. Granodiorite of the State Farm Gneiss comprises coarse mesoperthite with interstitial quartz and clinopyroxene. The pyroxene has been replaced almost completely by coronas of hornblende around clusters of hornblende + quartz ± biotite ± garnet.

These replacement textures grade outward, with increased deformation, from the granulite lenses into typical amphibolite-grade assemblages of enclosing units. Progressive deformation forms a strong S_1 foliation which destroys coronas of the State Farm granodiorite, leaving mesoperthite augen in a foliated groundmass of plagioclase + quartz + microcline + hornblende + garnet + biotite. This is the typical assemblage described for this unit by Poland (1976) and Reilly (1980). Progressive hydration of clinopyroxene + garnet granulite of the Sabot amphibolite results in hornblende + plagioclase amphibolite with minor relict

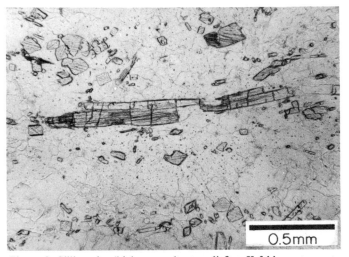

Figure 8. Sillimanite (high to moderate relief) + K-feldspar + quartz. Typical of dry, pelitic assemblages produced in the granulite event. Sample W9-33.

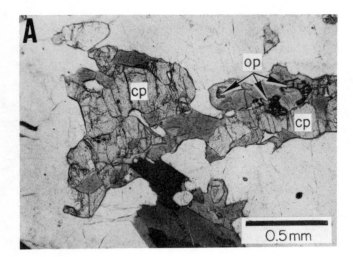

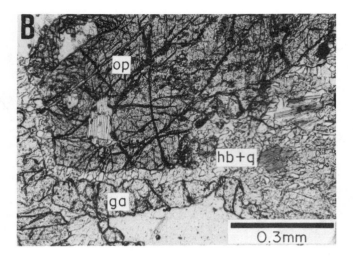

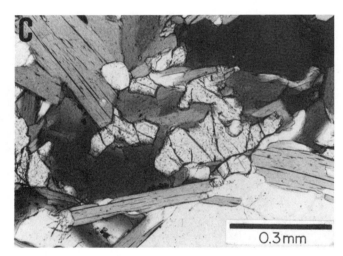

Figure 9. Partial reaction of granulite to amphibolite-facies assemblages has occurred even in the best preserved granulite gneisses. (a) Relict M_g orthopyroxene (op) and clinopyroxene (cp) have continuous M_1 hornblende rims. Metanorite (FP6-1); plane light. (b) M_g orthopyroxene (op) with M_1 coronas of hornblende + quartz (hb+q), and garnet (ga). Metanorite (R8-633); plane light. (c) M_1 biotite almost completely replaces M_g clinopyroxene in a K-feldspar-bearing granulite. Metagranodiorite (FP5-328); plane light.

clinopyroxene, as described by Reilly (1980). Progressive hydration of K-feldspar-bearing granulites of the Maidens gneiss results in typical Maidens granodiorite gneiss, with relict K-feldspar augen in a groundmass of plagioclase + biotite + quartz ± garnet ± hornblende. Rare relict clinopyroxene or orthopyroxene in biotite-rich layers confirm that the granulite event affected the entire area. None of the granitic gneisses of the terrane has been found to preserve granulite assemblages. However, relict K-feldspar augen surrounded by a strongly foliated biotite-rich groundmass of biotite + microcline + quartz + plagioclase is a common texture in granitic Maidens gneiss, suggesting that any granulite assemblages that existed have been hydrated to biotite granitic gneiss.

Pelitic Gneisses and Schists. Pelitic rocks, like the intermediate to mafic rocks, preserve textural evidence of two metamorphic events. Sillimanite + K-feldspar-bearing assemblages of the granulite event were partially to completely replaced by muscovite + quartz ± kyanite ± staurolite. In most of the central, eastern, and southern portions of the Goochland terrane the reaction involves sillimanite + K-feldspar + H_2O going to muscovite + quartz + kyanite (Fig. 7; appendix 2). In nearly dry pelitic rocks

minor muscovite occurs along grain boundaries between sillimanite + K-feldspar. Progressive hydration results in relicts of sillimanite in muscovite clots, and no direct contact of sillimanite and K-feldspar. Finally, sillimanite is either: (1) completely consumed; (2) remains as grains isolated from reaction in quartz layers; or (3) remains where protected from deformation and accompanying hydration in embayments or inclusions in garnet (Fig. 10). Kyanite generally replaces sillimanite as part of the reaction producing muscovite + quartz, destroying preexisting textures, and suggesting that the reaction progresses most readily in the presence of fluid. In rare cases kyanite directly replaces individual sillimanite grains, even retaining the diagonal sillimanite cleavage. In approximately the western third of the Goochland terrane, the sillimanite + K-feldspar pelitic gneiss has been hydrated to muscovite + quartz schist with no kyanite produced (Fig. 7; appendix 2). Relict sillimanite is common as inclusions in large muscovite clots in these schists, and K-feldspar augen are preserved in some. These schists commonly have what at first appear to be flattened quartz cobbles as much as 6-8 cm long. Closer examination shows these to be quartz + muscovite pseudomorphs of K-feldspar augen. Thin, quartz-rich layers around

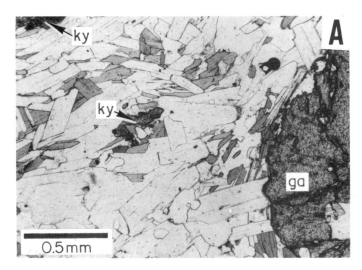

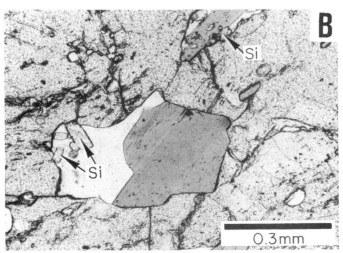

Figure 10. (a) Amphibolite-grade M_1 pelitic assemblage comprises kyanite (ky) + muscovite + biotite + quartz + garnet (ga). Relict M_g sillimanite is preserved only where protected with quartz + biotite as an inclusion in garnet (lower right). Sample SF1-254; plane light. (b) Enlargement of part of garnet in (a), showing M_g assemblage sillimanite (si) + quartz + biotite in garnet inclusions.

these augen contain excess sillimanite that remained when the K-feldspar was consumed. Along the westernmost edge of the Goochland terrane (Fig. 7; appendix 2) staurolite + muscovite + quartz replaces sillimanite + K-feldspar + garnet. Although this is the only area of staurolite occurrence in the mapped Goochland terrane, the assemblage staurolite + sillimanite + biotite + garnet has been reported (with no comment as to textural relations) along the western edge of the Po River metamorphic suite to the north (Bobyarchick and others, 1981).

In the northeastern part of the Goochland terrane, sillimanite is replaced directly by kyanite with only minor, or no, muscovite produced (Fig. 7; appendix 2). Some of the replacement is pseudomorphic and some forms coarse kyanite laths, but the very minor muscovite produced, coupled with abundant relict(?)

K-feldspar, indicate that little H_2O was introduced in this area during M_1.

The late Precambrian to early Paleozoic rocks west of the Goochland terrane show textural evidence of only one metamorphic event. Pelitic rocks, with sedimentary layering preserved, have assemblages including muscovite + kyanite ± staurolite ± garnet ± fibrolitic sillimanite (Fig. 7; appendix 3). This regional, amphibolite-grade event appears to correlate with the amphibolite-grade event in the Goochland terrane. Textures indicate that kyanite ± fibrolite ± staurolite replaces muscovite + chlorite in prograde reactions. Here, no evidence exists of the earlier granulite assemblages which are used to define the Goochland terrane to the east.

DEFORMATION

A minimum of three deformational events affected the Goochland terrane. D_g and penetrative S_g foliation, which resulted in isoclinal F_g folds, occurred during the granulite metamorphic event. These D_g structures are preserved in granulite rocks which suffered the least subsequent deformation. The F_g folds are observed only at the outcrop scale, or smaller. The second event (D_1) was an intense deformational event which occurred near the peak of the amphibolite-grade metamorphic event. It produced isoclinal F_1 folds and an amphibolite-grade, penetrative axial planar foliation (S_1). S_1 is the penetrative foliation in those rocks which were reduced to amphibolite grade, but it is nearly absent in those rocks which preserve granulite assemblages. The granulite rocks apparently were preserved as large lenses of rock least deformed during D_1. The third event (D_2) formed open to tight F_2 folds which have a weak to very weak S_2 foliation. Minerals, including biotite and muscovite, which form the S_2 foliation suggest that D_2 occurred during cooling from the amphibolite-grade event.

F_2 folds dominate the map pattern of the Goochland terrane. Small scale F_g and F_1 folds have various orientations as a result of the later F_2 folding, but S_1 generally has a shallow to moderate dip. F_2 folds at all scales are consistently overturned to the west (Poland, 1976). Attitude of S_2 is generally N10E, 25-50E, and F_2 fold axes plunge gently north and south (Poland, 1976; Farrar, Reilly, and Wehr, unpub. data). The map patterns of lithologic units, and S_g and S_1 foliations define regional-scale, elongated, F_2 domes and tight F_2 basins as a result of interference with preexisting structures.

The apparent continuity of the second (M_1) metamorphic event from the Goochland terrane into the Late Precambrian-Early Paleozoic rocks to the west (Fig. 6), suggests that these terranes were juxtaposed during D_1, which occurred near the thermal maximum of M_1. D_2 folding, and mylonitization along the Lakeside zone during D_2 or later, modified the form of the terrane. This interpretation essentially agrees with that of Glover and others (1982) which suggested that these two terranes were juxtaposed early in their Paleozoic tectonic history.

DISCUSSION

Pressure and Temperature of Metamorphism

The Goochland terrane was metamorphosed under granulite-facies conditions, as indicated by orthopyroxene + plagioclase, two-pyroxene + plagioclase, orthopyroxene + garnet + plagioclase, and clinopyroxene + garnet + plagioclase granulite gneisses. Dry, sillimanite + K-feldspar-bearing pelitic gneiss interlayered with granulite gneiss indicates that the granulite event was in the sillimanite stability field. The interlayered nature of the different granulite assemblages and their occurrences across the terrane suggest that the occurrence of orthopyroxene, versus clinopyroxene + garnet, is controlled by composition rather than P-T variation. No evidence was found to suggest significant P-T variation across the terrane during the granulite event. These are typical granulite assemblages of sillimanite-bearing massif granulite terranes, which according to Newton and Perkins (1982) generally form in the range 7.5-9.0 kb and 750-850°C.

The second, amphibolite-facies, metamorphic event, accompanied by intense deformation, remobilized the granulite terrane. Hydration of granulite assemblages to amphibolite-facies assemblages was pervasive. Granulite assemblages are preserved only as isolated lenses in intermediate to mafic rocks. No charnockites are preserved, if, indeed, they had ever formed during the granulite event. Retrogressive textures dominate the remaining granulite assemblages, with partial replacement of pyroxenes and feldspar by hornblende and biotite in intermediate to mafic rocks and replacement of plagioclase by scapolite in some of the mafic rocks. In pelitic rocks sillimanite + K-feldspar was replaced by muscovite + quartz ± kyanite ± staurolite, leaving relict sillimanite and/or K-feldspar in muscovite or muscovite + kyanite schists. P-T conditions of this second event, as roughly indicated by pelitic assemblages were probably in the kyanite stability field throughout the terrane. Variation in P-T across the area is suggested by (1) staurolite occurrence only at the western edge of the terrane, (2) muscovite + quartz schists lacking kyanite in the western third of the terrane, (3) kyanite + muscovite in the south and center, and (4) kyanite + K-feldspar in the northeast. This distribution of assemblages suggests some combination of increase in T and/or decrease in P_{H_2O} from southwest to northeast across the terrane. The general range of P_{total}-T for this second event, as indicated by the pelitic assemblages, is 5-7 kb and 550-650°C.

Areal Extent of the Goochland Terrane

The extent of the Goochland terrane is defined in this study by the areal extent of the granulite-facies metamorphic event. Thus, two major assumptions are involved: (1) that lenses of granulite assemblages indicate the former extent of a continuous granulite terrane; and (2) that in this terrane sillimanite + K-feldspar-bearing pelitic assemblages were produced in the granulite event and that their distribution can be used in addition

to granulite assemblages to define the extent of the terrane. Textural evidence supports these two assumptions. Throughout the terrane granulite assemblages and sillimanite + K-feldspar are replaced by amphibolite facies assemblages, whereas to the west of the Goochland terrane the amphibolite-facies assemblages are prograde with no evidence of an earlier, higher-grade event. Data presented by Bobyarchick and others (1981) for the Ladysmith quadrangle suggest that the Po River metamorphic suite has also been subjected to both sillimanite-grade and kyanite-grade metamorphic events. It is suggested here that these are probably the same events recorded in the Goochland area, and that the Po River metamorphic suite is a probable northern continuation of the Goochland terrane. Reconnaissance mapping (by the writer) in the Raleigh belt of North Carolina indicates that similar textural relations exist in that area. The boundaries showing the probable extent of the Goochland terrane (Fig. 1) reflect these interpretations.

Age of the Goochland Terrane

The Goochland terrane, as defined here by the extent of the granulite facies metamorphic assemblages, includes the State Farm Gneiss, Sabot amphibolite, Maidens gneiss and Montpelier metanorthosite. The State Farm Gneiss has a Grenville Rb-Sr whole-rock age of 1031 ± 94 m.y. (Glover and others, 1982), and it is petrographically similar to plutonic gneisses of the Blue Ridge and other Grenville terranes of the southern Appalachians. Because the overlying Sabot and Maidens units are not similar to other southern Appalachian Grenville lithologies, Poland (1976) and Reilly (1980) proposed that they might comprise a nonconformably overlying, Late Precambrian-early Paleozoic metavolcanic-metasedimentary sequence, equivalent to sequences overlying Grenville rocks of the Blue Ridge, Pine Mountain belt, Sauratown Mountains, and Baltimore Gneiss domes. Glover and others (1978), however, pointed out that the Maidens gneiss does not compositionally, nor in facies succession, resemble other post-Grenville successions in the region, and that the entire sequence of State Farm, Sabot, and Maidens may be of Grenville age.

Three lines of evidence support the inference of Glover and others (1978) that the entire area, the Goochland terrane of this paper, is Grenville or older. First, the State Farm is dated as Grenville in age. Second (the main contribution of this paper), the entire terrane was subjected to granulite-facies metamorphism, requiring either that the entire terrane is at least Grenville in age or that the granulite event occurred in the Paleozoic. Third, the Montpelier metanorthosite, which is interpreted to have intruded the other members of the sequence, is nearly identical to the Roseland metanorthosite of the Blue Ridge, as first noted by Watson and Tabor (1913), and very distinct from most other anorthosites. Field relations and radiometric dating of surrounding rocks support a Grenville or older age for the Roseland (Herz, 1968; Force and Herz, 1982). With one unit (State Farm) definitely Grenville, the youngest (Montpelier) probably Grenville, and the entire area subjected to granulite facies metamorphism of probable Grenville age, the conclusion is warranted that the entire Goochland terrane is Grenville in age or older.

Comparison to Other Granulite Terranes

If granulite metamorphism in the Appalachians was restricted to Grenville rocks, the assumption of a Grenville age for the Goochland terrane would probably not be questioned. However, along strike, approximately 200 km to the north, lies the Wilmington Complex of Delaware and Pennsylvania (Fig. 1). Zircon (Grauert and Wagner, 1975) and Rb/Sr (Foland and Muessig, 1978) studies of the Wilmington Complex indicate that the granulite-grade metamorphism in that area occurred in the Taconic, at about 440 m.y. ago. However, in contrast to the Goochland terrane no radiometric evidence exists for Grenville rocks in the Wilmington Complex. It has been proposed that Wilmington Complex rocks are a continuation of the James Run volcanics (Higgins, 1972), and an intrusive charnockite has a whole-rock Rb/Sr age of 502 m.y. (Foland and Muessig, 1978). Also, petrologic studies indicate that the Wilmington Complex had a significantly different metamorphic history than the Goochland terrane. The Wilmington event was apparently relatively low pressure, lacking garnet in the granulite assemblages (Ward, 1959; Crawford and Crawford, 1980), in contrast to the Goochland terrane, which has plentiful clinopyroxene-garnet assemblages.

The Goochland terrane would appear to correlate more closely in metamorphic history with the West Chester Prong of Pennsylvania (Fig. 1), as described by Wagner and Crawford (1975). As in the Goochland terrane, the West Chester Prong is interpreted to have been metamorphosed to granulite facies during the Grenville event, producing two--pyroxene granulites, clinopyroxene-garnet granulites, and sillimanite in aluminous rocks. In the West Chester Prong this was followed by a Taconic amphibolite-grade event during which much of the terrane was reduced to amphibolite-grade assemblages by hydration reactions, and most sillimanite was replaced by kyanite. In contrast to the Goochland terrane, much of the West Chester Prong was not hydrated. Garnet coronas around mafic minerals apparently grew in these dry rocks during the Taconic event (Wagner and Crawford, 1975). Equivalent coronas are rare (but see Fig. 9b) in the Goochland terrane where the hydration was much more complete. Wagner and Crawford (1975) propose that this second event took place at about 440 m.y. ago in the West Chester Prong, while the Wilmington Complex was at granulite grade.

In contrast to the West Chester Prong, amphibolite grade metamorphism in the Goochland terrane was probably Acadian-Alleghanian, although it is possible from regional considerations (Glover and others, 1983) that Taconic metamorphism also affected the Goochland terrane. The Arvonia and Quantico formations west of the Goochland terrane are Late Ordovician or younger, as indicated by fossils (Tillman, 1970; Pavlides and others, 1980). Also, the Columbia granite, which is nonconformably overlain by the Arvonia, has an Rb/Sr whole-rock age of

454 ± 9 (Mose and Nagel, 1982), again supporting a Late Ordovician or younger age for the Arvonia. The Arvonia and Quantico were metamorphosed to kyanite + staurolite ± fibrolite assemblages (Fig. 7) in what appears to be the same amphibolite grade event that affected the Goochland terrane. $^{40}Ar/^{39}Ar$ cooling ages (Durrant and others, 1980) on hornblende and biotite in the State Farm Gneiss indicate that this region was cooling in the time period 280-260 m.y. ago. This suggests an Alleghanian age for the latest metamorphic event.

Absolute age of the granulite metamorphism of the Goochland terrane is thus limited only by the age of the State Farm Gneiss (1031 ± 94 m.y.) and the regional cooling age (280 m.y.). However, because in character these granulite assemblages match other Grenville terranes of the Appalachians, with sillimanite as the stable aluminum silicate and clinopyroxene + garnet granulites associated with orthopyroxene granulites, it seems reasonable to conclude that this was a Grenville event. Indeed, these assemblages characterize other Precambrian massif granulite terranes of the world (Newton and Perkins, 1982), and contrast with the variable character of the rare younger granulite terranes.

The Goochland terrane is typical of remobilized basement, or internal massifs, of orogenic belts, although the Goochland differs most from other such terranes in its greater extent of recrystallization in the remobilizing event. In most terranes, such as the West Chester Prong discussed above, or the Lewisian of Scotland (Evans and Lambert, 1974), coherent basement terranes are described first and then extended into areas where remobilization dominates. In the Goochland terrane no coherent, non-recrystallized terrane has been found. In other terranes of amphibolite-grade recrystallization, such as the Lewisian (Evans and Lambert, 1974), hydration reactions go to completion first in the granitic rocks, because they contain fewer mafic minerals to be hydrated, whereas granulite assemblages are preserved in intermediate to mafic rocks. This appears also to be the case in the Goochland terrane.

Of the internal massifs of the southern Appalachians, including the Pine Mountain belt, Sauratown Mountains, Baltimore Gneiss domes, and West Chester Prong, the Goochland terrane is the largest and easternmost of the group. In spite of its eastern location, however, no strong reasons exist to consider this to be African basement, as suggested for part of this terrane by Rankin (1975). The Grenville age of the State Farm, the similarity of the Montpelier metanorthosite to the Roseland of the Blue Ridge, and the apparent similarities in Grenville metamorphic history of the Goochland terrane to other southern Appalachian basement rocks, strongly suggest that this is another slice of North American Grenville basement.

ACKNOWLEDGEMENTS

This research was supported by U.S. Nuclear Regulatory Commission contract no. NRC-04-75-237 to Lynn Glover, III and John K. Costain, and National Science Foundation Grant no. EAR-800 9449-2 to Lynn Glover, III and John K. Costain.

Many of the samples used in this study were collected by F. B. Poland, J. M. Reilly, A. R. Bobyarchick, W. C. Bourland, and F. L. Wehr. I thank J. A. Speer and L. Glover, III for their helpful suggestions.

APPENDIX 1. GRANULITE MINERAL ASSEMBLAGES OF THE GOOCHLAND TERRANE

Sample	opx	cpx	gar	plg*	Ksp	bio	hbd	qtz	scp	opq	tit	apt	zir	rut	mus	cac	epd
Opx+plg																	
R9-90	+			+	+		+			+		+	+	+			
SF1-111	+			+	+	+	+			+		+	+				
SF1-280-2	+		50		+		+	+		+		+					
Opx+cpx+plg																	
FP5-328	+	+	45	+	+	+	+				+						
FP6-1	+	+	51		+	+	+			+	+	+					
R8-633	+	+	+	57	+	+	+			+	+		+	+			
R9-110	+	+		+	+	+	+			+	+	+					
R9-113	+	+		+	+	+	+			+	+	+	+				
R9-205	+	+		+	+	+	+			+	+	+					+
SFO-111-4	+	+	82		+	+	+			+	+	+	+				
SFO-111-5	+	+		+	+	+	+			+	+	+					
SFO-112	+	+		+	+	+	+			+	+	+					
Opx+gar+plg																	
SFO-111-3	+		+	+		+	+			+		+					
Cpx+gar+plg																	
AB5-152		+	+	+		+				+	+	+					
FP5-2		+	+	+	+	+				+	+	+					
FP5-14		+	+	+	+	+				+	+	+			+		
FP5-339		+	+	+	+	+		+	+	+	+			+	+	+	
FP5-345		+	+	+	+	+				+	+	+					
FP5-371		+	+	+	+	+				+	+	+					
FP5-457		+	+	+	+	+				+	+	+					
R8-318		+	36	+	+	+				+	+	+			+		+
R9-219		+	+	+	+	+				+	+	+					
SF1-161		+	+	+		+				+	+	+			+		
SF1-169		+	+	+		+	+	+	+	+	+	+					+
SF1-182-4		+	21	+		+				+	+	+					
SF1-263		+	+			+		+	+	+	+	+					
SF1-280		+	+		+	+	+	+		+	+	+			+		
W9-22		+	43			+		+	+	+	+	+					

Abbreviations used: opx, orthopyroxene; cpx, clinopyroxene; gar, garnet; plag, plagioclase; Ksp, K-feldspar; bio, biotite; hbd, hornblende; qtz, quartz; scp, scapolite; opq, opaque; tit, titanite; apt, apatite; zir, zircon; rut, rutile; mus, muscovite; cac, calcite; epd, epidote.
*Plagioclase An compositions determined by microprobe.

APPENDIX 2. PELITIC MINERAL ASSEMBLAGES OF THE GOOCHLAND TERRANE

Sample	Ksp	sil	kya	sta	mus	gar	bio	plg	qtz	opq	tit	apt	zir	hem	chl	rut	gph
Sil+Ksp±mus																	
CB5-298	+	+			+	+	+	+	+								
FP5-279	+	+			+		+	+									
FP5-282	+	+			+		+	+									
SFO-74	+	+			+	+	+	+	+			+	+	+			
SFO-82	+	+			+	+	+	+	+			+	+				
SFO-111-6	+	+			+	+	+	+	+			+		+			
SF1-263-5	+	+			+	+	+	+	+		+	+	+	+	+		
W9-33	+	+			+	+	+	+		+		+		+			
P*	+	+				+	+		+					+			
Sil+mus																	
CB5-197		+			+	+	+	+					+	+			
R8-379		+			+	+	+	+					+	+			
SF1-200		+			+	+	+	+									
SF1-205-2		+			+	+	+	+					+				
SF1-214-3		+			+	+	+	+					+				
SF1-215		+			+	+	+		+		+						
SF1-247		+		+	+	+	+					+					
W9-123		+			+	+	+	+									
Sil+kya+mus																	
FP5-29		+	+		+	+	+	+	+	+		+					
FP5-254		+	+		+	+	+	+	+	+							
FP5-261		+	+		+	+	+	+	+	+							
FP5-309		+	+		+	+	+	+	+		+		+				
FP5-432	+	+	+		+	+	+	+	+	+		+					
R8-94		+	+		+	+	+	+	+								+
SFO-54		+	+		+	+	+	+	+	+							+
SF1-179-2	+	+	+		+	+	+	+									+
SF1-254	+	+	+		+	+	+	+				+					+
Sil+mus+sta																	
SF1-204		+		+	+	+		+	+			+					
Kya+Ksp±mus																	
AB6-34	+		+			+	+	+	+								
SF1-224	+		+	+	+	+	+	+			+	+			+	+	
SF1-282-2	+		+	+	+	+	+	+			+	+	+				
SF1-283-2	+		+		+	+	+	+			+	+					
Kya+mus																	
FP5-166			+		+	+	+	+					+				
FP5-211			+		+	+	+	+	+				+				
FP5-253			+		+	+	+	+	+								
FP5-448			+		+	+	+	+	+		+	+	+				
R8-97			+		+	+	+	+		+							
R9-42			+		+	+	+	+	+	+							
SF1-284			+		+	+	+	+									+

Abbreviations used: Ksp, K-feldspar; sil, sillimanite; kya, kyanite; sta, staurolite; mus, muscovite; gar, garnet; bio, biotite; plg, plagioclase; qtz, quartz; opq, opaque; tit, titanite; apt, apatite; zir, zircon; hem, hematite; chl, chlorite; rut, rutile; gph, graphite.
* From Pegau (1932).

APPENDIX 3. PELITIC MINERAL ASSEMBLAGES IN THE CHOPAWAMSIC AND ARVONIA, WEST OF THE GOOCHLAND TERRANE

Sample	Ksp	sil	fib	kya	sta	mus	gar	bio	plg	qtz	opq	tit	apt	zir	hem	chl	rut
Kya+mus±sta±fib																	
FP6-86		+	+			+	+	+		+	+						
FP7-40			+		+	+	+	+		+	+			+		+	
FP7-42			+		+	+	+	+		+	+		+		+		
FP7-176		+	+	+	+	+	+	+		+	+						
FP7-188			+			+	+			+	+				+		+

Abbreviations used: Ksp, K-feldspar; sil, sillimanite; fib, fibrolitic sillimanite; kya, kyanite; sta, staurolite; mus, muscovite; gar, garnet; bio, biotite; plg, plagioclase; qtz, quartz; opq, opaque; tit, titanite; apt, apatite; zir, zircon; hem, hematite; chl, chlorite; rut, rutile.

REFERENCES CITED

Albee, A. L. and Ray, L., 1970, Correction factors for electron-probe microanalysis of silicates, oxides, carbonates, phosphates, and sulfates: Analytical Chemistry, v. 42, p. 1408–1414.

Bence, A. E. and Albee, A. L., 1968, Emperical correction factors for electron microanalysis of silicates and oxides: Journal of Geology, v. 76, p. 382–403.

Bice, K. L. and Clement, S. C., 1982, A study of the feldspars of the Montpelier andesine anorthosite, Hanover County, Virginia: Geological Society of America Abstracts with Programs, v. 14, p. 5.

Bobyarchick, A. R., 1976, Tectogenesis of the Hylas zone and eastern Piedmont near Richmond, Virginia [M.S. thesis]: Blacksburg, Virginia Polytechnic Institute and State University, 168 p.

Bobyarchick, A. R. and Glover, L., III, 1979, Deformation and metamorphism in the Hylas zone and adjacent parts of the eastern Piedmont in Virginia: Geological Society of America Bulletin, Part I, v. 90, p. 739–752.

Bobyarchick, A. R., Pavlides, L., and Wehr, K., 1981, Piedmont geology of the Ladysmith and Lake Anna East quadrangles and vicinity, Virginia: United States Geological Survey Miscellaneous Investigations Series Map I-1282.

Bourland, W. C., 1976, Tectogenesis and metamorphism of the Piedmont from Columbia to Westview, Virginia, along the James River [M.S. thesis]: Blacksburg, Virginia Polytechnic Institute and State University, 113 p.

Brown, C. B., 1937, Outline of the geology and mineral resources of Goochland County, Virginia: Virginia Geological Survey Bulletin, v. 48, 68 p.

Clement, S. C. and Bice, K. L., 1982, Andesine anorthosite in the eastern Piedmont of Virginia: Geological Society of America Abstracts with Programs, v. 14, p. 10.

Crawford, M. L. and Crawford, W. A., 1980, Metamorphic and tectonic history of the Pennsylvania Piedmont: Journal of the Geological Society of London, v. 137, p. 331–320.

Durrant, J. M., Sutter, J. F., and Glover, L., III, 1980, Evidence for an Alleghanian (Hercynian?) metamorphic event in the Piedmont province near Richmond, Virginia: Geological Society of America Abstracts with Programs, v. 12, p. 176.

Evans, C. R. and Lambert, R.St.J., 1974, The Lewisian of Lochinver, Sutherland; the type area for the Inverian metamorphism: Journal of the Geological Society of London, v. 130, p. 125–150.

Foland, K. A. and Muessig, K. W., 1978, A Paleozoic age for some charnockitic-anorthositic rocks: Geology, v. 6, p. 143–146.

Force, E. R. and Herz, N., 1982, Anorthosite, ferrodiorite, and titanium deposits in Grenville terrane of the Roseland district, central Virginia: in Lyttle, P. (ed.), Central Appalachian Geology, NE—SE GSA Field Trip Guide 1982, p. 109–119.

Glover, L., III, Mose, D. G., Costain, J. K., Poland, F. B., and Reilly, J. M., 1982, Grenville basement in the Eastern Piedmont of Virginia: a progress report: Geological Society of America Abstracts with Programs, v. 14, p. 20.

Glover, L., III, Mose, D. G., Poland, F. B., Bobyarchick, A. R., and Bourland, W. C., 1978, Grenville basement in the eastern Piedmont of Virginia: implications for orogenic models: Geological Society of America Abstracts with Programs, v. 10, p. 169.

Glover, L., III, Speer, J. A., Russell, G. S., and Farrar, S. S., 1983, Ages of regional metamorphism and ductile deformation in the central and southern Appalachians: Lithos, v. 16, p. 223–245.

Glover, L., III and Tucker, R. D., 1979, Map showing Virginia Piedmont geology along the James River from Richmond to the Blue Ridge: in, Glover, L., III and Read, J. F. (eds.), Guides to Field Trips 1-3 for Southeastern Section Meeting, Geological Society of America, p. 1–41.

Goodwin, B. K., 1970, Geology of the Hylas and Midlothian quadrangles, Virginia: Virginia Division of Mineral Resources Report of Investigations 23, 51 p.

Grauert, B. and Wagner, M. E., 1975, Age of the granulite facies metamorphism of the Wilmington Complex, Delaware-Pennsylvania Piedmont: American Journal of Science, v. 275, p. 683–691.

Herz, N., 1969, The Roseland anorthosite massif, Virginia: in, Isachsen, Y. I. (ed.), Origin of Anorthosite and Related Rocks, New York State Museum and Science Service Memoir 18, p. 357–367.

Higgins, M. W., 1972, Age, origin, regional relations, and nomenclature of the Glenarm Series, central Appalachian Piedmont: a reinterpretation: Geological Society of America Bulletin, v. 83, p. 989–1026.

Leake, B. E., 1978, Nomenclature of amphiboles: American Mineralogist, v. 63, p. 1023–1052.

Mose, D. G. and Nagel, M. S., 1982, Plutonic events in the Piedmont of Virginia: Southeastern Geology, v. 23, p. 25-39.

Newton, R. C. and Perkins, D., III, 1982, Thermodynamic calibration of geobarometers based on the assemblages garnet-plagioclase-orthopyroxene (clinopyroxene)-quartz: American Mineralogist, v. 67, p. 203–222.

Pavlides, L., 1980, Revised nomenclature and stratigraphic relationships of the Fredericksburg complex and Quantico formation of the Virginia Piedmont: United States Geological Survey Professional Paper 1146, 29 p.

Pavlides, L., 1981, The Central Virginia volcanic plutonic belt: an island arc of Cambrian(?) age: United States Geological Survey Professional Paper 1231-A, 34 p.

Pavlides, L., Pojeta, J., Gordon, M., Parsley, R. L., and Bobyarchick, A. R., 1980, New evidence for the age of the Quantico formation in Virginia: Geology, v. 8, p. 286–290.

Pegau, A. A., 1932, Pegmatite deposits of Virginia: Virginia Geological Survey Bulletin, v. 33, 123 p.

Poland, F. B., 1976, Geology of the rocks along the James River between Sabot and Cedar Point, Virginia [M.S. thesis]: Blacksburg, Virginia Polytechnic Institute and State University, 98 p.

Rankin, D. W., 1975, The continental margin of eastern North America in the southern Appalachians: the opening and closing of the Proto-Atlantic Ocean: American Journal of Science, v. 275-A, p. 298–336.

Reilly, J. M., 1980, A geologic and potential field investigation of the central Virginia Piedmont [M.S. thesis]: Blacksburg, Virginia Polytechnic Institute and State University, 111 p.

Rucklidge, J. C., 1971, Specifications of FORTRAN program SUPERRECAL: Department of Geology, University of Toronto, Toronto, Ont.

Tillman, C. G., 1970, Metamorphosed trilobites from Arvonia, Virginia: Geological Society of America Bulletin, v. 81, p. 1189–1200.

Wagner, M. E. and Crawford, M. L., 1975, Polymetamorphism of the Precambrian Baltimore gneiss in southeastern Pennsylvania: American Journal of Science, v. 275, p. 653–682.

Ward, R. F., 1959, Petrology and metamorphism of the Wilmington Complex, Delaware, Pennsylvania, and Maryland: Geological Society of America Bulletin, v. 70, p. 1425–1458.

Watson, T. L. and Tabor, S., 1913, Geology of the titanium and apatite deposits of Virginia: Virginia Geological Survey Bulletin 3A, 308 p.

Weems, R. E., 1974, Geology of the Hanover Academy and Ashland quadrangles, Virginia [M.S. thesis]: Blacksburg, Virginia Polytechnic Institute and State University, 98 p.

Williams, H., 1978, Tectonic-lithofacies map of the Appalachians orogen: St. John's Newfoundland, Canada, Memorial University.

Ziebold, T. O., and Oglivie, R. E., 1964, An empirical method for electron microanalysis: Analytical Chemistry, v. 36, p. 322–327.

MANUSCRIPT ACCEPTED BY THE SOCIETY AUGUST 2, 1983

Geological Society of America
Special Paper 194
1984

Evolution of Grenville massifs in the Blue Ridge geologic province, southern and central Appalachians

Mervin J. Bartholomew*
Virginia Division of Mineral Resources Office
Department of Geological Sciences
Virginia Polytechnic Institute and State University
Blacksburg, Virginia 24061

Sharon E. Lewis*
Department of Geology
Eastern Kentucky University
Richmond, Kentucky 40475

ABSTRACT

Within the southern and central Appalachians, Grenville-age basement rocks are found in major massifs in the Blue Ridge and Sauratown Mountains anticlinoria and in the vicinity of the Grandfather Mountain window. These massifs are, respectively, Pedlar and Lovingston Massifs in the Blue Ridge anticlinorium, Sauras Massif in the Sauratown Mountains anticlinorium, and Watauga, Globe, and Elk River Massifs near the Grandfather Mountain window. In central Virginia the Lovingston Massif is juxtaposed against the Pedlar Massif, and in northwestern North Carolina–southwestern Virginia, the Elk River Massif is thrust over the Globe and Watauga Massifs, all along faults of the Fries fault system, which includes the Rockfish Valley, Fork Ridge, Devil's Fork, and Linville Falls faults, as well as the Fries fault *per se.* The Pedlar Massif is a deeper granulite facies country-rock terrane intruded by charnockite plutonic suites. The Lovingston Massif primarily is a shallower granulite/amphibolite facies terrane intruded by biotite dioritoid plutonic suites containing bodies of charnockite. Country rocks of the Watauga Massif were subjected to metamorphic conditions similar to those of the Lovingston Massif, but were intruded by a plutonic suite of biotite dioritoid, biotite granitoid, and granitoid. The Elk River, Globe, and Sauras Massifs all are terranes metamorphosed to amphibolite facies and intruded by granitoid/dioritoid suites containing some porphyritic biotite dioritoid phases. A suite of late Precambrian (post-Grenville) peralkaline granitoid plutons intruded all of the massifs except the Pedlar. These plutons presumably are related to upper Precambrian volcanic rocks that were associated with a rifting environment and that were later metamorphosed and deformed along with overlying sedimentary rocks to form part of the Appalachian orogenic belt.

*Present address: Bartholomew—Montana Bureau of Mines and Geology, Montana Tech, Butte, Montana 59701. Lewis—Department of Geological Engineering, Montana Tech, Butte, Montana 59701.

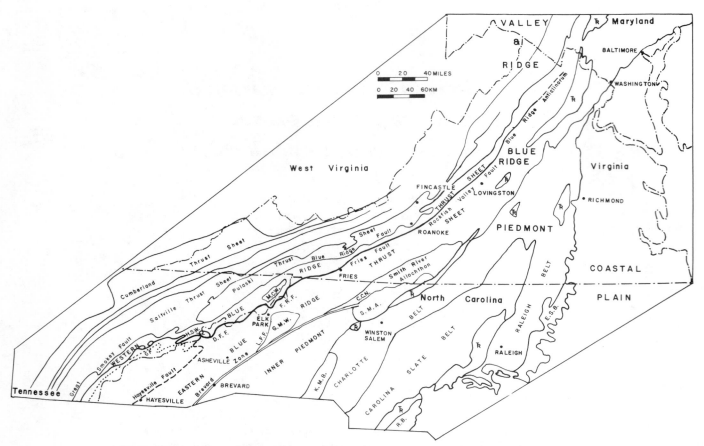

Figure 1. Tectonic map of a portion of the central and southern segments of the Appalachian orogenic belt. E.S.B.—Eastern Slate Belt; R.B.—Raleigh Belt; S.M.A.—Sauratown Mountains anticlinorium; C.C.N.—Camp Creek Nappe; K.M.B.—Kings Mountain Belt; G.M.W.—Grandfather Mountain window; M.C.W.—Mountain City window; H.S.W.—Hot Springs window; G.F.—Greenbrier fault; D.F.F.—Devils Fork fault; F.R.F.—Fork Ridge fault; L.F.F.—Linville Falls fault; dotted line—trace of Greenbrier fault; dashed line—trace of Hayesville fault.

INTRODUCTION

This paper elucidates major relationships among, and the internal geology of, basement massifs of the Blue Ridge province in Virginia and northwestern North Carolina. These Grenville-age massifs served as the cratonic platform upon which upper Precambrian metavolcanic and metasedimentary rocks were deposited nonconformably. The massifs occur as basement primarily in two major thrust sheets (Figs. 1 and 2). The western and eastern Blue Ridge thrust sheets contain the Pedlar and Watauga Massifs and the Lovingston and Elk River Massifs, respectively. Basement rocks of the Globe Massif are found within the Grandfather Mountain window in a tectonic slice originally (before Paleozoic deformation) between the Watauga and Elk River Massifs. Farther to the southeast the Sauras Massif forms the core of the Sauratown Mountains anticlinorium (Figs. 1 and 2).

All of these massifs lie northwest of the traditional boundary between the Blue Ridge geologic province and the various provinces of the Piedmont (Fig. 1). Grenville-age rocks also occur

within the Baltimore Gneiss domes and State Farm Gneiss dome farther northeast and east, respectively (Figs. 1 and 2), but are outside the scope of this paper. Mentioned only briefly is the enigmatic Moneta gneiss, which occurs at the late Precambrian nonconformity along the southeastern flank of the Lovingston Massif and in the Moneta domes (Fig. 2).

The Fries fault system of late Devonian age is a major zone of ductile deformation that separates the western and eastern Blue Ridge thrust sheets (Fig. 1). Because of the increasing amount of displacement southwestward along the Fries fault system, contrast between juxtaposed massifs on either side of the fault system becomes greater to the southwest and provides a tectonic basis for distinction and discussion of the various plutonic and layered rocks of the massifs.

Discussions are keyed to the three 1:500,000 geologic maps (Figs. 3, 4, and 5) of the massifs shown on Fig. 2. These maps represent the first attempt to regionally map and relate the various

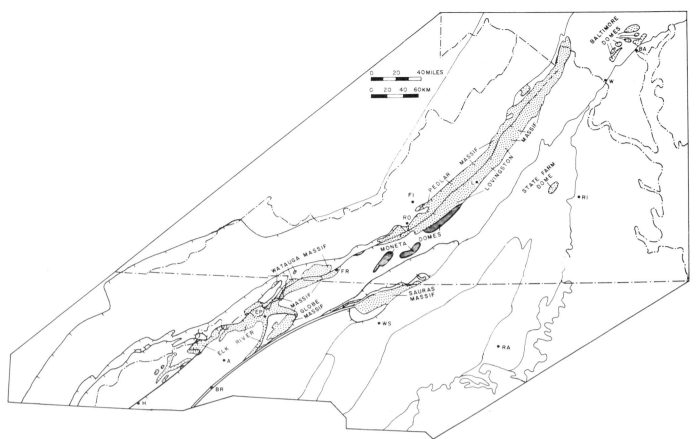

Figure 2. Tectonic map showing the distribution and names of Grenville-age basement massifs and domes relative to the tectonic boundaries shown in Figure 1. BA—Baltimore; W—Washington; RI—Richmond; L—Lovingston;FI—Fincastle; RO—Roanoke; FR—Fries; RA—Raleigh; WS—Winston-Salem; EP—Elk Park; A—Asheville; BR—Brevard; H—Hayesville.

plutonic and country rocks of Grenville age in Virginia and North Carolina. In order to simplify the discussion of these Grenville-age rocks, a number of new names are introduced here for what are considered to be regionally important plutonic suites, plutons, and country rocks. Type localities for these new units, as well as stratigraphic revisions of previously named units, are included in Appendix 1.

The term "Massif" is capitalized here when used with proper names to emphasize its use to define specific geographic-geologic entities within the Blue Ridge geologic province, and not to imply any stratigraphic ranking. As such, it is used to show that each massif contains rocks that are spatially and/or genetically related to each other but are distinctively different from those in adjacent massifs.

THE BASEMENT MASSIFS

The purpose of distinguishing and naming the basement massifs is to provide clear and unique identification of crystalline terranes that were subjected to Grenville orogenesis so that they may be referred to or discussed without further specific reference

to either their cover-rock sequences or later structures produced during repeated episodes of Paleozoic and younger deformation. Bartholomew (1977) first introduced the terms "Pedlar" and "Lovingston" for massifs (Fig. 2), which subsequently were defined by Bartholomew and others (1981). They defined the Pedlar Massif as the deeper (higher P and T conditions) granulite facies terrane within which the term "Pedlar Formation" (Bloomer and Werner, 1955) had been applied to a wide variety of rock types. Similarly, they defined the Lovinston Massif as the shallower (lower P and T conditions) granulite facies terrane within which the term "Lovingston Formation" (Jonas, 1928) had received similar reconnaissance use for a variety of rock types.

In addition to the Pedlar and Lovingston Massifs, this paper focuses on four other massifs that are defined herein. The Watauga, Globe, and Elk River Massifs are found in the Grandfather Mountain window area (Figs. 1, 2, and 4), and the Sauras Massif is located in the Sauratown Mountains anticlinorium (Figs. 1, 2, and 5). The Watauga Massif is a terrane, northwest of the Fries-- Fork Ridge fault, that was mainly subjected to shallower granulite facies and amphibolite facies conditions during Grenville

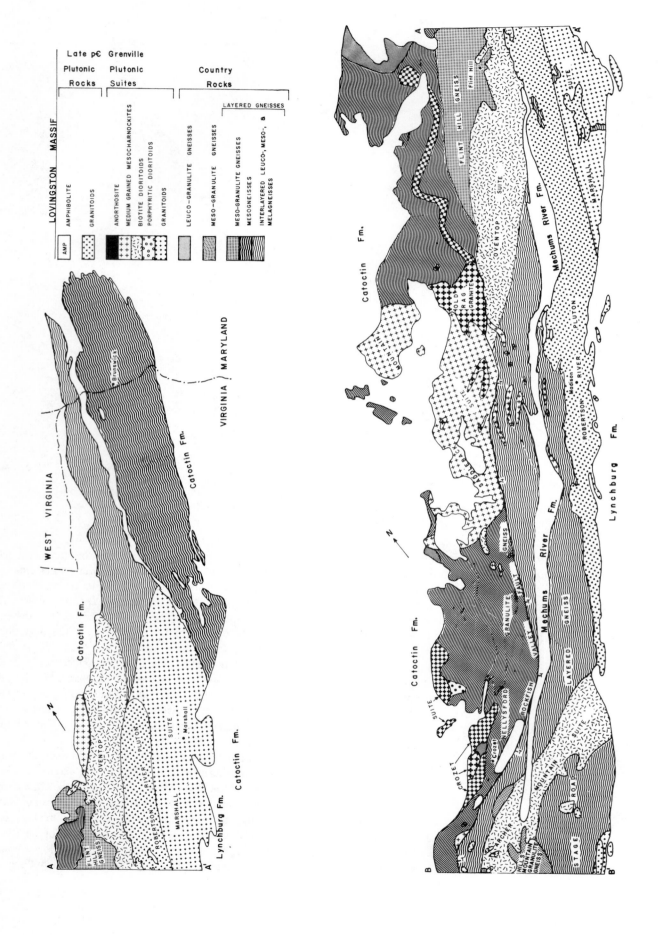

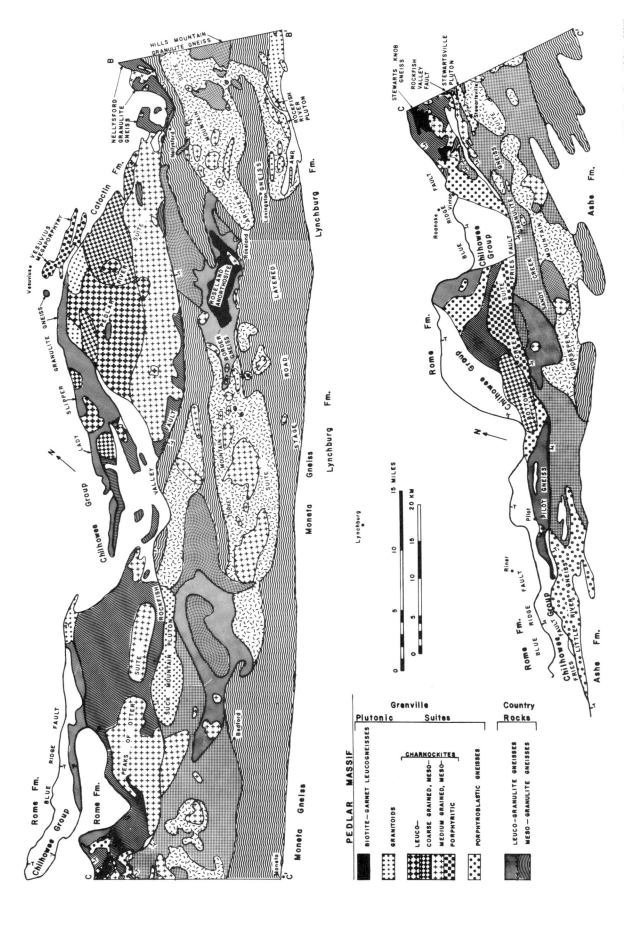

Figure 3 (this and facing page). Geologic map of the Pedlar and Lovingston Massifs in the Blue Ridge anticlinorium of Virginia and Maryland, approximate scale 1:500,000; geology compiled from state geologic maps of Virginia and Maryland (Virginia Division of Mineral Resources, 1963; Maryland Geological Survey, 1968) and from Allen (1963); Bailey (1983); Bartholomew (1977); Bartholomew (1981); Bartholomew and Hazlett (1981); Bloomer and Werner (1955); Brock (1981); Brown (1958); Davis (1974); Deitrich (1959); Diggs (1955); Furcron (1939); Gathright (1976); Gathright and Nys-trom (1974); Gathright and others (1977); Hamilton (1964); Herz (1969); Henika (1981); Hillhouse (1960); Hudson (1981); Kaygi (1979); Lewis and others (1984); Lukert and others (1977); Lukert and Halladay (1980); Lukert and Nuckols (1976); Moore (1940); Nelson (1962); Parker (1968); Truman (1976); Werner (1966); unpublished data from J. F. Conley, E. Force, T. M. Gathright II, W. S. Henika, N. Herz, and A. K. Sinha.

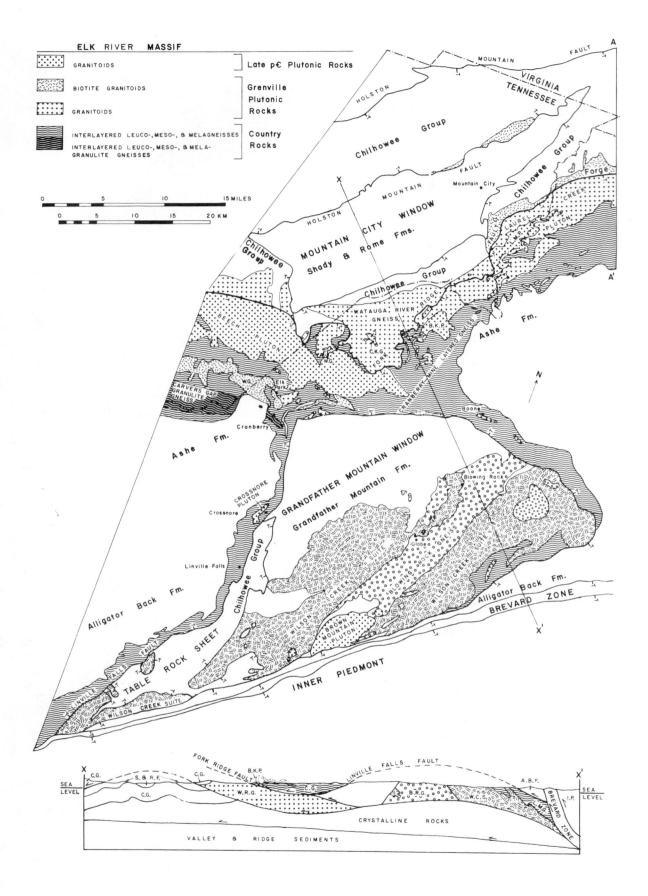

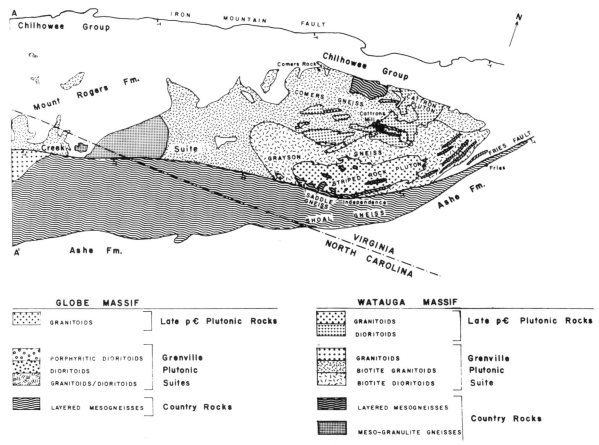

Figure 4 (this and facing page). Geologic map of the Watauga, Globe, and Elk River Massifs in the Grandfather Mountain window area, North Carolina, Virginia, and Tennessee; approximate scale 1:500,000. C.K.G.—Crossing Knob Gneiss; B.K.P.—Buckeye Knob Pluton; L.M.P.—Leander Mountain Pluton; W.G.—Whaley Gneiss. Cross section X–X' is along line of cross section of Rankin and others (1972). W.R.G.—Watauga River Gneiss; B.R.G.—Blowing Rock Gneiss; W.C.S.—Wilson Creek Suite; L.M.G.—layered meso-gneiss; A.B.F.—Alligator Back Formation; I.P.—Inner Piedmont; C.G.—Chilhowee Group; S.&R.F.—Shady and Rome Formations; C.G. (wavy pattern)—Cranberry—Mine Layered Gneiss. Geology compiled from: Bartholomew (1983a); Bartholomew and Gryta (1980); Bartholomew and Wilson (1984); Bryant and Reed (1970b); Conley and Drummond (1979); Jones (1976); Keith (1903); Lewis and Bartholomew (1984); Lewis and Butler (1984); Rankin and others (1972); Rodgers (1953); Stose and Stose (1957).

metamorphism. The Globe Massif is the basement terrane within the Grandfather Mountain window, and the Elk River Massif forms the basement of the eastern Blue Ridge thrust sheet near the Grandfather Mountain window (Fig. 1). The Globe and Elk River Massifs, as well as the Sauras Massif, mainly were subjected to upper amphibolite facies conditions of metamorphism during Grenville orogenesis.

The names "Watauga," "Globe," and "Sauras" were chosen from (1) the Indian name now used for Watauga County, North Carolina; (2) the community of Globe, North Carolina; and (3) the Sauras Indians of North Carolina. The Elk River Massif is named for the Elk River which flows across the massif a few miles north of the town of Elk Park. The name "Elk Park" was originally used for the Elk River Massif by Bartholomew (1983b)

and Bartholomew and others (1983) but because of the desirability of retaining "Elk Park" as a formal stratigraphic term, the massif name is here changed to Elk River rather than Elk Park.

All of the basement massifs are composed of pre-Grenville or early Grenville layered gneisses that served as country rock into which the Grenville-age intrusive suites were emplaced. These plutonic suites typically are composed of several genetically related rock types that were either intruded as separate bodies from common magma sources or derived from one or more large magmas that underwent differentiation after emplacement. At the present time, insufficient petrology and geochronology exist for most of the suites to give more specific interpretations.

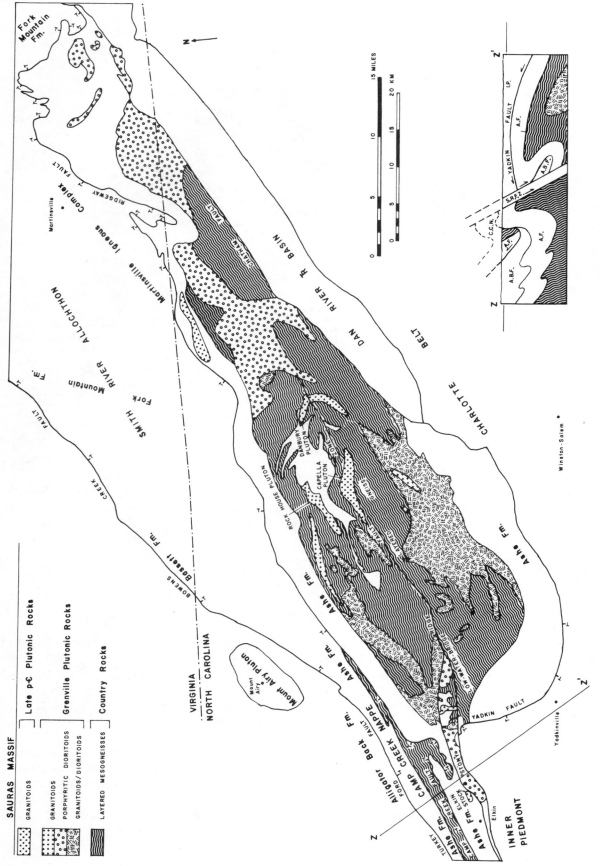

Figure 5. Geologic map of the Sauras Massif in the Sauratown Mountains anticlinorium, North Carolina and Virginia; approximate scale 1:500,000. Cross section Z–Z′ is along line of cross section of Espenshade and others (1975). A.B.F.—Alligator Back Formation; A.F.—Ashe Formation; C.C.N.—Camp Creek Nappe; I.P.—Inner Piedmont; S.R.F.Z.—Stony Ridge Fault Zone. Geology compiled from: Conley (1978); Dunn and Weigand (1969); Espenshade and others (1975); Fullagar and Butler (1980); Lewis (1980a); Price and others (1980a, 1980b).

During late Precambrian (post-Grenville) time, all of the massifs, except the Pedlar Massif, were intruded by a suite of peralkaline (Rankin, 1975) granitoid plutons commonly lumped into the plutonic suite of the Crossnore Plutonic-Volcanic Complex of Rankin and others (1983). Recently, Hudson (1981) reported a very little granitic body associated with the Irish Creek tin deposit near the western flank of the Pedlar River Suite. If this granitoid rock proves to be similar to the granites of the Crossnore suite, then it will be the first such body located in the Pedlar Massif.

FRIES FAULT SYSTEM

A major key to unraveling stratigraphic and tectonic relationships of the basement is an understanding of post-Grenville structural displacement, particularly along what has been called the Fries fault. The Fries fault system is one continuous zone of ductile deformation, which encompasses a series of nomenclatural variations along strike. The fault system is one of the longest known tectonic features in the southern and central Appalachians and separates the western Blue Ridge thrust sheet from the structurally superjacent eastern Blue Ridge thrust sheet (Fig. 1). The Fries thrust dies out in the core of the Blue Ridge anticlinorium in northern Virginia (Bartholomew and others, 1981). From there southward to the Roanoke area (Fig. 1) it has been mapped as the Rockfish Valley fault, and the overall relationships are described by Bartholomew and others (1981). North of Roanoke, the Rockfish Valley fault separates two basement massifs, the Lovingston Massif to the southeast and the Pedlar Massif to the northwest.

Southwestward from Roanoke to the Virginia/North Carolina boundary, the Fries fault originally was mapped and described by Stose and Stose (1957) near the state boundary and extended northeastward by Dietrich (1954). Dietrich incorrectly thought that the Fries fault either merged with or was cut off by the Blue Ridge fault just southwest of Roanoke. Lewis (1975) recognized that the ductile deformation zone separated two distinctly different types of Grenville-age rocks and correctly located the trace of the Fries fault in the area where Dietrich (1954, 1959) had worked. Bartholomew and Lewis (1977) recognized that the Fries fault could be extended northward from the region mapped by Dietrich totally within the Grenville-age basement, to at least the Roanoke area. They first proposed that a single large zone of ductile deformation extended from the core of the Blue Ridge anticlinorium in northern Virginia at least to the vicinity of the Grandfather Mountain window in northwestern North Carolina.

In southwestern Virginia, Grenville-age rocks of the herein named Watauga Massif were described first by Stose and Stose (1957). The Watauga Massif is that part of the western Blue Ridge thrust sheet southwest of Fries, Virginia (Fig. 1), that extends into the North Carolina/Tennessee area where it forms the footwall of the Fork Ridge fault (Figs. 1 and 3). This portion of the fault zone was mapped and named by Bartholomew and

Gryta (1980) and only recently has been traced northeastward to link up with the Fries fault proper. Rankin (1970), Rankin and others (1972), and Espenshade and others (1975) previously had recognized the striking contrast between late Precambrian/early Paleozoic cover rocks across the Fries fault. However, they followed earlier interpretations by Stose and Stose (1957), Bryant and Reed (1970a, 1970b), and Keith (1903) and extended the fault around the northern side of, and structurally above, the Linville Falls fault, the border fault of the Grandfather Mountain window (Figs. 1 and 2). This interpretation led to the conclusion that the Ashe Formation was deposited on thin continental crust or oceanic crust (Hatcher, 1978). Recent detailed mapping (Bartholomew and Gryta, 1980; Bartholomew, 1983a; Bartholomew and Wilson, 1984; Conley and Drummond, 1979; Lewis and Bartholomew, 1984; Lewis and Butler, 1984) indicates that the Fork Ridge fault borders the herein named Elk River Massif on the northwest, and the Linville Falls fault borders the same massif around the window (Figs. 1, 2, and 4), thus implying that the Linville Falls fault is an integral part of the Fries fault system (Fig. 4) and not a structurally lower fault as proposed by previous workers (Rankin and others, 1972). Hence, a minimum horizontal displacement on the Fries fault system can be estimated across the window. This minimum displacement is about 50 km if the leading edge of the Fork Ridge fault were originally just southeast of the eastern edge of the window (Bartholomew, 1983b).

The Watauga and Elk River Massifs, juxtaposed across the fault, are terranes of the granulite/amphibolite and amphibolite facies, respectively. Locally within the Elk River Massif at Roan Mountain (Fig. 4), deeper granulite facies rocks are described by Gulley (1982) and Monrad and Gulley (1983). Both the Grenville-age rocks of the Globe Massif and its cover rocks within the Grandfather Mountain window originated prior to thrusting from a geographic position between the Watauga and Elk River Massifs (Bartholomew, 1983b) and hence have lithological differences from the structurally adjacent massifs. The cover rocks of the Globe Massif more closely resemble the cover rocks of the Watauga Massif in both lithology and grade of Paleozoic metamorphism, whereas the Grenville rocks of the Globe Massif proper more closely resemble rocks of the Elk River, Lovingston, and Sauras Massifs in lithology and grade of Grenville metamorphism.

Extension of the Fries fault system southwestward from the vicinity of the Hot Springs window (Fig. 1) is uncertain because of a lack of detailed mapping, particularly of basement rocks. Hatcher (1978, 1981) suggested that the Fries is the northeastward extension of the Hayesville fault. However, inasmuch as the Late Devonian-age Fries system postdated Taconic metamorphism and the Hayesville fault has been interpreted (Hatcher and Odom, 1980) as premetamorphic to synmetamorphic (Taconic), this correlation seems untenable unless the fault was reactivated. For similar reasons the Fries also is not likely to be the extension of the premetamorphic Greenbrier fault. Thus, the Fries fault probably lies to the west of the Greenbrier fault and probably truncates either or both of the older faults just southwest of the Hot Springs window, as depicted in Fig. 1.

The age of movement on the Fries fault system was established at about 345 m.y. ago by Dietrich and others (1969). The placement of the age of faulting at about 345 m.y. (Late Devonian) is consistent with our interpretation that the Fries fault system postdates both Taconic metamorphism, placed at about 450 m.y. ago, and emplacement of the Spruce Pine Pegmatites, placed at about 380 m.y. ago (Butler, 1972, 1973). Moreover, a Late Devonian age for the faulting is consistent with truncation of several related zones of ductile deformation in the Roanoke region by the Alleghanian-age Blue Ridge fault, which is characterized by brittle deformation (Bartholomew and others, 1982). Bartholomew and others (1981) also argued for a pre-Alleghanian age for movement on the Rockfish Valley fault. Possible pre-Late Devonian and/or multiple times of movement along all or portions of the Fries fault system has not been investigated.

REGIONAL GEOLOGY

The Grenville-age massifs are each nonconformably overlain by upper Precambrian to Lower Cambrian clastic sedimentary and volcanic rocks. These cover rocks, along with the massifs themselves, were subjected to several episodes of metamorphism and/or deformation between middle and late Paleozoic time. The terms "Taconic" and "Alleghanian" as used here refer to orogenic pulses that occurred approximately during Middle to Late Ordovician and late Paleozoic time, respectively. The three principal areas of occurrence of the Grenville-age massifs (Figs. 1 and 2) are the Blue Ridge anticlinorium (Fig. 3), the Grandfather Mountain window area (Fig. 4), and the Sauratown Mountains anticlinorium (Fig. 5).

Blue Ridge Anticlinorium

In northern Virginia the Pedlar Massif (Fig. 2) is nonconformably overlain by the late Precambrian Catoctin Formation, metavolcanic and associated metasedimentary rocks, which pinch out in central Virginia (Bartholomew and others, 1981; Bloomer and Werner, 1955). From there southwestward the Pedlar Massif is overlain by younger clastic metasedimentary rocks of the Chilhowee Group of Eocambrian/Cambrian age. The Lovingston Massif also is overlain partially by Catoctin metavolcanic rocks but only around the northern terminus of the Blue Ridge anticlinorium in northern Virginia and Maryland. By contrast, along most of the eastern flank of the anticlinorium, the Lovingston Massif is directly and nonconformably overlain by the eastwardly thickening sequence of terrigenous and marine metasedimentary rocks of the Lynchburg Formation (including the Mechums River metasedimentary rocks), which stratigraphically underlies the Catoctin metavolcanic rocks. In the southern portion of the anticlinorium, the enigmatic Moneta Gneiss (Figs. 1, 2, and 3), with its associated pegmatites, lies between recognizable basement rocks of the Lovingston Massif and the late Precambrian Lynchburg Formation. The presence of the Moneta domes to the southwest suggests that Grenville-age rocks are probably nearer the surface there than in other adjacent areas. Southwest of the Roanoke oroclinal bend, the southern end of the Lovingston Massif is overlain by metasedimentary rocks of the Ashe Formation, which is the lateral equivalent of the Moneta Gneiss and/or Lynchburg Formation.

These late Precambrian to Cambrian-age cover rocks of the Blue Ridge anticlinorium were metamorphosed during the Paleozoic. The presence of clasts from metasedimentary rocks of the Chilhowee Group within the Fincastle Conglomerate of Middle Ordovician-age implies that the metamorphism was part of the Taconic event (Bartholomew and others, 1981; Bartholomew and others, 1982). The cover rocks over the Pedlar Massif typically reached biotite grade, whereas those overlying the Lovingston Massif are commonly at or above the transition from upper greenschist facies to lower amphibolite facies. Both eastward and southeastward from the Lovingston Massif, the metamorphic grade of the cover rocks progressively increases to middle or upper amphibolite facies. Because basement rocks were relatively "dry," due to dehydration during Grenville-age metamorphism at higher grades, they show few retrograde effects of the Taconic metamorphism, except locally in the northern terminus of the anticlinorium and near the Moneta Gneiss boundary.

By contrast, during the Late Devonian ductile deformation and associated retrograde metamorphism synchronous with movement along faults of the Fries fault system, rocks of the basement massifs were altered extensively to greenschist facies assemblages presumably due to introduction of water during thrusting. The cover rocks, however, show few metamorphic effects as a result of latest Devonian orogenesis, except locally near the thrusts where the earlier foliation is commonly transposed into the plane of the mylonitic fabric (Kaygi, 1979). Finally, during Alleghanian orogenesis in the late Paleozoic, both massifs and cover rocks were cut by the Blue Ridge thrust and related faults that are characterized by brittle deformation (Bartholomew and others, 1982). A locally pervasive crenulation cleavage that postdates the earlier metamorphic and mylonitic fabrics probably developed synchronously with Alleghanian thrusting and folding in the adjacent Valley and Ridge province. The development of this cleavage is strongly lithology-dependent.

The Blue Ridge anticlinorium in central and northern Virginia differs from the Blue Ridge farther southwest in the Grandfather Mountain window area in regional orientation of multiply formed structural elements. In the latter region, deformational events produced different structural elements that typically are oriented in different directions, whereas in the Blue Ridge anticlinorium orogenic events apparently produced structural elements that are nearly coaxial and coplanar but not readily distinguishable. This change in structural orientations follows the approximate radiometric separation boundary of Fullagar and Dietrich (1976), where to the south, Late Devonian ages (350–380 m.y.) are well documented. In the Sauratown Mountains structural elements produced during both Taconic and younger metamorphism are distinguishable, even though they are often coplanar and coaxial (Lewis, 1980a).

Grandfather Mountain Window Area

The Watauga Massif, which lies northwest of the Fries fault (Figs. 2 and 4), is nonconformably overlain by clastic metasedimentary rocks of the Chilhowee Group (Fig. 4), except locally where interlayered metavolcanic and metasedimentary rocks of the older Mount Rogers Formation lie disconformably beneath the Chilhowee Group (Stose and Stose, 1957; Rankin, 1970). Chilhowee metasedimentary rocks, along with the conformably overlying Shady Dolomite, also are found on the Table Rock thrust sheet (Fig. 4) within the Grandfather Mountain window. Basement rocks of the Globe Massif, nonconformably overlain by interlayered clastic metasedimentary and metavolcanic rocks of the Grandfather Mountain Formation of late Precambrian age, lie structurally beneath the Table Rock sheet, as well as below the Linville Falls fault that surrounds the window (Bryant and Reed, 1970b). The Elk River Massif (Figs. 2 and 4) is overlain by a thick sequence of mafic metavolcanic rocks that pinch out eastward and southeastward (Rankin and others, 1972; Espenshade and others, 1975). These mafic metavolcanic rocks form the basal portion of the Ashe Formation. Stratigraphically above, as well as laterally east and southeast of the type area, clastic metasedimentary rocks and some interlayered metavolcanic rocks form the Ashe Formation. As the Ashe Formation pinches out southeastward, the Elk River Massif is overlapped by younger clastic metasedimentary rocks of the Alligator Back Formation (Fig. 4). Abundant mafic dikes cut the basement gneisses beneath the metavolcanic pile, and both the basal portion of the pile and the basement beneath are intruded by dikes and plutons of the Bakersville Metagabbro (Bartholomew, 1983a; Bartholomew and Wilson, 1984; Bartholomew and others, 1983; Lewis and Bartholomew, 1984). Contacts between metagabbro and basement gneisses are truncated by the Beech Pluton (Bartholomew and others, 1983; Lewis and Bartholomew, 1984). Thus, the age of about 700 m.y. for the Beech Pluton (Davis and others, 1962; Odom and Fullagar, 1984) provides a minimum age for both the Bakersville Metagabbro and the basal portion of the metavolcanic pile. Because of the difficulty of obtaining radiometric dates that represent the time of original deposition for metasedimentary rocks, such as the Ashe Formation, most workers (cf., Rankin and others, 1973) accept a late Precambrian age for time of deposition of the Ashe Formation metasedimentary and metavolcanic rocks.

The Watauga Massif and its cover rocks were subjected to Taconic folding and metamorphism to greenschist facies followed by Late Devonian ductile deformation and retrograde metamorphism of the basement rocks during movement along the faults of the Fries fault system. Subsequent folding and development of locally widespread crenulation cleavage probably represent structural features formed during the Alleghanian event. At this time,

the massifs and their cover rocks were thrust over the Cambrian-age Shady and Rome formations that are now exposed both within the Mountain City window and along the leading edge of the frontal faults of the Blue Ridge fault system (Fig. 4).

The rocks within the Grandfather Mountain window experienced a similar sequence of deformation and metamorphism, but differ in that the metamorphic conditions reached during the Taconic event were upper greenschist facies and lower amphibolite facies in the western and eastern portions of the window, respectively (Bryant and Reed, 1970b; Rankin and others, 1972). By contrast, the Elk River Massif and its associated cover rocks were subjected to metamorphic conditions of the middle to upper amphibole facies (Rankin and others, 1972) during the Taconic event, placed by Butler (1973) at about 450 m.y. ago. A second period (Late Devonian) of close to isoclinal folding of the cover rocks appears to be nearly concurrent with the Spruce Pine Pegmatites, which were emplaced about 380 m.y. ago (Butler, 1973). Inasmuch as (1) upper amphibolite facies cover rocks overlie the Elk River Massif, which is thrust over the Watauga Massif with its greenschist facies cover rocks, as well as over the greenschist facies rocks of the Grandfather Mountain window; and (2) Spruce Pine Pegmatites are found both north and south of the window but are solely confined to the eastern Blue Ridge thrust sheet (Figs. 1 and 4), the 345-m.y. age (Late Devonian) assigned to the Fries fault system is consistent with the ages of metamorphism and emplacement of pegmatites that necessarily preceded the thrusting.

Sauratown Mountains Anticlinorium

The Sauras Massif forms the basement core of the Sauratown Mountains anticlinorium at the northern terminus of the Brevard zone (Figs. 1, 2, and 5). This massif, like the Elk River Massif, is nonconformably overlain by the Ashe Formation (Rankin and others, 1972; Espenshade and others, 1975). However, as pointed out by Lewis (1980a; 1983), the Ashe Formation here differs significantly from the same unit in its type region, just described above, where it is dominantly a volcanic pile (Bartholomew and others, 1983). Around the anticlinorium the formation consists of interbedded lenses of metaquartzite (mainly near the base), metapelite, metagraywacke, and some mafic metavolcanic rocks and associated ultramafic rocks higher in the sequence. Scattered mafic dikes and several late Precambrian granitoid stocks intrude the metasedimentary sequence.

The anticlinorium structurally is flanked on the northwest by the Camp Creek nappe/Smith River allochthon and on the southwest and southeast by the Inner Piedmont and Charlotte belt, respectively. The contact with the Charlotte belt is largely concealed beneath a Triassic basin (Fig. 1). The classic Brevard zone characteristics begin at the confluence of three major structural elements (Lewis, 1980a, 1980b); these are (1) the Smith River allochthon/Camp Creek nappe on the northwest, (2) the Yadkin thrust, and (3) the younger Stony Ridge Fault zone that approximately bisects the Sauras Massif (Figs. 1, 2, and 5). "Camp Creek nappe" is used here for allochthonous probable basement and cover rocks northwest of the anticlinorium that have strong lithological affinities with both basement and cover rocks of the anticlinorium and, hence, probably were derived from its southeastern flank (Lewis, 1980a, 1980b). By contrast, Conley and Henika (1973) and Conley (1978) consistently have suggested an Inner Piedmont origin for the Smith River allochthon proper, which does contain some notably different lithologies from those of the Camp Creek nappe. Whether the Camp Creek nappe and Smith River allochthon proper are a single tectonic sheet derived from the Blue Ridge geologic province or two sheets derived from different provinces as yet remains unresolved. Lewis (1980a, 1980b) recently summarized the deformational history of the Sauratown Mountains area as follows.

Taconic isoclinal folding and metamorphism to upper amphibolite facies (kyanite and sillimanite grade) was followed by emplacement of the allochthonous Camp Creek nappe. Devonian(?) metamorphism, possibly as high as amphibolite facies and associated with tight to isoclinal folding of the allochthonous nappe, was followed by thrusting of the Inner Piedmont over the anticlinorium along the Yadkin fault. To the southwest, Brevard zone faults truncate the Table Rock and Linville Falls faults of the Fries fault system along the southeastern flank of the Grandfather Mountain window (Fig. 1); hence, movement on the Yadkin fault is also interpreted to be Late Devonian or possibly even Alleghanian. Much younger, large-scale, dip-slip and/or strike-slip movement along the high-angle Stony Ridge fault zone truncates the Yadkin fault and imparts a remarkably linear appearance to this northern segment of the Brevard zone. Brittle deformation characterizes the Stony Ridge fault zone, which has a Rb-Sr date of about 180 m.y. (Fullagar and Butler, 1980).

Consistent with Hatcher's (1978) interpretation that his Hayesville-Fries thrust sheet contains lower Precambrian rocks deposited on oceanic crust (or very thin continental crust), Hatcher and others (1983) recently invoked an inordinately complicated explanation of truncated "premetamorphic" and younger folded thrusts to explain the presence of metasedimentary rocks of the Ashe Formation above older "metagraywacke" in the Sauratown Mountains anticlinorium. The presence of continental crust on the eastern Blue Ridge thrust sheet (Figs. 1, 2, 3, and 4) and the relationship of the Fries and Hayesville faults have already been discussed in a previous section of this paper.

Our interpretation of their older "metagraywacke" is based on our comparison of these rocks with those in the other Grenville massifs where more geochronological data are available. We prefer the interpretation of Lewis (1980a, 1983) that these older rocks are all part of the Grenville-age sequence within the Sauras Massif and thus are below the late Precambrian unconformity. However, until more geochronological data are available for rocks in this region, the structural evolution of the Sauras Massif and its cover rocks is open to speculation.

GRENVILLE OROGENESIS

Although the massifs were subjected to repeated episodes of Paleozoic metamorphism, these basement rocks typically contain relict textures and mineral assemblages developed during Grenville orogenesis. From these relict mineral assemblages, preserved in a variety of rock types within each massif, the general Grenville metamorphic facies may be gleaned. Rocks of the Pedlar Massif all were subjected to the deeper (higher P and T conditions) granulite facies, as indicated by widespread orthopyroxene + garnet assemblages, whereas the Lovingston Massif primarily was subjected to conditions of the shallower (lower P and T) granulite facies, as indicated by widespread orthopyroxene assemblages (Bartholomew and others, 1981). Locally within the Lovingston Massif around the Roseland Anorthosite the preserved mineral assemblages are indicative of the deeper-seated granulite facies. the higher grade of metamorphism around the anorthosite is suggestive of a metamorphic aureole related to emplacement of the intrusion. Along much of the eastern flank of the Lovingston Massif, the preserved mineral assemblages (biotite ± amphibole) in the Stage Road Gneiss and similar rocks indicate that metamorphic conditions reached only the upper amphibolite facies. Shallower granulite facies conditions prevailed over most of the Watauga Massif, except in the eastern portion where amphibolite facies conditions were attained. The Globe, Elk River, and Sauras Massifs all were subjected to amphibolite facies conditions, except for a small area west of the town of Cranberry, North Carolina, where gneisses subjected to deeper granulite facies conditions (Gulley, 1982; Monrad and Gulley, 1983) are preserved in the Elk River Massif (Fig. 4).

Grenville-Age Plutonic Rocks

Within the Pedlar Massif, Grenville-age rocks are found in a series of plutonic charnockitic suites (see Appendix 1). these are, from north to south, the Saddleback Mountain, Crozet, Pedlar River, Peaks of Otter, and Bottom Creek suites (Fig. 3). All of these composite charnockitic plutonic suites consist of one or more of the following rock types: (1) massive, medium-grained, green charnockite with or without garnet; (2) massive, coarse-grained, green charnockite; (3) massive, coarse-grained, light-gray leucocharnockite; (4) massive to crudely foliated, very coarse-grained, green charnockite with feldspar megacrysts up to about 10 cm across; and (5) crudely to well-foliated, very coarse-grained, green to grayish green porphyroblastic gneiss with or without garnet and with feldspar porphyroblasts up to 10 cm in diameter.

The large Pedlar River Suite is typical of these composite suites and is discussed here as a representative example (see also Sinha and Bartholomew, 1984). A change in grain size and mineralogy of rocks within this suite was noted by Bartholomew and others (1981) who observed that grain size coarsens away from the margin of the suite toward its interior. They did not, however, recognize that the suite actually consisted of discretely mappable rock bodies that may represent either individual intrusions or portions of a single magma that underwent *in situ* differentiation after emplacement due to different cooling histories.

From the margin to the interior of the main portion of the

suite, the rock changes from medium-grained charnockite (± garnet), to coarse-grained charnockite, to coarse-grained leucocharnockite. This change is suggestive of a single magma cooling from the margin inward with progressive depletion of the mafic phase during cooling. The Crozet and Saddleback Mountain suites appear to be similar, and as part of the latter suite, the Old Rag Granite may represent the final granitic composition, devoid of a mafic phase, produced by differentiation.

The very coarse-grained charnockite (Vesuvius Megaporphyry), which is crudely foliated, probably represents a slightly earlier, but still comagmatic, pluton that crystallized more slowly at fewer nucleation sites. The presence of large magmatic hornblende crystals with pyroxene cores suggests that the partial pressure of H_2O remained higher in this magma than in the later emplacements which contain only orthopyroxene. The porphyritic charnockite of the Bottom Creek Suite differs from the Vesuvius Megaporphyry in that it is more foliated and contains a high percentage of white feldspar (rather than green feldspar); similarly, the garnet-bearing porphyroblastic gneiss, mapped by Bartholomew and Hazlett (1981), between the Bottom Creek and Peaks of Otter suites is interpreted as being closely related to the megaporphyries. This gneiss is clearly cut by younger stocks of medium-grained charnockite (Bartholomew, 1981). The presence of garnet, the development of a foliation, and the cross-cutting relationships near Stewartsville all suggest that the megaporphyries represent earlier charnockitic intrusions, which were followed by the younger plutons that generally underwent more differentiation during their cooling histories.

The overall petrologic and chemical similarities of the suites and country rocks suggest some type of common bond (Sinha and Bartholomew, 1984); probably the latter served as source rock from which the charnockitic magmas were derived during anatexis at crustal levels not far below that at which they were emplaced.

Grenville-age intrusive rocks, other than the charnockitic suites, are limited to scattered minor occurrences of mafic granulites (Bartholomew, 1977, 1981; Hudson, 1981); nelsonite dikes (Hillhouse, 1960; Hudson, 1981; Force and Herz, 1982; Herz and Force, 1984; Bartholomew, 1981); small hornblende granitoid bodies in the Pilot Gneiss; and the anomalous Stewarts Knob Gneiss.

As deduced from its cross-cutting relationships (Bartholomew, 1981) with both porphyroblastic granulite gneiss and stocks of massive charnockite, the small bodies of Stewarts Knob Gneiss just northeast of Roanoke, Virginia, are the youngest Grenville intrusions in this portion of the Pedlar Massif. This distinctive light gray to white porphyroblastic quartzo-feldspathic gneiss contains abundant garnet and biotite and most resembles rocks typically associated with anorthosite suites such as the Border Gneiss of the Roseland Anorthosite (Hillhouse, 1960; Herz and Force, 1984) or the Honey Brook Anorthosite of Pennsylvania (Crawford and Hoersch, 1984). Similarly, nelsonite dikes are associated with both the Stewarts Knob Gneiss and the Roseland Anorthosite.

Charnockitic rocks also are found in the Lovingston Massif where small bodies of massive, medium-grained, gray charnockite form part of the Oventop, Archer Mountain, Turkey Mountain, and Horsepen Mountain plutonic suites. Mapping in much of the Oventop Suite and Little River Gneiss is insufficient to tell if charnockite is present. In the other suites, most of these small charnockite bodies occur within or near the main bodies of biotite dioritoid (dominantly quartz monzonite), which suggests that they are genetically related. As interpreted by other workers, the charnockite bodies represent portions of magmas that crystallized early. Thus, the charnockite bodies are interpreted as more anhydrous phases that did not react subsequently with the magma as it became relatively enriched in H_2O. However, the general small size of the charnockites, their scattered distribution, relatively sharp boundaries, probable cross-cutting relationships, and the lack of relict pyroxenes in the biotite-dioritoids are used to infer that crystallization of the magma that produced both the biotite-dioritoid and charnockite probably began with water added from the country rock. Incomplete mixing and/or insufficient water as crystallization progressed produced less hydrous magma resulting in small stocks of charnockite as the final crystallization products. These conditions would be more compatible with the observed shallower granulite or amphibolite facies of the country rock into which the suites were emplaced.

On the other hand, the Roseland Anorthosite is surrounded by the Border Gneiss (emplaced as an integral part of the anorthosite diapir) and portions of the Turkey Mountain Suite, both with orthopyroxene + garnet mineral assemblages indicative of the deeper granulite facies. As stated previously, this is suggestive of a metamorphic aureole, which would imply that emplacement of the Roseland Anorthosite diapir was (1) younger than and (2) intruded at a much higher temperature than the nearby biotite-dioritoid suites. This younger Grenville age also is consistent with our inference that the late Grenville-age Stewarts Knob Gneiss is more closely related to anorthositic rocks than to other rocks of the Pedlar Massif. The young charnockites associated with the Border Gneiss and Roseland Anorthosite probably are essentially *in situ* melts of the Border Gneiss and hence not genetically related to formation of charnockite within the biotite dioritoid suites.

Biotite dioritoid also forms the coarse-grained core facies (Grayson Gneiss, Stose and Stose, 1957) of the Forge Creek Suite in the Watauga Massif (Fig. 4). Here it is associated with medium-grained biotite granitoid (Comers Gneiss, Stose and Stose, 1957) that probably represents the marginal facies of the suite. The biotite granitoid is gradational into granitoid (Watauga River Gneiss) lacking abundant biotite. Biotite granitoid that is very similar to the Comers Gneiss also is found near Elk Park in the Elk River Massif (Fig. 4). There, cross-cutting relationships (Lewis and Bartholomew, 1984) indicate it is the younger of the two Grenville-age intrusive rocks of the Cranberry Suite.

Rocks of the Forge Creek Suite and similar rocks of the Elk River Massif differ mineralogically from the intrusive suites of the Pedlar and Lovingston Massifs. Forge Creek Suite rocks contain

both pink and green feldspar, whereas the charnockitic suites of
the Pedlar Massif have only green or both green and white feldspar and the dioritoid suites of the Lovingston Massif only white
feldspar. Rocks of the Forge Creek Suite rarely contain mafic
minerals other than biotite, whereas suites in the Lovingston Massif have both biotite- and pyroxene-bearing rocks, and the Pedlar
Massif suites are charnockitic.

Intrusive suites of the Globe and Sauras Massifs most resemble those of the Lovingston Massif in that they contain
abundant biotite granitoid/dioritoid with white feldspars. Porphyritic dioritoid plutons also characterize both the Globe (Blowing Rock Gneiss, Keith, 1903) and Sauras Massifs but only occur
at the extreme southwestern end of the Lovingston Massif (Little
River Gneiss, Dietrich, 1959).

The remaining rocks that form part of the intrusive suites are
granitoids which contain white feldspar and lack abundant biotite. Such granitoids form the small stocks of the Sauras Massif
(Fig. 5) as well as the Marshall Suite (Fig. 3) of the Lovingston
Massif and the Laurel Creek Pluton (Fig. 4) and related stocks of
the Cranberry Suite of the Elk River Massif.

Country Rocks

The country rocks of the Pedlar Massif consist of mesogranulite and leucogranulite gneisses. The Nellysford Granulite Gneiss
(Bartholomew and others, 1981) is predominantly a uniformly
dark green gneiss of intermediate composition with well developed millimetre- or centimetre-scale segregation layering. Nellysford-type gneisses appear to form the basal portion of a country
rock sequence that grades upward (westward) into the less
uniform light-greenish-gray sequence of interlayered massive and
layered granulite gneisses as characterized by the more leucocratic Lady Slipper Granulite Gneiss (Sinha and Bartholomew,
1984). Garnet is ubiquitous in the Nellysford-type gneiss, but
occurrence is selective with lithology within the Lady Slipper--
type gneisses.

Within the Lovinston Massif, both mesogranulite and leucogranulite gneisses also occur, but from the size, distribution, and
apparent relationships (Fig. 3) with adjacent rocks, some of these
granulite gneisses probably represent small to moderate sized
plutons, whereas others like the garnet-graphite gneisses of the
Border Gneiss are probably paragneisses. Bartholomew and others (1981) drew attention to the textural and mineralogical similarities of the Hills Mountain and Nellysford granulite gneisses
and suggested that they may be facies equivalents. However, the
Hills Mountain Granulite Gneiss of the Lovingston Massif appears to be a group of intrusive bodies, and thus only could be
related genetically to the much more widespread mesogranulite
and leucogranulite gneisses of the Pedlar Massif if the latter represented a volcanic equivalent.

Within the Lovingston Massif both this equivalent group of
mesogranulite and leucogranulite gneisses, as well as the
Grenville-age suites, intrude an older layered gneiss sequence
characterized best by the Stage Road Layered Gneiss (Sinha and

Bartholomew, 1984), which consists of interlayered biotite augen
gneisses, garnet biotite gneisses, well-layered biotite gneisses, and
feldspathic gneisses. The Stage Road and related biotite-rich
meso-gneisses of the northern end of the anticlinorium probably
are amphibolite facies equivalents of layered mesogranulite
gneisses like the Flint Hill Gneiss (Lukert and others, 1977) and
the herein-named Sandy Creek Granulite Gneiss, although
neither appears to have the wide variety of lithologic layers that
characterizes the Stage Road Layered Gneiss. The contact, as
mapped by Lukert and Nuckols (1976), between the Flint Hill
Gneiss and Nellysford-type gneisses probably is the garnet isograd, within the granulite faices, which we use to separate the
shallower and deeper granulite facies (Figs. 3 and 6).

The Saddle Gneiss (Stone and Stose, 1957) of the Watauga
Massif; the Cranberry-Mine Layered Gneiss and Shoals Gneiss
(Stose and Stose, 1957) of the Elk River Massif; and the herein-
named Low-Water-Bridge Layered Gneiss of the Sauras Massif
are all similar to the Stage Road Layered Gneiss and, in similar
fashion, all represent the country rocks that were intruded by
Grenville suites. All of these gneisses lack relict granulite facies
assemblages but do contain relict amphibolite facies assemblages
characterized by abundant biotite. One of the two known occurrences of granulite facies rocks in these other massifs is found in a
small area of the Watauga Massif where shallower granulite facies gneisses form part of the country rock of the Forge Creek
Suite. The other is the Carvers Gap Granulite Gneiss described by
Gulley (1982) and Monrad and Gulley (1983), which appears to
be the deeper granulite facies equivalent of the Cranberry-Mine
Layered Gneiss in the Elk River Massif.

Near Cranberry (Fig. 4), both amphibole and orthopyroxene-bearing gneisses have been reported (Bryant and Reed,
1970b; Lewis and Bartholomew, 1984). Thus, within the region
from Cranberry to Carvers Gap, an entire transition from amphibolite to shallower granulite to deeper granulite facies may be
preserved.

The Cranberry-Mine Layered Gneiss is a migmatitic amphibolite facies rock with abundant pegmatoid and small granitoid bodies contained within the more migmatized zones (Lewis
and Bartholomew, 1984). The nearby granulite facies rocks,
however, are lacking both the migmatized appearance of the
Cranberry-Mine Layered Gneiss and the granitoid intrusions. If
the migmatization here is a characterization of upper amphibolite
facies metamorphism and is produced by dehydration reactions
in the gneisses coupled with release of hydrous phases in the
granitoid melts, then granitoid plutons logically would be expected near the amphibolite/granulite transition, but not within
the relatively "dry" granulite facies.

Late Precambrian Plutonic Rocks

The Crossnore Complex (Rankin and others, 1983), contains a suite of peralkaline granites, representing the youngest
series of major granitoid intrusions cutting the Blue Ridge basement massifs. The larger bodies (Figs. 3, 4, and 5) are the Robertson River Pluton (Allen, 1963), the herein-named Suck

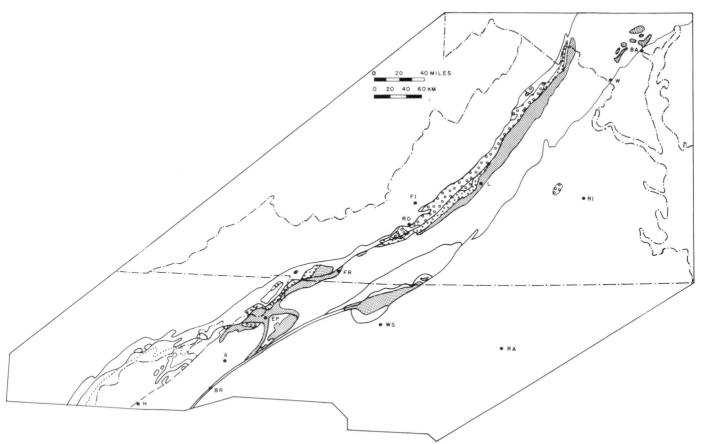

Figure 6. Metamorphic facies of the Grenville massifs of the Blue Ridge geologic province; large circles—deeper granulite facies terrane; small circles—shallower granulite facies terrane; stippled—amphibolite facies terrane. BA—Baltimore; W—Washington; RI—Richmond; L—Lovingston; FI—Fincastle; RO—Roanoke; FR—Fries; RA—Raleigh; WS—Winston-Salem; EP—Elk Park; A—Asheville; BR—Brevard; H—Hayesville.

Mountain Pluton, the associated Striped Rock and Cattron plutons (Stose and Stose, 1957), and the Beech Pluton (Keith, 1903). All of these major late Precambrian granites are composite plutons typically composed of (1) the major rock type, a medium- to coarse-grained facies containing aegerine-augite or a sodic amphibole, (2) a coarse-grained biotite-rich facies, and/or (3) a fine-grained facies also with augite or amphibole, and/or (4) a fine-grained dioritoid facies usually found near the margins. The Robertson River Pluton (Allen, 1963; Lukert and Nuckols, 1976; Lukert and Halladay, 1980) is composed of facies (1), (3), and (4); the Suck Mountain Pluton of facies (1) and (2); the Striped Rock Pluton of facies (1) and (3); the Cattron Pluton of (4) (Stose and Stose, 1957); and the Beech Pluton of (1), (2), and (3) (Lewis and Bartholomew, 1984). Most of the smaller plutons are typically composed of facies (1). Among the better-mapped and/or better-studied smaller bodies are the Rockfish River Pluton (Davis, 1974), the Stewartsville Pluton (Bartholomew, 1981), and the Mobley Mountain Pluton (Brock, 1981) in the Lovingston Massif; the Leander Mountain Pluton (Bartholomew, 1983a) and the Buckeye Knob Pluton (Bartholomew and Gryta,

1980) in the Elk Park Massif; the Brown Mountain Pluton in the Globe Massif (Bryant and Reed, 1970b); and the Elkin Stock (Harper, 1977; Lewis, 1980a) in the Sauras Massif.

Rankin (1970, 1975; Rankin and others, 1973; 1983) defined and named the Crossnore Plutonic-Volcanic Complex to include the granitoid plutons with peralkaline affinities that were similar to felsic volcanic rocks of the Mount Rogers Formation and thus represented the felsic end of related, bimodal plutonic and volcanic rocks. He (Rankin, 1976) speculated that their distribution was related to the occurrence of salients in the Appalachian orogen formed during an initial breakup of the North American continent to form Iapetus. He suggested that the silicic intrusive/extrusive rocks were excluded from recesses, which represented aulacogens. However, at that time detailed mapping was lacking in the Roanoke recess area. Recent work there has shown an abundance of these granitoid rocks (Fig. 3), including the Suck Mountain Pluton which is one of the larger bodies. The concept that these granitoid rocks are part of related, bimodal plutonic and volcanic rocks likely remains valid with those granites in the Roanoke area related to the Moneta Gneiss (Fig. 2) which probably represents a small volcanic pile that accumulated along the present flank of the Lovingston Massif (Fig. 2).

Age Relationships

Geochronology of Blue Ridge units is still in its nascency. No large areas have been mapped geologically in detail and then systematically examined from a geochemical, petrological, and/or geochronological standpoint. Preliminary results on one such traverse based on reconnaissance mapping are presented in this volume (Sinha and Bartholomew, 1984).

In general, the age relationships that have been worked out for the entire Blue Ridge (Table 1) support an interpretation that the early Grenville-age country rocks are generally about 1100 to 1200 m.y. old but probably contain older (detrital?) 1400- to 1800-m.y.-old zircons. The Grenville-age intrusive suites typically are slightly younger than the country rocks, about 1050 to 1150 m.y. old, and the metamorphic peak appears to be still younger, about 900 to 1000 m.y. old.

DISCUSSION

The rock unit symbols on the geologic maps (Figs. 3, 4, and 5) have been used in an attempt to portray lithological similarities of rock types among the different massifs. In the same way, the metamorphic grade and age relationships among the plutonic suites and country rocks are indicated on Figure 6 and Table 1. If the country rocks are interpreted (Sinha and Bartholomew, 1984) as largely a volcanic or volcaniclastic sequence, then most of the rocks of the basement massifs can be interpreted as part of a single orogenic cycle in which a eugeosynclinal sequence accumulated along the North American craton during early Grenville time. Subsequently, during later Grenville time, the deeply buried sequence was metamorphosed to upper amphibolite and granulite facies and was intruded by plutonic suites. The interpretation of country rocks as largely volcanic or volcaniclastic is suggested by (1) the absence of pelitic rocks (sillimanite was reported solely in the Saddle Gneiss by Stose and Stose, 1957); (2) a dearth of rocks representing possible shallow-water carbonates or quartzites; (3) an absence of abundant detrital zircon suites which would be expected in sediments derived from cratonic source terranes; and (4) the scarcity of mafic rocks which would be expected if the country rocks were largely oceanic and/or part of the upper mantle.

One of the more intriguing relations is the apparent association of different types of Grenville-age plutonic suites with different grades of Grenville metamorphism. Of course, one explanation is that the different plutonic suites simply reflect different types of magmatic sources. However, if the country rocks are all part of a stratified sequence buried to depths where anatectic melts are likely to be generated, then different magmatic sources (upper mantle versus lower crust) for the plutonic suites seem a less likely explanation than linking anatectic magma compositions with the grade of metamorphism of the source rocks. Admittedly, the massifs preserve the relationships where the intrusive suites were emplaced, but the inference seems warranted that regionally these suites represent nearly *in situ* lower crustal melts. Of course, some plutons, as yet unrecognized, may be generated in the upper mantle during the crustal heating event. In any case, the general association of (1) a charnockitic supersuite with deeper granulite facies terrane, (2) a biotite dioritoid supersuite with shallower granulite facies terrane, and (3) granitoid and mixed granitoid/dioritoid suites with amphibolite facies terrane tentatively is interpreted as genetic.

CONCLUSIONS

The Fries fault system is a major tectonic boundary between distinctive Grenville terranes extending all the way from the northern end of the Blue Ridge anticlinorium southwestward to the Grandfather Mountain window area of northwestern North Carolina. Our broad-scale study has resulted in distinguishing Grenville plutonic suites from the surrounding Precambrian country rocks and associating the type of Grenville plutonic suites with the grade of Grenville metamorphism in adjacent country rocks.

The Watauga, Globe, and Elk River Massifs comprise the Grenville rocks of the Grandfather Mountain window area, the Sauras Massif forms the crystalline core of the Sauratown Mountains anticlinorium, and the Pedlar and Lovingston Massifs form the core of the Blue Ridge anticlinorium from Maryland to southwestern Virginia.

Grenville metamorphic grade reached shallower granulite/amphibolite facies in the Watauga Massif, amphibolite facies in the interposed Globe Massif, as well as in the Elk River and Sauras Massifs. Overprinting by Paleozoic metamorphism is more intense in these three massifs than farther northeast in Virginia where Grenville metamorphism reached the deeper granulite facies in the Pedlar Massif and shallower granulite/amphibolite facies in the Lovingston Massif.

ACKNOWLEDGMENTS

During the course of our investigation, much of the work done by Bartholomew in the State of Virginia was funded by the Virginia Division of Mineral Resources. The work done by Lewis in the Sauratown Mountains area of North Carolina was funded partially by National Science Foundation Grant DES 73-06495-A01 awarded to J. R. Butler and P. D. Fullagar of the University of North Carolina, Chapel Hill, and partially by the North Carolina Division of Land Resources. The work by both of us in the Grandfather Mountain window area was funded partially by the North Carolina Division of Land Resources.

TABLE 1. GEOCHRONOLOGICAL DATA FROM PUBLISHED SOURCES CURRENTLY AVAILABLE ON BLUE RIDGE CRYSTALLINE ROCKS

	Country rocks	Grenville intrusive suite	Late Precambrian units
PEDLAR MASSIF			
Nellysford Granulite Gneiss	1100 m.y. (Pb-Pb)		
Lady Slipper Granulite Gneiss	1130 m.y. (U-Pb)		
Flint Hill Gneiss	1080 m.y. (U-Pb)		
Saddleback Mtn. Suite			
mesocharnockite		1075 m.y. (U-Pb)	
Old Rag Granite		1115 m.y. (U-Pb)	
Pedlar River Suite			
leucocharnockite		1075 m.y. (U-Pb)	
mesocharnockite		1040 m.y. (Rb-Sr)	
Vesuvius Megaporphyry (metamorphism)		940 m.y. (K-Ar)	
Irish Creek greisen			635 m.y. (^{40}Ar/^{39}Ar)
LOVINGSTON MASSIF			
Stage Road Layered Gneiss			
augen gneiss (detrital zircons)	1420+ m.y. (U-Pb)		
(metamorphism)	915 m.y. (U-Pb)		
augen gneiss	1065 m.y. (U-Pb)		
biotite gneiss	1120 m.y. (U-Pb)		
Archer Mtn. Suite			
biotite dioritoid		1100 m.y. (U-Pb)	
pegmatite		1100 m.y. (U-Pb)	
mesocharnockite		1060 m.y. (U-Pb)	
Little River Gneiss		1130 m.y. (U-Pb)	
Catoctin volcanic rocks			685 m.y. (Pb-Pb)
Robertson River Pluton			730 m.y. (U-Pb)
riebeckite granite			650 m.y. (U-Pb)
Rockfish River Pluton			730 m.y. (U-Pb)
Mobley Mountain Pluton			650 m.y. (Rb-Sr)
WATAUGA MASSIF			
Forge Creek Suite			
Grayson Gneiss		1150 m.y. (Rb-Sr)	
Watauga River Gneiss		1175 m.y. (Rb-Sr)	
Comers Gneiss (metamorphism)		780 m.y. (Rb-Sr)	
Striped Rock Pluton			680 m.y. (Rb-Sr)
" " "			695 m.y. (Rb-Sr)
Mount Rogers volcanic rock			820 m.y. (U-Pb)
GLOBE MASSIF			
Wilson Creek Suite		1020 m.y. (Pb-Pb)	
Crossing Knob Gneiss (metamorphism)		945 m.y. (Rb-Sr)	
Blowing Rock Pluton		1060 m.y. (Pb-Pb)	
" " " (metamorphism)		1005 m.y. (Rb-Sr)	
Grandfather Mtn. volcanic rocks			820 m.y. (U-Pb)
Brown Mtn. Pluton			735 m.y. (U-Pb)
ELK RIVER MASSIF			
Cranberry-Mine Layered Gneiss			
(metamorphism)	1020 m.y. (Rb-Sr)		
Carvers Gap Granulite Gneiss	1815 m.y. (Rb-Sr)		
(metamorphism)	805 m.y. (Rb-Sr)		
Beech Pluton			705 m.y. (Rb-Sr)
" "			700 m.y. (Pb-Pb)
Crossnore Pluton			645 m.y. (Rb-Sr)
Lansing Pluton			700 m.y. (Rb-Sr)
SAURAS MASSIF			
Granitoid/Dioritoid		1190 m.y. (Pb-Pb)	
Capella Pluton			675 m.y. (Rb-Sr)
Rock House Pluton			570 m.y. (Rb-Sr)
Danbury Pluton			625 m.y. (Rb-Sr)

Note: Correlation of these lithologies with country rock or intrusive suites is, in some cases, tentative. Geochronologic data from: Bartholomew (1971); Davis (1974); Davis and others (1962); Dietrich and others (1969); Fullagar and Bartholomew (1983); Fullagar and Butler (1980); Fullagar and others (1979); Fullagar and Odom (1973); Herz and others (1981); Hudson and Dallmeyer (1981); Lukert (1973, 1982); Lukert and Banks (1984); Lukert and others (1977); Monrad and Gulley (1983); Odom and Fullagar (1973, 1984); Rankin (1976); Rankin and others (1969); Rankin and others (1973); Sinha and Bartholomew (1984); Tilton and others (1960); Truman (1976). Ages rounded off to nearest 5 m.y.

APPENDIX 1. NEW AND REVISED FORMAL STRATIGRAPHIC NOMENCLATURE

Unit name	7.5-min. quad. long. and lat	Location of representative or type locality
Archer Mountain Suite (biotite dioritoid and charnockite)	–	See Sinha and Bartholomew (1984)
Blue Ridge Complex	–	Abandoned as formal stratigraphic unit
Bottom Creek Suite (charnockite)	Elliston 80°11'W 37°08'N	Named for Bottom Creek in south-western Virginia
1. mesocharnockite	Check 80°11'W 37°06'N	Crops out along state roads 637, 669, and 607 near Bottom Creek in the Elliston and Check quadrangles
2. megaporphyry	Bent Mountain 80°04'W 37°11'N	Crops out along Blue Ridge Parkway about 2 mi northeast of Slings Gap
3. porphyroblastic gneiss	Roanoke 79°53'W 37°16'N	Outcrop from which sample R-8444 (Bartholomew, 1981) was collected near Virginia Avenue in Vinton, Virginia
Carvers Gap Granulite Gneiss	Carvers Gap 82°06'W 36°06'N	Representative exposures are described by Gulley (1982) along the access road to the top of Roan Mountain for about one mile west of intersection with Tennessee State Road 143 (North Carolina State Road 261) at Carvers Gap on the North Carolina-Tennessee border
Cranberry Suite	Elk Park 81°58'W 36°09'N	This name is derived from Keith's (1903) Cranberry Granite which he describes in part as follows: "... In its principal area is situated Cranberry, N.C., from which its name is derived. The formation consists of granite, of varying texture and color, and of schists and granitoid gneisses derived from the granite..." "The granite is an igneous rock. Especially in the coarse varieties, the feldspar is by far the most prominent mineral and gives a prevailing light-gray or white color to the rock." Keith (1903) also included a "variety" of granite which has "...a marked red appearance..." due to the abundant pink feldspars. He describes these elsewhere as the "red feldspathic granites near Cranberry..." As thus described by Keith, the Cranberry Suite clearly included both light-gray granite (which forms the Laurel Creek Pluton and numerous smaller igneous bodies just north of the town of Cranberry) and pink granite (Whaley Gneiss) which also forms numerous small igneous bodies within the Elk River Massif near Cranberry. Rankin (1970), and Rankin and others

APPENDIX 1. (continued)

Unit name	7.5-min. quad. long. and lat	Location of representative or type locality
		(1973, 1983) followed this usage but applied the name Cranberry to a variety of rocks in the Watauga Massif which were mapped, described, and named by Stose and Stose (1957) and Hamilton (in King and Ferguson, 1960). We are restricting the term "Cranberry Suite" to the plutonic rocks of the Elk River Massif.
1. Laurel Creek Pluton (granitoid)	Baldwin Gap 81°40'W 36°29'N	Representative exposures (Bartholomew, 1983a) are found along the Big Laurel Creek road for about half a mile southwest of the junction with Soup Bean Branch road, near Big Laurel and Roaring Fork churches
2. Whaley Gneiss	Elk Park 82°56'W 36°13'30"N	Representative exposures are found along North Carolina State Road 1316 for about half a mile north of Cannon Gap which is located about one mile south of the community of Whaley
Cranberry-Mine Layered Gneiss	Elk Park 81°58'W 36°08'N	In Keith's (1903) description of his "Cranberry Granite" he states "this is the most extensive formation in the quadrangle..." and "consists of granite...and of schists and granitoid gneisses derived from the granite." In his description of the Cranberry mines Keith (1903) stated "...the ore and gangue occur as a series of great lenses...about parallel to the planes of...schistosity in the granite..." As thus described and mapped by him, Keith (1903) did not distinguish the older layered gneiss sequence from the plutonic rocks (and mylonitic gneisses derived therefrom) of the Cranberry Suite. Likewise, on recent maps (Bryant and Reed, 1970a; Rankin and others, 1972) the country rocks were not distinguished from the plutonic Cranberry Suite. On still more recent detailed 7 1/2-minute quadrangles (including the Elk Park quadrangle within which is situated both the town of Cranberry and the nearby Cranberry mine), the country rocks have been mapped separately from the plutonic Cranberry Suite. Because of Keith's (1903) inclusion of the layered country rocks (which specifically occur at both the town of Cranberry and the Cranberry mine) in his "Cranberry Granite" as well as his indication that it is the "most extensive" unit in the area, this older layered gneiss sequence was initially mapped as "Cranberry Gneiss" by Bartholomew

APPENDIX 1. (continued)

Unit name	7.5-min. quad. long. and lat	Location of representative or type locality
		(1983a) and Bartholomew and Gryta (1980). This same older gneiss was also referred to as "Cranberry Gneiss" by Bartholomew and others (1983) and Fullagar and Bartholomew (1983) in articles in the 1983 Carolina Geological Society Guide-book.
		As a result of discussions on that field trip, we feel it is desirable to clarify the following stratigraphic units. First, the term "Cranberry Suite" is retained for the plutonic rocks originally described by Keith (1903) as "Cranberry Granite" and which occur near the town of Cranberry, N.C. Second, in order to clearly distinguish that the layered gneisses at the Cranberry mine are not part of that plutonic suite, but rather are part of the more extensive layered gneiss sequence which forms the country rock into which the Cranberry suite was intruded, the term "Cranberry-Mine Layered Gneiss" is here designated to replace the term "Cranberry Gneiss" as used by Bartholomew and coworkers in the abovementioned articles. The Cranberry mine is the type locality for the Cranberry-Mine Layered Gneiss; however, because of extensive alteration of the gneiss at this iron ore mine, a representative locality nearby is herein chosen as a substitute more typical of the gross lithology (Lewis and Bartholomew, 1984). This outcrop is located along North Carolina State Road 194 approximately a quarter of a mile east of its junction with U.S. Highway 19E and about 1 mile northeast of the Cranberry mine.
Crossing Knob Gneiss (dioritoid)	Sherwood 81° 52'30"W 36° 16' N	Type outcrop (Bartholomew and Gryta, 1980) is along Bethel Church Road half a mile south of the junction with U.S. Highway 321 on west flank of Crossing Knob
Plutonic suite of Crossnore Complex (granitoid and dioritoid)	-	Crossnore Plutonic Volcanic Complex of Rankin (1970) and Rankin and others (1973; 1983) includes newly named Rockfish River, Suck Mountain, Stewartsville, Leander Mountain, and Buckeye Knob plutons and Elkin Stock, as well as previously named Robertson River (Allen, 1963), Striped Rock and Cattron (Stose and Stose, 1957), Beech (Keith, 1903), Crossnore (Rankin, 1970), Brown Mountain (Bryant and Reed, 1970b), and Rock House, Danbury, and Capella plutons (Fullagar and Butler, 1980)

Unit name	7.5-min. quad. long. and lat	Location of representative or type locality
1. Rockfish River Pluton	-	See Sinha and Bartholomew (1984)
2. Suck Mountain Pluton	Peaks of Otter 79° 33'W 37° 27'N	Typical exposures are found along Virginia State Road 643 about a quarter of a mile west of Suck Springs Church; coarse-grained biotite-rich facies crop out along Virginia State Road 639 about half a mile south of Curtis
3. Stewartsville Pluton	Stewartsville 79° 48' W 37° 15'30"N	Type outcrop is from where sample R-8431 was collected (Bartholomew, 1981) along Virginia State Road 619 about 1-1/2 mi south of town of Stewartsville in southwestern Virginia
4. Leander Mountain Pluton	Baldwin Gap 81° 43'W 36° 27'N	Representative exposures of the massive facies of this gneiss are found along crest of Leander Mountain (Bartholomew, 1983a) for approximately 1 mi west of the crossing of the mountain by the Tennessee-North Carolina boundary
5. Buckeye Knob Pluton	Sherwood 81° 47'30"W 36° 19' N	Representative exposures are found on Buckeye Knob on the crest and southern flank of Buckeye Knob (Bartholomew and Gryta, 1980)
6. Elkin Stock	Elkin North 80° 49' W 36° 15'30"N	Type exposures are at Vulcan Materials Elkin quarry along North Carolina State Road 1142 approximately a quarter of a mile north of junction with North Carolina State Road 268 (Harper, 1977; Lewis, 1980a)
Crozet Suite (charnockite)	Crozet 78° 38'W 38° 04'N	Named for Town of Crozet, central Virginia, and informal useage of Crozet granite (Nelson, 1962) for leucocharnockite of this suite
1. leucocharnockite	Browns Cove 78° 44'W 38° 08'N	Below dam for Charlottesville Reservoir on Moormans River along Virginia State Road 614
2. mesocharnockite	Crozet 78° 39'W 38° 05'N	Junction of Virginia State Roads 684 and 788 at northwest edge of Town of Crozet
3. porphyroblastic gneiss	Browns Cove 78° 41'W 38° 10'N	Adjacent to Virginia State Road 673 for 1-1/4 mi west of junction with Virginia State Road 810
Elk Park Supersuite	-	Informally and then formally named Elk Park Plutonic Suite by Rankin and others (1973, 1983) and here elevated to Elk Park Supersuite and specifically designated to include the Grenville-age suites and plutons of granitoid/dioritoid rocks of the Elk River, Watauga and Globe Massifs. This supersuite does include both the Cranberry Suite

APPENDIX 1. (continued)

Unit name	7.5-min. quad. long. and lat	Location of representative or type locality
		and the Forge Creek Suite. As thus defined the broad-scale intent of the use of "Elk Park" for these "orogenic calc-alkaline" rocks by Rankin and others (1973, 1983) is retained. However, the charnockite supersuite of the Pedlar Massif and the dioritoid/charnockite super-suite of the Lovingston Massif are specifically excluded from the Elk Park Supersuite.
Forge Creek Suite	Baldwin Gap 81°44'W 36°28'N	Named for Forge Creek in Tennessee (Bartholomew, 1983a) and for prior informal usage by Hamilton (in King and Ferguson, 1960); includes herein-named Watauga River Gneiss as well as the Comers and Grayson Gneisses of Stose and Stose (1957)
1. Watauga River Gneiss	Sherwood 81°52'W 36°17'N	Typical exposures (Bartholomew and Gryta, 1980) of the massive facies are found (1) along the Bethel Church road about half a mile north of where it crosses the Watauga River; (2) outcrops are found along the road adjacent to the west side of old Bethel School; and (3) along Rube Creek road about 1-1/4 mi north of the community of Bethel, North Carolina
Horsepen Mountain Suite (biotite dioritoid and charnockite)	Stewartsville 79°49'W 37°17'N	Crops out along Virginia State Highway 24 on the southern flank of Horsepen Mountain in south-western Virginia where samples R-8433 to R-8435 were collected (Bartholomew, 1981)
Lady Slipper Granulite Gneiss	–	See Sinha and Bartholomew (1984)
Lovingston Formation	–	Abandoned by Bartholomew and others (1981) as a formal strati-graphic unit; now used as name for terrane-Lovingston Massif
Low-Water-Bridge Layered Gneiss	Copeland 80°39'W 36°16'N	Type outcrop (Lewis, 1980a) is about half a mile north along river road north of intersection of North Carolina State Roads 1527 and 1510 on southeast side of the low-water bridge over the Yadkin River approximately half a mile south of Rockford
Oventop Suite (biotite dioritoid)	Flint Hill 78°02'W 38°50'N	Named for Oventop Mountain in northern Virginia; representative exposure along Virginia State Road 635 about 1 mi west of Hume (Lukert and Nuckols, 1976)
Peaks of Otter Suite (charnockite)	Peaks of Otter 79°35'W 37°27'N	Named for Peaks of Otter in central Virginia; typical expo-sures of mesocharnockite are found there and in the adjacent Montvale quadrangle (Henika, 1981)

APPENDIX 1. (continued)

Unit name	7.5-min. quad. long. and lat	Location of representative or type locality
Pedlar Formation	-	Abandoned by Bartholomew and others (1981) as a formal strati-graphic unit; now used as name for terrane-Pedlar Massif
Pedlar River Suite (charnockite)	Buena Vista 79°17'W 37°42'N	Named for Pedlar River in central Virginia and prior informal useage of Pedlar River pluton (Bartholomew and others, 1981) for this suite
1. granitoid	Montebello 79°08'W 37°52'N	Mapped by Hudson (1981) around town of Montebello, Virginia
2. leucocharnockite	Montebello 79°10'W 37°46'N	Crops out along Virginia State Road 634 about 2 mi east of Salt Log Gap
3. mesocharnockite	Sherando 78°58'W 37°55'N	Crops out along Virginia State Road 664 about 2 mi south of Reeds Gap on Blue Ridge Parkway
4. Vesuvius megaporphyry	Vesuvius 79°10'W 37°54'N	Crops out along Virginia Highway 56 about 2 mi east of the town of town of Vesuvius and half a mile east of old unakite quary
Pilot Gneiss (leuco-granulite gneiss)	Pilot 80°22'W 37°03'N	Named for town of Pilot in south-western Virginia; type outcrop behind Pilot Community Center (Dietrich, 1959) and prior infor-mal useage by Lewis (1975) and Kaygi (1979)
Saddleback Mountain Suite (charnockite)	Swift Run 78°30'W 38°22'N	Named for Saddleback Mountain, northern Virginia
1. Old Rag Garnite (leucocharnockite)	Old Rag Mountain 78°19'W 38°33'N	Named for Old Rag Mountain (Furcron, 1934)
2. mesocharnockite	Swift Run 78°32'W 38°21'N	Crops out along U.S. Highway 33 for about 2 mi east of Swift Run Gap on Skyline Drive (south-western flank of Saddleback Mountain)
Sandy Creek Granulite Gneiss (meso-granulite gneiss)	Stewartsville 79°49' W 37°15'30"N	Type outcrop is along Virginia State Road 619 from where sample R-8438 was collected (Bartholomew, 1981) along Sandy Creek in south-western Virginia
Stage Road Layered Gneiss	-	See Sinha and Bartholomew (1984)
Stewarts Knob Gneiss (biotite-garnet leucogneiss)	Stewartsville 79°52' W 37°17'30"N	Type outcrop is at Reference Locality 21 (Bartholomew, 1981) along the Blue Ridge Parkway about half a mile north of junction with Virginia State Highway 24
Turkey Mountain Suite (biotite dioritoid and and charnockite)	Piney River 78°03'W 37°40'N	See Herz and Force (1984)

REFERENCES CITED

Allen, R. M., Jr., 1963, Geology and mineral resources of Greene and Madison counties: Virginia Division of Mineral Resources Bulletin 78, 102 p.

Bailey, W. M., 1983, Geology of the northern half of the Horseshoe Mountain quadrangle, Nelson County, Virginia [M.S. thesis]: Athens, University of Georgia, 100 p.

Bartholomew, M. J., 1971, Geology of the Humpback Mountain area of the Blue Ridge in Nelson and Augusta counties, Virginia [Ph.D. dissertation]: Blacksburg, Virginia Polytechnic Institute and State University, 159 p.

—— 1977, Geology of the Greenfield and Sherando quadrangles, Virginia: Virginia Division of Mineral Resources Publication 4, 43 p.

—— 1981, Geology of the Roanoke and Stewartsville quadrangles, Virginia: Virginia Division of Mineral Resources Publication 34, 23 p.

—— 1983a, Geologic map and mineral resources summary of the Baldwin Gap quadrangle, North Carolina and Tennessee: North Carolina Division of Land Resources, GM 220-NW.

—— 1983b, Palinspastic reconstruction of the Grenville terrane in the Blue Ridge Geologic Province, southern and central Appalachians, U.S.A.: Geological Journal, v. 18, p. 241–253.

Bartholomew, M. J., and Gryta, J. J., 1980, Geologic map of the Sherwood quadrangle, North Carolina and Tennessee: North Carolina Division of Land Resources, GM 214-SE.

Bartholomew, M. J., and Hazlett, W. H., Jr., 1981, Geologic map of the Roanoke quadrangle, Virginia *in* Bartholomew, M. J., Geology of the Roanoke and Stewartsville quadrangles, Virginia: Virginia Division of Mineral Resources Publication 34, 23 p.

Bartholomew, M. J., and Lewis, S. E., 1977, Relationships of the Fries thrust to the basement complex and Paleozoic deformation in the Blue Ridge of northwestern North Carolina and southwestern Virginia: Geological Society of America Abstracts with Programs, v. 9, no. 2, p. 116–117.

Bartholomew, M. J., and Wilson, J. R., 1984, Geologic map of the Zionville quadrangle, North Carolina and Tennessee: North Carolina Division of Land Resources GM 220-SW (in press).

Bartholomew, M. J., Gathright, T. M., II, and Henika, W. S., 1981, A tectonic model for the Blue Ridge in central Virginia: American Journal of Science, v. 281, p. 1164–1183.

Bartholomew, M. J., Schultz, A. P., Henika, W. S., and Gathright, T. M., II, 1982, Geology of the Blue Ridge and Valley and Ridge at the junction of the central and southern Appalachians *in* Lyttle, P. T., ed., Central Appalachian geology, NE-SE GSA 1982 Field Trip Guidebook: American Geological Institute, p. 121–170.

Bartholomew, M. J., Lewis, S. E., Wilson, J. R., and Gryta, J. J., 1983, Deformational history of the region between the Grandfather Mountain and Mountain City windows, North Carolina and Tennessee, *in* Lewis, S. E., ed., Geological Investigations in the Blue Ridge of Northwestern North Carolina: 1983 Guidebook for the Carolina Geological Society, North Carolina Division of Land Resources, Article I, 30 p.

Bloomer, R. O., and Werner, H. J., 1955, Geology of the Blue Ridge in central Virginia: Geological Society of America Bulletin, v. 66, p. 579–606.

Brock, J. C., 1981, Petrology of the Mobley Mountain Granite, Amherst County, Virginia [M.S. thesis]: Athens, University of Georgia, 130 p.

Brown, W. R., 1958, Geology and mineral resources of the Lynchburg quadrangle, Virginia: Virginia Division of Mineral Resources Bulletin 74, 99 p.

Bryant, B. H., and Reed, J. C., Jr., 1970a, Structural and metamorphic history of the southern Blue Ridge, *in* Fisher, G. W. and others, eds., Studies of Appalachian geology—central and southern: New York, Interscience Publishers, p. 213–225.

—— 1970b, Geology of the Grandfather Mountain window and vicinity, North Carolina and Tennessee: U.S. Geological Survey Professional Paper 615, 190 p.

Butler, J. R., 1972, Age of Paleozoic regional metamorphism in the Carolinas, Georgia, and Tennessee, southern Appalachians: American Journal of Science, v. 272, p. 319–333.

—— 1973, Paleozoic deformation and metamorphism in part of the Blue Ridge thrust sheet, North Carolina: American Journal of Science, v. 273-A, Cooper Volume, p. 72–88.

Conley, J. F., 1978, Geology of the Piedmont of Virginia—interpretations and problems, *in* Contributions to Virginia geology—III: Virginia Division of Mineral Resources Publication 7, p. 115–149.

Conley, J. F., and Drummond, K. M., 1979, Geologic map of the Marion-NE 7.5 minute quadrangle, North Carolina: North Carolina Division of Land Resources, GM 210-NE.

Conley, J. F., and Henika, W. S., 1973, Geology of the Snow Creek, Martinsville East, Price, and Spray quadrangles, Virginia: Virginia Division of Mineral Resources Report of Investigations 33, 71 p.

Crawford, W. A., and Hoersch, A. L., 1984, The geology of the Honey Brook Upland, southeastern Pennsylvania *in* Bartholomew, M. J., and others, eds., The Grenville event in the Appalachians and related topics: Geological Society of America Special Paper 194, (this volume).

Davis, G. L., Tilton, G. R., and Wetherill, G. W., 1962, Mineral ages from the Appalachian province in North Carolina and Tennessee: Journal of Geophysical Research, v. 67, p. 1987–1996.

Davis, R. G., 1974, Pre-Grenville ages of basement rocks in central Virginia: a model for the interpretation of zircon ages [M.S. thesis]: Blacksburg, Virginia Polytechnic Institute and State University, 47 p.

Dietrich, R. V., 1954, Geology of the Pilot Mountain area, Virginia: Bulletin of the Virginia Polytechnic Institute, Engineering Experiment Station Series no. 91, 32 p.

—— 1959, Geology and mineral resources of Floyd County of the Blue Ridge upland, southwestern Virginia: Bulletin of the Virginia Polytechnic Institute, Engineering Experiment Station Series no. 134, 160 p.

Dietrich, R. V., Fullagar, P. D., and Bottino, M. L., 1969, K/Ar and Rb/Sr dating of tectonic events in the Appalachians of southwestern Virginia: Geological Society of America Bulletin, v. 80, p. 307–314.

Diggs, W. E., 1955, Geology of the Otter River area, Bedford County, Virginia: Bulletin of the Virginia Polytechnic Institute, Engineering Experiment Station Series no. 101, 23 p.

Dunn, D. E., and Weigand, P. W., 1969, Geology of the Pilot Mountain and Pinnacle quadrangles, North Carolina: North Carolina Division of Mineral Resources Map Series I, scale 1:24,000.

Espenshade, G. H., Rankin, D. W., Shaw, K. W., and Neuman, R. B., 1975, Geologic map of the east half of the Winston-Salem quadrangle, North Carolina–Virginia: U.S. Geological Survey Miscellaneous Investigations Map I-709B, text and 1:250,000 scale map.

Force, E. R., and Herz, N., 1982, Anorthosite, ferrodiorite, and titanium deposits in Grenville terrane of the Roseland district, central Virginia *in* Lyttle, P. T., ed., Central Appalachian geology, NE-SE GSA '82 Field Trip Guidebooks: American Geological Institute, p. 109–120.

Fullagar, P. D., and Bartholomew, M. J., 1983, Rubidium-Strontium ages of the Watauga River, Cranberry, and Crossing Knob Gneisses, Northwestern North Carolina *in* Lewis, S. E., ed., Geological Investigations in the Blue Ridge of Northwestern North Carolina: 1983 Guidebook for the Carolina Geological Society, North Carolina Division of Land Resources, Article II, 29 p.

Fullagar, P. D., and Butler, J. R., 1980, Radiometric dating in the Sauratown Mountains area, North Carolina, *in* Price, V., Jr., and others, eds., Geological investigations of Piedmont and Triassic rocks, central North Carolina and Virginia: 1980 Guidebook for the Carolina Geological Society, North Carolina Division of Land Resources, Article II, 10 p.

Fullagar, P. D., and Dietrich, R. V., 1976, Rb-Sr isotopic study of the Lynchburg and probably correlative formations of the Blue Ridge and western Piedmont of Virginia and North Carolina: American Journal of Science, v. 276, p. 347–365.

Fullagar, P. D., and Odom, A. L., 1973, Geochronology of Precambrian gneisses in the Blue Ridge province of northwestern North Carolina and adjacent parts of Virginia and Tennessee: Geological Society of America Bulletin, v. 84, p. 3065–3080.

Fullagar, P. D., Hatcher, R. D., Jr., and Merschat, C. E., 1979, 1200-m.y.-old gneisses in the Blue Ridge province of North and South Carolina: Southeastern Geology, v. 20, p. 69–77.

Furcron, A. S., 1934, Igneous rocks of the Shenandoah National Park area: Journal of Geology, v. 42, p. 400–410.

——1939, Geology and mineral resources of the Warrenton quadrangle, Virginia: Virginia Division of Mineral Resources Bulletin 54, 94 p.

Gathright, T. M., II, 1976, Geology of the Shenandoah National Park, Virginia: Virginia Division of Mineral Resources Bulletin 86, 93 p.

Gathright, T. M., II, and Nystrom, P. G., Jr., 1974, Geology of the Ashby Gap quadrangle, Virginia: Virginia Division of Mineral Resources Report of Investigation 36, 55 p.

Gathright, T. M., II, Henika, W. S., and Sullivan, J. L., 1977, Geology of the Waynesboro East and Waynesboro West quadrangles, Virginia: Virginia Division of Mineral Resources Publication 3, 53 p.

Gulley, G. L., Jr., 1982, The petrology of the granulite-facies metamorphic rocks on Roan Mountain, western Blue Ridge province, NC-TN: [M.S. thesis]: Chapel Hill, University of North Carolina, 163 p.

Hamilton, C. L., 1964, Geology of the Peaks of Otter area, Bedford and Botetourt counties, Virginia [Ph.D. dissertation]: Blacksburg, Virginia Polytechnic Institute and State University, 129 p.

Harper, S. B., 1977, The age and origin of granitic gneisses of the Inner Piedmont northwestern North Carolina: [M.S. thesis]: Chapel Hill, University of North Carolina, 92 p.

Hatcher, R. D., Jr., 1978, Tectonics of the western Piedmont and Blue Ridge, southern Appalachians: Review and speculation: American Journal of Science, v. 278, p. 276–304.

——1981, Thrusts and nappes in the North American Appalachian orogen: Journal of the Geological Society of London, v. 138, p. 491–499.

Hatcher, R. D., Jr., and Odom, A. L., 1980, Timing of thrusting in the southern Appalachians, USA: Model for orogeny: Journal of the Geological Society of London, v. 137, p. 321–327.

Hatcher, R. D., Jr., McConnell, K. I., and Heyn, T. 1983, Preliminary results from detailed geologic mapping studies in the western Sauratown mountains anticlinorium, North Carolina, *in* Lewis, S. E., ed., Geological investigations in the Blue Ridge of northwestern North Carolina: 1983 Guidebook for the Carolina Geological Society, North Carolina Division of Land Resources, Article VI, 12 p.

Henika, W. S., 1981, Geology of the Villamont and Montvale quadrangles, Virginia: Virginia Division of Mineral Resources Publication 35, 18 p.

Herz, N., 1969, The Roseland alkalic anorthosite massif, Virginia, *in* Isachsen, Y. W., ed., Origin of anorthosite and related rocks: New York State Museum and Science Service Memoir 18, p. 357–367.

Herz, N., and Force, E. R., 1984, Rock suites in Grenvillian terrane of the Roseland district, Virginia, *in* Bartholomew, M. J., and others, eds., The Grenville event in the Appalachians and related topics: Geological Society of America Special Paper 194 (this volume).

Herz, N., Mose, D. C., and Nagel, M. S., 1981, Mobley Mountain granite and the Irish Creek tin district, Virginia: A genetic and temporal relationship: Geological Society of America Abstracts with Programs, v. 13, p. 472.

Hillhouse, D. N., 1960, Geology of the Piney River–Roseland titanium area, Nelson and Amherst counties, Virginia [Ph.D. dissertation]: Blacksburg, Virginia Polytechnic Institute and State University, 129 p.

Hudson, T. A., 1981, Geology of the Irish Creek Tin District, Virginia Blue Ridge [M.S. thesis]: Athens, University of Georgia, 144 p.

Hudson, T. A., and Dallmeyer, R. D., 1982, Age of mineralized greisens in the Irish Creek tin district, Virginia Blue Ridge: Economic Geology, v. 77, p. 189–192.

Jonas, A. I., ed., 1928, Geologic map of Virginia: Virginia Division of Mineral Resources, Charlottesville, Virginia, scale 1:500,000.

Jones, T. Z., 1976, Petrography, structure and metamorphic history of the Warrensville and Jefferson quadrangles, southern Blue Ridge, northwestern North Carolina [Ph.D. dissertation]: Oxford, Miami University, 130 p.

Kaygi, P. B., 1979, The Fries fault near Riner, Virginia: An example of a polydeformed, ductile deformation zone [M.S. thesis]: Blacksburg, Virginia Polytechnic Institute and State University, 165 p.

Keith, A., 1903, Cranberry folio: U.S. Geological Survey, Geologic Atlas of the United States, Folio 90, 9 p.

King, P. B., and Ferguson, H. W., 1960, Geology of northeasternmost Tennessee *with a section on the* Description of the basement rocks by Warren Hamilton: U.S. Geological Survey Professional Paper 311, 136 p.

Lewis, S. E., 1975, Geology of the southern part of the Riner quadrangle, Montgomery and Floyd counties, Virginia [M.S. thesis]: Raleigh, North Carolina State University, 106 p.

——1980a, Geology of the Brevard zone, Smith River allochthon, and Inner Piedmont in the Sauratown Mountains anticlinorium, northwestern North Carolina [Ph.D. dissertation]: Chapel Hill, University of North Carolina, 131 p.

——1980b, An examination of the northwestern terminus of the Brevard zone and relationships with the Stony Ridge fault zone, the Sauratown Mountains anticlinorium, the Smith River allochthon, and the Inner Piedmont *in* Price V., Jr. and others, eds., Geological investigations of Piedmont and Triassic rocks, central North Carolina and Virginia: 1980 Guidebook of the Carolina Geological Society, North Carolina Division of Land Resources, Article VIII, 12 p.

Lewis, S. E., 1983, Distribution and relationships of late Precambrian and underlying Grenville(?) age rocks, Sauratown Mountains area, North Carolina, *in* Lewis, S. E., ed., Geological investigations in the Blue Ridge of northwestern North Carolina: 1983 Guidebook for the Carolina Geological Society, North Carolina Division of Land Resources, Article VII, 35 p.

Lewis, S. E., and Bartholomew, M. J., 1984, Geologic map of the Elk Park quadrangle, North Carolina and Tennessee: North Carolina Division of Land Resources, GM 215-NW (in press).

Lewis, S. E., and Butler, J. R., 1984, Geologic map of the Little Switzerland quadrangle, North Carolina: North Carolina Division of Land Resources, GM 209-SE (in press).

Lewis, S. E., Bartholomew, M. J., and Kaygi, P. B., 1984, Geology of the Riner quadrangle, Virginia: Virginia Division of Mineral Resources Publication (in press).

Luket, M. T., 1973, The petrology and geochronology of the Madison area, Virginia [Ph.D. dissertation]: Cleveland, Case Western Reserve University, 218 p.

——1982, Uranium-lead isotope age of the Old Rag Granite, northern Virginia: American Journal of Science, v. 282, p. 391–398.

Lukert, M. T., and Banks, P. O., 1984, Geology and age of the Robertson River Pluton, *in* Bartholomew, M. J., and others, eds., The Grenville event in the Appalachians and related topics: Geological Society of America Special Paper 194, (this volume).

Lukert, M. T., and Halladay, C. R., 1980, Geology of the Massies Corner quadrangle, Virginia: Virginia Division of Mineral Resources Publication 17, text and 1:24,000 scale map.

Lukert, M. T., and Nuckols, E. B., III, 1976, Geology of the Linden and Flint Hill quadrangles, Virginia: Virginia Division of Mineral Resources Report of Investigations 44, 33 p.

Lukert, M. T., Nuckols, E. B., and Clarke, J. W., 1977, Flint Hill Gneiss—a definition: Southeastern Geology, v. 19, p. 19–28.

Maryland Geological Survey, 1968, Geologic map of the state of Maryland: Baltimore, Maryland, scale 1:250,000.

Monrad, J. R., and Gulley, G. L., Jr., 1983, Age and P-T conditions during metamorphism of granulite-facies gneisses, Roan Mountain, NC-TN, *in* Lewis, S. E., ed., Geological investigations in the Blue Ridge of Northwestern North Carolina: 1983 Guidebook for the Carolina Geological Society, North

Carolina Division of Land Resources, Article IV, 29 p.

Moore, C. H., 1940, Origin of the nelsonite diks of Amherst County, Virginia: Economic Geology, v. 35, p. 629–645.

Odom, A. L., and Fullagar, P. D., 1973, Geochronologic and tectonic relationships between the inner Piedmont, Brevard zone, and Blue Ridge belts, North Carolina: American Journal of Science, Cooper Volume 273-A, p. 133–149.

——1984, Rb-Sr whole-rock and inherited zircon ages of the plutonic suite of the Crossnore Complex, southern Appalachians, and their implications regarding the time of opening of the Iapetus ocean, *in* Bartholomew, M. J., and others, eds., the Grenville event in the Appalachians and related topics: Geological Society of America Special Paper 194, (this volume).

Parker, P. E., 1968, Geologic investigation of the Lincoln and Bluemont quadrangles, Virginia: Virginia Division of Mineral Resources Report of Investigations 14, 23 p.

Price, V., Jr., Conley, J. F., Piepul, R. G., Robinson, G. R., Thayer, P. A., and Henika, W. S., 1980a, Geology of the Whitmell and Brosville quadrangles, Virginia: Virginia Division of Mineral Resources Publication 21, text and 1:24,000 scale map.

——1980b, Geology of the Axton and Northeast Eden quadrangles, Virginia: Virginia Division of Mineral Resources Publication 22, text and 1:24,000 scale map.

Rankin, D. W., 1970, Stratigraphy and structure of Precambrian rocks in northwestern North Carolina, *in* Fisher, G. W., and others, eds., Studies of Appalachian Geology—central and southern: New York, Interscience Publishers, p. 227–245.

——1975, The continental margin of eastern North America in the southern Appalachians: The opening and closing of the proto-Atlantic Ocean: American Journal of Science, v. 275-A, p. 298–336.

——1976, Appalachian salients and recesses: Late Precambrian continental breakup and the opening of the Iapetus Ocean: Journal of Geophysical Research, v. 81, p. 5605–5619.

Rankin, D. W., Stern, T. W., Reed, J. C., Jr., and Newell, M. F., 1969, Zircon ages of felsic volcanic rocks in the upper Precambrian of the Blue Ridge, Appalachian Mountains: Science, v. 166, p. 741–744.

Rankin, D. W., Espenshade, G. H., and Neuman, R. B, 1972, Geologic map of the west half of the Winston-Salem quadrangle, North Carolina, Virginia, and Tennessee: U.S. Geological Survey Miscellaneous Geological Investigations Map I-709A, text and 1:250,000 scale map.

Rankin, D. W., Espenshade, G. H., and Shaw, K. W., 1973, Stratigraphy and structure of the metamorphic belt in northwestern North Carolina and southwestern Virginia; a study from the Blue Ridge across the Brevard fault zone to the Sauratown Mountains anticlinorium: American Journal of Science, v. 273-A, Cooper Volume, p. 1–40.

Rankin, D. W., Stern, T. W., McLelland, J., Zartman, R. E., and Odom, A. L., 1983, Correlation chart for Precambrian rocks of the eastern United States, *in* Harrison, J. E. and Peterman, Z. E., eds., Correlation of Precambrian rocks of the United States and Mexico: U.S. Geological Survey Professional Paper 1241-E, 18 p.

Rodgers, J., 1953, Geologic map of east Tennessee with explanatory text: Tennessee Division of Geology Bulletin 58, pt. 2, 168 p.

Sinha, A. K., and Bartholomew, M. J., 1984, Evolution in the Grenville terrane in the central Virginia Appalachians, *in* Bartholomew, M. J., and others, eds., The Grenville event in the Appalachians and related topics: Geological Society of America Special Paper 194, (this volume).

Stose, A. J., and Stose, G. W., 1957, Geology and mineral resources of the Gossan Lead District and adjacent areas in Virginia: Virginia Division of Mineral Resources Bulletin 72, 233 p.

Tilton, G. R., Wetherill, G. W., Davis, G. L., and Bass, M. N., 1960, 1000-million-year-old minerals from the eastern United States and Canada: Journal of Geophysical Research, v. 65, p. 4173–4179.

Truman, W. E., 1976, Geology of the Blue Ridge Front near Riner, Virginia [M.S. thesis]: Blacksburg, Virginia Polytechnic Institute and State University, 102 p.

Virginia Division of Mineral Resources, 1963, Geologic map of the Commonwealth of Virginia: Charlottesville, Virginia, scale 1:500,000.

Werner, H. J., 1966, Geology of the Vesuvius quadrangle, Virginia: Virginia Division of Mineral Resources Report of Investigations 7, 53 p.

MANUSCRIPT ACCEPTED BY THE SOCIETY AUGUST 2, 1983

Geological Society of America
Special Paper 194
1984

Rb-Sr whole-rock and inherited zircon ages of the plutonic suite of the Crossnore Complex, southern Appalachians, and their implications regarding the time of opening of the Iapetus Ocean

A. Leroy Odom
Department of Geology
The Florida State University
Tallahassee, Florida 32306

Paul D. Fullagar
Department of Geology
University of North Carolina
Chapel Hill, North Carolina 27514

ABSTRACT

New U-Pb and Rb-Sr isotopic data on the suite of the Crossnore Complex, herein referred to as the Crossnore plutonic suite (CPS), indicate that these plutons crystallized between 680 and 710 m.y. ago; the Crossnore Pluton itself may be as young as approximately 650 m.y. Bulk zircon separates from these rocks contain an older xenocrystic component due to contamination of the CPS magmas by older gneisses.

During late Precambrian time, the ancestral Atlantic Ocean basin (Iapetus) formed and separated crustal units which are now part of the Caledonian-Appalachian mountains. These new CPS isotopic data indicate that continental rifting, which preceded the actual formation of the Iapetus Ocean, occurred approximately 690 m.y. ago. The Iapetus Ocean formed 690 to 570 m.y. ago.

INTRODUCTION

Basement rocks of the southern Appalachian Blue Ridge province consist of 1200 to 1000 m.y.-old peraluminous ortho- and paragneisses with lesser amounts of amphibolites and schists (Fig. 1). Granitoid gneisses dominate this terrain, a product of the Grenville orogeny. Stratigraphically between these Grenville gneisses and strata of the early paleozoic miogeoclinal sequence are upper Precambrian to lower Cambrian igneous and sedimentary rocks.

These upper Precambrian to lower Cambrian igneous rocks include the Catoctin, Mount Rogers, and Grandfather Mountain volcanics, plus the Crossnore, Beech, Striped Rock, Brown Mountain, and Lansing plutons. On stratigraphic, mineralogic, chemical, and isotopic-geochronologic grounds, Rankin (1968, 1970, 1975, 1976) and Rankin and others (1969) presented evidence that these rocks constitute a single bimodal and anorogenic magmatic suite (suite of the Crossnore Complex plus correlative

volcanic units herein referred to as the Crossnore plutonic suite (CPS)) in which felsic members have mildly to distinctly peralkaline affinities. Major members of this CPS suite and correlative volcanic rocks, except for the Catoctin volcanics, are shown in Fig. 1.

Rankin and others (1969) reported zircon ages from felsic members of the Mount Rogers, Grandfather Mountain, and Catoctin formations. These zircons plus two earlier analyses (one from the Beech Pluton and one from the Crossnore Pluton) by Davis and others (1962) defined a discordia curve indicating a crystallization age of approximately 820 m.y.

As did Rodgers (1972), we find such an old age disturbing. We concur with Rankin and others (1969) in questioning:

so high an age as 820 m.y. for the following reasons: (i) the most conspicuous hiatus in the stratigraphic section is at the base of the Mount

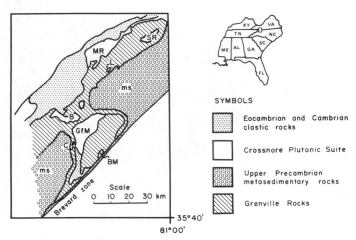

Figure 1: Generalized geologic map showing location of suite of the Crossnore Complex and correlative Mount Rogers and Grandfather Mountain formations in southern Appalachians. Inset map of southeastern United States shows location (rectangle) of field area. CPS symbols: B, Beech Pluton; BM, Brown Mountain Pluton; SR, Striped Rock Pluton; C, Crossnore Pluton; L, Lansing Pluton; correlative volcanic and sedimentary units are MR, Mount Rogers Fm., and GfM, Grandfather Mountain Fm. Eocambrian and Cambrian clastic rocks are primarily Chilhowee Group, but include Rome Fm. and Shady Dolomite. Upper Precambrian metasedimentary rocks (ms) are those above the Fries thrust, and include probable metavolcanic rocks.

Rogers, (ii) the Mount Rogers-Chilhowee contact appears conformable on a local scale, and (iii) although the lowest reported Cambrian fossils are near the top of the Chilhowee, the suite of clastic rocks comprising the Chilhowee probably represents a relatively short interval of time.

However, Rankin and others (1969) and Rankin (1970) mentioned what they considered evidence for an erosional interval between the deposition of Mount Rogers and Chilhowee clastic sediments, primarily felsic and mafic dikes of probably late Precambrian age (and probably belonging to the CPS) which intrude Grenville basement but are truncated by a nonconformity at the base of the Chilhowee. Obviously an age of 820 m.y. for Mount Rogers volcanic rocks implies a major hiatus nearly as long as the entire Paleozoic Era between the volcanic rocks and Chilhowee. However, the time significance of nonconformities should be evaluated in the context of their tectonic setting. In continental rift zones, erosion, sedimentation, and volcanism are concomitant processes, and time intervals are best estimated by local conformities—not local nonconformities.

Preliminary Rb-Sr data caused us to question the validity of the 820 m.y. age for the CPS (Odom, 1971; Odom and Fullagar, 1971). We now present additional Rb-Sr whole-rock and U-Pb zircon ages for plutonic members of the CPS. These data require re-evaluation of the time of continental rifting which preceded the formation of the Iapetus Ocean.

GEOCHRONOLOGY

Rb and Sr analyses were made using standard isotope dilu-

tion and mass spectrometric techniques. All sample $(^{87}Sr/^{86}Sr)_N$ values are reported relative to a $(^{87}Sr/^{86}Sr)_N$ ratio of 0.70800 for the Eimer and Amend $SrCO_3$ standard. One-standard-deviation (1σ) analytical uncertainties are estimated to be no more than 1 percent for $^{87}Rb/^{86}Sr$ and 0.05 percent for $(^{87}Sr/^{86}Sr)_N$. Replicate analyses (10) of the NBS-70a K-feldspar standard give 24.81 $\pm$ 0.10 (1σ) for $^{87}Rb/^{86}Sr$ and 1.2006 $\pm$ 0.0003 (1σ) for $(^{87}Sr/^{86}Sr)_N$. Ages and initial $^{87}Sr/^{86}Sr$ ratios were calculated using the regression method of York (1969). The decay constant used for ^{87}Rb is 1.42×10^{-11} yr^{-1}. All age and initial ratio errors represent analytical uncertainties of one-standard-deviation (1σ).

U and Pb analyses were made using essentially the same procedures given in Krogh (1973). The zircon samples analyzed weighed 10 to 32 mg. Decay constants and ratios for common lead corrections are listed in Table 2. The ratios for common lead corrections were obtained from analyses of blanks; if 700 m.y.-old model lead were used instead, the correction difference would be insignificant. Analytical uncertainties (1σ) for the ages reported in Table 2 are estimated to be no greater than 1 percent.

As noted in Table 1, most of the zircon samples analyzed in this study were extracted from the same rock samples from which splits were taken for Rb-Sr analyses.

Results of Rb-Sr whole-rock analyses are given in Table 1 and in the isochron diagrams (Fig. 2 and 3). The isochron ages of the Beech and Lansing plutons are indistinguishable and indicate an age of approximately 695 to 710 m.y., with MSWD values of 4.8 and 4.1 for the Beech and Lansing isochrons, respectively. The Striped Rock Pluton isochron indicates a younger age of 681 $\pm$ 5 m.y., with an MSWD of 1.3. The Crossnore Pluton has limited surface exposure; hence, the best samples we could obtain came from a single, small outcrop. All four samples are moderately weathered with altered feldspars and biotite. Because the effect of weathering seems to lower Rb-Sr ages (Fullagar and Ragland, 1975), and because only four samples were analyzed (Fig. 3), little confidence is placed in the 646 $\pm$ 9 m.y. date

TABLE 1. Rb-Sr DATA FOR CPS WHOLE-ROCK SAMPLES

Pluton and location	Sample	$(^{87}Sr/^{86}Sr)$	$^{87}Rb/^{86}Sr$	Rb(ppm)	Sr(ppm)
Striped Rock					
36°40'37"N,81°05'27"W	832(S1)*	0.7934	8.86	159.4	52.56
36°40'37"N,81°05'27"W	833(S2)	0.7891	8.26	171	58.57
36°40'43"N,81°05'02"W	836(S3)	1.2323	53.3	213	11.77
36°40'43"N,81°05'02"W	837(S4)	1.3379	65.5	237	10.76
36°39'02"N,81°10'37"W	842	3.302	266.9	267	3.63
Beech					
~36°12'N,~81°52'30"W	071(B3)	0.7800	7.46	160	61.00
36°12'12"N,82°04'56"W	072B	0.7787	7.06	177	70.6
36°12'12"N,82°04'56"W	072C	0.7634	5.50	123	63.5
36°12'11"N,82°05'58"W	074A(B2)	1.2427	52.23	334	18.4
36°12'11"N,82°05'58"W	074B	1.2695	56.63	317	16.5
Crossnore					
36°01'48"N,81°56'18"W	807	0.8036	9.85	80.4	23.0
36°01'48"N,81°56'18"W	816	0.7247	1.262	167	370
36°01'48"N,81°56'18"W	041(C2)	0.7375	2.61	53.3	57.17
36°01'48"N,81°56'18"W	041A(C3)	0.7424	3.10	55.4	50.3
Lansing					
36°29'34"N,81°30'16"W	1427	1.1542	43.82	169.9	11.72
36°29'34"N,81°30'16"W	1428	1.0349	32.40	131.2	12.10
36°29'34"N,81°30'16"W	1429	0.9513	24.15	163.3	20.04
36°29'34"N,81°30'16"W	1789	0.8260	11.09	159.5	42.14
36°29'34"N,81°30'16"W	1790	1.1572	44.90	147.1	11.72
36°29'34"N,81°30'16"W	1791	1.1870	46.58	183.3	11.93
36°29'34"N,81°30'16"W	1792	0.8739	16.66	116.7	20.61

*Number of zircon sample (Table 2, Fig. 4) which was extracted from same rock sample used for Rb-Sr analyses.

TABLE 2. U-Pb DATA FOR CPS ZIRCONS

Sample and Location*	ppm by weight		observed isotopic ratios			ages in 10^6 years		
	U	Pb	$\dfrac{^{208}\text{Pb}}{^{206}\text{Pb}}$	$\dfrac{^{207}\text{Pb}}{^{206}\text{Pb}}$	$\dfrac{^{204}\text{Pb}}{^{206}\text{Pb}}$	$\dfrac{^{206}\text{Pb}}{^{238}\text{U}}$	$\dfrac{^{207}\text{Pb}}{^{235}\text{U}}$	$\dfrac{^{207}\text{Pb}}{^{206}\text{Pb}}$
Crossnore Pluton								
C2 (lmz)	431	58.2	0.3223	0.07580	0.000752	802	719	770
C3 (mmz)	519	58.3	0.3838	0.07450	0.000606	562	604	762
Beech Pluton								
B2 (>200m)	816	98.1	0.2950	0.07432	0.000622	640	671	784
B3 (<200m)	1093	129	0.3044	0.07011	0.000476	624	644	714
Brown Mt. Pluton								
BM1 (bz) 35°54'07"N,81°43'07"W	1121	133	0.3876	0.06815	0.000321	587	616	724
BM2 (lmz,>300m) 35°54'02"N,81°43'03"W	567	75.4	0.4004	0.07186	0.000575	656	666	704
Striped Rock Pluton								
S1 (bz)	780	107	0.4296	0.07350	0.000626	662	684	754
S2 (lmz,>200m)	625	88.2	0.3906	0.06915	0.000417	694	697	712
S3 (>200m)	716	90.8	0.4101	0.07289	0.000541	622	656	776
S4 (<200m)	1256	142	0.3222	0.06559	0.0001538	586	614	720
Grandfather Mtn. Fm.								
G1 (bz) 36°03'55"N,81°53'48"W	394	43.0	0.2613	0.07979	0.001056	606	638	756

Note: Values used: $\lambda 238U = 1.5512 \times 10^{-10}$/yr, $\lambda 235U = 9.8485 \times 10^{-10}$/yr, $238U/235U = 137.8$.
Common lead corrections: 208/204 = 38.1, 207/204 = 15.7, 206/204 = 18.5 (see text for discussion of values)
lmz = less magnetic zircons; mmz = more magnetic zircons; >200m = zircons larger than 200 mesh; bz = split of bulk zircon extract.
*Location not repeated if it is given in Table 1.

(MSWD = 0.9), and it is best regarded a minimum age. Attempts to obtain a Rb-Sr isochron on the Brown Mountain Pluton where only rather severely weathered samples could be found, were unsuccessful. Attempts at obtaining Rb-Sr whole-rock isochrons on rhyolite members of the Mount Rogers and Catoctin have been unsuccessful, probably because they experienced low-grade Paleozoic metamorphism.

Analyses of separated magnetic and size fractions of zircons (Table 2) extracted from single rock samples reveal a complex pattern of "ages," very suggestive of mixed components of different ages. (In contrast, the Rb-Sr whole-rock isochron dating of four granite members of the CPS yielded fairly consistent and significantly younger ages.)

The concordia diagram (Fig. 4) includes eleven new analyses of CPS zircons as well as the two analyses reported by Davis and others (1962); these two analyses are labeled B1 (Beech Pluton) and C1 (Crossnore Pluton). Of the new analyses, one sample is from a volcanic member of the Grandfather Mountain Formation; the other ten samples are from plutons of the CPS. The host rock and analytical data are given in Table 2. The 820-240 m.y. chord is that obtained for bulk zircon separates from rhyolite members of the correlative Mount Rogers, Grandfather Mountain and Catoctin formations (Rankin and others, 1969). This chord is shown here for reference, though the indi-

vidual data points are not shown. Samples designated BM1, S1 and G1 in Fig. 4 represent splits of bulk zircon extracted from a single rock. The remaining samples are fractions of a bulk zircon extraction, separated by size and magnetic properties (Table 2).

Most of the new analyses fit closely to the 820-240 m.y. chord. However, several zircon fractions depart significantly from the chord, and taken together these analyses reveal a more complex pattern of ages than do the bulk zircon samples of the rhyolites reported by Rankin and others (1969).

The four samples of zircons from the Striped Rock Pluton do not fit any single chord; a chord through S1, S2 and S3 has a lower intercept much less than zero. The three samples of the Beech Pluton, B1, B2 and B3, also do not fit any single chord. A chord between the two Brown Mountain Pluton samples, BM1 and BM2, cannot have an upper intercept greater than 735 m.y. One fraction of zircons from the Crossnore Pluton (C3) plots on the "old side" of the chord, and one fraction each of Beech (B3), Striped Rock (S2) and Brown Mountain (BM2) zircons plot on the "young side." Most significantly, Striped Rock Pluton sample S2 is, within analytical uncertainty, concordant at approximately 695 m.y.

The complex pattern of U-Pb zircon ages is evidence that the zircons within the granites are mixed populations of at least two components of different ages. Most likely these compo-

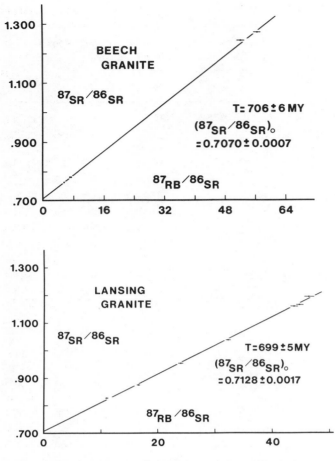

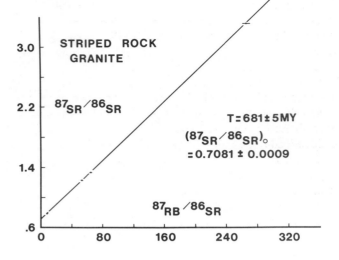

Figure 2: Rb-Sr isochron plots for Beech and Lansing plutons.

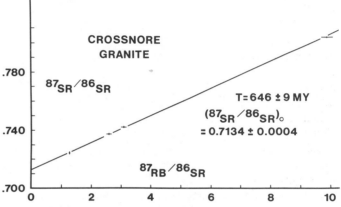

Figure 3: Rb-Sr isochron plot for Striped Rock and Crossnore plutons.

nents are (1) approximately 1000-1200 m.y. old xenocrysts derived from incorporation of the intruded Grenville age gneisses, and (2) younger, primary zircons which crystallized from the CPS magmas. In the case of the Striped Rock Pluton, this younger magmatic component is closely approximated by the concordant fraction, S2, at 695 m.y. though even this might contain a small older component.

We have seen evidence of stoping and assimilation of Grenville gneisses by the CPS magmas, as have Riecken (1966) and Bryant and Reed (1970). Therefore, at least some small portion of the CPS zircons probably are xenocrysts. Approximately 20 percent of the zircons from the Crossnore Pluton are rounded, and many others possess visible cores and overgrowths (Eckelmann and Kulp, 1956; Davis and others, 1962). One of us (A.L.O.) separated about 1 mg of each of two types of zircon from the Crossnore Pluton. Type A is euhedral with rare synneusis twins, essentially colorless, >200 mesh, and contains no inclusions or cores. Type B is anhedral, somewhat rounded or ovoid, darker, <200 mesh, and somewhat metamict. These two types of zircon from the Crossnore Pluton were analyzed for Zr/Hf ratios by proton-induced X-ray emission (PIXE). The analytical procedures and results are discussed in Van Grieken and others (1975);

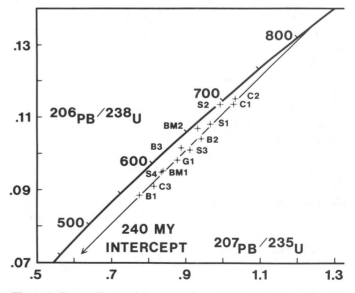

Figure 4: Concordia plot for samples from CPS. Symbols explained in text. 820 to 240 m.y. reference chord is from Rankin and others (1969).

analytical uncertainties are estimated to be 4 percent or less. Type A zircons had a Zr/Hf ratio of 41, Type B a ratio of 50. Based on the known behavior of Zr and Hf, it would be surprising to find zircons with such a wide range of Zr/Hf ratios produced from one magma. Ecklemann and Mose (1981) reported that the Crossnore Pluton contains two distinctly different zircon populations. One population is subhedral to euhedral, clear and colorless, and elongate; the other is anhedral to subhedral, dark, ovoid, and small. Eckelmann and Mose also examined zircons from the Beech, Striped Rock and Robertson River (Virginia) plutons and found petrographically complex populations which may indicate the presence of different components in the zircon population. We also observed these complex zircon populations in the samples we analyzed from the CPS plutons.

From the above considerations it appears that the best estimate of the age of the CPS granites is that obtained from the Rb-Sr isochrons. In particular the crystallization of the Striped Rock Pluton is accurately dated. The concordant zircon age (fraction S2) is approximately 695 m.y.; the isochron age is 681 ± 5 m.y. The difference is small and perhaps insignificant.

Our zircon analyses contain only one sample from a volcanic member of equivalent age to the CPS. Our sample G1 is a bulk zircon separate from the Grandfather Mountain Formation, and it plots on the 820-240 m.y. chord with those of Rankin and others (1969). We were unable to obtain adequate separation of magnetic or size fractions from this sample. Morphologically, zircons in the rhyolites appear to represent a single population as noted by Rankin and others (1969). Nonetheless, an older isotopic influence appears to be present in the volcanic units. The following reasons (in addition to stratigraphic relations) lead us to conclude that the previously reported 820 m.y. age of the volcanic units is erroneously high:

1. We find the arguments of Rankin (1968, 1970, 1975, 1976) and Rankin and others (1969) compelling that these plutons and volcanic rocks are consanguineous.

2. The plutons of the CPS yield bulk zircon ages essentially the same as the volcanic units; the bulk zircon ages for the plutons are affected by an older zircon component.

3. The zircon age for the Brown Mountain Pluton, from data reported here, cannot be older than 735 m.y. (and is too high because of inherited, older zircons). Rounded clasts of Brown Mountain granite are incorporated in flows of the Grandfather Mountain Formation (Bryant and Reed, 1970). Unless these clasts were rounded by transport in lava, the time interval separating the crystallization of the Brown Mountain granite and deposition of the Grandfather Mountain Formation was long enough to allow for the exhumation of the pluton prior to eruption of the flows.

UPPER PRECAMBRIAN-LOWER CAMBRIAN ROCKS OF THE BLUE RIDGE

Sixteen years ago Tuzo Wilson (1966) boldly suggested that during the early Paleozoic an ancestral Atlantic Ocean existed separating crustal units now incorporated within the Caledonian-Appalachian mountain belt. In more recent years the concept of an ancestral Atlantic or Iapetus ocean opening in the latest Precambrian time and closing during the Paleozoic Era has dominated tectonic models of the orogen. The closing of the Iapetus is recorded throughout the orogen at various places in rocks ranging in age from Cambrian to Permian. Judging from evidence relating to the opening of the Atlantic Ocean, the opening of the Iapetus probably also spanned a considerable length of time, and was initiated at different times in different places.

Several years ago we presented (Odom and Fullagar, 1973) reasons for the interpretation that uppermost Precambrian rocks of the Blue Ridge Province in the southern Appalachians were formed within and along a late Precambrian rift zone, and, that together with locally conformable and overlying lower Paleozoic miogeoclinal deposits, reflect the opening and spreading (growth) of the pre-Appalachian (Iapetus) Ocean basin. Since then this interpretation has been advanced and expanded by Rankin (1975, 1976) and subsequently been incorporated in many tectonic interpretations. Our new geochronological data provide information on the time of rifting and initiation of the Iapetus Ocean.

In the southern Appalachian Blue Ridge Province, upper Precambrian-lower Cambrian igneous and sedimentary rocks are stratigraphically between Grenville gneisses and the strata of the early Paleozoic miogeoclinal sequence (Fig. 1). In places the Lower Cambrian rocks rest nonconformably on Grenville basement, but elsewhere older clastic and volcanic rocks intervene. Several general aspects are important to understanding the late Precambrian-early Cambrian geology of the Blue Ridge Province.

The great clastic deposits of the Ocoee Supergroup, the Grandfather Mountain and Mount Rogers formations (probably portions of the Lynchburg and Ashe formations of the eastern Blue Ridge should be included [Rankin, 1975]) are generally poorly sorted, locally derived, of extremely variable and locally immense thickness (up to several kilometers), and for the most part are suggestive of rapid deposition in deeply subsiding isolated basins or troughs. From place to place various parts of the section can be found in contact with the basement; Odom and Fullagar (1973) and Rankin (1975) interpreted these deposits as terrigenous rift-basin-fills analogous to clastic rocks of the Mesozoic basins of the Piedmont associated with the opening of the Atlantic Ocean. Possible correlations among these rocks were discussed by many workers including Hadley (1970), Rodgers (1972), Rankin and others (1973), and Rankin (1975).

Interlayered within most of these clastic sequences (the exception being Ocoee) are lavas and tuffs. Most notable are the extensive flows of the Catoctin basalts (rhyolites are present at the base of the Catoctin in the northern Blue Ridge). Volcanic rocks of the Mount Rogers and Grandfather Mountain formations are predominantly basalt but include rhyolite. These volcanic rocks are correlative with the Crossnore plutonic suite (Rankin, 1968, 1970, 1975, 1976; Rankin and others, 1969).

Superjacent and (at least) locally conformable to the Catoctin, Mount Rogers, and Grandfather Mountain formations is the Chilhowee Group. The lowest part of the Chilhowee, the Unicoi Formation, contains extensive basaltic flows (Rogers, 1953; Bloomer and Werner, 1955; Stose and Stose, 1957), and coarse, poorly sorted sediments. Though some of these basalts occur stratigraphically above the base of the Unicoi Formation, the intervening clastics are fluvial deposits (Schwab, 1972), probably representing a short period of time. The Chilhowee also overlies, without stratigraphic break, terrigenous clastic rocks of the Mount Rogers which also contain basaltic flows. The Chilhowee Group begins as a terrigenous volcanic-clastic sequence, becomes increasingly more mature upwards and is clearly of coastal to shallow marine origin (Schwab, 1972; Brown, 1970). In Tennessee, the upper part of the Chilhowee, the Murray Shale, contains *Olenellus* (Walcott, 1890; Keith, 1903), and is conformably overlain by the Lower Cambrian to Middle Ordovician carbonate sequence which Rodgers (1968) pointed out was deposited on and along the pre-Appalachian continental margin of North America.

If a pre-Appalachian (Iapetus) ocean basin, whose closure spawned the orogeny did exist, then it must have opened before deposition of the Upper Chilhowee Group and perhaps just prior to or during deposition of the Lower Chilhowee.

Judging from the nature of modern continental rift zones (such as East Africa) and experimental studies such as the early studies of Cloos (1939), initial arching and distension of the continental lithosphere resulted in tilting, down-dropping, and rotation of large crustal rocks. Lower surfaces of tilted blocks became depositional basins of Mount Rogers, Grandfather Mountain, Ocoee and Mechum clastics, while higher surfaces were the sources. Extension fractures served as avenues for magmas. In basins, where sedimentation was continuous, volcanic rocks were interlayered with clastic sediments, while elsewhere they erupted onto Grenville basement. These basement highs continued to be eroded, however, and their volcanic deposits removed (leaving today only dikes and larger intrusions), while basins were continuously filled. The process continued into the time of deposition of the Chilhowee Group as indicated by the volcanic rocks and alluvial fans of the lowest Chilhowee, the Unicoi Formation (Brown, 1970). As separation of the continent was complete, relief diminished and the source terrane became more distant, hence the Chilhowee sediments matured into coastal and continental shelf deposits. Thus, the Chilhowee sediments were deposited conformably on sediments in rift basins while Grenville basement was onlapped and covered, and exposed dikes in former highlands were truncated by these same Chilhowee sediments.

This situation is analogous to the stratigraphic relationships between the eastern North America Triassic-Jurassic rift deposits with associated volcanic rocks and dikes and the overlying Cretaceous transgressive sequence (Sheridan, 1974). In the subsurface of the Coastal Plain, Cretaceous and younger strata overlie with apparent conformity the sedimentary and volcanic sequence

of the Mesozoic rift basins, while truncating dikes outside of basins. Care must be taken to draw inferences regarding time from such relationships.

The above considerations suggest that the time of initial rifting prior to the formation of the Iapetus Ocean can be inferred from isotopic ages of the upper Precambrian-lower Cambrian volcanic and related plutonic rocks.

CONCLUSIONS

(1) Plutons of the CPS crystallized between 680 and 710 m.y. ago. Because the Crossnore Pluton might be as young as approximately 650 m.y., and because we studied only a few plutons of the suite, the actual age range might be much greater. (Work in progress on a small CPS pluton near Warrensville, North Carolina, indicates that this pluton has a Rb-Sr whole-rock age of about 580 m.y.) Similar rocks of the White Mountain-Monteregian Hills Supersuite span a time interval of about 135 m.y. (Foland and Faul, 1977). Based on the agreement between the Rb-Sr whole-rock age and the zircon data from the Striped Rock Pluton, an age of 690 ± 10 m.y. seems a good estimate for the crystallization of the CPS.

(2) Bulk zircon separates from these rocks contain an older xenocrystic component, and thus previously reported U-Pb data, while yielding fairly consistent discordant ages, are deceiving.

(3) The CPS magmas were contaminated with material from older gneisses which they assimilated. The effects of this contamination is seen in both the zircon data and the somewhat high initial $^{87}Sr/^{86}Sr$ ratios. The original, uncontaminated magma was probably more strongly peralkaline.

(4) At least some of the uppermost Precambrian igneous activity related to the CPS occurred about 690 m.y. ago. The geochronology, stratigraphy, and interpretations of field relationships imply that this dates a period of continental rifting which preceded actual formation of the Iapetus Ocean. A Paleozoic continental shelf and margin had developed by the time the Chilhowee group was deposited and was superceded by a carbonate bank (Rodgers, 1968). The Murray Shale in the Upper Chilhowee contains *Olenellus* (Walcott, 1890; Keith, 1903). Regardless of where one places the base of the Cambrian in the southern Appalachians, or the local age of the *Olenellus* (which is certainly not below Lower Cambrian), surely the Murray Shale of the Chilhowee Group is not older than 600 m.y.

The base of the Cambrian is considered to be approximately 570 m.y. (Cowie, 1964). Thus the opening of the Iapetus Ocean occurred during the interval 690 to 570 m.y., or approximately 630 ± 60 m.y. ago.

ACKNOWLEDGMENTS

We thank J. R. Butler, T. Z. Jones, S. A. Kish, and R. E. Zartman for their assistance and suggestions during this study. Winston Russell performed some of the isotopic analyses. Early drafts of the manuscript were reviewed by M. J. Bartholomew,

J. R. Butler, and H. E. Gaudette. Financial assistance was provided by National Science Foundation grant GA-18448 and by the North Carolina Department of Natural Resources and Community Development.

REFERENCES CITED

Bloomer, R. O., and Werner, H. J., 1955, Geology of the Blue Ridge region in central Virginia: Geological Society of America Bulletin, v. 66. p. 579–606.

Brown, W. R., 1970, Investigations of the sedimentary record in the Piedmont and Blue Ridge of Virginia, *in* Fisher, G. W., and others, eds., Studies of Appalachian Geology: Central and Southern: New York, Interscience Publishers, p. 335–349.

Bryant, B. H., and Reed, J. C., Jr., 1970, Geology of the Grandfather Mountain window and vicinity, North Carolina and Tennessee: U.S. Geological Survey Professional Paper 615, 190 p.

Cloos, H., 1939, Hebung-Spaltung-Vulkanismus; Elemente einer geometrischen Analyse irdischer Grossformen: Geologische Rundschau, Bd. 30, p. 405–527.

Cowie, J. M., 1964, The Cambrian Period, *in* Harland, W. B., and others, eds., The Phanerozoic Time-scale: Quarterly Journal of the Geological Society of London, v. 120S, p. 255–258.

Davis, G. L., Tilton, G. R., and Wetherill, G. W., 1962, Mineral ages from the Appalachian province in North Carolina and Tennessee: Journal of Geophysical Research, v. 67, p. 1987–1996.

Eckelmann, F. D., and Kulp, J. L., 1956, The sedimentary origin and stratigraphic equivalence of the so-called Cranberry and Henderson granite in western North Carolina: American Journal of Science, v. 254, p. 288–315.

Eckelmann, F. D., and Mose, D. G., 1981, Search for xenocrystic zircon in latest Precambrian plutonic rocks of the Blue Ridge: Geological Society of America Abstracts with Programs, v. 13, p. 7.

Foland, K. A., and Faul, H., 1977, Ages of the White Mountain Intrusives—New Hampshire, Vermont, and Maine, USA: American Journal of Science, v. 277, p. 888–904.

Fullagar, P. D., and Ragland, P. C., 1975, Chemical weathering and Rb-Sr whole-rock ages: Geochimica et Cosmochimica Acta, v. 39, p. 1245–1252.

Hadley, J. B., 1970, The Ocoee Series and its possible correlatives, *in* Fisher, G. W., and others, eds., Studies of Appalachian Geology: Central and Southern: New York, Interscience Publishers, p. 247–259.

Keith, Arthur, 1903, Description of the Cranberry Quadrangle (North Carolina–Tennessee): U.S. Geological Survey Geological Atlas of the United States, Folio 90, 9 p.

Krogh, T. E., 1973, A low-contamination method for hydrothermal decomposition of zircon and extraction of U and Pb for isotopic age determinations: Geochimica et Cosmochimica Acta, v. 37, p. 485–494.

Odom, A. L., 1971, A Rb-Sr isotopic study: implications regarding the age, origin, and evolution of a portion of the southern Appalachians, western North Carolina, southwestern Virginia, and northwestern Tennessee: Ph.D. dissertation, University of North Carolina, Chapel Hill, 92 p.

Odom, A. L., and Fullagar, P. D., 1971, A major discordancy between U-Pb zircon ages and Rb-Sr whole-rock ages of Late Precambrian granites in the Blue Ridge Province: Geological Society of America Abstracts with Programs, v. 3, p. 663.

——1973, Geochronologic and tectonic relationships between the Inner Piedmont, Brevard Zone, and Blue Ridge belts, North Carolina: American Journal of Science, v. 273-A (Cooper volume), p. 133–149.

Rankin, D. W., 1968, Magmatic activity and orogeny in the Blue Ridge province of the southern Appalachian Mountain system in northwestern North Carolina and southwestern Virginia (abs.): Geological Society of America Special Paper 115, p. 181.

Rankin, D. W., 1970, Stratigraphy and structure of Precambrian rocks in northwestern North Carolina, *in* Fisher, G. W., and others, eds., Studies of Appalachian Geology: Central and Southern: New York, Interscience Publishers, p. 227–245.

Rankin, D. W., 1975, The continental margin of eastern North America in the southern Appalachians: The opening and closing of the proto-Atlantic ocean: American Journal of Science, v. 275-A, p. 298–336.

Rankin, D. W., 1976, Appalachian salients and recesses: Late Precambrian continental breakup and the opening of the Iapetus Ocean: Journal of Geophysical Research, v. 81, p. 5605–5619.

Rankin, D. W., Stern, T. W., Reed, J. C., Jr., and Newell, M. F., 1969, Zircon ages of felsic volcanic rocks in the upper Precambrian of the Blue Ridge central and southern Appalachian Mountains: Science, v. 166, no. 3906, p. 741–744.

Rankin, D. W., Espenshade, G. H., and Shaw, K. W., 1973, Stratigraphy and structure of the metamorphic belt in northwestern North Carolina and Southwestern Virginia: a study from the Blue Ridge across the Brevard fault zone of the Sauratown Mountains anticlinorium: American Journal of Science, v. 273-A (Cooper volume), p. 1–40.

Riecken, C. C., 1966, Petrology of the Striped Rock Granite and surrounding rocks, Grayson County, Virginia [Ph.D. dissertation]: Blacksburg, Virginia Polytechnic Institute, 161 p.

Rogers, J., 1953, Geologic map of east Tennessee with explanatory text: Tennessee Division of Geology, Bull. 58, 168 p.

Rodgers, John, 1968, The eastern edge of the North American continent during the Cambrian and Early Ordovician, *in* Zen, E-an, and others, eds., Studies of Appalachian Geology: Northern and Maritime: New York, Interscience Publishers, p. 141–149.

——1972, Latest Precambrian (post-Grenville) rocks of the Appalachian Region: American Journal of Science, v. 272, p. 507–520.

Schwab, F., 1972, The Chilhowee group and the late Precambrian–early Paleozoic sedimentary framework in the central and southern Appalachians, *in* Appalachian structures: Origin, evolution and possible potential for new exploration frontiers: West Virginia Geologic and Economic Survey Report, p. 59–101.

Sheridan, R. E., 1974, Atlantic margins of North America, *in* Burk, C. A., and Drake, C. L., eds., The Geology of Continental Margins: Springer-Verlag, New York, p. 391–407.

Stose, A. J., and Stose, G. W., 1957, Geology and mineral resources of the Gossan Lead District and adjacent areas in Virginia: Virginia Division of Mineral Resources, Bull. 72, 233 p.

Van Grieken, R. E., Johansson, T. B., Winchester, J. W., and Odom, A. L., 1975, Micro-determination of zirconium/hafnium ratios in zircons by proton induced x-ray emission: Zeitschrift für Analytische Chemie, v. 275, p. 343–348.

Walcott, C. D., 1890, The fauna of the Lower Cambrian or *Olenellus* zone: U.S. Geological Survey 10th Annual Report, 1888-1889, pt. 1, p. 509–763.

Wilson, J. T., 1966, Did the Atlantic close and then re-open?: Nature, v. 211, no. 5050, p. 676–681.

York, D., 1969, Least squares fitting of a straight line with correlated errors: Earth and Planetary Science Letters, v. 5, p. 320–324

MANUSCRIPT ACCEPTED BY THE SOCIETY AUGUST 2, 1983

Geological Society of America
Special Paper 194
1984

Basement–cover rock relationships along the western edge of the Blue Ridge thrust sheet in Georgia

Keith I. McConnell
John O. Costello*
Georgia Geologic Survey
19 Martin Luther King, Jr., Drive
Atlanta, Georgia 30334

ABSTRACT

The southwesternmost exposures of Grenville-age basement in the Appalachian Blue Ridge are present in rootless anticlinoria that lie along the western margin of the Blue Ridge thrust sheet in north-central Georgia. The southernmost of these massifs, the Corbin Gneiss Complex, lies in the core of the Salem Church anticlinorium, while farther to the north, the Fort Mountain Gneiss is exposed in the core of a smaller, unnamed anticlinorium. Both the Corbin Gneiss Complex and the Fort Mountain Gneiss are mantled by thick sequences of predominantly clastic rocks of the Ocoee Supergroup. Those clastic rocks lying nonconformably on basement gneisses, the Pinelog and Parr Branch Formations, respectively, clearly were derived from the gneisses themselves and are probably lithostratigraphic equivalents of the Snowbird Group, basal member of the Ocoee Supergroup. Conformably overlying these coarse clastics are graphitic phyllites, metaconglomerates, and sandy marbles of the Wilhite Formation.

While relict textures related to Grenville-age granulite facies metamorphism still persist locally in the basement gneisses, no evidence of this event is apparent in the cover rocks. However, all aforementioned rocks show evidence of an episode of mid-Paleozoic regional metamorphism that retrograded earlier-formed, higher temperature mineral assemblages in basement rocks. Coincident with Paleozoic metamorphism was development of overturned to recumbent isoclinal folds (F_1) with well-developed axial-planar schistosity. Subsequent deformational events (1) fold earlier structures, (2) deform isograds formed during the Paleozoic metamorphism, and (3) are at least partially responsible for the arcuate trace of the Great Smoky fault[1] in this area.

INTRODUCTION

In north-central Georgia, rocks that compose the Fort Mountain Gneiss and Corbin Gneiss Complex represent the southwesternmost exposures of Grenville basement in the Blue Ridge portion of the Appalachian orogen. As with gneisses of similar age to the northeast in Tennessee and Virginia, the Corbin and Fort Mountain gneisses are nonconformably overlain by a thick succession of predominantly clastic rocks. However, unlike those gneisses to the northeast, very little published information is available concerning the lithologic complexity of the gneisses, the nature of the metasedimentary pile that overlies them, or the deformation that affected them. Therefore, their role in the evolution of the orogen is uncertain, and models for tectonic development of the southern Appalachians are speculative as to their significance.

---

*Present address: Department of Geology, University of South Carolina, Columbia, South Carolina 29208.

[1]Since this report was written our interpretation regarding relationship of the Great Smoky and Cartersville fault systems in northwest Georgia has changed. The major fault separating Ocoee Supergroup rocks from Chilhowee Group rocks is now interpreted to be only the Cartersville fault. Where exposed, the Great Smoky fault lies farther to the west. See Costello and McConnell (1983) for a further discussion of these relationships.

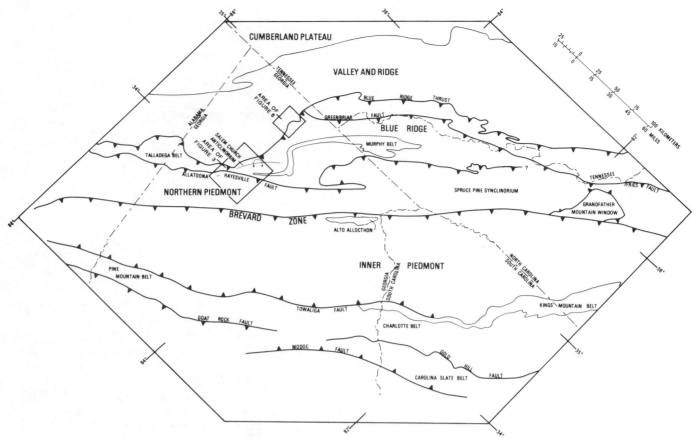

Figure 1. Generalized map of the major subdivisions and tectonic features of the southern Appalachians (modified after Williams, 1978; Hatcher and Odom, 1980; and Bartholomew and Lewis, 1984).

The purpose of this report is to present information regarding the nature of the Corbin and Fort Mountain gneisses, as well as the stratigraphic sequence above them. In addition, interpretations regarding timing of metamorphism, folding, and faulting are presented. The basis for these interpretations is derived from detailed mapping of 7.5-minute quadrangles in the Salem Church anticlinorium, a major structural feature just east of the trace of the Blue Ridge thrust in north-central Georgia (Fig. 1). In this area, the nature of the contact between basement gneiss and its cover rocks is discernible, and isotopic ages (Odom and others, 1973; Dallmeyer, 1975) provide insight into timing of folding and faulting. Reconnaissance mapping northward to Fort Mountain, just east of Chatsworth (Fig. 2), indicates that similar stratigraphic and structural relationships are present around basement gneiss exposed there (i.e., the Fort Mountain Gneiss).

STRATIGRAPHY OF THE WESTERNMOST BLUE RIDGE IN GEORGIA

Rocks described in this report lie east of the exposed trace of the Great Smoky fault, the frontal fault of the Blue Ridge thrust

system, and south of the apparent trace of the Greenbrier fault (Fig. 1). The study area is bounded on the east by the Murphy belt, which contains in its core Paleozoic(?) metasedimentary rocks (McLaughlin and Hathaway, 1973). To the south, rocks of the northern Piedmont (Fig. 1) are thrust over basement gneisses and rocks of the Ocoee Supergroup along the Allatoona-Hayesville fault. Although small parts of the study area were subjects of short reports, guidebooks, or master's theses (Furcron and Teague, 1947; Smith, 1959; Bentley and others, 1966; Needham, 1972; Cressler and others, 1979; McConnell and Costello, 1980), no comprehensive report was previously published. The most detailed information available on the stratigraphy of the study area was done by C. W. Hayes and his assistants in the late 1880s. In two short reports (Hayes, 1891, 1900), Hayes defined the Cartersville fault and outlined the stratigraphy within the western part of the Salem Church anticlinorium. Unfortunately, the large mass of Hayes's data remains unpublished but is accessible through the U.S. Geological Survey's archives, which has both manuscripts and maps that he prepared for the Cartersville and Dalton 30-minute quadrangles. This unpublished information is the base on which stratigraphic and structural models for this report were built.

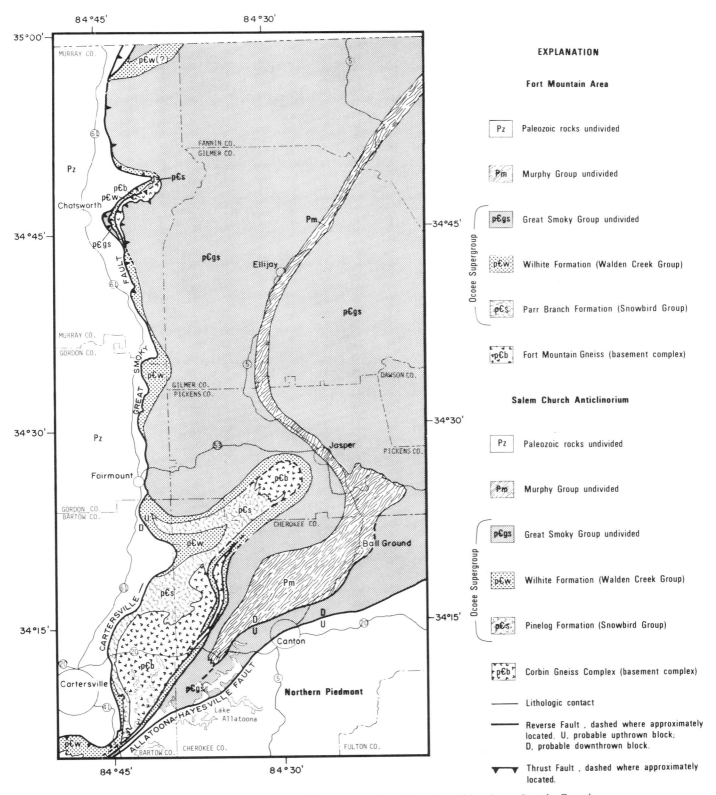

Figure 2. Generalized geologic map of the western edge of the Blue Ridge thrust sheet in Georgia (modified after Hayes, unpub. data; Cressler and others, 1979; Furcron and Teague, 1947; and McConnell and Costello, 1980).

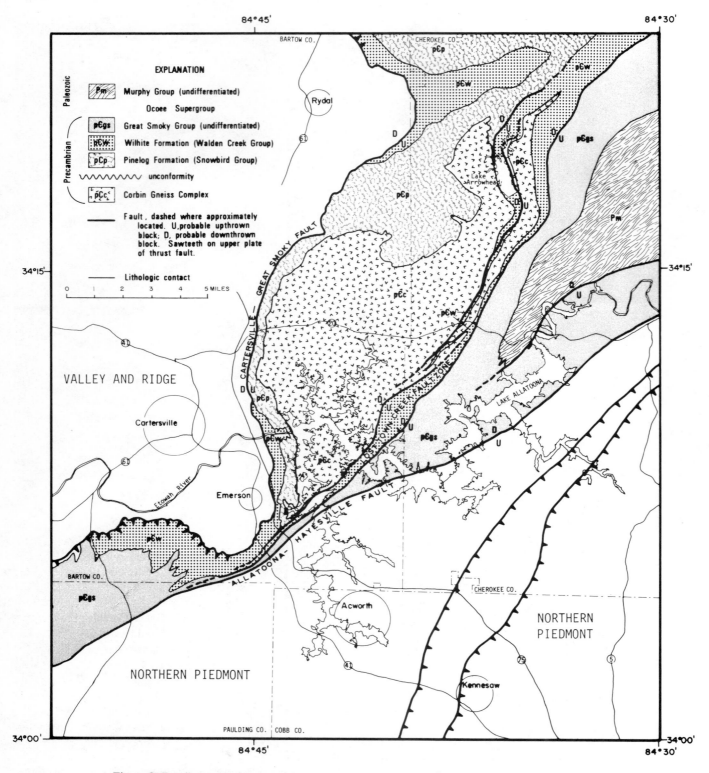

Figure 3. Detailed geologic map of the southwestern terminus of the Salem Church anticlinorium (modified after McConnell and Costello, 1980; and Cressler and others, 1979).

Figure 4. Slab section cut through the porphyroblastic phase of the Corbin Gneiss Complex showing the initial stages of ductile deformation.

Stratigraphy of the Salem Church Anticlinorium

The Salem Church anticlinorium (Fig. 1) is approximately 8 km east of Cartersville and lies east of the Great Smoky fault and north of the Allatoona-Hayesville fault. Although the axial trace of the anticlinorium was mapped for approximately 30 km, the relationships between basement gneiss and its cover rocks (Fig. 3) are seen best in exposures along the shore of Allatoona Lake east of Cartersville.

Corbin Gneiss Complex. Rocks of the Corbin Gneiss Complex occur in the core of the Salem Church anticlinorium. The term "Corbin Gneiss Complex" was informally introduced by McConnell and Costello (1980) to describe rocks previously defined as Corbin granite (Hayes, 1900) and Salem Church granite (Bayley, 1928). Although areally separated within the Salem Church anticlinorium, they are believed to be exposed parts of a single basement complex. This conclusion is based not only on apparent similarities in mineralogy and texture but also on similarities of stratigraphic sequences overlying both gneisses. One phase of the Corbin Gneiss Complex was dated isotopically by Odom and others (1973) and was determined to be in excess of 1 b.y. old on the basis of a lead-lead age. Later work by Dallmeyer (1975) using $^{40}Ar/^{39}Ar$ techniques reaffirmed a Grenville age for the Corbin. We suggest that the present paper be a formal definition of the Corbin Gneiss Complex.

Rocks composing the Corbin Gneiss Complex represent an intricate association of orthogneisses and paragneisses, the dominant phase being a coarse-grained quartz-monzonitic gneiss containing megacrysts of microcline (Fig. 4). This is the phase from which samples were taken by Odom and others (1973) for their isotopic age date. However, this rock cross-cuts small bodies of paragneiss composed of graphite bearing meta-arkose (Costello, 1978; Pl. VI). The age of these paragneisses is unknown, but they may represent remnants of an earlier orogenic belt largely remobilized during the Grenville event. Compositions within the meta-igneous phase of the Corbin vary from granite to granodiorite (Table 1; Fig. 5). Variation diagrams plotting weight percent silica versus weight percent of other oxides (Harker diagrams) are given for the analyses presented in this report (Fig. 6). They show well-defined relationships with silica for FeO and TiO_2, while the

McConnell and Costello

TABLE 1.   WHOLE ROCK CHEMISTRY AND NORMATIVE ANALYSES FOR THIRTY SAMPLES
FROM THE CORBIN GNEISS COMPLEX

Sample No.	19	25	28	30	34	40	49	51	53	58	62B	64A	67	68	76
SiO_2	69.12	68.90	67.80	70.14	73.60	70.60	63.46	68.60	69.00	71.60	66.20	62.24	71.80	57.84	68.00
Al_2O_3	15.00	15.80	15.00	15.30	13.00	14.30	14.50	14.80	13.80	12.50	14.80	15.50	13.80	17.40	14.80
Fe_2O_3	1.13	1.46	1.88	0.89	1.09	1.38	1.15	0.47	1.76	0.99	1.18	1.65	0.79	1.58	1.26
FeO	2.04	2.11	2.99	2.26	1.09	1.82	5.98	2.55	2.92	2.62	4.88	6.70	2.62	7.58	2.92
MgO	0.40	0.77	0.83	0.68	0.24	0.20	0.60	0.25	0.35	0.40	0.75	1.00	0.30	1.45	1.00
CaO	1.72	1.00	1.16	1.62	2.04	1.84	4.06	2.80	2.60	3.04	2.60	3.10	2.10	4.50	0.76
Na_2O	2.56	2.70	2.00	3.37	2.63	2.70	3.64	2.96	2.63	2.70	2.70	2.70	2.00	3.10	2.70
K_2O	6.00	5.06	4.82	4.10	5.42	5.30	3.00	6.00	5.06	4.02	4.22	3.97	4.82	2.77	5.42
H_2O	0.88	1.10	2.07	0.56	0.40	0.83	0.13	0.45	0.58	0.28	0.43	0.25	0.85	0.34	1.30
TiO_2	0.70	0.78	1.10	0.80	0.30	0.78	1.75	0.80	1.00	0.80	1.60	2.00	0.60	2.50	1.50
MnO	0.04	0.04	0.04	0.02	0.01	0.02	0.05	0.02	0.04	0.02	0.05	0.08	0.02	0.09	0.04
TOTAL	99.59	99.72	99.69	99.74	99.73	99.77	98.32	99.70	99.74	98.97	99.41	99.19	99.70	99.15	99.70
Q	26.20	29.88	33.00	29.39	33.08	29.94	19.17	21.65	27.61	30.69	25.18	19.66	35.31	13.46	27.56
CO Z	1.17	4.06	4.38	2.37	0.00	0.78	0.00	0.00	0.00	0.00	1.06	1.13	1.47	1.12	3.11
OR	35.46	29.90	28.48	24.23	32.03	31.32	17.73	35.46	29.90	28.48	24.92	23.46	28.48	16.37	32.03
AB	21.66	22.85	16.92	28.52	22.25	22.85	30.80	25.05	22.25	22.85	22.85	22.85	16.92	20.56	22.85
AN	8.53	4.96	5.75	8.04	7.66	9.13	14.37	9.38	10.91	7.75	12.90	15.38	10.42	22.32	3.77
DI					2.04		5.01	3.92	1.70	6.26					
NE															
EN	1.00	1.92	2.07	1.69	0.10	0.50	1.05	0.27	0.63	0.13	1.87	2.49	0.75	3.61	2.49
FS	1.73	1.45	2.20	2.13	0.11	0.95	5.08	1.33	1.70	0.37	5.44	7.79	3.20	8.65	1.92
MT	1.64	2.12	2.73	1.29	1.58	2.00	1.67	0.68	2.55	1.44	1.71	2.39	1.15	2.29	1.83
IL	1.33	1.48	2.09	1.52	0.57	1.48	3.32	1.52	1.90	1.52	3.04	3.80	1.14	4.75	2.85

Sample No.	78	79	81	82	88	92	94	95	97	99	104	108	110	112	113
SiO_2	68.40	68.36	65.20	69.60	67.80	67.20	63.20	67.20	66.40	63.54	70.80	68.64	68.02	66.16	73.26
Al_2O_3	15.30	16.00	14.30	14.20	12.80	14.80	14.20	13.50	15.00	16.70	14.70	15.00	15.80	16.00	13.50
Fe_2O_3	1.67	1.09	2.73	1.16	0.85	0.78	1.00	1.35	1.10	1.98	0.92	2.01	0.97	0.80	0.68
FeO	2.55	1.90	5.10	2.92	4.37	3.35	7.29	4.00	4.23	4.52	1.60	1.97	2.55	3.06	1.82
MgO	0.58	0.87	1.95	0.72	0.98	1.45	0.98	1.52	1.16	0.54	0.94	1.30	0.72	0.80	0.33
CaO	1.76	0.50	2.66	1.86	3.36	2.56	3.44	2.96	4.66	3.88	0.10	1.60	1.74	2.44	2.08
Na_2O	2.43	2.00	2.43	2.17	2.70	2.96	2.70	2.70	2.83	2.43	1.62	2.00	2.83	2.96	2.00
K_2O	5.30	6.62	2.65	5.30	4.58	5.06	4.34	4.22	3.61	3.73	6.50	4.58	5.90	6.26	5.06
H_2O	0.90	1.54	0.62	0.48	0.45	0.44	0.29	0.75	0.43	0.64	1.80	1.59	0.24	0.30	0.22
TiO_2	0.80	0.80	1.75	1.32	1.50	1.00	1.66	1.32	1.00	1.50	0.80	1.09	0.90	0.78	0.80
MnO	0.02	0.02	0.08	0.02	0.04	0.02	0.09	0.04	0.04	0.09	0.00	0.02	0.02	0.02	0.03
TOTAL	99.71	99.70	99.47	99.75	99.43	99.62	99.19	99.56	99.56	99.55	99.78	99.80	99.69	99.58	99.78
Q	28.43	28.43	30.33	30.55	25.26	21.32	18.21	25.38	23.20	24.03	34.50	34.02	23.07	16.92	36.63
CO Z	2.37	4.64	2.60	1.51	0.00	0.00	0.00	0.00	0.00	1.61	4.82	3.84	1.59	0.00	0.95
OR	31.32	39.12	15.66	31.32	27.06	29.90	25.65	24.94	21.33	22.04	38.41	27.06	34.86	36.99	29.90
AB	20.56	16.92	20.56	18.36	22.85	25.05	22.85	22.85	23.95	20.56	13.71	16.92	23.95	25.05	16.92
AN	8.73	2.48	13.20	9.23	9.28	12.15	13.81	12.25	17.56	19.25	0.50	7.94	8.63	11.88	10.32
DI					6.26	.45	2.81	2.02	4.69						
NE														0.19	
EN	1.44	2.17	4.86	1.79	1.39	3.50	2.16	3.31	2.05	1.34	2.34	3.24	1.79	1.96	0.82
FS	2.02	1.30	4.37	2.39	2.80	3.78	8.82	3.60	3.75	4.35	0.86	0.19	2.43	3.65	1.50
MT	2.42	1.58	3.96	1.45	1.23	1.13	1.45	1.96	1.59	2.87	1.33	2.91	1.41	1.16	0.99
IL	1.52	1.52	3.32	2.51	2.85	1.90	3.15	2.51	1.90	2.85	1.52	2.07	1.71	1.48	1.52

Note: Data modified after Martin (1974). Analyses performed in the laboratory of the Georgia Geologic Survey.

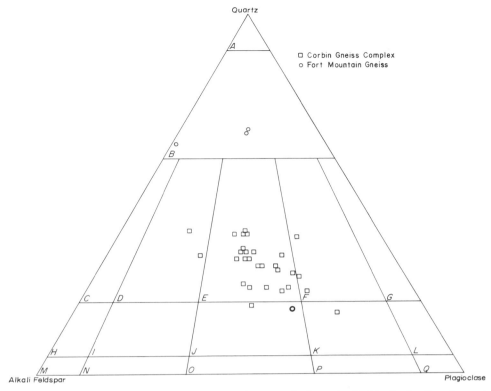

Figure 5. Quartz-alkali-feldspar-plagioclase plot of 30 samples from the Corbin Gneiss Complex and five samples of the Fort Mountain Gneiss (modified after Lyons, 1976) B = quartz-rich granitoids, C = alkali-feldspar granite, D = two-feldspar granite, E = quartz monzonite, F = granodiorite, G = quartz diorite, H = alkali-feldspar-quartz syenite, I = quartz syenite, J = quartz-rich monzonite, and K = quartz monzodiorite.

relationship between silica and other oxides is more diffuse. However, Martin (1974) reported that a differentiation trend is apparent for all oxides except CaO and Al_2O_3 when the analyses are plotted on Larson-type variation diagrams. Parts of the Corbin Gneiss Complex contain a significant amount of pyroxene. This phase was described originally as an augite granite by Hayes (unpub. data) and most recently mapped as a metagabbro by Crawford (*in* Cressler and others, 1979). The rock is composed of graphite, ilmenite, microcline, blue quartz, biotite, pyroxene, and andesine. Microprobe analyses of pyroxenes indicate that ferrohypersthene dominates augite (K. Gillon, 1980, written commun.).

Ocoee Supergroup. In his unpublished report on the Cartersville 30-minute quadrangle, Hayes included rocks above the Corbin in the Ocoee series and termed them the Pinelog conglomerate and Wilhite slate. Since Hayes's work, but prior to isotopic age determinations on the Corbin, rocks lying directly on the Corbin were termed Weisner Formation (Bentley and others, 1966), Walden Creek Group (Hadley, 1970), or Murphy Group equivalents (Hurst, 1973). Since determination of the age of the Corbin, these rocks generally are correlated with the Ocoee (Cressler and others, 1979; McConnell and Costello, 1980).

With minor modifications, McConnell and Costello (1980) redefined Hayes's Pinelog conglomerate and Wilhite slate to Pine-

log Formation and Wilhite Formation. While the Wilhite Formation was defined previously by King and others (1958) in the Great Smoky Mountains National Park area, we suggest that the present paper serve as a formal definition of the Pinelog Formation. The Pinelog is composed of an interlayered sequence of metaconglomerates, metasandstones, and locally carbonaceous metasiltstone. The basal units that lie nonconformably on the Corbin Gneiss Complex consist of well-sorted quartz-pebble metaconglomerates, metasandstone, and graphitic phyllite. These, in turn, are overlain by thickly bedded and graded metaconglomerates, cross-bedded metasandstones, graphitic phyllites, and thin lenses of immature, poorly sorted lithic metaconglomerates (diamictite). The greatest portion of the Pinelog Formation was derived directly from the Corbin. This was noted by Hayes (1900, p. 406), who stated that "This area of Corbin granite at one time probably formed an island, since it is surrounded, in part at least, by rocks derived from its waste." The best exposures documenting the relationships between the Corbin and its cover rocks are present along the western limb of the Salem Church anticlinorium. Although stratigraphic relationships along this limb are overturned due to later folding, an erosional unconformity was observed between the Corbin and its cover rocks at two locations (Fig. 7).

Gradationally above the Pinelog Formation is a relatively thin sequence of carbonaceous and noncarbonaceous, locally cal-

McConnell and Costello

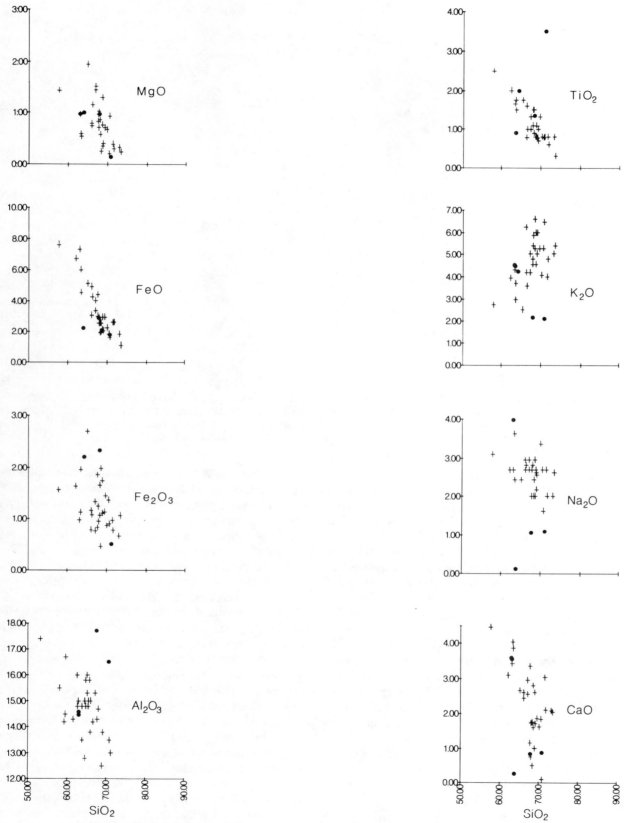

Figure 6. Variation diagrams showing the relationship of SiO$_2$ to other oxides in weight percent.

Figure 7. Photograph of nonconformable contact between the Corbin Gneiss Complex and Pinelog Formation near Lake Arrowhead (Fig. 3). At this location the units are overturned due to subsequent deformation. Pick end of hammer points to contact.

careous metasiltstone, metasandstone, and sandy marble. The lithologic character of this unit led Hayes (unpub. data) to equate it with the Wilhite slate that he mapped in the Cleveland, Tennessee, area (Hayes, 1901). According to Hayes (unpub. data), of the units mapped in the Cleveland area only the Wilhite retains its characteristics to the south. In later reports in the Great Smoky Mountains National Park, King and others (1958) redefined stratigraphic relationships of the Ocoee Supergroup and included the Wilhite Formation in their Walden Creek Group. In order to conform to King and others' (1958) stratigraphy, McConnell and Costello (1980) modified Hayes's Wilhite slate in the Salem Church area and renamed it the Wilhite Formation. It is the intention of this report to formally designate these rocks as Wilhite Formation (Fig. 3; Table 2). Similar to the Wilhite in the Great Smoky Mountains, the Wilhite of the study area contains most of the carbonate rocks in the Ocoee sequence.

The Pinelog Formation is known to nonconformably overlie the Grenville-age Corbin Gneiss complex and conformably underlie the Wilhite Formation of the Walden Creek Group. This

relationship, as well as favorable lithologic comparisons with the Snowbird Group of King and others (1958), prompted McConnell and Costello (1980) to equate the two (Table 2). While the aforementioned stratigraphic relationship (i.e., Snowbird Group rocks conformably overlain by Walden Creek Group rocks) occurs north of the Greenbrier fault, south of the fault, Great Smoky Group rocks lie conformably on the Snowbird Group (King and others, 1958; and Table 2). No evidence presently available suggests that the Greenbrier fault passes through the Salem Church area, and the apparent conflict with the relationships to the northeast is not resolvable at this time.

On the eastern limb of the Salem Church anticlinorium, the Pinelog and a large part of the Wilhite Formation are truncated by a series of *en echelon* faults grouped together into the Lost Town Creek fault zone. The Lost Town Creek and Salem Church anticlinorium resemble a duplex fault system similar to that described by Boyer (1976) in which a broad open fold forming the head of the duplex is paralleled to the southeast by a series of imbricate faults. In this area, rocks of the Wilhite Formation lie in fault contact with the Corbin Gneiss Complex and structurally are overlain by rocks of the Great Smoky Group (for a breakdown of the Great Smoky Group in this area, the reader is referred to McConnell and Costello, 1980). Although this faulting thrusts younger rocks over older rocks, Elliott and Johnson (1980) have shown that this is not an uncommon occurrence in previously folded rocks. On the northwestern limb of the Salem Church anticlinorium, rocks of the Wilhite Formation are overlain in apparent stratigraphic conformity by rocks of the Great Smoky Group.

The exact thickness of the Ocoee Supergroup in this area is unknown owing to complications resulting from multiple penetrative deformations. However, map distribution suggests that the Ocoee is less than 5,000 m thick. This is in contrast to a 16,200-m thickness for the Ocoee in the Great Smoky Mountains (King, 1964).

Stratigraphy of the Fort Mountain Area

At Fort Mountain, Grenville basement (Russell, 1976) is exposed in the core of an unnamed anticlinorium and is nonconformably overlain by a metamorphosed succession of clastic rocks (Fig. 8). Stratigraphic relationships, however, are not as clear in this area due to complex faulting, paucity of unweathered exposures and lack of detailed mapping.

Fort Mountain Gneiss. The Fort Mountain Gneiss, like the Corbin Gneiss Complex, is an assemblage of lithologies rather than a single rock type. Both coarse-grained, porphyroblastic and medium-grained, equigranular varieties are defined. Uranium-lead isotopic age determinations on zircons from the Fort Mountain Gneiss are concordant at 1030 m.y. (Russell, 1976), while a 368 ± 9 m.y. Rb-Sr age on the gneiss reflects homogenization of isotopes during ductile deformation (Russell, 1976).

Hayes (unpub. data) originally defined the Fort Mountain Gneiss as dominantly chlorite schist intruded by felsic and mafic

TABLE 2. CHART COMPARING THE VARIOUS STRATIGRAPHIES USED IN THE SALEM CHURCH
ANTICLINORIUM AND FORT MOUNTAIN AREA, AS WELL AS THEIR RELATIONSHIP TO
THE STRATIGRAPHIC SEQUENCE DEFINED IN GREAT SMOKY NATIONAL PARK

SALEM CHURCH ANTICLINORIUM		GREAT SMOKY MOUNTAINS NATIONAL PARK	
Hayes (1900)	This Report	after King and Others (1958)	
		North of Greenbrier fault	South of Greenbrier fault
Gilmer formation	Great Smoky Group	Walden Creek Group	Great Smoky Group
Wilhite slate	Wilhite Formation (Walden Creek Group)	Cades Sandstone	
Pinelog conglomerate	Pinelog Formation (Snowbird Group)	Snowbird Group	Snowbird Group
Corbin granite	Corbin Gneiss Complex	Granitic and gneissic rocks	Granitic and gneissic rocks

FORT MOUNTAIN AREA

Hayes (unpub. data)	Furcron and Teague (1947)	This report
Cohutta formation		
Gilmer formation	Wilhite Slate	Great Smoky Group
Citco conglomerate	Ocoee Series	Wilhite Formation
Wilhite slate	Corbin Granite	Parr Branch Formation
Fort Mountain gneiss	Fort Mountain Gneiss	Fort Mountain Gneiss (includes Corbin Granite of Furcron and Teague, 1947)
	Cohutta Schist	

rocks. He believed that the chlorite schist represented altered "augite" granite similar to parts of the Corbin granite. Hayes also included a coarse augen granite, similar to parts of the Corbin granite in the Fort Mountain Gneiss. Furcron and Teague (1947) redefined the Fort Mountain Gneiss to be primarily a biotite augen gneiss of granitic and granodioritic composition, which was intruded by what they termed Corbin Granite, a coarse-grained granite with little or no biotite. In this report, the Fort Mountain Gneiss includes the granitic and granodioritic augen gneisses of Furcron and Teague (1947) as well as their "Corbin Granite." Table 3 presents chemical analyses and norms for five samples of Fort Mountain Gneiss. The results of these analyses are plotted in Figures 5 and 6 along with the results from the Corbin Gneiss Complex. Although there are apparent differences in chemistry between the Fort Mountain Gneiss and the Corbin Gneiss Complex, they probably are partially the result of the disparity in the number of samples analyzed (i.e., 30 from the Corbin Gneiss Complex and 5 from the Fort Mountain Gneiss). In any event, both the "Corbin Granite" of Furcron and Teague and the granodioritic and granitic phases of the Fort Mountain Gneiss have textural and compositional similarities to phases of the Corbin Gneiss Complex. These similarities suggest that the Corbin Gneiss Complex and the Fort Mountain Gneiss represent isolated exposures of the same basement gneiss complex; how-ever, because they occur in different "domes," both terms are retained in this report.

An enigmatic problem of the Fort Mountain area is the presence of chlorite schist and talc associated with the Fort Mountain Gneiss. Furcron and Teague (1947) interpreted these rocks as metamorphosed sedimentary rocks representing the oldest units present in this area and subsequently intruded by the Fort Mountain Gneiss. Later, Needham (1972) concluded that the talc deposits represented an altered ultramafic body that was intrusive into the Fort Mountain Gneiss. While a metasedimentary origin for these deposits is suggested by the substantial amounts of dolomite associated with them (Furcron and Teague, 1947), the origin of these rocks is still open to speculation.

Ocoee Supergroup. Rocks that overlie the Fort Mountain Gneiss generally are grouped into either the Ocoee Series (Furcron and Teague, 1947) or Great Smoky Group (Needham, 1972). These reports concentrated more on the Fort Mountain Gneiss and associated talc deposits than on the overlying metasedimentary rocks. In this report, we modify the original stratigraphy developed for the Fort Mountain area by Hayes in his unpublished map of the Dalton 30-minute quadrangle, adding additional data derived by Furcron and Teague (1947) and Needham (1972), as well as our own mapping.

One of the most distinctive rock types present in the Fort

TABLE 3. WHOLE-ROCK CHEMISTRY AND NORMATIVE ANALYSES
FOR FIVE SAMPLES FROM THE FORT MOUNTAIN GNEISS

Sample No.	238-1A*	245-1*	301-1*	FT-1+	FT-2+
SiO_2	70.92	64.02	67.96	63.18	63.25
Al_2O_3	16.50	20.70	17.70	14.51	14.58
Fe_2O_3	0.52	2.23	2.36	4.63	4.63
FeO	1.70	2.16	2.83	ND	ND
MgO	0.14	1.02	0.98	1.00	1.01
CaO	0.88	0.28	0.84	3.59	3.57
Na_2O	1.07	0.13	1.07	4.11	4.06
K_2O	2.10	4.22	2.18	4.51	4.47
H_2O	2.64	2.86	2.31	ND	2.40
TiO_2	3.53	2.00	1.35	0.92	0.92
MnO	0.00	0.04	0.10	ND	ND
Total	100.00	99.66	99.68	96.45	98.89
Q	54.82	45.59	50.08	14.99	15.39
CO	11.08	15.92	12.43	26.65	26.41
Z					
OR	12.41	24.94	12.76	26.65	26.41
PL	12.83	1.10	12.24	42.60	42.71
DI	0.00	0.00	0.00	5.28	4.79
HY	0.35	2.54	3.64	0.04	0.30
MT	0.00	1.30	3.42	0.00	0.00
IL	3.59	3.80	2.56	0.00	0.00
RU	1.64	0.00	0.00	0.00	0.00
HM	0.52	1.34	0.00	4.63	4.63
SP	0.00	0.00	0.00	2.26	2.26
AP	0.21	0.50	0.36	0.00	0.00

Note: Analyses performed in the laboratory of the Georgia Geologic Survey.
* Samples from Fort Mountain provided by Bruce O'Connor.
+ After Furcron and Teague (1947).

Mountain area is a coarse, bouldery lithic conglomerate that is known to overlie the Fort Mountain Gneiss, although the contact between the two units is not exposed. Hayes noted the presence of these conglomerates (Hayes, unpub. data) and included them in his Cohutta Formation. He believed that the Cohutta Formation was one of the youngest members of the Ocoee in this area and that it was deposited nonconformably on the Fort Mountain Gneiss. The only other known occurrence of this lithology in this part of the Blue Ridge is in the Pinelog Formation, which non-conformably overlies Grenville basement and underlies the Wilhite Formation. The boulder conglomerate in the Fort Mountain area (see pЄs in Fig. 8) apparently lies in a similar stratigraphic position. This part of Hayes's Cohutta Formation is considered in this report to be a lithostratigraphic equivalent of the Pinelog Formation but is renamed the Parr Branch Formation to eliminate confusion regarding the name "Cohutta" (Table 2). Hayes described several other locations within the Dalton quadrangle where this type of lithology was observed; however, mapping in

these areas is not of sufficient detail to determine the exact relationships of these rocks.

As in the Salem Church anticlinorium, the Wilhite Formation in the Fort Mountain area is composed primarily of a calcareous, gray to black, carbonaceous phyllite and contains beds of siliceous marble (Hayes, unpub. data). It lies along the western side of Fort Mountain and for the most part is fault bounded. However, along Holly Creek (Fig. 9), approximately 7 km to the northeast of Fort Mountain, coarse, lithic conglomerate is bounded on either side by carbonaceous phyllite. These outcrops represent the coarse conglomerate of the Parr Branch Formation conformably overlain by the Wilhite Formation in the northeastern closure of an anticlinorium that is cored by the Fort Mountain Gneiss. Thus, the Wilhite-Parr Branch relationship is a lithostratigraphic twin to the Wilhite-Pinelog sequence observed to the south in the Salem Church anticlinorium.

Structurally above the rocks of the Wilhite in the Fort Mountain area are metamorphosed arkoses, siltstones, and conglomerates of what Hayes termed Gilmer Formation. These rocks are stratigraphically and lithologically similar to Great Smoky Group rocks.

METAMORPHISM OF THE BASEMENT GNEISS-COVER ROCK SEQUENCE

Basement gneisses exposed in the Salem Church anticlinorium and Fort Mountain area contain evidence of Grenville age granulite facies metamorphism. Ferro-hypersthene and Mg-rich garnet-bearing phases of the Corbin Gneiss Complex (Martin, 1974) are indicative of this granulite facies event; an event not observed in the cover rocks. Textures related to this metamorphism, for the most part, have a metamorphic overprint caused by Paleozoic events, although some rocks do retain a marked discordance between the planar fabric present in the Corbin and the more dominant planar fabric in the cover rocks. Evidence for the timing of the Grenville metamorphic event that formed these structures was provided by $^{40}Ar/^{39}Ar$ work of Dallmeyer (1975). Ages of 702 and 735 m.y. from biotite concentrates of the coarse-grained phase of the Corbin were interpreted to represent cooling ages following Grenville metamorphism (Dallmeyer, 1975).

Ocoee Supergroup rocks overlying basement gneisses show no evidence of Grenville metamorphism but do contain metamorphic textures and mineral assemblages of a later, lower grade event. Figure 10 is a revised isograd map for this part of Georgia. It modifies the earlier maps of Smith and others (1968) and Dallmeyer (1975) and was derived from published data on metamorphic index minerals and from detailed mapping in the northeastern part of the Salem Church anticlinorium. Metamorphic mineral assemblages from the Corbin Gneiss Complex were not used in the construction of Figure 10 because of the presence of relict assemblages from Grenville metamorphism. Although Dallmeyer (1975) suggested that parts of the Corbin Gneiss Complex were affected by garnet or higher grade metamorphism,

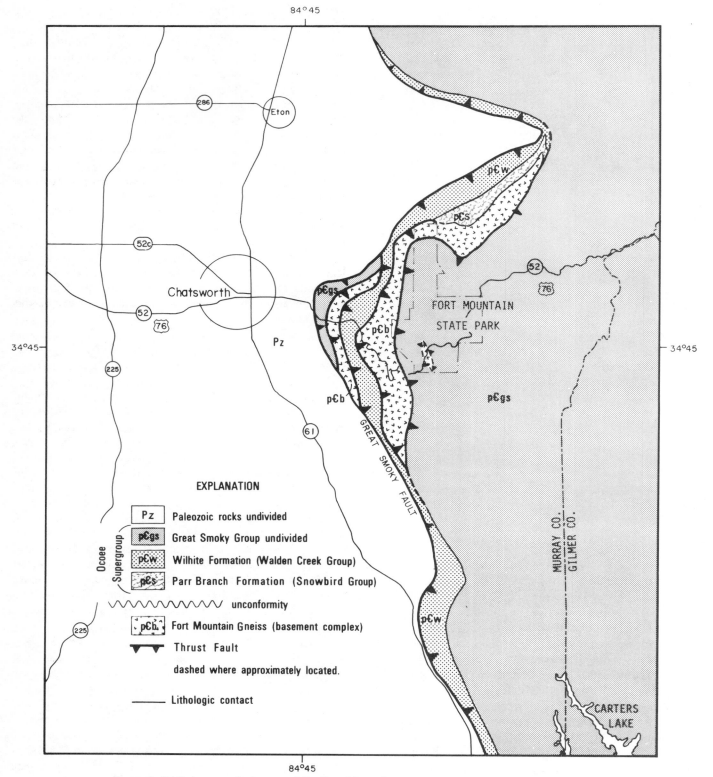

Figure 8. Preliminary geologic map of the Fort Mountain area (modified after Hayes, unpub. data; Furcron and Teague, 1947; and Needham, 1972).

Figure 9. Photograph of an outcrop along Holly Creek northeast of Chatsworth showing the boulder bearing lithic conglomerate of the Parr Branch Formation.

his basis for this conclusion was the resetting of the argon system in one sample of Corbin. However, the age derived for that sample (379 ± 15 m.y.) is similar to the Rb-Sr age determined on mylonites in the Fort Mountain Gneiss (368 ± 9 m.y.) (Hatcher and Odom, 1980). This suggests that the argon system in that sample of Corbin Gneiss was reset by ductile shearing common along the eastern side of the Corbin Gneiss Complex (Costello, 1978). In general, metamorphic grade increases from lower greenschist facies in the west to upper amphibolite facies to the east (Fig. 10). Deformation after the peak of metamorphism is suggested by apparent folding of isograds on map scale by the Salem Church anticlinorium (Fig. 10). The exact timing of this second event is uncertain. While Mose and Stransky (1973) indicated that the peak of metamorphism in the Blue Ridge during the Paleozoic occurred between 350 and 380 m.y. ago, most workers in the Blue Ridge believe it occurred between 430 and 450 m.y. ago, as suggested by Butler (1972).

DEFORMATION OF THE BASEMENT-COVER ROCK SEQUENCE

Deformation in the western Blue Ridge of Georgia is widespread and varies in intensity and style from south to north. Multiple deformation is evident in rocks along the western edge of the Blue Ridge thrust sheet in Georgia and contrasts with a single major deformational event present along the leading edge of the thrust sheet approximately 20 km north in the Ocoee Gorge area of Tennessee (Holcombe, 1973). Thus, correlation of deformational events, particularly later ones, across the orogen may not be possible.

Faults

Faults within the study area are characterized by both pre-metamorphic and post-metamorphic movement. Premetamor-

phic faults within the cover sequence, namely, the Allatoona-Hayesville on the south and Greenbrier fault on the north, border the study area (Fig. 1). Hatcher (1978) defined the Hayesville fault and suggested that it followed the trace of Hurst's (1973) Allatoona fault. He also indicated that the Hayesville was a probable cryptic suture separating a volcanic, ultramafic, granite-bearing terrane on the southeast from a nonvolcanic, abundant, basement-bearing terrane on the northwest (Hatcher and Odom, 1980). Mylonites and phyllonites along the trace of the Hayesville fault southwest of Canton, as well as retrograded mineral assemblages in this same area, indicate postmetamorphic movement along this part of the fault zone; hence the term "Allatoona" is retained as a qualifier (McConnell and Costello, 1980). Movement along the Allatoona-Hayesville fault has thrust amphibolites, mica schists, and felsic gneisses of the northern Piedmont over rocks of the Ocoee Supergroup in the area of this report.

The Greenbrier fault separates two zones within the Blue Ridge thrust sheet that have different stratigraphic sequences within the Ocoee Supergroup; namely, on the north side of the fault, Walden Creek Group rocks are believed to conformably overlie Snowbird Group rocks, while to the south of the fault, only the Great Smoky Group overlies the Snowbird (King and others, 1958). This is of particular significance because the Greenbrier fault is not traceable south of where it abuts the Great Smoky fault (Fig. 1). In the study area, the Snowbird Group (i.e., Pinelog and Parr Branch Formations) is overlain by Wilhite Formation (Walden Creek Group). These rocks apparently are overlain conformably by rocks of the Great Smoky Group.

Postmetamorphic faults in this area are characterized by an early ductile stage and a later brittle stage. Both basement gneisses and cover rocks commonly contain zones of ductile shearing. In the Corbin Gneiss Complex, these shear zones were at one time mistakenly believed to be metsedimentary rocks (Kesler, 1950), but more recent investigations established their true identity (Costello, 1978). Ductile deformation is most commonly observed along the eastern limb of the Corbin Gneiss Complex and along the western contact of the Fort Mountain Gneiss. Mylonites associated with the Corbin have not been dated isotopically, but Russell (1976) and Hatcher and Odom (1980) reported Rb-Sr ages for mylonites in the Fort Mountain Gneiss of 368 ± 9 m.y. They suggest that this age represents homogenization of strontium isotopes during ductile deformation in Acadian times (Russell, 1976). The orientation of mylonitic foliation in ductile shear zones present in the basement gneisses is parallel to the axial planes of F_1 folds observed in the cover rocks and was folded by subsequent deformations. Parallelism between the orientation of the mylonitic foliation in the basement gneiss shear zones and the S_1 schistosity in the cover rocks suggests that they are a reflection of intense shearing related to transposition of original layering during F_1 folding.

Late-stage brittle faulting in the study area is evident in breccias and gouge zones along the Great Smoky fault and Lost Town Creek fault zone. The southwestern continuation of the Great Smoky fault thrusts rocks of the Talladega belt over rocks

TABLE 4. POST GRENVILLE-AGE STRUCTURAL ELEMENTS
OF THE SOUTHWESTERN BLUE RIDGE OF GEORGIA

Generation	Fold Style	Lineations Generation	Type	Timing	Event
D_1				Pre-metamorphic	Allatoona-Hayesville & Greenbriar faults
D_2 (F_1)	Isoclinal, overturned; NE trending, NW vergence	L_{1a} L_{1b}	Elongation Intersection	Mid-Paleozoic (350-480 m.y.) Mid-Paleozoic (370 m.y.)	Low grade metamorphism Ductile shearing in basement gneisses
D_3 (F_2)	Tight, upright to overturned; NE trending, NW vergence	L_2	Intersection	Mid-Paleozoic	Salem Church anticlinorium
D_4				Late Carboniferous	Brittle faulting (Lost Town Creek fault zone); emplacement of Blue Ridge thrust sheet.
D_5 (F_3)	Open, upright to overturned; SE trending, SW vergence	L_3	Intersection	Hercynian(?)	Bends plane of Great Smoky fault
D_6 (F_4)	Open, upright; NW trending			(?)	

as young as Mississippian-age Fort Payne Chert (Crawford, 1977).

Folds

Except for relict textures locally present in the Corbin Gneiss Complex, no evidence remains in this area for Grenville-age deformation or folding related to movement on the Greenbrier fault. Therefore, the numerical sequence used to refer to deformational events in this report excludes the Grenville deformation. (D_g).

F_1 folding (D_2) (Table 4) in this area is characterized by overturned isoclinal folds (Fig. 11) that developed contemporaneously with Paleozoic metamorphism. A well-developed axial-planar schistosity is parallel to axial planes of F_1 folds. Original layering was transposed subparallel to the plane of the S_1 schistosity, and in zones of high strain, a mylonitic foliation was imparted to the rocks. Intersections with bedding (L_{1a}) and a well-formed clast elongation in the conglomerates (L_{1b}) (Fig. 12) are the dominant lineations formed during this event.

Second-generation folding (F_2/D_2) postdated the peak of metamorphism. Isograds formed during D_2 (Table 4) related metamorphism were folded during D_3. Folds formed during this event were superimposed on recumbent F_1 folds in the study area. Interference between F_1 and F_2 folds are responsible for the geometry of outcrop patterns of various units, in particular configuration of the rocks in the Murphy syncline (F_1) near the northeastern terminus of the Salem Church anticlinorium. The Salem Church anticlinorium is a second-generation (F_2) structure.

Folds formed during third-generation folding (F_3/D_5) represent one of the most important and least understood structural events in the western Blue Ridge of Georgia in that they play an important role in the configuration of the trace of the Great Smoky fault and the relationship between rocks of the Ocoee Supergroup and Talladega belt. F_3 folds trend northwest-southeast and plunge moderately to steeply to the southeast. These folds are open and vergent to the southwest. Fairley (1965) recognized folds of this orientation in the Tate quadrangle and suggested that they were responsible for the outcrop pattern of the Murphy syncline (Fairley, 1965, p. 62). However, interference between southeast-trending folds and an earlier northeast-trending fold pattern should produce a Ramsay Type 1 or 2 interference pattern, depending on the orientation of the axial surface of the first fold event (Ramsay, 1967). This is not the case in this area (Fig. 2). The arcuate trace of the Great Smoky fault southward from the Georgia-Tennessee line is an example of the effect of F_3 folds. In particular, southeast-trending folds are responsible for the arching of units in the Fort Mountain area (Fig. 8), while in the Salem Church area, F_3 folds interfere with earlier-formed structures to form a dome and basin pattern resulting in the separation of the exposed masses of the Corbin Gneiss Complex (Fig. 2). Finally, the most important effect that F_3 folds have is in the area southeast of Emerson (Fig. 3) where Ocoee

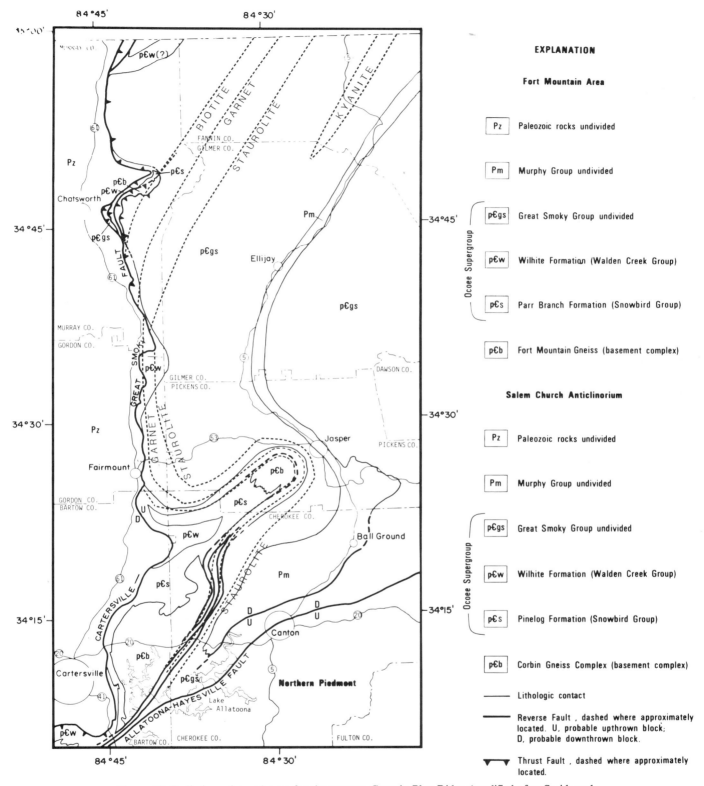

Figure 10. Preliminary isograd map for the western Georgia Blue Ridge (modified after Smith and others, 1968; and Dallmeyer, 1975).

Figure 11. Recumbent isoclinal folds in the Pinelog Formation related to F_1 folding.

Supergroup rocks trend into the Talladega belt. In this area, F_3 folds rotate rock units and earlier-formed planar structures from a northerly trend northeast of Emerson to a southwesterly trend south of Emerson (McConnell and Costello, 1981). Both the Wilhite Formation and rocks of the Great Smoky Group are traceable around the reentrant in the Great Smoky fault at Emerson and, therefore, this part of the Talladega belt is composed of Ocoee Supergroup rocks.

Deformation of the plane of the Great Smoky fault by F_3 folds places constraints on the timing of this event. Because rocks of Mississippian age form part of the footwall of the Great Smoky fault, D_5 deformation probably occurred during the late Carboniferous.

DISCUSSION

In his early work, Hayes indicated that gneisses here termed Corbin Gneiss Complex and Fort Mountain Gneiss represented rocks of Precambrian age nonconformably overlain by rocks of the Ocoee Series. After Hayes's original efforts, subsequent workers in this area proposed various and often contradictory interpretations regarding the age and stratigraphic relationships of the Corbin and Fort Mountain gneisses. Isotopic dating of the gneisses in the early 1970s and mapping related to this report resurrect, to a large degree, Hayes's original interpretation.

In the western Blue Ridge of Georgia, the Corbin Gneiss Complex and Fort Mountain Gneiss lie at the base of the stratigraphic sequence. Distinct lithologic and compositional similarities are present between the two gneissic bodies; thus, they probably represent exposed parts of a single basement complex. Both the Corbin and Fort Mountain gneisses are overlain nonconformably by a unit composed predominantly of rocks derived from the gneisses themselves. These rocks are termed the Pinelog and Parr Branch Formations, respectively. Conformably overly-

Figure 12. Stretched quartz clasts (L_{1b}) in the Great Smoky Group near Carters Dam.

ing both the Pinelog and Parr Branch Formation are graphitic phyllites and sandy marbles of the Wilhite Formation. The Pinelog-Wilhite and Parr Branch–Wilhite sequences exposed in the area of this report probably represent lithostratigraphic equivalents to the Snowbird Group–Walden Creek Group sequence observed in the Great Smoky Mountains of Tennessee.

While relict textures and mineral assemblages indicative of granulite facies metamorphism occur locally in the basement gneisses, this Grenville-age event was overprinted by Paleozoic metamorphism. Coincident with Paleozoic metamorphism was development of isoclinal folds vergent to the northwest. Subsequent doming of these isoclinal folds by second-generation folds formed the Salem Church anticlinorium with its core of basement gneiss. A third major folding event in this area folded the earlier events and is responsible for the dome and basin interference patterns in the Salem Church area as well as the arcuate trace of the Great Smoky fault. Finally, this third major fold event is responsible for the bend in rock units south and east of Emerson where the Talladega belt is composed of Ocoee Supergroup lithologies.

ACKNOWLEDGMENTS

We thank R. D. Hatcher Jr., R. D. Dallmeyer, and M. J. Bartholomew for constructive reviews of this report. We are also grateful to C. E. Abrams for assistance in preparation of the manuscript, and J. Barrett and M. Foley for drafting the figures.

REFERENCES CITED

Bartholomew, M. J., and Lewis, S. E., 1984, Evolution of Grenville massifs in the Blue Ridge geologic province, southern and central Appalachians, *in* Bartholomew, M. J., ed., The Grenville event in the Appalachians and related topics: Geological Society of America Special Paper 194 (this volume).

Bayley, R. D., Fairley, W. M., Fields, H. H., Power, W. R., and Smith, J. W.,

1966, The Cartersville fault problem: Georgia Geological Society Guidebook no. 4, 38 p.

Boyer, S. E., 1976, Formation of the Grandfather Mountain Window, North Carolina, by duplex thrusting: Geological Society of America Abstracts with Programs, v. 8, p. 788–789.

Butler, J. R., 1972, Age of Paleozoic regional metamorphism in the Carolinas, Georgia and Tennessee, southern Appalachians: American Journal of Science, v. 266, p. 865–894.

Costello, J. O., 1978, Shear zones in the Corbin gneiss of Georgia: Georgia Geologic Survey Bulletin 93, p. 32–37.

Costello, J. O., and McConnell, K. I., 1983, Relationship of the Cartersville fault and Great Smoky fault in the southern Appalachian orogen: A reinterpretation: Geological Society of America Abstracts with Programs, v. 15, no. 2, p. 94.

Crawford, T. J., 1977, The Cartersville fault, *in* Chowns, T. M., ed., Stratigraphy and economic geology of Cambrian and Ordovician rocks in Bartow and Polk Counties: Georgia Geological Society, 12th Annual Field Trip, 21 p.

Cressler, C. W., Blanchard, H. E., Jr., and Hester, W. G., 1979, Geohydrology of Bartow, Cherokee, and Forsyth Counties, Georgia: Georgia Geologic Survey Information Circular 50, 45 p.

Dallmeyer, R. D., 1975, $^{40}Ar/^{39}Ar$ age spectra of biotite from Grenville basement gneisses in northwest Georgia: Geological Society of America Bulletin, v. 86, p. 1740–1744.

Elliott, D., and Johnson, M.R.W., 1980, Structural evolution in the northern part of the Moine thrust belt, NW Scotland: Transactions of the Royal Society of Edinburgh, Earth Sciences, v. 71, p. 69–96.

Fairley, W. M., 1965, The Murphy syncline in the Tate quadrangle, Georgia: Georgia Geologic Survey Bulletin 75, 71 p.

Furcron, A. S., and Teague, K. H., 1947, Talc deposits of Murray County, Georgia: Georgia Geologic Survey Bulletin 53, 75 p.

Hadley, J. B., 1970, The Ocoee series and its possible correlatives, *in* Fisher, G. W., Pettijohn, F. J., Reed, J. C., and Weaver, K. N., eds., Studies in Appalachian geology: Central and Southern: New York, Wiley Interscience, p. 247–260.

Hatcher, R. D., Jr., 1978, Tectonics of the western Piedmont and Blue Ridge, southern Appalachians: Review and speculation: American Journal of Science, v. 278, p. 276–304.

Hatcher, R. D., Jr., and Odom, A. L., 1980, Timing of thrusting in the southern Appalachians, USA: Model for orogeny? Journal of the Geological Society of London, v. 137, p. 321–327.

Hayes, C. W., 1891, The overthrust faults of the southern Appalachians: Geological Society of America Bulletin, v. 2, p. 141–154.

—— 1900, Geological relations of the iron-ores in the Cartersville district, Georgia: American Institute of Mining Engineers Transactions, v. 30, p. 403–419.

—— 1901, Cleveland folio: U.S. Geological Survey Atlas 20, 12 p.

Holcombe, R. J., 1973, Mesoscopic and microscopic analysis of deformation and metamorphism near Ducktown, Tennessee [Ph.D. thesis]: Stanford, Stanford University, 225 p.

Hurst, V. J., 1973, Geology of the southern Blue Ridge belt: American Journal of Science, v. 273, p. 643–670.

Kesler, T. L., 1950, Geology and mineral deposits of the Cartersville district, Georgia: U.S. Geological Survey Professional Paper 224, 97 p.

King, P. B., 1964, Geology of the central Great Smoky Mountains, Tennessee: U.S. Geological Survey Professional Paper 349-C, 148 p.

King, P. B., Hadley, J. B., Newman, R. B., and Hamilton, W., 1958, Stratigraphy of Ocoee Series, Great Smoky Mountains, Tennessee and North Carolina: Geological Society of America Bulletin, v. 69, p. 947–966.

Lyons, P. C., 1976, I.U.G.S. classification of granitic rocks: A critique: Geology, v. 4, p. 425–426.

Martin, B. F., Jr., 1974, The petrology of the Corbin Gneiss [M.S. thesis]: Athens, University of Georgia, 133 p.

McConnell, K. I., and Costello, J. O., 1980, Guide to geology along a traverse through the Blue Ridge and Piedmont provinces of North Georgia, *in* Frey, R. W., ed., Excursions in southeastern geology: American Geological Institute, v. 1, p. 241–258.

—— 1981, The relationship between the Talladega Group and Ocoee Supergroup near Cartersville, Georgia: Geological Society of America Abstracts with Programs, v. 13, p. 30.

McLaughlin, R. E., and Hathaway, D. J., 1973, Fossils in the Murphy Marble: Geological Society of America Abstracts with Programs, v. 5, no. 5, p. 418–419.

Mose, D. G., and Stransky, T., 1973, Radiometric age determinations in the southern Appalachian Blue Ridge: Geological Society of America Abstracts with Programs, v. 5, p. 422.

Needham, R. E., 1972, The geology of the Murray County, Georgia, Talc District [M.S. thesis]: Athens, University of Georgia, 107 p.

Odom, A. L., Kish, S., and Leggo, P. J., 1973, Extension of "Grenville basement" to the southern extremity of the Appalachians: U-Pb ages of zircons: Geological Society of America Abstracts with Programs, v. 5, p. 425.

Ramsay, J. G., 1967, Folding and fracturing of rocks: New York, McGraw-Hill,

568 p.

Russell, G. S., 1976, Rb-Sr evidence from cataclastic rocks for Devonian faulting in the southern Appalachians: Geological Society of America Abstracts with Programs, v. 8, p. 1081.

Smith, J. W., 1959, Geology of an area along the Cartersville fault near Fairmount, Georgia [M.S. thesis]: Atlanta, Emory University, 41 p.

Smith, J. W., Wampler, J. M., and Green, M. A., 1968, Isotopic dating and metamorphic isograds of the crystalline rocks of Georgia, *in* Precambrian-Paleozoic Appalachian problems: Georgia Geologic Survey Bulletin 80, p. 121–139.

Williams, H., 1978, Tectonic lithofacies map of the Appalachian orogen: St. John's Newfoundland, Canada, Memorial University of Newfoundland, scale: 1:1,000,000.

Manuscript Accepted by the Society August 2, 1983

Geological Society of America
Special Paper 194
1984

An overview of the Grenville basement complex of the Pine Mountain window, Alabama and Georgia

James W. Sears
Department of Geology
University of Montana
Missoula, Montana 59812

Robert B. Cook, Jr.
Department of Geology
Auburn University
Auburn, Alabama 36849

ABSTRACT

Grenville basement rocks and their metasedimentary cover are exposed in the Pine Mountain window of Georgia and Alabama. The Grenville complex of the window includes charnockite, charnockite gneiss, feldspathic and migmatitic augen gneiss, and mylonite. The metasedimentary cover, the Pine Mountain Group, consists of an approximately 2 km-thick sequence of aluminous and graphitic schist, orthoquartzite and dolomitic marble.

Schistosity of the Grenville rocks conforms to that of the enclosing Pine Mountain Group, is parallel to compositional layering and the basement-cover interface, and is the axial surface of the Auburn, Memory Hill, and Thomaston nappes. In Alabama, the basement-cover interface was refolded coaxially about a N70E axis so that the axial surfaces of the major fold-nappes are folded isoclinally. To the northeast, complex interference folding resulted in development of broad basins and domes and curvilinear axial traces of folds of foliation. An apparent increase in strain to the southwest closely corresponds to the appearance of thick sheets of basement-derived mylonite and an increase in metamorphic grade from kyanite-garnet-muscovite-staurolite-biotite in Georgia to sillimanite-biotite-muscovite-anatectite in Alabama.

INTRODUCTION

The Pine Mountain window of Alabama and Georgia (Fig. 1 and 2) lies at the southwestern extremity of the Appalachian Mountain belt. The window has been eroded through the crystalline Piedmont allochthon, and contains the southernmost recognized Grenville massif in the orogenic system. Zircons from charnockitic rocks and felsic gneisses of the massif yield U-Pb ages of more than 1 b.y. (Odom and others, 1973). The massif is exposed over more than 1000 km², and lies 150 km south of the nearest Grenville rocks of the Blue Ridge Province. The massif is similar in structural configuration to other internal massifs along the Appalachian orogen—the West Chester dome, Berkshire Massif, Philadelphia-Baltimore Gneiss domes,

and Salem Church anticlinorium, in that it is deformed complexly with its younger metasedimentary cover sequence, and was metamorphosed retrogressively over much of its outcrop extent.

Despite its strategic importance at the southwestern flank of the orogen, no systematic, detailed study has focused on the Grenville massif of the window. This may be partly because of the poor exposures, generally deep weathering, heavy vegetative cover and the low topographic relief of the area. Structural data is obtainable from saprolite exposures along most roads and streams, but mapping is largely based upon weathered residuum, and outcrops fresh enough for petrographic study are confined to quarries, local exfoliation pavements and major drainage chan-

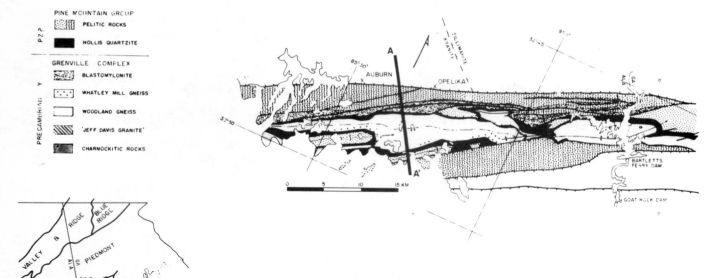

Figure 1. Geologic map of the western part of the Pine Mountain window. Modified after Sears and others, 1981b; Schamel and others, 1980; Bentley and others, 1985. Vertical ruling on Pine Mountain Group pelites indicates sillimanite zone.

nels. Outcrops of felsic rocks are particularly poor and because many of the Grenville rocks in the windows are felsic to some degree, the megascopic structural configuraiton of the basement is, to some extent, conjectural and based upon relations with the less weathered, more quartzose cover sequence.

This paper is a general overview based upon literature review and a combined total of 11 years of study of the window and its environs by the authors. Our work in Georgia largely has been of a reconnaissance nature, while ongoing mapping in Alabama is at a scale of 1:24,000. This paper is presented with the hope of encouraging further work in this interesting but difficult area.

STRUCTURAL CONFIGURATION

The Pine Mountain window is eroded through the Piedmont allochthon along the axis of the Pine Mountain antiform, a multiply folded and faulted, elongate structural culmination extending for 165 km from Auburn, Alabama, east-northeastward to Forsyth, Georgia. The antiform is the consequence of the vertical stacking and warping of three relatively thin nappes involving Grenville rocks and their cover (Fig. 3). The Grenville massif is polydeformed and metamorphosed and over large areas is dominated by S-surfaces and mineral assemblages that appear, on the basis of their structural configuration, to be of post-Grenville age.

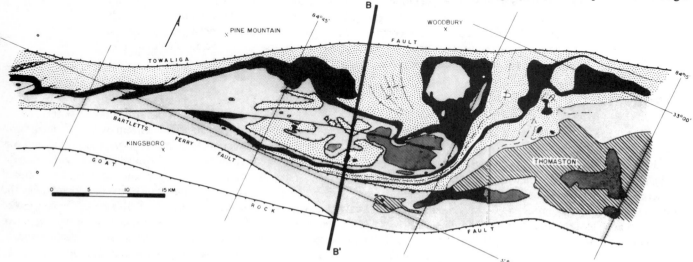

Figure 2. Geologic map of the eastern part of the Pine Mountain window. See Fig. 1 for and symbols. Modified after Sears and others, 1981b; Schamel and others, 1980; Schamel and Bauer, 1980, Hewett and Crickmay, 1937, Clarke, 1952.

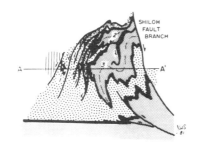

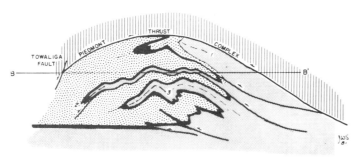

Figure 3. Composite Geologic cross sections through western (A-A') and eastern (B-B') parts of Pine Mountain window. (1) Thomaston Nappe, (2) Memory Hill Nappe, (3) Auburn Nappe. No vertical exaggeration intended.

In Alabama, less than 1 percent of the rocks assigned to the Grenville contain relict Precambrian textures. In Georgia, the percentage increases gradually eastward with decreasing post-Grenville structural complexity.

The Pine Mountain Group is the metasedimentary cover sequence nonconformably overlying the massif within the window. The group is estimated to have a structural thickness of about 2 km and is composed of a sequence of aluminous and graphitic schists, schistose quartzite, orthoquartzite and local dolomitic marble. Pervasive shearing, complex deformation and amphibolite facies metamorphism have obscured original thicknesses and lithofacies distributions, and no fossils have been recovered. However, the Pine Mountain Group is dominated by miogeoclinal lithologies that could represent metamorphosed equivalents of Upper Proterozoic-Lower Cambrian sequences of the Blue Ridge and Valley and Ridge provinces exposed 110 to 150 km to the northwest (Sears and others, 1981b).

The Grenville massif and its cover sequence contrast with the regionally metamorphosed graywackes, cherts, volcaniclastic and volcanic rocks of the structurally superjacent Piedmont allochthon in the Inner Piedmont (Bentley and Neathery, 1970; Brown and Cook, 1981; Sears and others, 1981a) and Uchee belts (Schamel and others, 1980). If the Pine Mountain Group is the distal equivalent of Valley and Ridge strata, the Piedmont allochthon has a minimum displacement of 150 km relative to the rocks in the window (the distance from the southern margin

of the window to the northern margin of the Piedmont allochthon along a transect passing through Atlanta).

The northern boundary of the window is the Towaliga fault (Bentley and Neathery, 1970; Schamel and Bauer, 1980). This fault has a moderately straight trace, truncates foliations and map units in the Inner Piedmont and within the window, and juxtaposes rocks of different metamorphic grade. It is post-metamorphic and is a normal fault with the north side downthrown in central Georgia where observed by Schamel and Bauer (1980) along the Towaliga River. Elsewhere along its trace, the fault is not exposed.

The southern margin of the window in Alabama and western Georgia is a 3-4 km thick zone of mylonite apparently derived from comminution of felsic gneiss and amphibolite of the Uchee belt along the sole of the Piedmont allochthon. The mylonite zone diminishes in thickness eastward in central Georgia. Near Forsyth, Georgia, the window is bounded by a pre-metamorphic fault (Hatcher and others, 1981).

The most persistent map unit in the Pine Mountain Group is the Hollis Quartzite, which is up to 325 m thick and is relatively resistant to weathering. It supports ridges up to 180 m high and is traceable for 150 km along the strike of the window. It is the key to the internal structure of the window.

In the western part of the window, the structural plunge is consistently to the ENE between 9 and 15 degrees, with relatively few plunge reversals. First-and second-generation folds are coaxial and parallel a pervasive intersection lineation. A generalized downplunge projection is thus possible with the present data base in this area (Fig. 3, A-A').

Three nappes are recognized in a transect near Auburn, Alabama. The structurally lowest of these is the Auburn Nappe, which underlies a large segment of the western part of the window. Its axial surface is tightly refolded into two major antiforms in which the inverted limb of the nappe is exposed. In the structural core of the Auburn Nappe, feldspathic gneiss is in direct contact with a massive orthoquartzite facies of the Hollis Quartzite. The Auburn Nappe is a fold-nappe with the inverted limb intact but highly sheared. The upright limb is the locus of a folded fault zone. The nappe is truncated on the south by a fault. The Auburn Nappe contains the majority of the Grenville rocks exposed in Alabama. These rocks have a well-developed foliation, in a few localities, and this foliation is structurally conformable to that in the enclosing Pine Mountain Group over most areas.

Structurally overlying the Auburn Nappe is the Memory Hill Nappe. This nappe was not recognized in earlier discussions of the Pine Mountain window (Sears and others, 1981b). It lies upon a folded thrust which merges with a major porphyroclastic mylonite zone. It is tightly refolded along with the Auburn Nappe. The folds of its axial surface also deform an overlying porphyroclastic mylonite zone. The Memory Hill Nappe is relatively small, and its Grenville core is very poorly exposed in Alabama. The Hollis Quartzite occupying its upright limb is traceable continuously into Georgia, where the underlying Grenville rocks are better exposed.

The Memory Hill Nappe is structurally overlain by a thick zone of mylonite derived by comminution of Grenville rocks and metasedimentary rocks of the Pine Mountain Group. Structural relationships are complex in this zone, but the basic pattern is of repetitions of sheared gneisses and porphyroclastic mylonites derived from the basement and mylonitic quartzites and button sillimanite-muscovite schists derived from the cover. The mylonites form a belt up to 5 km wide, of mostly northeasterly dipping rocks, along the north margin of the window.

A synformal body of Pine Mountain quartzite and schist structurally overlies the mylonite zone near Auburn, Alabama. This body is interpreted to be the hinge area of the refolded Thomaston Nappe, which was thrust over the Memory Hill Nappe along the mylonite zone. It contains no basement rocks in Alabama and is truncated on the north by the Towaliga Fault.

In summary, the Grenville rocks of Alabama are of limited and poor exposure, are confined to the cores of two tightly appressed and refolded nappes, and have few recognized vestiges of Precambrian fabric.

The structural style of the basement and cover in Georgia strongly differs from that in Alabama. Structural plunge is shallow and variable, the nappes are not coaxially refolded, foliation dips are gentle to moderate, and the basement and cover are less highly strained. Thus, large areas of the basement contain relict Precambrian fabrics.

Much of the western part of the Pine Mountain window in Georgia is underlain by an upright sequence of Grenville basement, Hollis Quartzite and overlying Manchester Schist, here interpreted to occupy the upper limb of the Memory Hill Nappe. This sequence is deformed into a broad, open anticlinorium with domes and basins having curved axial traces. In the core of the anticlinorium are irregular areas of Sparks Schist (Hewett and Crickmay, 1937) structurally overlain by sheared Grenville gneiss. These are interpreted as windows through the Grenville core of the Memory Hill Nappe, exposing the underlying Auburn Nappe. This interpretation differs from previous ones in which these areas were considered parts of the upright cover sequence underlying the Hollis Quartzite (Schamel and Bauer, 1980; Sears and others, 1981).

The Memory Hill Nappe is structurally overlain by a tight, overturned syncline with the Pine Mountain Group in its core. This is in turn structurally overlain by the Thomaston Nappe, which has Grenville rocks in its core. This nappe underlies the remainder of the Pine Mountain window to the east. The Thomaston Nappe is truncated on the south by the Goat Rock and Bartletts Ferry faults, which form the sole faults of the Piedmont allochthon.

REGIONAL METAMORPHISM OF THE PINE MOUNTAIN GROUP

Near Manchester, Georgia, aluminous schists of the Pine Mountain Group contain the assemblage kyanite-garnet-biotite-muscovite-quartz, and locally kyanite-garnet-staurolite-biotite-

muscovite-quartz. In Alabama, kyanite was observed only in the central part of the window near the Chattahoochee River. Flanking the kyanite-bearing rocks at a structurally higher position, in Alabama, are sillimanite-biotite-muscovite-garnet-quartz schists that contain pods and lenses of pegmatite suggestive of anatexis. These schists are contorted, sheared and associated with thick porphyroclastic mylonite zones derived from the Grenville basement. Sillimanite schists also are found in the Thomaston Nappe near Auburn, Alabama. Sillimanite also occurs at the eastern end of the window near Forsyth, Georgia (Hatcher and others, 1981).

Metamorphic conditions in the window thus vary from the kyanite-staurolite zone of middle amphibolite facies to the sillimanite-anatectite zone of upper amphibolite facies. Relationships are not yet definitive, but the higher grade rocks appear confined to the Thomaston Nappe, while the lower grade, higher pressure (?) rocks occupy the Memory Hill Nappe. Thus, the isograds may be inverted in the window. Over much of the length of the window the metamorphic grade of the cover sequence is similar to or *lower* than that of the superjacent Inner Piedmont.

The Grenville rocks were subjected to the metamorphic conditions described above for the Pine Mountain Group, and all show metamorphic recrystallization to some extent. Primary Grenville mineral assemblages only can be documented where relict grains of some minerals are preserved.

PETROLOGY OF THE GRENVILLE COMPLEX

This section discusses the major lithologies recognized within the Grenville complex. All rocks so far examined show some evidence of retrograde metamorphism (Schamel and others, 1980), and most are dominated by foliations thought to have formed during Paleozoic orogenesis.

The Grenville complex of the window contains five main lithologic types: (1) charnockite with mafic xenoliths (Cunningham granite) and a weak foliation of probably Paleozoic age; (2) extensively foliated and retrograded leucocharnockitic gneiss (Jeff Davis "granite"); (3) feldspathic and migmatitic augen gneiss with zones of intense shearing (Woodland Gneiss); (4) mylonitic gneiss with large feldspar augen (Whatley Mill Gneiss); and (5) fine-grained, dark mylonite and porphyroclastic mylonite.

The general distribution of these rock types in the 3 nappes is shown in Table 1.

Table 2 shows mineral assemblages of probable Grenville and post-Grenville age for each major rock type. The distinction is made on the basis of texture.

Charnockite

Charnockitic rocks form the least altered parts of the Grenville complex. They recur in three major and several minor bodies in the eastern part of the window in the upper two nappes, but have not been recognized in Alabama in the lowermost nappe. The charnockitic rocks were first recognized in the Tho-

TABLE 1. GENERAL DISTRIBUTION OF MAJOR ROCK TYPES OF GRENVILLE COMPLEX

Nappe	Grenville Rocks	Charnockite	Jeff Davis "Granite"	Woodland Gneiss	Whatley Mill Gneiss	Mylonite
(highest)	(1) Thomaston	abundant	abundant	abundant	absent	dominant
	(2) Memory Hill	locally significant	absent	dominant	very restricted	in zone between (1) and (2)
(lowest)	(3) Auburn	absent	absent	dominant	locally significant	local

TABLE 2. DOMINANT GRENVILLE AND POST-GRENVILLE MINERAL ASSEMBLAGE OF MAJOR ROCK TYPES

	Charnockite	Jeff Davis "Granite"	Woodland Gneiss	Whatley Mill Gneiss	Mylonite
Grenville minerals					
Quartz	x	x	x	x	x
Hypersthene	x	x			
Microcline and/ or orthoclase	x	x	x	x	x
Plagioclase	x		x(?)		
Post-Grenville					
Garnet	x	x	x	x	x
Biotite	x	x	x	x	x
Muscovite			x	x	x
Plagioclase			x	x	x
Epidote				x	x
Sphene				x	

maston, Georgia, area of the Pine Mountain window by Clarke (1952), who described in detail a number of units ranging compositionally from "hypersthene gabbro" to "hypersthene granite." The areal extent of these granulite facies units has been extended to include the Cunningham "Granite" of Hewett and Crickmay (1937) in Fig. 2. All units within the charnockite grouping contain hypersthene and are retrograded, exhibiting ubiquitous coronas of garnet, hornblende, biotite and/or clinopyroxene or relict hypersthene (Clark, 1952; Schamel and others, 1980). Mafic xenoliths are locally conspicuous and may be angular and randomly oriented. Fabrics vary from massive to weakly foliated, with the foliation interpreted to be related to Appalachian disturbances. The charnockitic rocks typically weather into massive rounded boulders or develop wide pavement surfaces.

The charnockites are characterized by allotriomorphic fabrics and grain sizes averaging from 0.5 to 2.0 mm. Porphyroblastic phases of intermediate composition, containing subhedral perthitic microcline grains up to 3 cm in length and only minor hypersthene, occur in some localities.

The most common compositional variant is a charnockite containing pale-blue quartz. Microperthitic microcline occurs in amounts up to 50 percent. Oligoclase and quartz occur in variable amounts up to approximately 30 percent each. Hypersthene content ranges from less than one to 5 percent. Garnet is ubiquitous in amounts up to 7 percent. Accessory minerals include biotite, augite, apatite, zircon, rutile, pyrite and ilmenite. Piemontite occurs sparingly in the Cunningham "Granite."

The mesocharnockites are characterized by an increase in plagioclase content (up to 40%), with a coincident increase in anorthite content to An_{75}. Microcline and quartz are absent. Hornblende and actinolite occur in amounts up to 30 percent collectively, and the overall garnet content approaches 20 percent in the more mafic phases.

The Grenville granulite facies mineral assemblage was probably hypersthene, potassium feldspar, plagioclase and accessories. The post-Grenville retrograde products include garnet, biotite, muscovite, epidote and sphene.

Jeff Davis Leucocharnockitic Gneiss

The Jeff Davis "Granite" of Clarke (1952) is a textural variant of the Cunningham charnockite in which a moderate to strong foliation is present and mafic xenoliths are flattened (Schamel and Bauer, 1980). The foliation lies in the axial surface of the Thomaston Nappe and thus appears to be related to Paleozoic orogenesis (Schamel and Bauer, 1980). The Jeff Davis extends across a wide area in the core of the Thomaston Nappe. Locally it is strained into a porphyroclastic gneiss with mafic schlieren, but in many areas it contains relicts of massive, weakly deformed leucocharnockite. The Jeff Davis is mineralogically similar to the charnockite, containing dominant microcline and quartz with subordinate biotite and locally significant plagioclase, pyroxene, hornblende and opaques.

Woodland Gneiss

The Woodland Gneiss is the most widespread rock type of the Grenville complex. In this paper the name is extended into Alabama to embrace rocks formerly mapped as the Wacoochee Complex, because the units are structurally continuous and lithologically similar, and "Woodland" is already a formal name (Hewett and Crickmay, 1937).

In the vicinity of the type locality, the Woodland Gneiss consists of potassium feldspar augen in a fine- to medium-grained groundmass. Augen are typically microcline and rarely exceed 6 mm in maximum dimension. Locally, textures are transitional from a finely laminated mylonite, adjacent to the overlying Hollis Quartzite, to a more typical porphyroclastic biotite gneiss with increasing distance from the contact.

In Lee County, Alabama, similar gneisses that locally contain augen and exhibit banding on a scale of 2-8 mm are here assigned to the Woodland Gneiss. These rocks are characterized by microcline-quartz lenses exceeding 1 cm in thickness, bounded by layers of subhedral biotite aggregates. In the westernmost part of the window, this unit exhibits only very weak foliation of probable Paleozoic age, and appears to have been derived from a coarse-grained microcline granite.

Petrographic descriptions of Clarke (1952) indicate a generally allotriomorphic texture with only biotite occurring in euhedral grains. The average grain size within the groundmass varies from 0.1 mm to 3 mm with strained quartz as the dominant mineral. Both muscovite and biotite occur as foliated aggregates throughout. Garnets of unknown composition occur randomly as isolated grains. Accessory minerals are plagioclase (oligoclase-andesine), zircon, sphene, and ilmenite. The Grenville mineral assemblage may have been quartz, potassium feldspar, pyroxene and minor plagioclase.

In some areas of the Thomaston Nappe, rocks mapped as Woodland Gneiss may have been partly derived from the Jeff Davis lithology (Schamel and others, 1980), but usually the rock is more felsic. Because of the high feldspar content, the Woodland Gneiss is typically deeply weathered into a quartz-rich, light-red soil.

Whatley Mill Gneiss

The Whatley Mill Gneiss (Bentley and others, 1985) is the most restricted map-unit in the Grenville complex. It occurs near the contact with the Pine Mountain Group along the inverted limb of the Auburn Nappe in Alabama (Fig. 1).

The Whatley Mill Gneiss is characterized by potassium feldspar augen in a medium-grained felsic groundmass. Textures vary widely and are transitional between the apparent protolith, a porphyroblastic biotite granite that crops out in very restricted areas, to porphyroclastic mylonites of more general occurrence. Euhedral to subhedral porphyroblasts as much as 5 cm long occur in the protolith. Porphyroclasts derived from these porphyroblasts approach this maximum dimension at numerous locations along the outcrop belt.

Both microcline and orthoclase are present locally as individual augen. Tapered, crushed zones parallel with foliation typically occur at the ends of augen. Foliated aggregates of biotite and subordinate muscovite, together with quartz and minor oligoclase constitute a medium-grained matrix. Accessory minerals are epidote, zoisite, sphene, apatite, zircon, pyrite, and magnetite.

The original Grenville paragneses may have been quartz, potassium feldspar, plagioclase and pyroxene.

Fine-grained Mylonite and Porphyroclastic Mylonite

Fine-grained mylonitic rocks derived from the basement are present at many localities throughout the complex, but are especially abundant in Alabama in a belt of imbricate thrust slices between the Auburn and Thomaston nappes (Fig. 1 and 2).

As shown by detailed mapping within mylonitic zones, a major derivative of the basement complex units within the window is porphyroclastic mylonite. Wide variations reflecting original lithologies and degree of mylonitization exist. Porphyroclastic mylonites are developed gradationally from porphyroblastic units such as the Whatley Mill Gneiss through an intermediate phase of mylonitic gneiss. Other mylonites contain augen composed of abraded rock fragments of phases of the Woodland and related felsic gneisses. Flaser structure locally is well developed. Mylonite derived from charnockitic units appears to be confined to the central portions of the window near Thomaston, Georgia.

In general, mylonites contain relatively coarse-grained spherical to lenticular augen, abraded rock fragments, or flaser structures in a darker-colored, dense matrix composed of biotite and fine-grained partially recrystallized mylonitic debris. Biotite and, less commonly, muscovite foliation bends around porphyroclasts, imparting an overall wavy fabric. Mortar structures are commonly well developed. Porphyroclastic mylonites grade into finer-grained mylonites with fragments of larger crystals exhibiting extreme peripheral granulation with trails of crushed material extending into the sheared groundmass. Crushed garnets are conspicuous in many thin sections.

Basement-derived rocks exhibiting extreme comminution and partial or complete grain boundary recrystallization collectively are termed fine-grained mylonite for the purpose of this paper. Transition from porphyroclastic mylonite to fine-grained mylonite is in part controlled by the mineralogical and textural characteristics of the parent rock. Fine-grained mylonites formed from basement units are of less frequent occurrence and are less continuous in the window than are porphyroclastic mylonites.

In general, the fine-grained mylonites are somewhat lighter colored than porphyroclastic mylonites, and are streaked with narrow bands of quartz and mica. Mineralogical compositions vary widely, but, in general, quartz, plagioclase, microcline, orthoclase, and, less commonly, amphibole and pyroxene constitute the finer grained portions of the mylonites. Biotite and muscovite occur as isolated, somewhat larger, recrystallized grains or aggregates. Remnants of augen in some cases exhibit fracturing with reconcilable displacement of individual fragments.

CONCLUSIONS

Throughout much of the Pine Mountain window the Grenville massif contains foliations that are closely parallel to those of the nonconformably overlying Pine Mountain Group. The basement rocks were retrograded from granulite facies and in some areas were comminuted into mylonitic gneisses, porpphyroclastic mylonites, and fine-grained mylonites. The Pine Mountain Group was prograded to middle to upper amphibolite facies as it was deformed with its basement into tightly appressed recumbent isoclinal folds. Where the basement is most highly deformed in Alabama, the cover is at its highest grade. Thick mylonite zones derived from the basement are interleaved with quartz mylonites and button sillimanite schists with disrupted pegmatite pods derived from the cover. The mylonite zones are folded with the axial surfaces of the underlying nappes.

The window is much broader in Georgia where the Pine Mountain antiform is a broad, open fold of foliation with gently dipping limbs and a curved axial trace. In Alabama, the antiform is, in some transects, isoclinal and has a straight axial trace. Most structures in the window are asymmetric to the northwest.

The close spatial proximity of tightly refolded nappes, thick basement-derived mylonite zones, and anatectic sillimanite-grade metamorphism of the cover near the structural culmination in Alabama suggests a possible genetic interrelationship between structural complexity and metamorphism in the footwall of the Piedmont allochthon.

REFERENCES CITED

Bentley, R. D., and Neathery, T. L., 1970, Geology of the Brevard Fault Zone and related rocks of the Inner Piedmont of Alabama: Alabama Geological Society, Guidebook, 8th Annual, 119 p.

Bentley, R. D., Neathery, T. L., and Scott, J. C., (1985), Geology and mineral resources of Lee County, Alabama: Geological Survey of Alabama Bulletin 107, 110 p. (in press).

Brown, D. E., and Cook, R. B., Jr., 1981, Petrography of major Dadeville complex rock units in the Boyds Creek area, Chambers and Lee Counties, Alabama, *in* Sears, J. W., ed., Contrasts in Tectonic Style between the Inner Piedmont Terrane and the Pine Mountain Window: Alabama Geological Society, Guidebook, 18th Annual, p. 15–40.

Clark, J. W., 1952, Geology and mineral resources of the Thomaston Quadrangle, Georgia: Georgia Geological Survey Bulletin 59, 99 p.

Hatcher, R. D., Jr., Hopper, R. J., and Odom, A. L., 1981, Transition from the east end of the Pine Mountain Belt into the Piedmont, central Georgia: Preliminary results (Abstr.): Geological Society America, Abstracts with Programs, v. 13, no. 1, p. 9.

Hewett, D. F., and Crickmay, G. W., 1937, The warm springs of Georgia, their geologic relations and origin; a summary report: U.S. Geological Survey Water Supply Paper 819, 40 p.

Odom, A. L., Kish, S., and Leggo, P., 1973, Extension of "Grenville Basement" to the southern extremity of the Appalachians: U-Pb ages of zircons (Abstr.): Geological Society of America, Abstracts with Programs, v. 5, no. 5, p. 425.

Schamel, S. and Bauer, D., 1980, Remobilized Grenville basement in the Pine Mountain Window *in* Wones, D. R. (ed.), The Caledonides in the U.S.A.: International Geological Correlation Program project 27: Caledonite orogen; Virginia Sciences Memoir 2, p. 313–316.

Schamel, S., Hanley, T. B., and Sears, J. W., 1980, Geology of the Pine Mountain Window and adjacent terranes in the Piedmont Province of Alabama and Georgia: Geological Society of America Guidebook, 29th Annual Meeting, Southeastern Section, 69 p.

Sears, J. W., 1980, Nappe tectonics in the Pine Mountain Window of the Piedmont allochthon, Alabama (Abstr.): Geological Society of America, Abstracts with Programs, v. 12, no. 4, p. 280.

Sears, J. W., Cook, R. B., Jr., and Brown, D. E., 1981a, Tectonic evolution of western part of the Pine Mountain Window and adjacent Inner Piedmont Province *in* Sears, J. W., ed., Contrasts in Tectonic Style between the Inner Piedmont Terrane and the Pine Mountain Window: Alabama Geological Society Guidebook, 18th Ann., p. 1–13.

Sears, J. W., Cook, R. B., Gilbert, O. E., Jr., Carrington, T. J., and Schamel, S., 1981b, Stratigraphy and structure of the Pine Mountain Window in Georgia and Alabama: Georgia Geological Survey, Information Circular 54-A, p. 41–54.

MANUSCRIPT ACCEPTED BY THE SOCIETY AUGUST 2, 1983

Typeset by WESType Publishing Services, Inc., Boulder, Colorado
Printed in U.S.A. by Malloy Lithographing, Inc., Ann Arbor, Michigan